D1568160

Modern Physical Science

Holt, Rinehart and Winston, Inc.
Harcourt Brace Jovanovich, Inc.
Austin • Orlando • San Diego • Chicago • Dallas • Toronto

Authors

Harry E. Tropp
Former Science Supervisor
Hillsborough County Public Schools
Tampa, Florida

Alfred E. Friedl
Emeritus Professor of Education
Kent State University
Kent, Ohio

Safety Consultant
Juta Nolan
Montclair High School
Montclair, New Jersey

Reading Consultant
Patricia N. Schwab
University of South Carolina
Columbia, South Carolina

Reviewers

George Eblin
North High School
Downers Grove, Illinois

Richard H. Millett
Foxcroft Academy
Dover-Foxcroft, Maine

William T. Pawling
Warrensburg High School
Warrensburg, Missouri

Steven J. Rakow
University of Houston, Clear Lake
Houston, Texas

Cover: Plastic rods containing certain minerals, or phosphors, glow in a variety of colors after being exposed to energy in the form of light. Phosphors are substances that glow in specific colors when energized by light, heat, or electrical energy. Color television screens, for instance, have three types of phosphors; red, blue, and green, arranged in a pattern of dots. When energized by electrical signals, the phosphor dots produce color television images. Phosphors are discussed in Sections 25.3 and 28.3. Television is discussed in Section 28.3.

Cover Photo by Pelton and Associates/West Light

Printed in the United States of America

ISBN 0-03-005052-9

9 062 9

CONTENTS

UNIT 1

MATTER AND ENERGY

UNIT 2

MATTER INTERACTS

UNIT 5

WAVE MOTION AND ENERGY

USING MODERN PHYSICAL SCIENCE

Why do objects fall? What is light? Can we control pollution? How does a computer work? Physical science has the answers to these questions and explores many more unanswered questions about nature. ***Modern Physical Science*** is designed to guide you toward these answers and help you understand the questions that scientists are asking about nature today.

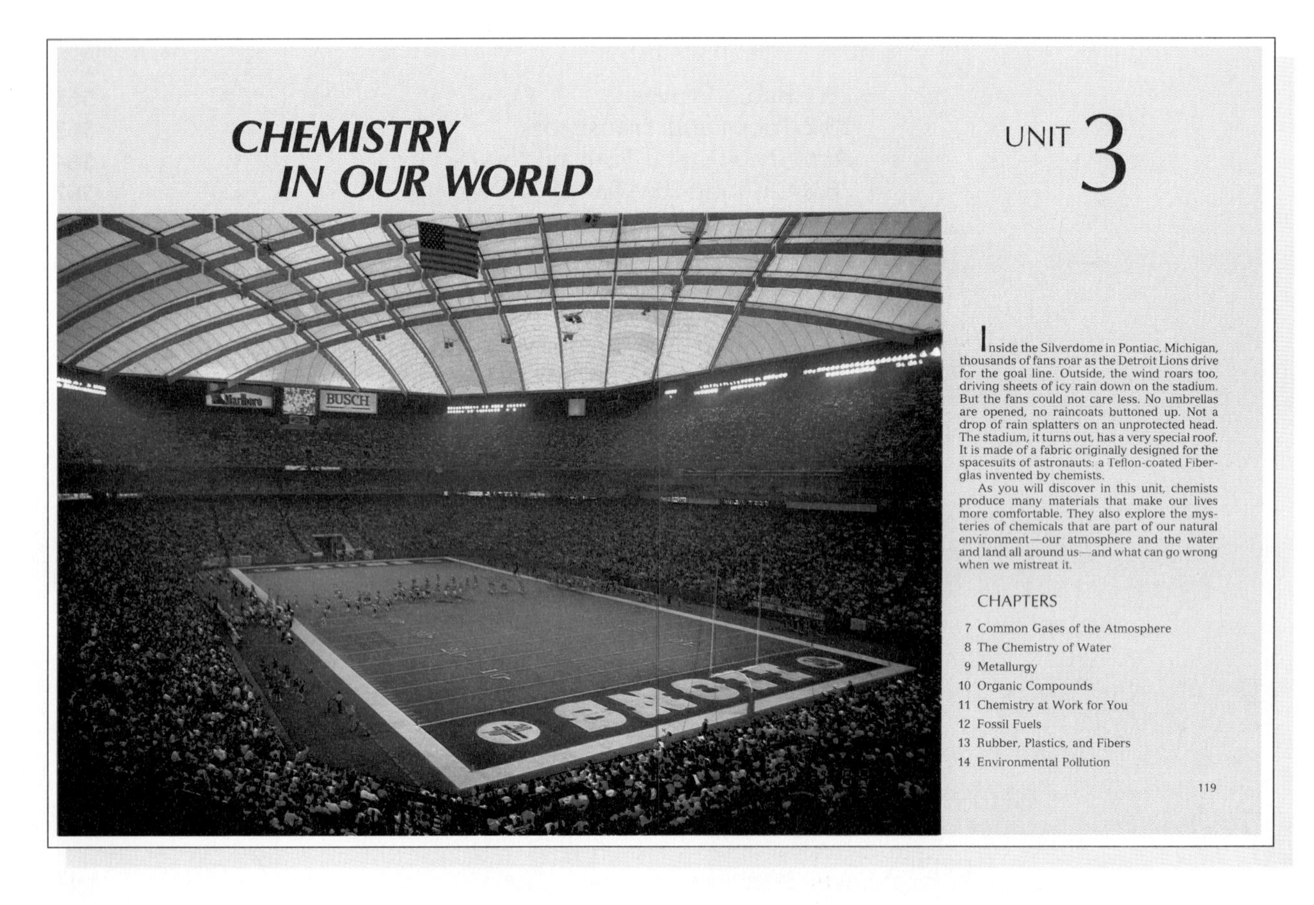

CHEMISTRY IN OUR WORLD

UNIT 3

Inside the Silverdome in Pontiac, Michigan, thousands of fans roar as the Detroit Lions drive for the goal line. Outside, the wind roars too, driving sheets of icy rain down on the stadium. But the fans could not care less. No umbrellas are opened, no raincoats buttoned up. Not a drop of rain splatters on an unprotected head. The stadium, it turns out, has a very special roof. It is made of a fabric originally designed for the spacesuits of astronauts: a Teflon-coated Fiberglas invented by chemists.

As you will discover in this unit, chemists produce many materials that make our lives more comfortable. They also explore the mysteries of chemicals that are part of our natural environment—our atmosphere and the water and land all around us—and what can go wrong when we mistreat it.

CHAPTERS

119

UNITS

Modern Physical Science is divided into units. Each unit acts as a short course, grouping chapters together that have common or connected scientific themes.

ORGANIC COMPOUNDS

CHAPTER 10

Trees, like this Baobab in Tanzania, are living organisms. All living things—whether they are plants, birds, insects, or humans—are made of organic compounds. How are these compounds different from those that make up some nonliving things, such as a quartz crystal or a glass of water?

SECTIONS

10.1 Organic Chemistry
10.2 Hydrocarbons
10.3 Addition and Substitution Reactions
10.4 Alcohols

OBJECTIVES

- ☐ Distinguish between organic compounds and inorganic compounds.
- ☐ Identify and give examples of five groups of hydrocarbons.
- ☐ List uses of organic compounds.
- ☐ Compare substituted hydrocarbons with their related alkanes.

10.1 ORGANIC CHEMISTRY

There are differences between organic compounds and inorganic compounds. Carbon compounds make up the structure of living things, or organisms. Thus, scientists classify carbon compounds as **organic compounds.** Other substances such as acids, bases, salts, and water do not contain carbon. These substances are classified as inorganic compounds. These two groups of compounds differ in several ways.

181

CHAPTERS

Units are divided into a total of 28 chapters. Each chapter introduces you to a main scientific subject or field, indicated by the chapter title, such as "Nuclear Reactions" or "Metallurgy."

SECTIONS
A list of objectives at the beginning of each chapter tells you what knowledge or skills to look for in the chapter. Keep these objectives in mind as you read. Ask yourself how the material you are reading will help you to reach those objectives.

OBJECTIVES
Chapters are divided into smaller, numbered sections to help you study that particular branch of physical science in a step-by-step manner.

HIGHLIGHTS
A boldface sentence tells you the main idea of that subsection. Important new terms appear in boldface in the sentence in which they are defined.

TABLES

Tables present information in a concise way for easy reference.

Not all glass has an index of refraction of 1.5. The exact value depends on the kind of glass. The index of refraction for several substances is shown in Table 22-2.

TABLE 22-2

Substance	Index of Refraction
Vacuum	1.0000
Air	1.0003
Ice	1.31
Water	1.33
Glass (crown)	1.52
Glass (flint)	1.61
Diamond	2.42

SAMPLE PROBLEM

What is the index of refraction of glass through which light travels at 200,000 km/sec?

SOLUTION:

Step 1: Analyze

You are given that the speed of light in glass is 200,000 km/sec. The index of refraction for the glass is the unknown.

Step 2: Plan

You can derive the following equation from the definition of the index of refraction:

$$\text{index of refraction} = \frac{\text{speed of light in vacuum or air}}{\text{speed of light in substance}}$$

Since you know the speed of light in air and in the substance (glass), you can use the equation to calculate the index of refraction.

Step 3: Compute and Check

Substitute all values in the equation.

$$\text{index of refraction} = \frac{300{,}000 \text{ km/sec}}{200{,}000 \text{ km/sec}} = 1.5$$

Since your answer is approximately equal to the index of refraction given for crown glass in Table 22-2, your solution is complete.

456 Unit 5 Wave Motion and Energy

Air can also cause refraction. Most often you can disregard the difference between the speed of light in air and its speed in a vacuum. However, this difference does result in a slight bending of light as it enters the atmosphere. This bending allows you to see the sun before it rises and after it sets. Figure 22-18 shows how this happens.

Suppose that you are at C, looking toward the setting sun. As the ray AB enters the air, it bends as shown. This refraction occurs because the ray is traveling into denser air and, as a result, slows down. As you can see, the effect of this refraction is to "lift" the sun above the horizon. Refraction lengthens the hours of daylight because the earth receives the sun's rays before the sun actually rises and after it sets.

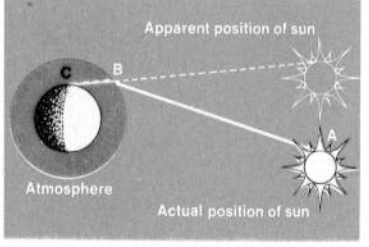

Figure 22-18 Why is the sun visible when it is still below the horizon?

You can see other ways in which air affects light. Look at a distant object using a line of sight that is directly above a heat source, such as a candle, gas burner, or hot plate. The dancing of heat rays on a hot metal roof, the shimmering of a hot roadway in the distance, and the twinkling of stars are all examples of distortions caused by the changing refraction of the light as it passes through air of varying densities. In working out star positions, astronomers must correct their readings to allow for the refraction of air.

Lenses refract light. A **lens** is a transparent substance having at least one curved surface. Although lenses differ in shape, size, and substance, the chief purpose of all lenses is to refract light. The amount of refraction produced by a lens depends both upon its shape and upon the index of refraction of the substance of which it is made. Lenses are used in telescopes, cameras, eyeglasses, microscopes, spectroscopes, movie projectors, and many other optical devices.

To understand how lenses affect light, first look at the path of light through a simple prism. Notice that in each prism in Figure 22-19A, the ray bends into the prism as it

Figure 22-19 In what ways are double convex and double concave lenses like prisms?

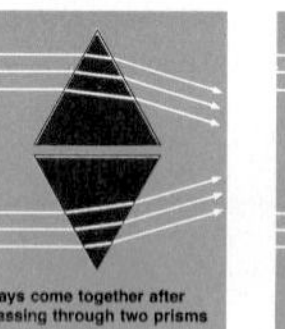

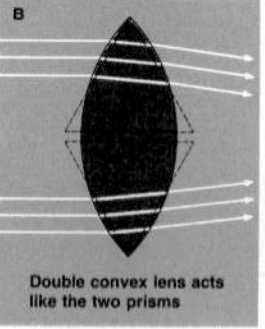

B
Double convex lens acts like the two prisms

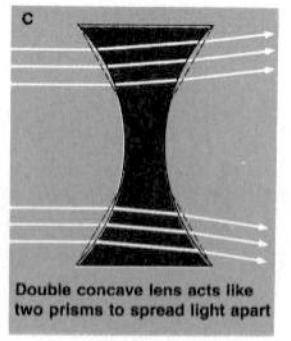

C
Double concave lens acts like two prisms to spread light apart

Chapter 22 Light 457

SAMPLE PROBLEMS

A logical three-step method is used to solve all sample problems.

ILLUSTRATIONS AND PHOTOGRAPHS

Illustrations and photographs clarify important concepts in physical science.

ACTIVITIES

Activities give you the chance to further explore physical science by using scientific methods.

CAUTION alerts you to possible hazards that may arise in the activities.

MAGNESIUM AND AN ACID

activity

OBJECTIVE: Determine how magnesium reacts with an acid.

PROCESS SKILLS
In this activity, you will *observe* the reaction of magnesium with acid and *classify* the products formed.

MATERIALS
magnesium ribbon test tube dilute hydrochloric acid

PROCEDURE
1. Place a 5-cm length of magnesium (Mg) ribbon in a test tube containing a small amount of dilute hydrochloric acid (HCl). CAUTION: *Magnesium ribbon is highly flammable. Wear safety goggles and an apron.*
2. Observe what happens.

OBSERVATIONS AND CONCLUSIONS
1. Describe your observations.
2. Identify the gas that evolves. What method did you use?
3. Write the balanced chemical equation for this reaction.

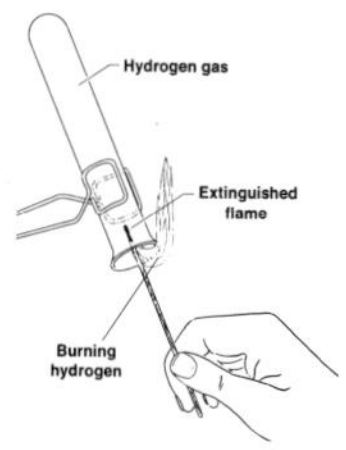

Figure 7-16 Burning a mixture of hydrogen and air. **CAUTION: Do not try this test for hydrogen without your teacher's supervision.**

bottle, nothing will happen because the hydrogen, which is less dense than air, has had time to escape from the open bottle.

If a flaming splint is placed inside an inverted bottle of hydrogen, the flame will go out, but the hydrogen will continue to burn at the mouth of the bottle (Figure 7-16). This is a test for the presence of hydrogen gas. While the hydrogen gas itself burns, it cannot support the burning of other substances such as the splint.

When hydrogen is added to cottonseed, coconut, or other food oils, it combines chemically to form solid cooking fats. This process is called **hydrogenation** (HIE-*drah-juh*-NAY-*shun*). Fine grains of nickel (Ni) are used as a catalyst to speed up the process. Peanut butter is hydrogenated to prevent the separation of the oils from the pulp.

Special welding and cutting devices use hydrogen to produce extremely hot flames. Hydrogen is also used to make high-grade gasoline from petroleum. Liquid hydrogen is used as fuel for rockets.

134 Unit 3 Chemistry in Our World

BIOGRAPHY

NEIL BARTLETT

Everyone said it was impossible to make a compound of a noble gas and another substance. As a matter of fact, the noble gases—helium, neon, argon, krypton, xenon, and radon—got their name because they did not react with other more "common" substances. The noble gases, like the royalty after which they were named, kept to themselves.

Then, in 1962, along came an English chemist by the name of Neil Bartlett. At the time, Bartlett was 30 and doing experimental chemistry in British Columbia, Canada. Bartlett's pet project involved the study of extremely reactive substances including one called platinum hexafluoride (PtF_6).

Based on theoretical calculations, Bartlett was convinced that PtF_6 was so reactive that it could get the noble gas xenon (Xe) to combine with it to form a substance never before seen. But was he correct? To find out, Bartlett built a special apparatus in which colorless xenon gas under pressure was suddenly mixed with deep red PtF_6 vapor. As Bartlett watched, the gases became transformed into something altogether new—a yellow solid! But what was this material? After careful chemical analysis, Bartlett uncovered the answer. The yellow solid was xenon hexafluoroplatinate ($XePtF_6$)—the "impossible" compound!

Bartlett's dramatic experiment revealed that what appears to be science "fact" one day may be science "fiction" the next. Soon other compounds of xenon, as well as of krypton and radon, were made. Perhaps Bartlett's most lasting contribution was to make scientists—and the rest of us—be more careful when using the word "impossible."

Chapter 7 Common Gases of the Atmosphere 135

SPECIAL FEATURES

Special features give you information about several career opportunities based on a knowledge of physical science, or introduce you to famous scientists or current scientific issues connected to physical science.

CHAPTER REVIEW

SUMMARY

1. All atoms are made of protons, neutrons, and electrons.
2. The atomic number of an element is equal to the number of protons in its nucleus. It also identifies an element in the periodic table.
3. In a neutral atom, the number of protons (+) in the nucleus equals the number of electrons (−) in the energy levels around the nucleus.
4. The mass number is equal to the total number of protons and neutrons in the nucleus of an isotope.
5. The charge an atom has in a particular molecule or ion is indicated by its oxidation number.
6. Ionic bonds are formed by the transfer of electrons from the outer energy level of one element (usually a metal) to the outer energy level of another element (usually a nonmetal).
7. Covalent bonds are formed when elements (usually nonmetals) share outer energy level electrons with each other.

VOCABULARY

Match the item in the left column with the best answer in the right column. Do not write in this book.

1. atomic mass	a. bond formed by sharing electrons
2. atomic number	b. number of neutrons plus protons
3. chemical bond	c. element on right side of periodic table
4. covalent bond	d. indicates the charge an atom has
5. electron	e. negatively charged particle in an atom
6. ionic bond	f. force holding atoms together
7. ions	g. positively charged particle in an atom
8. isotopes	h. number of protons
9. mass number	i. arrangement of elements according to their properties
10. metal	
11. neutron	j. element on left side of periodic table
12. nonmetal	k. bond formed by electron transfer
13. oxidation number	l. formed when electrons are lost or gained
14. periodic table	m. atomic particle with no charge
15. proton	n. atoms with the same number of protons and different numbers of neutrons
	o. average of the masses of an element's isotopes

REVIEW QUESTIONS

1. Name the three basic particles of an atom and the charge each has.
2. What is a scientific model?
3. Where in an atom is most of the mass found? Explain.
4. How does an atom of iron differ in structure from an atom of gold?

54 Unit 1 Matter and Energy

5. What is the maximum number of electrons in each of the first three energy levels of an atom?
6. Set up a table of 10 elements. List the following information about each: the number of protons, neutrons, and electrons.
7. Where in the periodic table are the most active elements located?
8. Where in the periodic table are the metals?
9. What is the oxidation number of Group 1 elements?
10. Name five elements that are included in the "nitrogen group."
11. What are the formulas for molecules of nitrogen and chlorine?
12. Write the formula for each of the following compounds: (a) calcium hydroxide, (b) ammonium sulfate, (c) hydrogen phosphate, and (d) copper(II) chloride.
13. Name the compound represented by each of the following formulas: (a) $KClO_3$, (b) NH_4OH, (c) Na_2CO_3, (d) $Pb(C_2H_3O_2)_2$, and (e) K_2S.
14. Describe two ways atoms combine to form compounds. Give examples.
15. Write a brief summary of the modern atomic theory of matter.
16. Explain the difference between the atomic mass of an element and the mass number of an atom.
17. Explain why some elements have positive oxidation numbers and other elements have negative oxidation numbers.
18. How has the periodic table been of value in advancing chemistry?
19. What type of chemical bonding would you expect to take place in the making of lithium fluoride? Explain.

CRITICAL THINKING

20. Sketch a diagram of the following atoms: (a) carbon-12, (b) magnesium-24, (c) nitrogen-14, and (d) sulfur-32.
21. The force of attraction holds the particles together within an oxygen atom. When the two oxygen atoms come near each other, both repellent and attractive forces act on the atoms. Describe the particles in each atom that produce these forces.
22. What evidence would you look for to tell that a chemical reaction has taken place?

PROBLEMS

1. Write the chemical formulas for each of the following compounds: (a) barium chloride, (b) calcium iodide, (c) copper(II) oxide, (d) lead(II) nitrate, and (e) aluminum hydroxide.
2. Write the names of each of the following compounds: (a) $MgSO_4$, (b) Na_2O, (c) $Ba(OH)_2$, (d) Li_2SO_4, (e) NH_4Cl, and (f) $PbCO_3$.
3. Give the oxidation number of the following underlined elements or groups: (a) $\underline{Ca}Cl_2$, (b) $\underline{Ag}_2S$, (c) $K_2\underline{SO_4}$, (d) $\underline{Cu}_2O$, and (e) $Al_2\underline{S}_3$.

FURTHER READING

Ardley, Neil. *Atoms and Energy*. New York: Franklin Watts, Inc., 1976.

Asimov, Isaac. *How Did We Find Out About Atoms?* New York: Walker, 1976.

Ley, Wily. *The Discovery of the Elements*. New York: Delacorte, 1983.

Chapter 3 Structure of Atoms and Molecules 55

CHAPTER REVIEW

SUMMARY
A summary at the end of each chapter lists key concepts covered in that chapter.

VOCABULARY
A vocabulary section helps you review boldface vocabulary words at the end of each chapter.

REVIEW QUESTIONS
Review questions serve as your self-test for each of the chapters.

CRITICAL THINKING
Critical Thinking questions ask you to apply and interpret what you have read.

PROBLEMS
Problems test your computing skills and understanding of quantitative concepts.

FURTHER READING
A list of books and articles at the end of each chapter provides suggestions for further reading.

INTRA-SCIENCE

Intra-Science unit features explore how physicists, chemists, biologists, and other scientists work together towards applying and integrating the sciences.

I N T R A - S C I E N C E

How the Sciences Work Together

The Process in Physical Science: Making a Computer Chip.

A good computer will take a second to do calculations an average person would spend weeks doing. Computerized controls on an airplane measure speed and position in heavy turbulence and make hundreds of adjustments per second to maintain a smooth ride. The brains behind these and other devices such as calculators and video games are integrated circuits—or computer chips.

Video games have become more sophisticated with improved computer chips.

Computer chips help maintain a smooth ride in today's airplanes.

Computer chips, often no larger than a few millimeters on a side, can contain thousands of transistors. Each transistor can process information by switching the flow of electrons on and off.

Most chips are made from silicon, which is a semiconductor. To make a chip, high purity silicon is mixed with impurities that either donate or accept electrons within the silicon crystal. These impurities are called "dopants." Silicon treated this way can be made to act like a conductor whose resistance depends on the voltage applied. This property can be used to amplify electrical signals or switch them on and off. Switching signals on and off is a key function of computer circuits.

Since their development in the early 1960's, chip circuitry has gone through a rapid miniaturization. This has reduced costs and increased performance. A tiny electronic circuit is cheaper to mass produce than a larger one that does the same things. Miniature circuits provide more computing power on a given surface area. Also, information is processed more quickly in smaller circuits since electrons have shorter distances to travel.

Businesses rely on information processed by computer chips.

578 Unit 6 Electricity and Magnetism

The Connection to Physics

Transistors were developed by physicists looking for ways to amplify (strengthen) electrical signals. They found that a small input current applied to two layers of doped silicon would create an output current through the adjoining layers. The output current was about equal to the input current but it flowed through a greater resistance. In effect, resistance was transferred from one circuit to another—thus the name "transistor" for transferred resistance. Since power is proportional to the resistance through which a current flows, the power in the output circuit was greater than in the input circuit. The energy for this increased power was obtained from voltage applied across certain layers of the transistor.

An engineer inspects a large scale circuit drawing used to design a computer chip.

The Connection to Chemistry

Several chemical processes are involved in making today's computer chips. First, extremely pure crystalline silicon is obtained by melting silicon dioxide (sand). Rods of pure silicon are then formed and sliced into wafers. Thousands of chips can then be manufactured on each wafer.

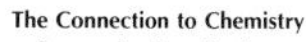

A pencil eraser is large compared to this computer chip.

Next, dopants are combined within the surface layer of pure silicon by heating the wafer in the presence of a gaseous dopant or bombarding the wafer with ion beams of the dopant.

Precise patterns and layers of differently doped silicon are then formed by applying a film of light sensitive material onto the wafer surface. An image of the microscopic circuit is then projected onto the wafer. The light sensitive film hardens where light strikes it. Lines of the soft film and underlying layers of doped silicon are removed by washing the wafer in chemical baths.

Once this is done, another chemical bath washes away the hardened portion of the film, exposing the rest of the chip surface. The chip is now ready for the process to be repeated until the required pattern of doped silicon layers is laid down.

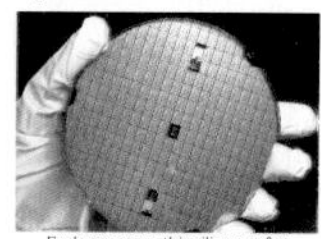

Each square on this silicon wafer contains a computer chip.

Continuing innovation in design and production has recently enabled chip manufacturers to put one million transistors on a 1.27 cm^2 chip. In the early 1960's, only four or five transistors could have been squeezed onto an area that small.

Intra-Science 579

INTEGRATED SCIENCE
Intra-Science features show how combining knowledge from various fields of science leads to innovation and a greater understanding of the physical world.

MATTER AND ENERGY

UNIT 1

Suppose you were the astronaut in this picture. What would be your first thought as you left your spacecraft and stepped upon the untouched surface of a distant moon or planet? Would you be dazzled by your surroundings and eager to see more? Would you marvel at your accomplishment and the distance you have traveled? Or would you suddenly be aware of the major role you have just played in scientific exploration?

Science is the search for answers, the charting of the unknown, and the thrill of discovery. People who study the planets, the weather, the soil, and the many forms of energy are all scientific explorers who gather information. They question the things around them and they look for answers. Their world of discovery extends from the tiny, invisible building blocks that make up all matter to the vast, unexplored universe.

In this unit, you will be introduced to what science is and why it is important to you. You will examine the smallest particles of matter, called atoms, and learn how these particles combine to build the many forms of matter that scientists seek to understand.

CHAPTERS

SCIENCE AND MEASUREMENT

In order to understand the world around us, scientists collect information. One method of getting information is taking careful measurements. How are the measurements scientists take similar to the ones you take when you look at your watch or step on a bathroom scale? What are some other examples of measurements that are useful to you each day?

CHAPTER 1

SECTIONS

OBJECTIVES

- ☐ Explain why it is helpful to study science.
- ☐ Define length, volume, force, and mass and identify their metric units of measure.
- ☐ Differentiate between the mass of an object and its weight.
- ☐ List various forms of energy.

1.1 SCIENCE IN TODAY'S WORLD

Science begins with curiosity. Human beings are naturally curious. They want to find out what things are made of and how they work. Scientists believe that everything in the universe behaves in an orderly way. They think that people can learn to understand this order. This desire to know about things around us is the key that unlocks our knowledge of the universe. **Science** is the orderly search for answers to questions about our world, our universe, and beyond.

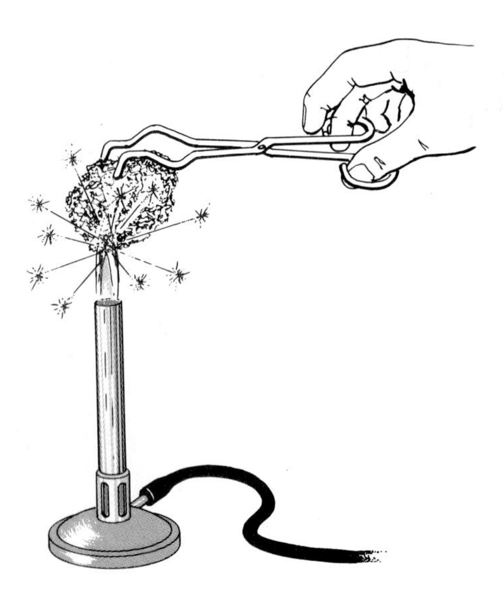

Figure 1-1 Science begins with careful observations. What observations can you make about the piece of steel wool being heated by the flame in the illustration? **CAUTION: Goggles should be worn when using a gas burner such as the one shown.**

Scientists seek to understand the world around them by making observations and asking questions. Look at Figure 1-1. When an ordinary piece of steel wool is held in a flame, the steel wool gives off sparks. Why do you think this happens? Why doesn't the steel wool melt, changing from a solid to a liquid? Are any new substances formed when the steel wool burns? The field of science helps you to find answers to questions like these.

Scientists use many methods and techniques to learn. A logical and systematic approach used by scientists to collect information is known as a **scientific method.** There are many different methods or approaches, but most include some or all of the following steps: (1) making observations, (2) identifying problems or questions, (3) forming hypotheses or explanations for observations or questions, (4) conducting experiments or testing hypotheses, (5) collecting and analyzing data, and (6) drawing conclusions. In applying the scientific method, scientists use a number of process skills, which are listed in the appendix. For example, when forming explanations for observations, scientists may use skills such as hypothesizing, predicting, and inferring. In order to collect and analyze data, scientists may use skills such as observing, measuring, and classifying.

Science is divided into branches. One branch consists of the biological sciences, which deal with the study of all living things. Another branch is made up of the physical sciences, which deal with matter, force, and energy. The two main areas of study in the physical sciences are chemistry and physics. Chemistry is the study of matter and its changes. Physics is the study of forces and energy that affect various forms of matter. You will study more about chemistry and physics later in this course.

Other areas of study in the physical sciences include geology (the study of the earth's surface and interior), meteorology (the study of the earth's atmosphere), astronomy (the study of the universe), and oceanography (the study of oceans). All of these sciences are related and often depend on one another. For example, a geologist must know about the chemistry of rocks and a meteorologist must know about the physics of air movements.

Science is a method for learning and applying knowledge. For example, scientific information is used to predict the outcome of certain events such as earthquakes or changes in weather patterns. Scientific information also leads to medical discoveries such as vaccines and organ

Figure 1-2 Robots at this automobile plant measure openings for doors and windshields. They work about 10 times as fast as human workers. What are some other advantages of these machines? What are some disadvantages?

transplants, and to technological advances such as solar energy and robots, like those shown in Figure 1-2. Unfortunately, scientists cannot ensure that their discoveries will be used wisely or safely. Scientific advances have also led to environmental pollution (Figure 1-3), energy shortages, and nuclear warfare.

Science helps us in many ways. When you drop a fork, does that mean you will have visitors? Do you carry a rabbit's foot for good luck? Are you afraid that if you break a mirror you will have seven years of bad luck? These may seem like silly questions, but many people still believe in such superstitions. Your study of science can help you to tell fact from superstition by explaining the mysteries of nature. Before the age of science, many people were frightened by comets, eclipses, thunder, and lightning. They thought that these events were supernatural. Once the natural causes for such events became known through scientific research, fewer people were frightened by them.

Besides explaining our mysteries, scientific developments affect our daily lives in other ways. The water you use in your home has probably been treated and purified. You use electric lights and enjoy the comforts of heating or air conditioning. Machines help you perform a job faster or more easily. Your clothing is durable and easy to care for because it is made of more resistant fabrics. Scientific research has also made it possible to produce more nourishing foods, better building materials, and more effective drugs and medicines. Without the benefit of scientific research, our life style would be very different.

Figure 1-3 Many modern conveniences are the result of scientific discoveries. Unfortunately, many of these conveniences will still be around long after they have outlived their usefulness.

Figure 1-4 Have you ever thought about where your drinking water comes from? Have you ever wondered where it goes once it enters the drain? Preparing water for home use and disposing of it once it has been used are some of the problems facing city planners.

Science can help you make wise decisions as a citizen. Officials of your city or town may have to decide whether or not to build new roads, bridges, parks, an airport, a water supply system (Figure 1-4), a sewage disposal plant, or a nuclear power plant. Much of what you learn in science class can help you deal with many of the problems that arise when these projects are considered by your community. A knowledge of science also helps you to answer questions about chemicals in your drinking water, air and water pollution, conservation of natural resources, new sources of food or energy, flood hazards, and pest control. By knowing and understanding the principles of science, you can make informed decisions regarding these issues. The future of our nation depends upon the decisions of its citizens.

Science is useful in many careers. A knowledge of science is necessary for many jobs in fields such as medicine, engineering, architecture, and agriculture. Science is also an important area of study for many sales, trade, and service occupations. Details about some of these jobs and how science relates to them is described in later chapters. This information will help you as you try to decide what type of work best fits your interests and abilities.

Table 1-1 lists several jobs related to one product in common use today, namely a school bus. As you read the table, note that some knowledge of science is required to answer each job-related question. What other questions about a school bus could arise that a knowledge of science would help to answer?

TABLE 1-1

Science and the School Bus	
Kind of Job	**Questions that Science Helps to Answer**
Designer	What is the best type of engine for the bus?
Materials Tester	What are the strongest and lightest construction materials that can be used?
Steel Worker	What is the best way to make and shape the steel used?
Salesperson	Is the product the best bus for the money spent?
Mechanic	What is needed to keep the bus running safely and smoothly?
Driver	What skills are needed to drive the bus safely?

1.2 MEASUREMENTS AND CONCEPTS OF SCIENCE

Scientists use exact measurements that can be reproduced. Weights and measures were among the earliest tools invented. Ancient civilizations based their units on body measurements. The Romans measured the length of objects by a "rule of thumb." What was later called an inch was in ancient times the width of a person's thumb! Twelve of these thumb widths roughly equaled the length of a person's foot, thus leading eventually to the standardized unit of measure known as the foot. Why are standards of measurement based on body measurements unreliable?

In 1793, the metric system of measurement was developed in France. This set of standards is known as the International System of Units (*Système International d'Unités* or SI). It is used by scientists all over the world. It is an exact, easy-to-use system, based on decimals. (The money system used in the United States is also based on decimals.) In most countries, it is also used as the system of weights.

Length is measured in meters. The basic unit of length in the metric system is the meter. A meter is about the length of a softball bat. Scientists have defined the meter exactly, and this definition has been agreed upon by countries throughout the world.

The meter is divided into 100 equal parts, called centimeters. See Figure 1-5. A centimeter is about the width of the tip of your little finger. The prefix centi- means 1/100. A centimeter is 1/100 part of a meter, just as a cent is 1/100 part of a dollar. A millimeter is 1/1,000 part of a meter. The prefix milli- means 1/1,000. A millimeter is about the thickness of a 10-cent coin (Figure 1-5). The fourth unit of length in the metric system is the kilometer. The prefix kilo- means 1,000, and so there are 1,000 meters in one kilometer. The metric units of length are summarized in Table 1-2 on the next page.

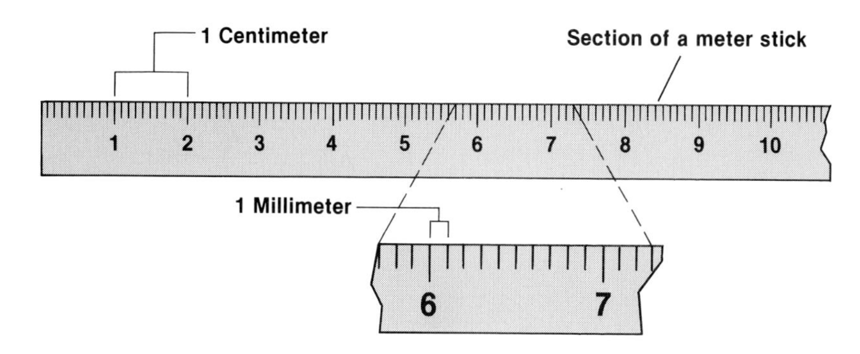

Figure 1-5 A centimeter and millimeter are marked on the metric ruler. How many millimeters are there in one centimeter?

TABLE 1-2

Metric Units of Length	
10 millimeters (mm)	= 1 centimeter (cm)
100 centimeters	= 1 meter (m)
1,000 meters	= 1 kilometer (km)

The basic unit for measuring volume is the liter. All material objects in the world occupy a certain amount of space. A car takes up more space than a book, and a school building takes up more space than a car. The measure of the space occupied by an object is called its **volume**. The volume of a box-shaped object is found by multiplying its length, width, and height:

$$\text{volume} = \text{length} \times \text{width} \times \text{height}$$

USING A METER STICK

activity

OBJECTIVES: Learn how to make measurements and convert numbers using the metric system.

PROCESS SKILLS

In this activity, you will *measure* and *collect data* in order to learn how to use the metric system.

MATERIALS

paper
pencil
meter stick

PROCEDURE

1. Measure the length and width of your classroom in meters.
2. Repeat Step 1 until you have measured the room three times. Take an average of the three readings for both length and width.
3. Convert both average readings to millimeters and then to kilometers.

OBSERVATIONS AND CONCLUSIONS

1. How do your results compare with those of your classmates? Explain any differences.
2. Why were you asked to take the average of three separate readings for both length and width?

The liter (LEE-*tur*) is the metric unit for measuring volume. The volumes of liquids and gases are often measured in units of liters. The same prefixes are used for parts of a liter as are used for parts of a meter. A cube measuring 10 cm on each side has a volume of 1,000 cubic centimeters, in other words, 10 cm × 10 cm × 10 cm = 1,000 cm^3. A volume of 1,000 cm^3 is defined as one liter (L). A box with a volume of 1,000 cm^3 would hold 1 L of liquid, as shown in Figure 1-6. Note that one milliliter (mL) is the same as one cubic centimeter (cm^3).

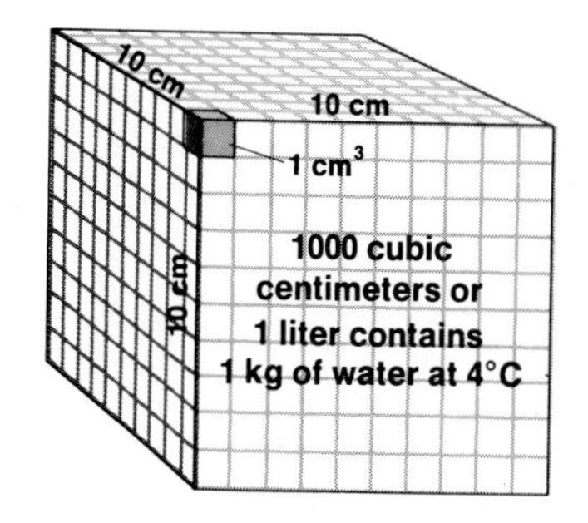

Figure 1-6 The large cube represents the volume of 1 L, or 1,000 cm^3. What part of a liter is represented by the small cube in the corner?

SAMPLE PROBLEM

A rectangular wooden box is 2,000 mm long, 0.50 m wide, and 72 cm high. Find the volume of the box.

SOLUTION

Step 1: Analyze

You are given that the box's length is 2,000 mm, width is 0.50 m, and height is 72 cm. Its volume is the unknown.

Step 2: Plan

Since you know the length, width, and height, you can use the following equation to find the volume:

volume = length × width × height

Since the three given values do not have the same units, you must convert them to meters.

Step 3: Compute and Check

$$2{,}000 \text{ mm} \times \frac{1 \text{ m}}{1{,}000 \text{ mm}} = 2.0 \text{ m}$$

$$72 \text{ cm} \times \frac{1 \text{ m}}{100 \text{ cm}} = 0.72 \text{ m}$$

Substitute all converted values in the equation.

$$\begin{aligned} \text{volume} &= 2.0 \text{ m} \times 0.50 \text{ m} \times 0.72 \text{ m} \\ &= 0.72 \text{ m}^3 \text{ (cubic meters)} \end{aligned}$$

(Note: $m \times m \times m = m^3$)

Since the unit m^3 can be converted to liters (the metric unit for volume) by multiplying by 1,000, your solution is complete.

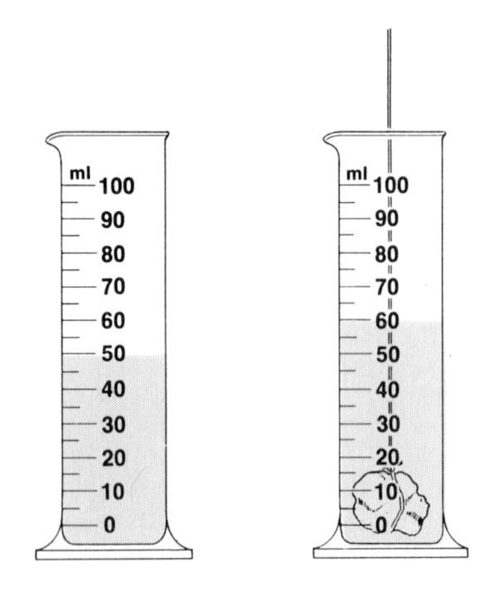

Figure 1-7 What is the volume of the rock in the graduated cylinder?

To find the volume of a small irregular object such as a rock or piece of metal, you can use the method of water displacement. Fill a graduated cylinder about half-full with water and note the level of the water (see Figure 1-7). Then place the irregular object in the water. Since the object displaces some of the water, the water level will rise. Now take a second reading of the water level. The difference between the two readings is equal to the volume of the object in milliliters or cubic centimeters. How accurate do you think this method is? Try it two more times on the same object to see whether there are any differences in volume. Would an average of these three readings give you a more accurate value for the volume than just one reading? Why?

The basic unit for measuring time is the second. For thousands of years, scientists have tried to find an accurate standard for measuring time. Early clocks like the sundial and the hourglass were off by as much as 30 minutes (min) in a single day. In 1894, astronomers developed a standard for keeping time. One second (sec) was defined as the time it took the earth to complete 1/86,400 of its daily rotation on its axis. In other words, an average 24-hour (hr) day was divided into 86,400 seconds (24 hr × 60 min/hr × 60 sec/min = 86,400 sec). However, the earth's exact rotation time varies slightly from day to day, partly due to the tilt of its axis, and thus making the earth's rotation an inaccurate time standard. The National Institute of Standards and Technology in Washington, D.C., currently maintains highly accurate atomic clocks for measuring time. These clocks measure time by counting the vibrations made by atoms of the element cesium, which vibrate 9,192,631,770 times per second.

Weight is a measure of force. A book falls to the floor because it is pulled to the earth. Any push or pull on an object is called a **force**. The unit for measuring force in the metric system is the newton (N). A measure of the pull between the earth and an object is called its **weight**.

The attraction of the earth for a given object is called the force of **gravity**. The pull of gravity is less on top of a mountain than it is in a valley. However, the difference in gravitational pull is so small that it would be hard to measure. The farther an object is from the center of the earth, the less the force of gravitational attraction and the less the object weighs.

The weight of an object depends on the following two factors: (1) the distance of the object above the surface of the earth (or other body such as the moon or planets) and (2) the amount of matter in the objects. The same object on the moon weighs less because the moon is less massive than the earth. Sugar, wood, water, air, the earth, and the moon are examples of matter. **Matter** is anything that takes up space and has weight.

The weight of a small object can be found by hanging it from a spring scale (Figure 1-8A). A spring scale consists of a pan or hook that hangs from a spring. The object to be weighed is placed in the pan or on the hook. Since the weight of the object stretches the spring, the reading on the scale shows the force of gravitational attraction between the earth and the object. The greater the stretch of the spring, the greater the force, and therefore, the more the object weighs.

Mass is a measure of the amount of matter. Look at a rubber sponge, a rock, and a book. You can see that each object takes up space and each has a given weight. Suppose you were to take one of the objects to the top of a high mountain or to the moon. You now know that the weight of the object would change. However, the amount of matter contained in the object would remain the same. A measure of the amount of matter in an object is called its **mass**.

Have you ever tried to push a friend in a go-cart or stop the same go-cart once you got it moving? If you have, then you know it is harder to start or stop the go-cart than it is to keep it in motion. This resistance to any change of motion or change in direction is called **inertia**. A soccer ball and a bowling ball of the same size do not have the same inertia. It takes more force to throw the bowling ball at a certain speed than it does to throw the soccer ball at the same speed. You can see that objects having a greater inertia (bowling ball) also have more mass than objects with less inertia (soccer ball). Inertia is a property of matter that opposes any change in the state of motion in matter.

Mass is measured in kilograms. The gram (g) is a unit of mass in the metric system. A paper clip has a mass of slightly less than 1 g. A United States 5-cent coin has a mass of about 5 g. The prefixes for the parts of a gram are similar to the parts of a meter. Thus, a centigram (cg) is 1/100 part of a gram and a milligram (mg) is 1/1,000 part of a gram. The kilogram (kg) is the basic metric unit for mass. It contains 1,000 times the mass of a gram. One kilogram equals 1,000 grams.

The kilogram standard of mass was defined and agreed upon by countries throughout the world. The standard kilogram mass is kept in a vault in France. An exact copy of the standard is stored in the National Institute of Standards and Technology in Washington, D.C.

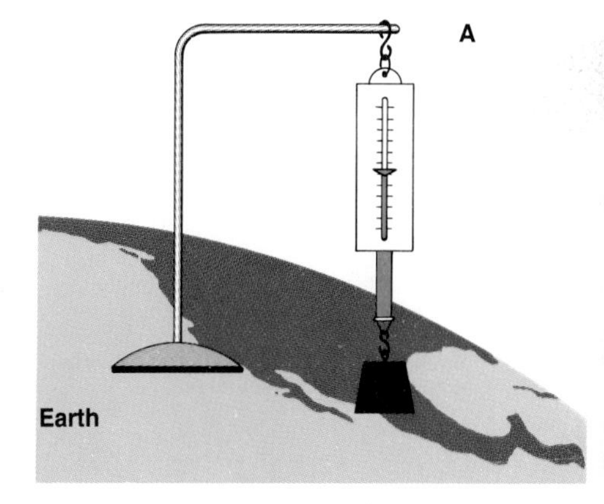

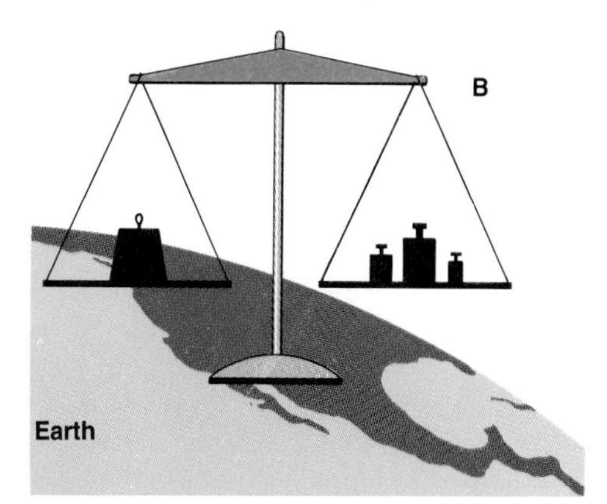

Figure 1-8 A spring scale (A) and a platform balance (B) are often used in the science laboratory. Describe how these devices are used to compare masses.

Figure 1-9 The student uses energy (does work) when she climbs the stairs against the force of gravity (F_g) through a vertical distance (d).

Masses can be compared by using a spring scale or a platform balance. See Figure 1-8. Since the pull of gravity is greater on a large mass than it is on a small one, the spring in the spring scale stretches farther for a large mass than for a small one. By reading the amount the spring stretches on a scale, comparisons between masses can be made.

At sea level on the earth, the weight of a 1-kg mass is about 9.8 N. A 2-kg mass weighs about 19.6 N (2 kg × 9.8 N/kg = 19.6 N), and a 60-kg student weighs about 588 N (60 kg × 9.8 N/kg = 588 N). You can see that, for a given location, the weight of an object is proportional to its mass. Since the pull of gravity on the moon is about 1/6 that of the earth, a 60-kg student would weigh only 98 N (1/6 × 588) on the moon. For objects near the earth's surface, you can use the following formula relating weight and mass:

$$\text{weight in newtons} = 9.8 \text{ N/kg} \times \text{mass in kilograms}$$

Using a platform balance (Figure 1-8B), you can also find the mass of a small object by placing the object on the left pan and one or more known masses on the right pan. A platform balance operates like a seesaw. When the pans balance, the mass of the object is equal to the total of the known masses. Would the pans still be in balance if the balance were taken to the moon? Explain.

Energy has several forms. The ability to do work is defined as **energy.** For example, you do work when you lift an object from the ground onto a truck against the pull of the earth's gravity. The amount of work done depends on the weight of the object and the vertical distance it is moved. The greater the weight or longer the distance, the greater the work involved. Walking up a set of stairs requires a form of energy called mechanical energy (Figure 1-9). Other forms of energy include heat, light, sound, electricity, magnetism, and nuclear energy. You will learn more about energy in later chapters.

The sun is our chief source of energy. It provides us with heat and light. The energy that the earth receives from the sun can be changed into other forms of energy. For example, sunlight can be changed directly into electricity by solar cells. A solar cell is a panel covered by a light-absorbing material such as silicon. Solar cells have been used to provide the energy to power many earth satellites. The sun's energy can also be used to provide heat and hot water for homes and other buildings (Figure 1-10). Large panels called solar collectors can be placed on the roof of a building to collect the sun's heat energy. This energy can then be stored for later use. Almost any kind of energy can be changed from one form to another. Solar energy may someday be an important source of many forms of energy.

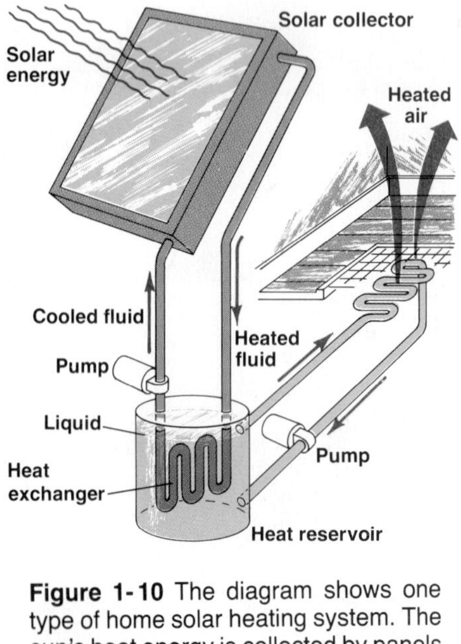

Figure 1-10 The diagram shows one type of home solar heating system. The sun's heat energy is collected by panels on the roof. This energy then heats a liquid that circulates through the house. In other solar heating systems, the sun's energy is stored in drums of water or in bricks and used when needed.

BIOGRAPHY

ALBERT EINSTEIN

the reluctant genius

Albert Einstein was one of the greatest scientists in history. In fact, many people think that Einstein had the most creative mind of all time! No matter how he is remembered, it is clear that his contributions to science will never be forgotten.

Einstein was born in Germany in 1879. Difficult though it is to imagine, as a youth, this scientific genius was a slow learner and a poor student. At the age of three, his parents doubted that he would ever learn to speak. In high school, Einstein was only interested in mathematics. Having received poor grades in geography, language, and history, he dropped out of school at age 15. Later, however, he completed his schooling in Switzerland.

Einstein is best known for his theory of relativity, which he first proposed when he was only 26 years old. One part of this theory says that when an object is moving, its mass increases. This new mass, which increases rapidly as the object approaches the speed of light, is called its relativistic mass. Most speeds, however, are well below that of light. Therefore, for most purposes, the relativistic mass is not used.

In 1921, Einstein was awarded the Nobel Prize in physics. He won this prize for providing evidence that suggests light is made up of a stream of tiny particles called photons. Shortly before World War II, he left his homeland to live and work in the United States.

Einstein lived a quiet life in Princeton, New Jersey. Toward the end of his life, he fought for world agreement to end the threat of nuclear warfare. Shortly after his death in 1955, element number 99—einsteinium—was discovered and named in his honor.

CHAPTER REVIEW 1

SUMMARY

1. Science is the study of the world around us; it is largely a process of learning by which new knowledge is found.
2. Physical science deals with matter, force, and energy.
3. The basic metric units for length, volume, time, weight, and mass are the meter, liter, second, newton, and kilogram, respectively.
4. Volume is the amount of space occupied by an object.
5. Matter is anything that occupies space and has a certain weight.
6. Weight is the force of attraction between the earth and an object.
7. Mass is a measure of the amount of matter in an object. Mass is also a measure of an object's inertia.
8. Energy is the ability to do work.

VOCABULARY

Match the item in the left column with the best answer in the right column. Do not write in this book.

1. energy
2. force
3. gravity
4. inertia
5. mass
6. matter
7. science
8. scientific method
9. volume
10. weight

a. orderly search for answers to questions about our world, our universe and beyond
b. measure of the space occupied by an object
c. any push or pull on an object
d. measure of the pull between the earth and an object
e. attraction of the earth for a given object
f. measure of the amount of matter in an object
g. resistance to any change of motion or change in direction
h. ability to do work
i. anything that takes up space and has weight
j. systematic approach used in collecting scientific information

REVIEW QUESTIONS

1. How are science and curiosity related?
2. In your own words, what is science? What are some reasons people study science?
3. What is meant by the physical sciences?
4. What are the basic units of length, volume, and mass in the metric system?
5. Describe how you could find the volume of a piece of coal?

6. How would you determine the mass of a small object?
7. What happens to a person's mass as the distance from the earth increases? What happens to a person's weight as the distance increases?
8. What is energy? Name five forms of energy.

CRITICAL THINKING

9. What is the connection between science and technology?
10. List some advantages and disadvantages of using the metric system in the United States for everyday purposes.
11. Using the information contained in Figure 1-6, find the mass of 1 mL of water.
12. What problem arises when you use water in the method shown in Figure 1-7 to determine the volume of a small piece of hard candy?
13. Suggest a way of finding the volume of an irregularly shaped sponge.
14. List the pros and cons associated with the following scientific developments: (a) nuclear power plants, (b) plastics, and (c) insecticides.

PROBLEMS

1. The distance from A to B is 3.5 km. Express this distance in (a) meters, (b) centimeters, and (c) millimeters.
2. How many liters are there in 4,000 mL?
3. A steel rod is 25 cm in length. Find its length in (a) meters, (b) millimeters, and (c) kilometers.
4. A rectangular box is 3.0 m long, 2.3 m wide, and 150 cm high. Find its volume in cubic meters (m^3).
5. An object has a mass of 3 kg. What is its mass in grams? What is its weight at sea level?

FURTHER READING

Ardley, Neil. *Making Metric Measurements.* New York: Franklin Watts, Inc., 1984.

Beller, J. *So You Want to Do a Science Project!* New York: Arco Publishing, Inc., 1982.

Parker, Barry. *Einstein's Dream: The Search for a Unified Theory of the Universe.* New York: Plenum Publishing Corp., 1986.

Shapiro, S. *Exploring Careers in Science.* New York: Rosen Publishing Group, 1989.

PROPERTIES, CHANGES, AND COMPOSITION OF MATTER

As the temperature drops, water on the surface of a pond freezes into tiny ice crystals. When this happens, how has the water changed? Is the composition of the matter still the same?

CHAPTER 2

SECTIONS

OBJECTIVES

- ☐ Explain the concept of density.
- ☐ Distinguish between physical and chemical properties, and between physical and chemical changes.
- ☐ Distinguish between atoms and molecules, and between elements, mixtures, and compounds.
- ☐ Identify common chemical symbols and chemical formulas, and use chemical equations to describe chemical reactions.
- ☐ Explain the law of constant proportions and the law of conservation of matter.

2.1 WAYS TO IDENTIFY MATTER

Matter ***has both physical and chemical properties.*** In day-to-day living, you learn to recognize many different materials. In order to tell one material from another, you look for certain traits that help identify a given material. These traits are called properties. Some properties are easy to see and to measure, whereas others are not. The properties of a material can be described as either physical or chemical.

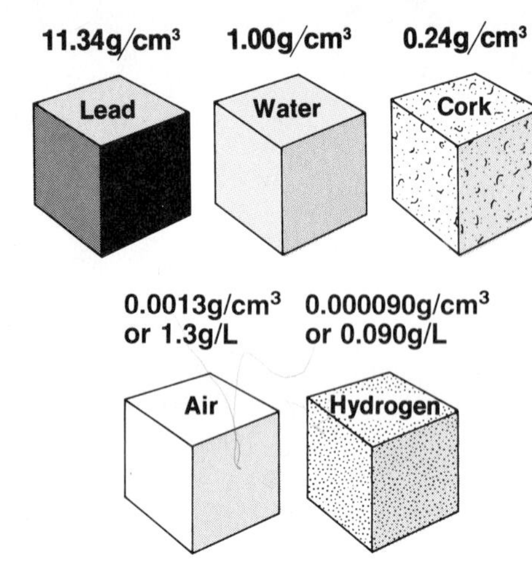

Figure 2-1 The substances in the diagram have equal volumes but different masses. Which material has the greatest density? Which material is the least dense?

Physical properties can be easily seen or measured without altering the identity of that substance. Physical properties of a material include color, hardness, solubility (how much of it will dissolve in a certain amount of liquid), freezing temperature (the temperature at which a material freezes), and boiling temperature (the temperature at which a material boils).

How would you describe how heavy or light a material is? Scientists describe this physical property by referring to the density of the material. The **density** of a substance is its mass per unit volume:

$$\text{density} = \frac{\text{mass}}{\text{volume}}$$

Density is a measure of how close the particles of matter are packed together. The density of solids and liquids is most often expressed in grams per cubic centimeter (g/cm^3). For example, the density of water is 1.0 g/cm^3. The density of gases is expressed in grams per liter (g/L). Compare the densities of the solids, liquids, and gases in Figure 2-1.

SAMPLE PROBLEM

A piece of brass has a mass of 32.0 g and a volume of 3.8 cm^3. Find the density of brass.

SOLUTION

Step 1: Analyze

You are given that the piece of brass has a mass of 32.0 g and a volume of 3.8 cm^3. Its density is the unknown.

Step 2: Plan

Since you know the mass and volume, you can use the following equation for density:

$$\text{density} = \frac{\text{mass}}{\text{volume}}$$

Step 3: Compute and Check

Substitute all values in the equation.

$$\text{density} = \frac{32.0\ \text{g}}{3.8\ \text{cm}^3}$$

$$= 8.4\ \text{g/cm}^3$$

Since density in solids is most often expressed in grams per cubic centimeter, your solution is complete.

The chemical properties of a material are not easy to see or measure. Chemical properties describe the way one substance can be changed into a different substance with an entirely different set of properties. One such property is whether or not a substance will burn when exposed to heat. For example, magnesium burns with a bright white flame as it joins with oxygen in the air to form a magnesium oxide. Another chemical property is whether a substance will react with other substances. The metal gold does not react with water or common acids. Chemical properties, as well as physical properties, are extremely useful in identifying an unknown substance.

2.2 CHANGES IN MATTER

Matter undergoes physical and chemical changes. Changes in matter are always taking place. Ice changes into water and then into steam when it is heated. Liquid gasoline evaporates quickly unless kept in a tightly closed container. A piece of chalk can be ground into a fine powder. All of these changes are alike in one way. They do not result in new substances. Such changes are called physical changes (Figure 2-2). Water is the same chemically whether it is in the form of ice or steam. Gasoline remains gasoline even after it evaporates. Chalk particles remain chalk no matter what the size or shape of the pieces.

Figure 2-2 The spice pepper comes from peppercorns that have been ground into a fine powder. Is the grinding of pepper an example of a chemical or a physical change?

Changes in matter that produce new substances are called chemical changes. The decay of dead plants, the souring of milk, the rusting of iron, and the digestion of food are examples of chemical changes. In each of these processes, new substances are formed. Their chemical makeup and properties are different from those of the original substances. See Figure 2-3. Magnesium burns to produce a white powder called magnesium oxide. Magnesium oxide does not have the color or metal luster of magnesium. Iron rust is chemically different from iron. Sweet milk is not the same chemically as sour milk.

Figure 2-3 New substances are formed when a tree burns. Is this an example of a chemical or a physical change?

Whenever a physical or chemical change takes place, it is always accompanied by an energy change. This change in energy is often detected by the absorption or release of heat, light, sound, or electricity. Examples of physical changes in which heat is absorbed are the melting of ice and the evaporation of water. The burning of wood is a chemical change in which heat and light are released.

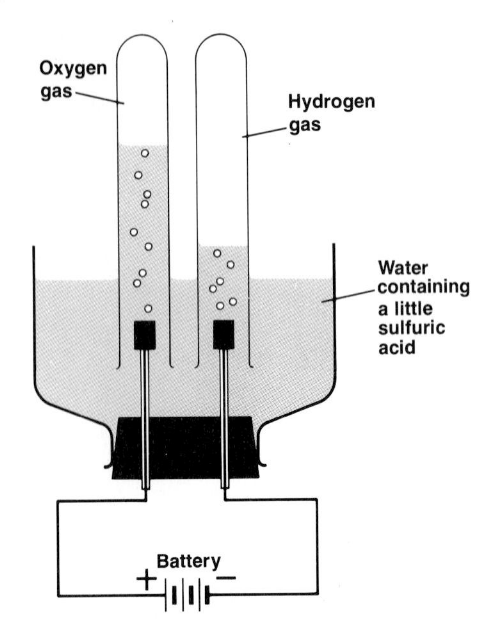

Figure 2-4 When an electric current is passed through water, the water breaks down into hydrogen and oxygen. **CAUTION: Sulfuric acid is corrosive. Only attempt this experiment in a well-ventilated room.**

Electrolysis is a chemical change. The laboratory setup pictured in Figure 2-4 can be used to demonstrate a chemical change. The experiment illustrates that water can be broken down into two different gases.

To begin the experiment, a jar and two test tubes are filled with water to which a small amount of sulfuric acid has been added. The acid is added to the water to make the liquid conduct an electric current. The current from a battery separates the water into two colorless gases, oxygen and hydrogen, that collect in the separate tubes. The chemical change that takes place when an electric current passes through a liquid is called **electrolysis** (*ih*-LEK-TRAHL-*uh*-*sis*). Electrolysis has many industrial uses, such as the purification of metals.

If the tubes in the diagram in Figure 2-4 were not labeled, how could you determine which gas is in each tube? Knowing some of the chemical properties of each gas would help you to identify them. Pure hydrogen burns with an almost colorless flame. When ignited, pure hydrogen combines with oxygen in the air to form water. This chemical change produces a popping sound. Pure oxygen, on the other hand, does not burn. However, it does support the burning of other substances. If you place a burning piece of wood into one of the test tubes and hear a pop, you can conclude that the tube contains hydrogen. If you place a glowing piece of wood into the other tube and it bursts into flame, you know the tube contains oxygen. You will learn more about hydrogen and oxygen in Chapter 7.

Matter
Element
Carbon (diamond)
Compound
Water (H_2O)
Mixture
Soil (sediments and organic matter)

Figure 2-5 Examples of the three classes of matter.

2.3 COMPOSITION OF MATTER

Matter can be divided into three classes. There are many different kinds of substances. Therefore, when scientists study substances, they must first classify them into similar groups. Scientists have found that all forms of matter can be classified on the basis of their properties into three major classes. These three classes are elements, mixtures, and compounds (Figure 2-5).

Elements are simple substances. Recall that water is made up of two gases, hydrogen and oxygen, that have different chemical properties. Hydrogen and oxygen cannot be broken down any further into simpler substances. A substance that cannot be broken down into simpler substances by chemical means is called an **element.**

Elements are nature's building blocks. All of the material in the universe is made up of elements or combinations of elements. There are 103 recognized elements on the earth. Some are common; others are rare. Scientists have been able to make a few of these elements under special conditions in the laboratory. You will learn about some of the laboratory-made elements in Chapter 6.

Table 2-1 contains a list of some of the more common elements found in the earth's crust and their relative concentrations. These elements are not concentrated evenly throughout the crust. Notice that many familiar elements, such as copper, lead, zinc, silver, and gold, are not included on this list. These important metals are rare in the earth's crust.

TABLE 2-1

Percent of Elements in the Earth's Crust

Element	Percent by Mass
Oxygen	49.5
Silicon	25.8
Aluminum	7.5
Iron	4.7
Calcium	3.4
Sodium	2.6
Potassium	2.4
Magnesium	1.9
Hydrogen	0.9
Titanium	0.6
All other elements	0.7

Mixtures are physical combinations of materials. An endless number of words can be created from the 26 letters in the alphabet. In the same way, an almost endless number of both physical and chemical combinations can be made from the 90 elements commonly found on the earth. Millions of these combinations exist in nature or have been created in the laboratory.

Substances that are combined physically are called mixtures. Air, for example, is a mixture of many gases including nitrogen and oxygen. In order for a combination of substances to be called a mixture, certain conditions must apply. Consider the following. When powdered iron and powdered sulfur are mixed, they form a grayish material.

Both iron and sulfur are elements. These two elements can be brought very close together by mixing, but they will not combine with each other to form a new substance. A strong magnet can separate the powdered iron from the sulfur. In other words, each element in this mixture maintains its own properties. In short, a **mixture** is a material containing different substances that have not been joined chemically. Since the substances in a mixture are not chemically combined, they can each be present in any amount.

MAKING AND SEPARATING A MIXTURE

activity

OBJECTIVE: Perform an experiment to determine whether a mixture can be separated.

PROCESS SKILLS

In this activity, you will *experiment* in order to create a method for separating a mixture and *communicate* your technique.

MATERIALS

- metric balance
- mortar and pestle
- table salt
- powdered charcoal
- distilled water
- 2 100-mL beakers
- stirring rod
- filter paper
- funnel
- gas burner
- safety goggles

PROCEDURE

1. Use a metric balance to measure 20 g of table salt and 15 g of powdered charcoal. CAUTION: *Be sure to keep powdered charcoal away from an open flame.*
2. Place both substances in a mortar. Grind them together using a pestle until you obtain a gray powder.
3. Using the equipment provided or additional materials, devise a way to separate the mixture into table salt and powdered charcoal. CAUTION: *Wear goggles when using the gas burner.*

OBSERVATIONS AND CONCLUSIONS

1. Explain how you separated the mixture into the two original substances.
2. Were the quantities of salt and charcoal significant to the mixing or separation? Explain your answer.

Compounds are chemical combinations of elements. If a sample of the iron-sulfur mixture is placed in a test tube and heated, the iron will unite chemically with the sulfur. (CAUTION: *Do not attempt this experiment yourself.*) The new substance formed is the compound iron sulfide. A **compound** is a substance composed of elements that are joined chemically. The physical and chemical properties of iron sulfide are different from those of the two original elements. Although mixtures can be separated by physical means, a compound can be broken down only by chemical means. See Figure 2-6.

Table 2-2 contains some examples of elements, mixtures, and compounds. Study the table carefully. Name three more examples that could be added under each heading.

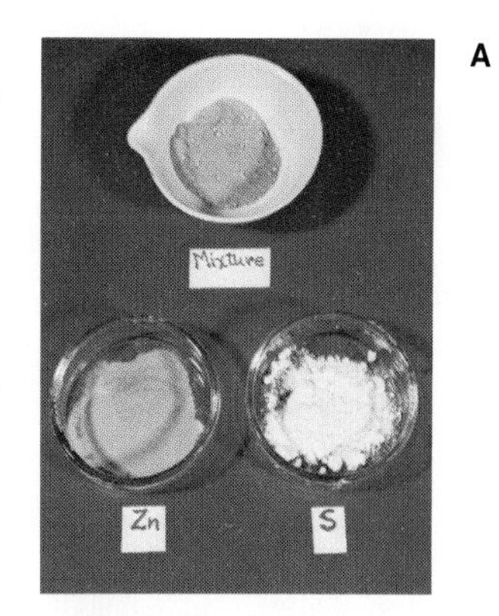

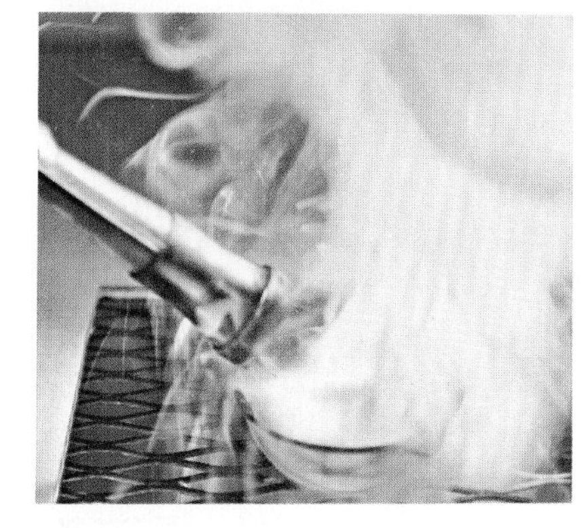

Figure 2-6 The elements zinc (Zn) and sulfur (S) do not react when mixed at room temperature, as shown in A. When they are heated, a chemical reaction takes place, as shown in B. In this reaction, heat is given off and the compound ZnS is formed. What forms of energy are released?

TABLE 2-2

Three Classes of Matter

Elements (symbols)	Mixtures	Compounds (formulas)
Tin (Sn)	Milk	Water (H_2O)
Iron (Fe)	Butter	Sugar ($C_{12}H_{22}O_{11}$)
Aluminum (Al)	Toothpaste	Table salt (NaCl)
Silver (Ag)	Baking powder	Baking soda ($NaHCO_3$)
Oxygen (O)	Paint	Ethyl alcohol (C_2H_5OH)
Copper (Cu)	Air	Carbon dioxide (CO_2)

A molecule is the smallest unit of a substance. Picture a lump of sugar. Imagine dividing it into two pieces, then dividing these two pieces into four, and the four into eight, and so on. You would eventually reach a point at which each particle is so small that, if you divided it again, the two halves would no longer be sugar. This small particle, or unit, would be one molecule of sugar. Any further division of this molecule would mean that its parts would no longer have the properties of sugar. The smallest unit of a substance that has all the properties of that substance is a **molecule.** Molecules are so small that even with the aid of an electron microscope only a few of the largest ones have even been seen.

An atom is the smallest unit of an element. You have learned that water is an example of a compound. Recall that each molecule of water contains particles of hydrogen and oxygen that have been joined chemically. These single particles of hydrogen and oxygen are called atoms. An **atom** is the smallest unit of an element that can combine chemically with other elements. All matter is made up of various combinations of atoms. In the case of water, one atom of oxygen combines with two atoms of hydrogen to form one molecule of water.

Each molecule of a compound is made of two or more atoms. Molecules of some gaseous elements, including oxygen and hydrogen, are made up of two identical atoms. Atoms of elements join in many combinations. It is estimated that there are over a million compounds known today. New compounds are being made almost daily in research laboratories around the world.

2.4 CHEMICAL SYMBOLS, FORMULAS, AND EQUATIONS

A chemical symbol identifies an element. Each element is designated by a symbol. A **chemical symbol** is a one- or two-letter abbreviation used to represent an element. For example, O is the symbol for oxygen, H is for hydrogen, C is for carbon, Al is for aluminum, Mg is for magnesium, S is for sulfur, Cl is for chlorine, and Ca is for calcium. See Figure 2-7.

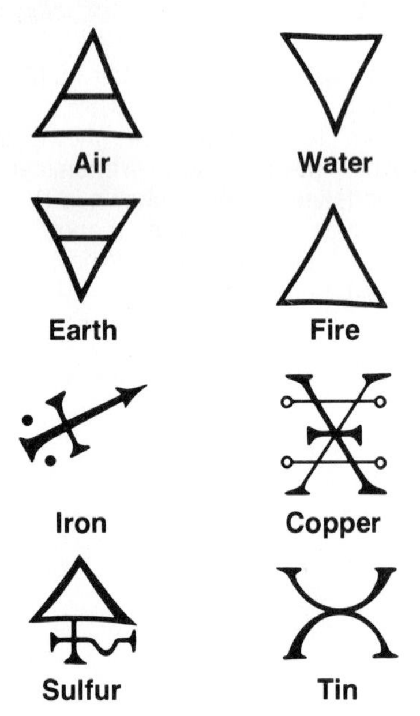

Figure 2-7 Chemists in ancient times believed that all matter was made up of the basic substances air, water, earth, and fire. They used triangles as symbols for these substances, as shown above. Other early symbols for several elements are also shown. How are today's chemical symbols an improvement over these earlier ones?

Many elements have been known since ancient times. The contemporary symbols for these elements were taken from the shortened form of their Latin names. Some examples of these symbols are Au (*aurum*) for gold, Ag (*argentum*) for silver, Pb (*plumbum*) for lead, Cu (*cuprum*) for copper, and Fe (*ferrum*) for iron. A list of common elements and their symbols appears in Table 3-2, page 48. The complete chart of elements appears in Figure 3-6 on pages 44–45.

Chemical formulas identify compounds. With so many known compounds, it would be very difficult to keep track of them all if each were referred to only by name. **Chemical formulas** are used to identify compounds by showing the kind and number of atoms that are present.

You probably know that H_2O is the chemical formula for the compound water. This formula can also be written as HOH. The formula indicates that water is a compound composed of the elements hydrogen and oxygen. It also indicates that one molecule of water consists of two atoms of hydrogen (2H) and one atom of oxygen (O). Two molecules of water are written as $2H_2O$.

Notice that chemical symbols are used in writing chemical formulas. A number written slightly below the line following a symbol indicates how many atoms of that element are found in the molecule. For instance, 1 molecule of table sugar, $C_{12}H_{22}O_{11}$, has 12 atoms of carbon (C), 22 atoms of hydrogen (H), and 11 atoms of oxygen (O). You would read this formula as C twelve, H twenty-two, O eleven. There are 45 atoms in 1 molecule of this kind of sugar. Table 2-3 contains a list of several examples of chemical compounds and their formulas.

TABLE 2-3

Some Chemical Compounds and Their Formulas

Name of Compound	Formula
Magnesium oxide	MgO
Sodium chloride (table salt)	$NaCl$
Sodium carbonate	Na_2CO_3
Sucrose (table sugar)	$C_{12}H_{22}O_{11}$
Sulfuric acid	H_2SO_4
Iron sulfide	FeS
Calcium hydroxide	$Ca(OH)_2$

Chemical equations represent chemical changes. One way to explain what takes place in a chemical change is to describe the change in words. This description is called a word equation. Another way of describing a chemical change is to use a chemical equation. Both equations state briefly what elements or compounds enter into a chemical change and what elements or compounds are produced by it. However, a word equation does not reveal as much information about a chemical change as does a chemical equation. Consider the following example. When heated, a mixture of iron and sulfur produces iron sulfide. The word equation for this reaction is written as

$$\text{iron} + \text{sulfur} \rightarrow \text{iron sulfide}$$

The arrow (→) in this case means "yields." This equation does not tell what was done to the iron-sulfur mixture to produce iron sulfide. Adding the fact that heat was applied, the equation can be rewritten as

$$\text{iron} + \text{sulfur} \xrightarrow{\text{heat}} \text{iron sulfide}$$

Even though it may seem complete, this word equation lacks certain critical information.

Chemical equations give greater detail about the arrangement of atoms taking part in a chemical change. The reaction involving iron and sulfur could be rewritten using the following chemical equation:

$$Fe(s) + S(s) \xrightarrow{\text{heat}} FeS(s)$$

This equation is read as one atom of iron and one atom of sulfur, when heated enough, will yield the simplest unit of iron sulfide. Of course, in an actual chemical change, many billions of particles are involved in the reaction. The symbols in parentheses tell you the state of matter of each component in the equation, which in this case is solid (s). The state of matter can also be gas (g), liquid (l), or aqueous (aq). Aqueous means that a substance exists as a mixture with water. You will learn more about the states of matter in Chapter 4.

You already know that electrolysis of water yields hydrogen and oxygen. This chemical change can be described simply by using the following word equation:

$$\text{water} \xrightarrow{\text{electricity}} \text{hydrogen} + \text{oxygen}$$

Note that in this word equation, some information is still unclear. How many atoms and molecules of each substance are involved in the reaction?

Tests reveal that a molecule of oxygen gas is made up of two atoms of oxygen. A molecule of oxygen is written as O_2. Similarly, a molecule of hydrogen gas is made up of two atoms of hydrogen. The hydrogen molecule is written as H_2. Since water can be broken down into two volumes of hydrogen and one volume of oxygen, the chemical equation is written as

$$2H_2O(l) \rightarrow 2H_2(g) + O_2(g)$$

This equation is read as two molecules of water yield two molecules of hydrogen and one molecule of oxygen. Notice that there are the same number of hydrogen and oxygen atoms on the left side of the chemical equation as there are on the right side. There are four hydrogen atoms and two oxygen atoms on each side.

A chemical equation reveals what occurs during a chemical reaction. To write a chemical equation, you must know the facts about the reaction. In other words, you must know all the reactants and all the products involved. **Reactants** are the substances that enter into a chemical reaction. Reactants are always written on the left hand side of the arrow. **Products** are substances resulting from a chemical reaction. They are always written on the right hand side of the arrow. You will learn more about writing chemical equations in the next section.

2.5 LAWS OF NATURE

Matter combines chemically in definite amounts. Scientists believe that everything in the universe behaves in an orderly way. They also believe that there are rules for these behaviors. A general statement that describes the behavior of nature is called a **scientific law** or principle.

To understand what a scientific law is, consider the following example. The compound iron sulfide (FeS) is always made of seven parts by mass of iron (Fe) and four parts by mass of sulfur (S). This means that the compound iron sulfide (FeS) contains 63.5 percent by mass of iron and 36.5 percent by mass of sulfur. Although the makeup of mixtures can vary, the same is not true for compounds. When elements unite in a given compound, they always unite in the same proportions. If there is too much iron in proportion to sulfur, the excess iron does not become a part of the compound. The same would apply if there were an excess of sulfur. This fact is expressed as the law of constant proportions. This law states that every compound always contains the same proportion by mass of the elements of which it is formed. The reason why this law holds true is explained in the next chapter.

In chemical changes, mass is neither gained nor lost. In ordinary physical and chemical changes, mass is not created nor is it lost. This idea can be demonstrated using the setup shown in Figure 2-8. (CAUTION: *All lead compounds are poisonous.*) In this chemical reaction, about 100 mL of sodium iodide (NaI) solution (15 g/L) is poured into a flask while 20 mL of lead nitrate [$Pb(NO_3)_2$] solution (33 g/L) is poured into a test tube. The test tube is placed with care into the flask in an upright position. The flask is stoppered and its mass is found, using a balance. The flask is then removed from the balance and turned upside down. When the solution in the test tube comes in contact with the solution in the flask, a chemical change occurs. A yellow solid is formed that is neither sodium iodide nor lead nitrate. This solid separates from the liquid and is called a **precipitate.** The formation of an insoluble precipitate is evidence that a chemical reaction has taken place. The new compound formed is lead iodide (PbI_2). If the flask is put back on the balance, you find that its mass remains the same. No mass has been gained or lost in the reaction. This experiment demonstrates the law of conservation of matter, which states that matter cannot be made or destroyed by ordinary chemical means.

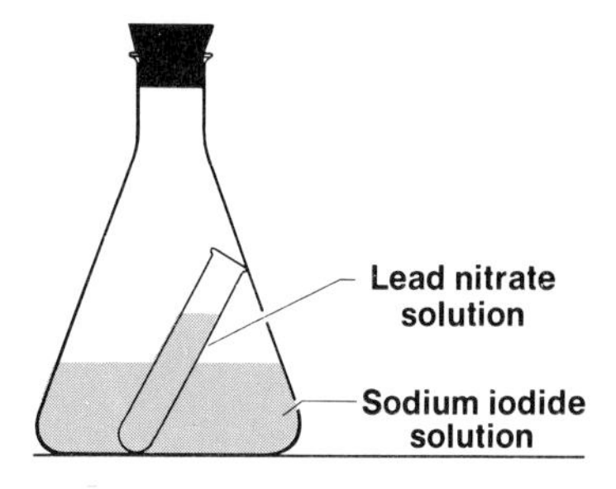

Figure 2-8 When the flask is turned upside down so that the two solutions come in contact, a chemical change takes place. How can you tell that a chemical reaction has taken place?

Now examine the following equation:

$$Mg(s) + HCl(aq) \rightarrow MgCl_2(aq) + H_2(g) \text{ (not balanced)}$$

Notice that on the left side of the equation there is one atom of hydrogen and one atom of chlorine. On the right, there are two atoms of hydrogen and two atoms of chlorine. According to the law of conservation of matter, chemical changes cannot create or destroy atoms. If you could put the reactants and products of this reaction on a double-pan balance, you would find that there is no gain or loss of mass during the reaction. An equation must be corrected, or balanced, to show this conservation of matter. It takes two parts of HCl to provide the additional atoms. This is indicated by the balanced equation

$$Mg(s) + 2HCl(aq) \rightarrow MgCl_2(aq) + H_2(g) \text{ (balanced)}$$

Now there are the same numbers and same kinds of atoms on both sides of the equation. The equation is balanced. Of course, in any chemical change, huge numbers of atoms and molecules take part. These atoms and molecules always combine in the proportion indicated by the whole numbers written in front of the symbols and formulas of the equation.

Thus, to write a balanced chemical equation, numbers may be needed in front of each formula. These numbers ensure that the numbers of atoms of each element are the same on both sides of the equation. This process is called balancing an equation. The smallest whole numbers are used to balance equations.

In writing chemical equations, three conditions must be satisfied:

1. Chemical equations must represent what is observed in the laboratory.
2. The formulas of compounds or the symbols of elements involved in the reaction must be written correctly in the equation.
3. The equation must be balanced. This means that the number of atoms of each element must be the same on both sides of the equation.

There are two things that a chemical equation does not indicate: (1) the rate (speed) at which a reaction takes place, and (2) the conditions under which a chemical reaction takes place.

The laws of conservation of matter and constant proportions, like all scientific laws, are based on many careful tests. However, keep in mind that science is a continuing process of discovery. Since scientific research is ongoing, new information is continually uncovered and even scientific laws may be changed to fit the newly discovered facts.

activity

CONSERVATION OF MATTER

OBJECTIVES: Initiate a reaction and determine whether matter has been created or destroyed.

PROCESS SKILLS

In this activity, you will *collect and analyze data* and *draw conclusions* about your results.

MATERIALS

metric balance
Erlenmeyer flask
balloon
vinegar
baking soda

PROCEDURE

1. Pour approximately 20 mL of vinegar into the flask. Put approximately 20 g of baking soda into the balloon. Carefully place the mouth of the balloon over the mouth of the flask. Do not let baking soda fall into the flask during this procedure. Lay the full end of the balloon against the side of the flask.
2. Use a metric balance to determine the mass of the flask, balloon, vinegar, and baking soda. Record your measurement.
3. Empty the baking soda into the flask by raising the end of the balloon and shaking gently. As the baking soda and vinegar come in contact, note the reaction that occurs.
4. When the reaction is complete, determine the mass of the flask, balloon, and their contents and compare this mass with the mass determined in Step 2.

OBSERVATIONS AND CONCLUSIONS

1. What kind of change occurred inside the beaker? Support your answer.
2. Did the substance undergo a change in mass?
3. Was matter created or destroyed in this activity? Explain your answer.

CHEMISTRY

from technician to engineer

Many exciting careers can be found in the field of chemistry. Jobs such as laboratory technician, chemist, and chemical engineer are just a few. Each of these jobs has a particular set of responsibilities.

Chemical laboratory technicians perform tests to find out if materials have and retain certain properties. They may test certain materials to find out their chemical content. Technicians also prepare chemical solutions used in making various products, such as drugs, textiles, soaps, foods, and beauty aids. Technicians also work with many laboratory instruments.

Chemists are most often engaged in research and development of new products. They set up tests and gather data. Chemists may make preparations of natural substances or combine existing elements to make new compounds. They sell consumer products, market new products, and consult with other chemists and business people. Many chemists teach in colleges and universities, or work for a range of industries and government agencies.

Chemical engineers apply their skills in chemistry to the design of equipment for new plants and to the development of chemical processes. They are involved in the making of paints, plastics, and hundreds of other items that are in daily use. Chemical engineers help to select new plant sites and to plan the layout of machines and equipment.

To pursue any of these careers, high school courses in physics, chemistry, and mathematics should be taken. Specialized technical training would also be required to become a laboratory technician. A degree program in a college or university is necessary to become a chemist or chemical engineer.

For more information on a career in chemistry, contact

American Chemical Society
1155 16th Street NW
Washington, DC 20036

CHAPTER REVIEW

SUMMARY

1. Density equals the mass of a substance divided by its volume.
2. A physical change does not produce a new substance.
3. A chemical change produces a new substance.
4. Every physical or chemical change in matter involves an energy change.
5. The smallest particle of an element is an atom.
6. A compound is formed by the chemical union of two or more elements. The smallest particle of a compound is a molecule.
7. A chemical equation shows the reactants and products involved in a chemical reaction.
8. The law of constant proportions states that a compound always contains the same proportion by mass of the elements that make up that particular compound.
9. The law of conservation of matter states that matter cannot be created or destroyed by ordinary chemical reactions.

VOCABULARY

Match the item in the left column with the best answer in the right column. Do not write in this book.

1. atom
2. chemical formula
3. chemical symbol
4. compound
5. density
6. electrolysis
7. element
8. mixture
9. molecule
10. precipitate
11. product
12. reactant
13. scientific law

a. smallest unit of an element that can combine chemically with other elements
b. substance that cannot be broken down into simpler substances by chemical means
c. mass per unit volume
d. smallest unit of a substance that has all the properties of that substance
e. general statement describing the behavior of nature
f. insoluble solid that separates from a solution
g. chemical change that occurs when an electric current passes through a liquid
h. material containing different substances that have not been joined chemically
i. one- or two-letter abbreviation for an element
j. expression used to identify compounds by showing the kind and number of atoms
k. substance entering into a chemical reaction
l. substance resulting from a chemical reaction
m. composed of chemically joined elements

REVIEW QUESTIONS

1. What is meant by properties of matter? Give five examples of physical properties.
2. What is density? How is it usually expressed?
3. What is the difference between a physical change and a chemical change? Give five examples of each.
4. Name three classes of matter.
5. What is an element? an atom?
6. Describe a mixture. Name five mixtures and their uses.
7. Define a chemical compound. What is the smallest particle of a compound called?
8. List five compounds and their uses.
9. Make a list of 10 chemical elements and their symbols.
10. What information do chemical formulas and chemical equations give?
11. What is meant by a scientific law or principle?
12. What compound is formed when hydrogen burns in air? Write its formula.

CRITICAL THINKING

13. If you were given a flask of an unknown gas, how would you test for oxygen? for hydrogen?
14. What is wrong with the following statement: "Lead is heavier than aluminum"? How can it be corrected?
15. Describe how you would find the density of a small irregular solid.
16. What does the formula for carbonic acid, H_2CO_3, tell about its composition? How many atoms are contained in a molecule of carbonic acid?
17. The reaction between sodium chloride (NaCl) and silver nitrate ($AgNO_3$) forms a white precipitate of silver chloride (AgCl). Write the complete chemical equation for this reaction.
18. Give an example of the law of constant proportions. Does it apply to mixtures or compounds? Explain.
19. During electrolysis, water is separated into hydrogen and oxygen. Describe a technique in which water can be formed.

PROBLEMS

1. The density of an unknown sample was found to be 11.2 g/cm^3. Could you find the volume of the sample from this information? What do you need to know to find its mass?
2. What is the mass in grams of water that fills a tank 100 cm long, 50 cm wide, and 30 cm high? What is the mass in kilograms?
3. The density of a piece of brass is 8.4 g/cm^3. If its mass is 500 g, find its volume.

4. A graduated cylinder is filled with water to a level of 40.0 mL. When a piece of copper is lowered into the cylinder, the water level rises to 63.4 mL. Find the volume of the copper sample. If the density of copper is 8.9 g/cm^3, what is its mass?
5. The results of an experiment showed that 24 g of magnesium (Mg) combined with 16 g of oxygen (O) to form magnesium oxide (MgO). Find the percent by mass of magnesium and oxygen in the sample. Will these percentages always be the same for magnesium oxide? Explain.

FURTHER READING

Allman, W.F. "The Journey to Inner Space." *U.S. News and World Report* 104 (May 9, 1988): 66–7.

Asimov, Isaac. *The Road to Infinity*. New York: Doubleday and Co., Inc., 1979.

Atkins, Peter William. *Molecules*. New York: W. H. Freeman and Co., 1988.

Berger, Melvin. *Atoms, Molecules, and Quarks*. New York: Putnam, 1986.

Neubauer, Alfred. *Chemistry Today*. New York: Arco Publishing Inc., 1983.

Waxter, Julia B. *Science Cookbook*. Belmont, Calif.: David S. Lake Publishers, 1981.

STRUCTURE OF ATOMS AND MOLECULES

These pyrite crystals are made up of vast numbers of iron and sulfur atoms. How is it possible that two different elements combine to form a new substance? Are iron and sulfur changed in the process?

CHAPTER 3

SECTIONS

OBJECTIVES

- ☐ Describe the structure of elements through the use of atomic models.
- ☐ Define atomic number, atomic mass, and oxidation number of an element.
- ☐ Explain the use of the periodic table of the elements.
- ☐ Explain how elements combine with other elements to form chemical compounds.

3.1 THE WORLD WITHIN THE ATOM

The atomic theory helps you to understand the concept of atoms. As early as 400 B.C., Greek scientists believed that matter could not be destroyed. Democritus (*deh*-MAHK-*rih-tus*), pictured in Figure 3-1, believed that matter was made up of particles. The Greeks called these particles atoms, which means cannot be divided. This particle theory was not widely accepted until the early nineteenth century when an English chemist, John Dalton (1766–1844), proposed the first useful theory of the nature and properties of matter. Although Dalton's theory included some of the old ideas about atoms, he was the first to describe the unchanging makeup of every compound.

Figure 3-1 Democritus (460–370 B.C.), a Greek thinker, was the first to suggest that all matter could be broken down into tiny particles called atoms.

The modern atomic theory was developed from the one proposed by Dalton. Today, scientists can provide evidence for proposing the following:

1. Matter is made up of very small particles called atoms.
2. Atoms of the same element are chemically alike and atoms of different elements are chemically different.
3. Single atoms of a given element may not all have quite the same mass. In any natural distribution, however, they do have a definite average mass that is a property of that element.
4. Atoms of different elements have different average masses.
5. Atoms do not break down in ordinary chemical changes.

Single atoms of any element are much too small to be seen even with the best optical microscopes. Yet, indirect observations of how atoms behave have provided scientists with evidence to support the modern atomic theory. Such indirect observations may be compared to concluding that a bullet was fired from an unseen gun because you heard a loud noise and then saw a hole in a nearby target.

Recall that one atom of oxygen joins with two atoms of hydrogen to form one molecule of water. A drop of water contains large numbers of these molecules. Each molecule of water in the drop is formed in the same way. Therefore, one water molecule will respond in the same way under the same conditions as all other water molecules.

An atom is made up of three basic particles. Scientists have gathered a great deal of evidence about the makeup of atoms. All elements (except hydrogen) contain three kinds of basic particles: electrons, protons, and neutrons.

Electrons and protons are electrically charged particles. The **electron** is a particle with a negative charge (−). The **proton** is a particle with a positive charge (+). The **neutron** is a particle with no charge (0). The mass of an electron is very small. Protons and neutrons also have small masses, but each has about 1,800 times the mass of an electron.

The arrangement of particles in an atom is largely the result of the electrical forces within the atom. The protons and neutrons of an atom are grouped closely together in a central core, called the nucleus. Since neutrons are neutral, the nucleus takes on the positive charge of the protons. Recall that electrons are negatively charged. They move rapidly around the nucleus. As you can see in Figure 3-2, unlike electrical charges attract each other while like charges repel each other. The shape of an atom is due to the attraction between the nucleus and its electrons.

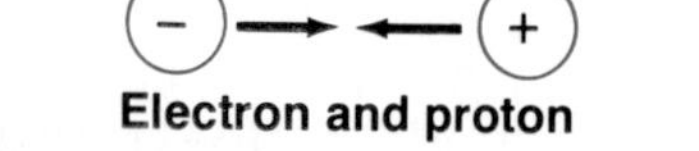

+ +
Proton and proton

− −
Electron and electron

− +
Electron and proton

Figure 3-2 Study the diagram of the atomic particles. What general statements can be made about like charges and unlike charges?

A model of an atom helps you imagine its structure. Because atoms are so small, and thus difficult to observe, what scientists know about them has been learned by observing the behavior of matter made up of atoms. From these observations, scientists make inferences and draw conclusions about the nature of atoms. Scientists find it useful to make diagrams or actual physical models to show what an atom is believed to look like. A model is a way of explaining how or why things behave the way they do. Models like the atomic model are subject to change as new evidence is discovered.

The electron cloud provides a good atomic model. Because an electron moves around the nucleus of an atom with such great speed, its position at any one time is uncertain. Thus it is hard to pinpoint an electron's exact position. Instead, all of the electrons are pictured as occupying a cloud-like region around the nucleus. Think of a spinning airplane propeller or electric fan. When the blades are moving very fast, their outline is no longer clear. Similarly, the fast-moving electrons also seem to form a hazy sphere around the nucleus. This spherical region in which electrons move is called an electron cloud.

Not all electrons in an atom move at the same distance from the nucleus. Instead, the electrons are in several different levels, each at a different distance from the nucleus. These levels are called energy levels (Figure 3-3). Electrons can occupy up to seven energy levels. The greater an electron's distance from the nucleus, the more energy the electron has.

According to the present atomic model, each energy level can hold only a certain number of electrons. The first energy level (level 1) can hold no more than 2 electrons. The second energy level (level 2) can hold no more than 8 electrons and the third energy level (level 3) can hold no more than 18 electrons. The lowest energy level must contain its full complement of electrons before additional electrons can be found in the next higher energy level.

Each element is different because no two neutral elements have atoms with the same number of electrons or protons. Chemists believe that the way atoms combine to form molecules is determined by the number of electrons in the outer energy levels of the atoms.

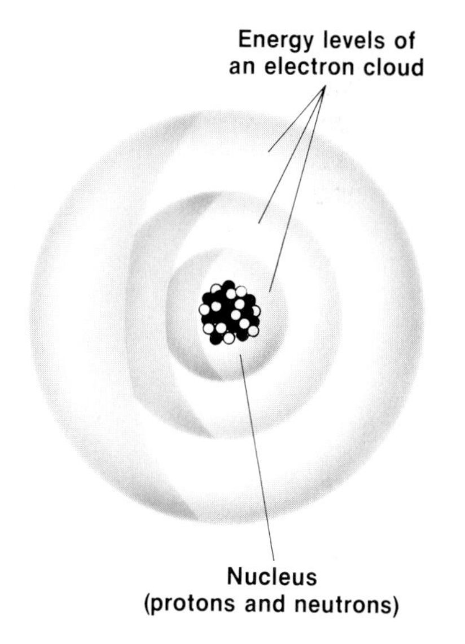

Figure 3-3 Protons and neutrons are found in the nucleus of an atom. Electrons circle around the nucleus at various energy levels within an electron cloud.

Different elements are made up of different numbers of protons. If we were to organize the natural elements in terms of atomic complexity, the simplest atom would be the most common form of hydrogen (H). It is different from the

DESCRIBING THINGS YOU CANNOT SEE

activity

OBJECTIVE: Use each of the five senses when making observations.

PROCESS SKILLS

In this activity, you will *observe and collect data* about an unknown object and *infer* what the object is without seeing it.

MATERIALS

objects of various size, shape, and texture
small box
masking tape
paper
pencil

PROCEDURE

1. Place an object in a small box. Use masking tape to seal the box completely. Write your name on the outside of the box.
2. Exchange boxes with someone else in the class.
3. Try to determine what unknown object is inside the sealed box without opening it. Move the box gently, tilting it from side to side.
4. Prepare a list of observations about the unknown object. Try to identify the object based upon your observations.
5. Return the box and your list of observations to its owner.

OBSERVATIONS AND CONCLUSIONS

1. Describe briefly the unknown object in your box.
2. Identify which of your senses were involved in making each of the observations.
3. How is the method you used to identify the unknown object similar to the method used by scientists to provide evidence to support the modern atomic theory?

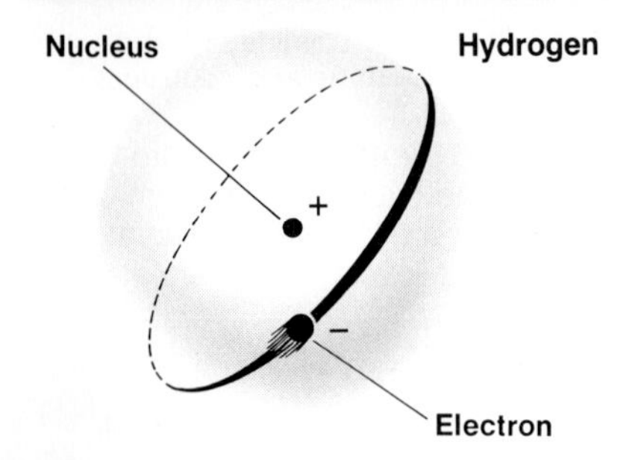

Figure 3-4 A diagram of an ordinary hydrogen atom. What kind of particle is found in the nucleus?

atoms of all other elements. What makes this atom unique is that it does not have a neutron. Hydrogen (H) has one proton and one electron (Figure 3-4).

The next simplest element after hydrogen is helium (He). This atom is made up of two protons, two neutrons, and two electrons. Lithium (Li) is the third atom in our sequence. The most common form of this element has three protons, four neutrons, and three electrons. Note that the electrons in lithium occupy two energy levels (Figure 3-5).

As you may have guessed, the number of electrons, protons, and neutrons increases from one element to the next in our sequence. However, all atoms of a given element have the same number of protons and electrons, and therefore are neutral.

The ninety-second and most complex of the natural elements in our sequence, uranium, has a total of 92 electrons in rapid motion around the nucleus. Uranium also has 92 protons and 146 neutrons in its nucleus. Like all other atoms, the uranium atom is electrically neutral. Because of its large number of protons and neutrons, it has a very high density.

Seventeen additional elements after uranium have been made by scientists to date, although not all of these have been officially accepted. The newest element contains 109 electrons in motion around a positively charged nucleus with 109 protons.

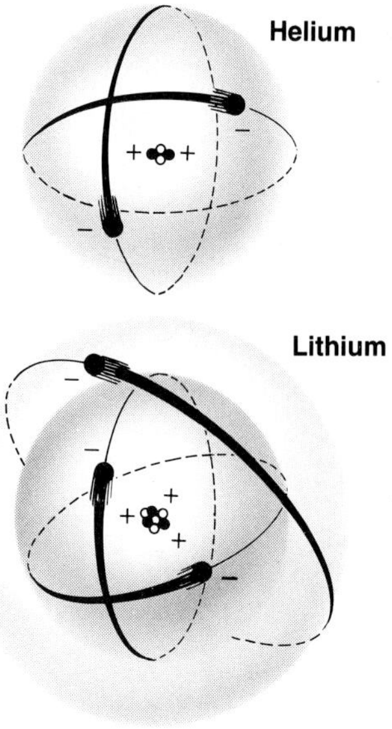

Helium

Lithium

Figure 3-5 Compare the diagrams of the two atoms. Explain why each atom is electrically neutral.

The atomic number identifies an element. The number of protons contained in the nucleus of an atom is called the **atomic number** of an element. The atomic number is also equal to the total number of electrons around the nucleus of a neutral atom. Table 3-1 contains a list of 18 elements and

TABLE 3-1

Elements, Atomic Numbers, and Electron Positions

Name of Element	Symbol	Atomic Number (protons)	Electrons Level 1	Level 2	Level 3
Hydrogen	H	1	1		
Helium	He	2	2		
Lithium	Li	3	2	1	
Beryllium	Be	4	2	2	
Boron	B	5	2	3	
Carbon	C	6	2	4	
Nitrogen	N	7	2	5	
Oxygen	O	8	2	6	
Fluorine	F	9	2	7	
Neon	Ne	10	2	8	
Sodium	Na	11	2	8	1
Magnesium	Mg	12	2	8	2
Aluminum	Al	13	2	8	3
Silicon	Si	14	2	8	4
Phosphorus	P	15	2	8	5
Sulfur	S	16	2	8	6
Chlorine	Cl	17	2	8	7
Argon	Ar	18	2	8	8

their protons and electrons. Notice that the elements are arranged in a definite order, according to increasing atomic numbers.

Each element has a different atomic number. The nucleus of common hydrogen, for example, contains one proton. The atomic number of hydrogen is 1. The nucleus of the helium atom has two protons, so the atomic number of helium is 2. Lithium, with its three protons, has an atomic number of 3. The number of protons keeps increasing by one proton for each element in the sequence.

The number of neutrons in the nucleus of an atom can vary. Recall that the atomic number is equal to the number of protons in an atom. Also recall that all atoms of a given element have the same number of protons. However, atoms of the same element do not necessarily contain the same number of neutrons. For example, the most common form of hydrogen has one proton and no neutrons in the nucleus of each atom. But each atom of another form of hydrogen has one proton and one neutron in its nucleus. A third form of hydrogen has one proton and two neutrons in the nucleus of the atom. Atoms that contain the same number of protons but different numbers of neutrons are **isotopes.**

Isotopes of the same element behave the same chemically. But because each atom has a different number of neutrons, each nucleus has a different mass. Each isotope is assigned a number called the **mass number,** which is equal to the total number of protons and neutrons in the nucleus. Each isotope is named according to its mass number. The isotope of hydrogen that has only one proton and no neutrons has a mass number of 1 and is called hydrogen-1. The isotope that has one proton and one neutron is hydrogen-2. How would you describe the nucleus of an atom of hydrogen-3?

To find the number of neutrons in an atom, subtract the atomic number (protons) from the mass number (protons + neutrons).

$$\text{neutrons} = \underset{\text{(protons + neutrons)}}{\text{mass number}} - \underset{\text{(protons)}}{\text{atomic number}}$$

The atomic mass unit indicates an element's relative mass. The actual mass of a single atomic particle is very small. These small values are not easy to use in calculations. As a result, chemists are mostly concerned with the relative masses of the various kinds of atoms. That is, they want to know how masses of different elements compare with each other.

SAMPLE PROBLEM

How many protons, electrons, and neutrons occur in a single atom of sodium-23?

SOLUTION

Step 1: Analyze

You are given the name of an isotope, sodium-23. The number of protons, electrons, and neutrons in an atom of this isotope are the unknowns.

Step 2: Plan

Since you know that isotopes of a given element have the same number of protons, and that the number of protons equals the atomic number, you can look up the number of protons in Table 3-1. The number of protons in sodium is 11. Since you know that the number of electrons around the nucleus equals the number of protons (or atomic number), you can conclude that there are also 11 electrons in sodium-23. Since you know the mass number (protons plus neutrons) of sodium-23 is 23, you can subtract the number of protons from this value to get the number of neutrons.

Step 3: Compute and check

$$\begin{aligned} \text{neutrons} &= \text{mass number} - \text{protons} \\ &= 23 - 11 \\ &= 12 \end{aligned}$$

In order to compare the masses of such small particles as atoms, a new unit had to be created. This unit is called the atomic mass unit (u). As with all units, an element's atomic mass unit is derived from a standard. Scientists have agreed to use the mass of the most common isotope of carbon (carbon-12) as the standard. Carbon-12 has 6 protons and 6 neutrons, and has been assigned a mass of exactly 12 u. Using this standard, hydrogen-1 has a mass of 1.0078 u and hydrogen-2 has a mass of 2.0014 u. This means that the mass of a hydrogen-1 atom is about twice that of a hydrogen-2 atom.

In making measurements, scientists do not work with single atoms. They work with samples of matter that contain many atoms. Instead of measuring the mass of a single atom, scientists measure the mass of a sample of an element. Any sample of an element will contain a mixture of different isotopes of that element.

The **atomic mass** of an element is defined as the average mass of atoms of all isotopes of that element and reflects the proportions in which the isotopes of that element appear in nature. The atomic mass and relative abundance of each isotope can be measured very precisely in devices called *mass spectrometers.*

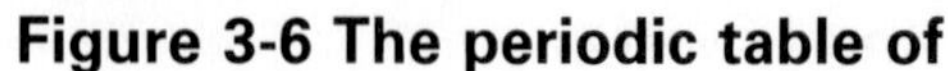

Figure 3-6 The periodic table of

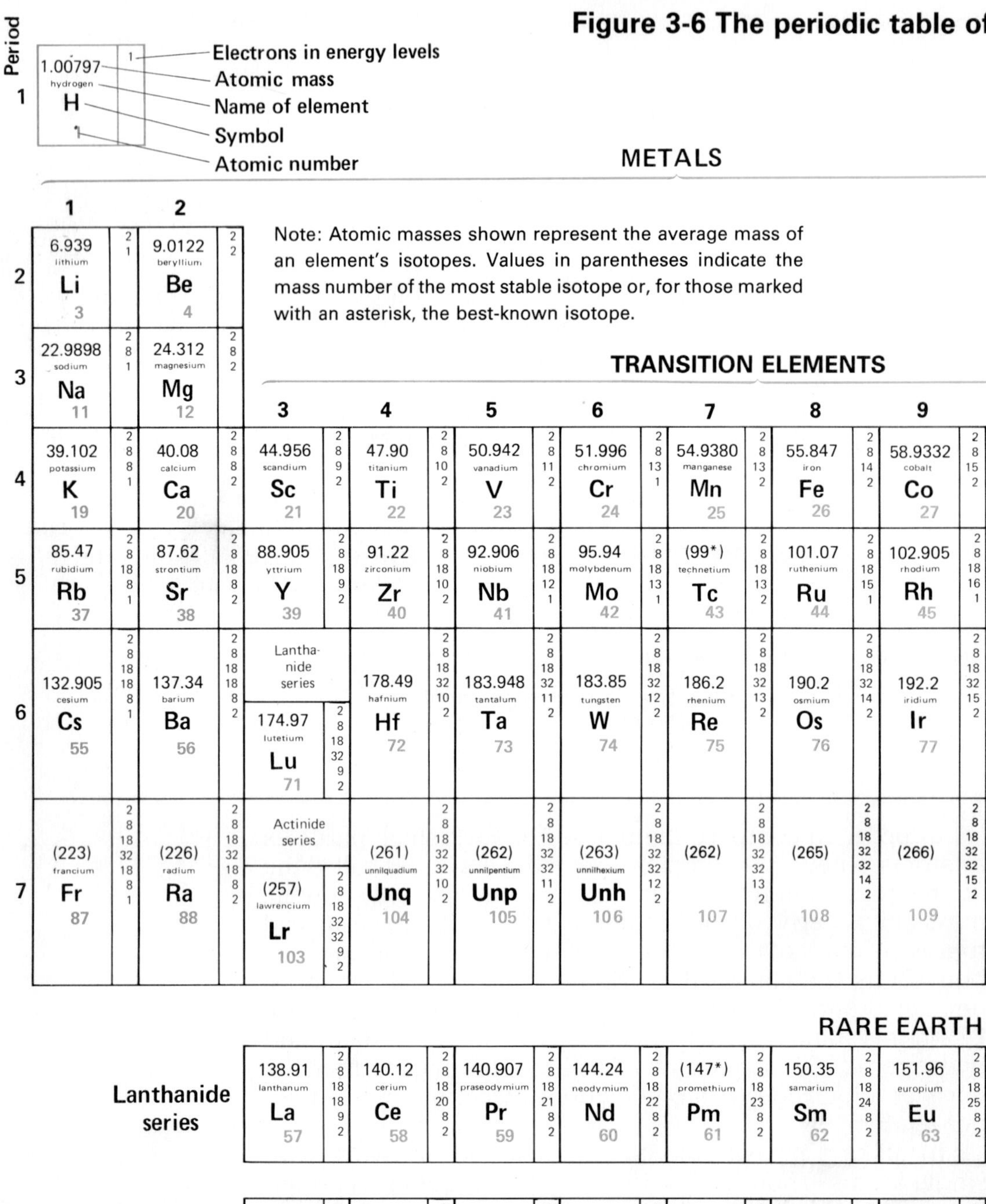

METALS

Note: Atomic masses shown represent the average mass of an element's isotopes. Values in parentheses indicate the mass number of the most stable isotope or, for those marked with an asterisk, the best-known isotope.

Period	1	2
2	6.939 lithium **Li** 3 (2, 1)	9.0122 beryllium **Be** 4 (2, 2)
3	22.9898 sodium **Na** 11 (2, 8, 1)	24.312 magnesium **Mg** 12 (2, 8, 2)
4	39.102 potassium **K** 19 (2, 8, 8, 1)	40.08 calcium **Ca** 20 (2, 8, 8, 2)
5	85.47 rubidium **Rb** 37 (2, 8, 18, 8, 1)	87.62 strontium **Sr** 38 (2, 8, 18, 8, 2)
6	132.905 cesium **Cs** 55 (2, 8, 18, 18, 8, 1)	137.34 barium **Ba** 56 (2, 8, 18, 18, 8, 2)
7	(223) francium **Fr** 87 (2, 8, 18, 32, 18, 8, 1)	(226) radium **Ra** 88 (2, 8, 18, 32, 18, 8, 2)

TRANSITION ELEMENTS

Period	3	4	5	6	7	8	9
4	44.956 scandium **Sc** 21 (2, 8, 9, 2)	47.90 titanium **Ti** 22 (2, 8, 10, 2)	50.942 vanadium **V** 23 (2, 8, 11, 2)	51.996 chromium **Cr** 24 (2, 8, 13, 1)	54.9380 manganese **Mn** 25 (2, 8, 13, 2)	55.847 iron **Fe** 26 (2, 8, 14, 2)	58.9332 cobalt **Co** 27 (2, 8, 15, 2)
5	88.905 yttrium **Y** 39 (2, 8, 18, 9, 2)	91.22 zirconium **Zr** 40 (2, 8, 18, 10, 2)	92.906 niobium **Nb** 41 (2, 8, 18, 12, 1)	95.94 molybdenum **Mo** 42 (2, 8, 18, 13, 1)	(99*) technetium **Tc** 43 (2, 8, 18, 13, 2)	101.07 ruthenium **Ru** 44 (2, 8, 18, 15, 1)	102.905 rhodium **Rh** 45 (2, 8, 18, 16, 1)
6	Lanthanide series; 174.97 lutetium **Lu** 71 (2, 8, 18, 32, 9, 2)	178.49 hafnium **Hf** 72 (2, 8, 18, 32, 10, 2)	183.948 tantalum **Ta** 73 (2, 8, 18, 32, 11, 2)	183.85 tungsten **W** 74 (2, 8, 18, 32, 12, 2)	186.2 rhenium **Re** 75 (2, 8, 18, 32, 13, 2)	190.2 osmium **Os** 76 (2, 8, 18, 32, 14, 2)	192.2 iridium **Ir** 77 (2, 8, 18, 32, 15, 2)
7	Actinide series; (257) lawrencium **Lr** 103 (2, 8, 18, 32, 32, 9, 2)	(261) unnilquadium **Unq** 104 (2, 8, 18, 32, 32, 10, 2)	(262) unnilpentium **Unp** 105 (2, 8, 18, 32, 32, 11, 2)	(263) unnilhexium **Unh** 106 (2, 8, 18, 32, 32, 12, 2)	(262) 107 (2, 8, 18, 32, 32, 13, 2)	(265) 108 (2, 8, 18, 32, 32, 14, 2)	(266) 109 (2, 8, 18, 32, 32, 15, 2)

RARE EARTH

Lanthanide series	138.91 lanthanum **La** 57 (2, 8, 18, 18, 9, 2)	140.12 cerium **Ce** 58 (2, 8, 18, 20, 8, 2)	140.907 praseodymium **Pr** 59 (2, 8, 18, 21, 8, 2)	144.24 neodymium **Nd** 60 (2, 8, 18, 22, 8, 2)	(147*) promethium **Pm** 61 (2, 8, 18, 23, 8, 2)	150.35 samarium **Sm** 62 (2, 8, 18, 24, 8, 2)	151.96 europium **Eu** 63 (2, 8, 18, 25, 8, 2)
Actinide series	(227) actinium **Ac** 89 (2, 8, 18, 32, 18, 9, 2)	232.038 thorium **Th** 90 (2, 8, 18, 32, 18, 10, 2)	(231) protactinium **Pa** 91 (2, 8, 18, 32, 20, 9, 2)	238.03 uranium **U** 92 (2, 8, 18, 32, 21, 9, 2)	(237) neptunium **Np** 93 (2, 8, 18, 32, 23, 8, 2)	(242) plutonium **Pu** 94 (2, 8, 18, 32, 24, 8, 2)	(243) americium **Am** 95 (2, 8, 18, 32, 24, 9, 2)

the elements

NONMETALS

10	11	12	13	14	15	16	17	NOBLE GASES 18
								4.0026 helium **He** 2 2
			10.811 boron **B** 5 2, 3	12.01115 carbon **C** 6 2, 4	14.0067 nitrogen **N** 7 2, 5	15.9994 oxygen **O** 8 2, 6	18.9984 fluorine **F** 9 2, 7	20.183 neon **Ne** 10 2, 8
			26.9815 aluminum **Al** 13 2, 8, 3	28.086 silicon **Si** 14 2, 8, 4	30.9738 phosphorus **P** 15 2, 8, 5	32.064 sulfur **S** 16 2, 8, 6	35.453 chlorine **Cl** 17 2, 8, 7	39.948 argon **Ar** 18 2, 8, 8
58.71 nickel **Ni** 28 2, 8, 16, 2	63.54 copper **Cu** 29 2, 8, 18, 1	65.37 zinc **Zn** 30 2, 8, 18, 2	69.72 gallium **Ga** 31 2, 8, 18, 3	72.59 germanium **Ge** 32 2, 8, 18, 4	74.9216 arsenic **As** 33 2, 8, 18, 5	78.96 selenium **Se** 34 2, 8, 18, 6	79.909 bromine **Br** 35 2, 8, 18, 7	83.80 krypton **Kr** 36 2, 8, 18, 8
106.4 palladium **Pd** 46 2, 8, 18, 18, 0	107.870 silver **Ag** 47 2, 8, 18, 18, 1	112.40 cadmium **Cd** 48 2, 8, 18, 18, 2	114.82 indium **In** 49 2, 8, 18, 18, 3	118.69 tin **Sn** 50 2, 8, 18, 18, 4	121.75 antimony **Sb** 51 2, 8, 18, 18, 5	127.60 tellurium **Te** 52 2, 8, 18, 18, 6	126.9044 iodine **I** 53 2, 8, 18, 18, 7	131.30 xenon **Xe** 54 2, 8, 18, 18, 8
195.09 platinum **Pt** 78 2, 8, 18, 32, 16, 2	196.967 gold **Au** 79 2, 8, 18, 32, 18, 1	200.59 mercury **Hg** 80 2, 8, 18, 32, 18, 2	204.37 thallium **Tl** 81 2, 8, 18, 32, 18, 3	207.19 lead **Pb** 82 2, 8, 18, 32, 18, 4	208.980 bismuth **Bi** 83 2, 8, 18, 32, 18, 5	(210*) polonium **Po** 84 2, 8, 18, 32, 18, 6	(210) astatine **At** 85 2, 8, 18, 32, 18, 7	(222) radon **Rn** 86 2, 8, 18, 32, 18, 8

ELEMENTS

157.25 gadolinium **Gd** 64 2, 8, 18, 25, 9, 2	158.924 terbium **Tb** 65 2, 8, 18, 27, 8, 2	162.50 dysprosium **Dy** 66 2, 8, 18, 28, 8, 2	164.930 holmium **Ho** 67 2, 8, 18, 29, 8, 2	167.26 erbium **Er** 68 2, 8, 18, 30, 8, 2	168.934 thulium **Tm** 69 2, 8, 18, 31, 8, 2	173.04 ytterbium **Yb** 70 2, 8, 18, 32, 8, 2
(247) curium **Cm** 96 2, 8, 18, 32, 25, 9, 2	(249*) berkelium **Bk** 97 2, 8, 18, 32, 27, 8, 2	(251*) californium **Cf** 98 2, 8, 18, 32, 28, 8, 2	(254) einsteinium **Es** 99 2, 8, 18, 32, 29, 8, 2	(253) fermium **Fm** 100 2, 8, 18, 32, 30, 8, 2	(256) mendelevium **Md** 101 2, 8, 18, 32, 31, 8, 2	(254) nobelium **No** 102 2, 8, 18, 32, 32, 8, 2

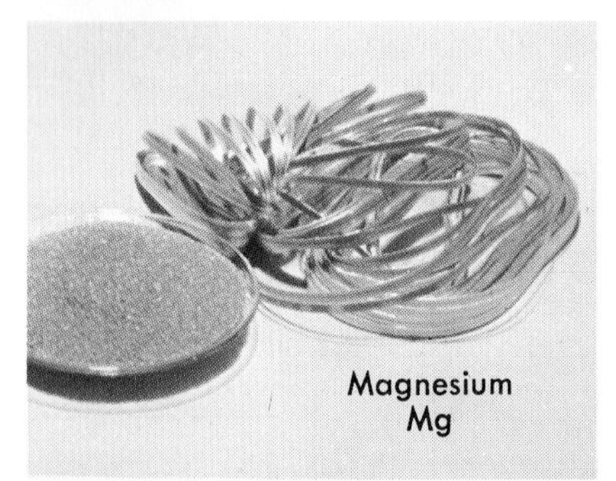

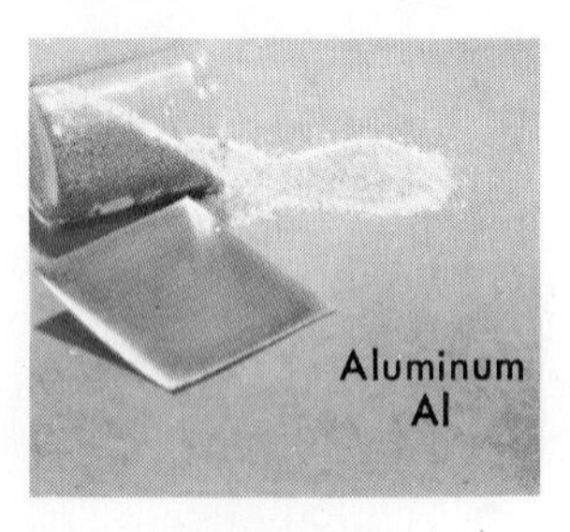

Figure 3-7 The elements pictured represent the first four elements in period 3 of the periodic table. What are the remaining elements of this period? **CAUTION: Sodium should only be stored in oil, as shown, since it reacts violently with water.**

3.2 THE PERIODIC TABLE

The periodic table contains useful data. To study the properties of the elements, scientists have grouped them in the order of their increasing atomic numbers in a chart called the **periodic table** of the elements. The periodic table is shown in Figure 3-6. In this arrangement, nuclei of the atoms of one element differ from the nuclei of atoms of the next element by the addition of one proton and usually by one or more neutrons.

All the elements are placed in seven numbered horizontal rows, called periods. Notice that the periodic table shows the electron arrangement of each element. Elements in the same period, or horizontal row, have electrons in the same energy level. The first four elements of the third period are shown in Figure 3-7.

The vertical columns numbered from 1 to 18 are called groups. Elements in each group have chemical and physical properties that are somewhat alike. They also have the same number of electrons in their outer energy level.

The six elements in Group 1 are very active chemically. That is, they combine readily with certain other elements. For instance, Group 1 elements react with water as follows: lithium reacts slowly; sodium, fast; potassium, very fast; and rubidium, violently. The six elements in Group 18 make up the family of noble gases. These elements are very stable; that is, they do not react readily with other elements. Scientists have been able to prepare only a few compounds containing these elements. You will learn more about noble gases in Chapter 7.

On the left side of the periodic table, the most active elements in a group are at the bottom. On the right side, the most active elements in a group are at the top.

Notice that the periodic table identifies metals and nonmetals. **Metals** are found at the left side of the periodic table. You are probably familiar with some of the metals, such as iron, zinc, lead, copper, tin, and gold. **Nonmetals** are found at the right side of the periodic table. Well-known nonmetals include nitrogen, oxygen, and chlorine.

Metals are described as having luster because they usually have a shiny surface. They are usually solid at room temperature and have relatively high densities. Metals can be easily hammered or formed into different shapes. Most metals are good conductors of heat and electric current.

Nonmetals lack luster. Most of them are poor conductors of heat and electric current. Nonmetals break or shatter when hammered or bent. They vary in color. For example, sulfur is yellow, chlorine is greenish-yellow, and carbon is black. Nonmetals have low densities.

3.3 HOW COMPOUNDS ARE FORMED

Electrons move between atoms. In Chapter 2, you learned that elements unite to form a compound. How does this occur? Consider a particle of sodium chloride (NaCl), which is common table salt. As you can see from the periodic table, sodium has an atomic number of 11. First, imagine what a sodium atom would look like if it could be seen: there would be 11 protons and 12 neutrons in its nucleus. Of the 11 electrons around its nucleus, 2 would be found in level 1, 8 in level 2, and 1 electron in level 3. See Figure 3-8. Notice that the electrons in the figure are shown in circles instead of in clouds. Scientists often use this kind of model to represent atoms because it clearly illustrates the position of the electrons.

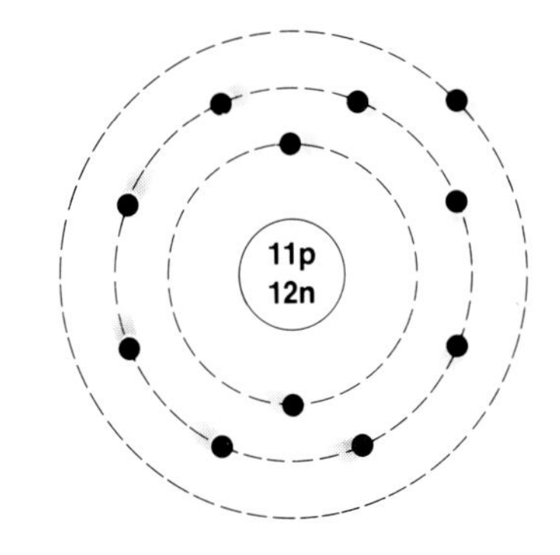

Figure 3-8 Atomic model of a sodium atom including the number of protons (p) and neutrons (n) in the nucleus.

Now, picture the chlorine atom. By looking at the periodic table, you can see that chlorine has an atomic number of 17. There are 17 protons and 18 neutrons tightly packed in the nucleus. A total of 17 electrons are in constant motion around the nucleus. Of these 17 electrons, 2 occur in level 1, 8 in level 2, and the remaining 7 are in level 3 (Figure 3-9).

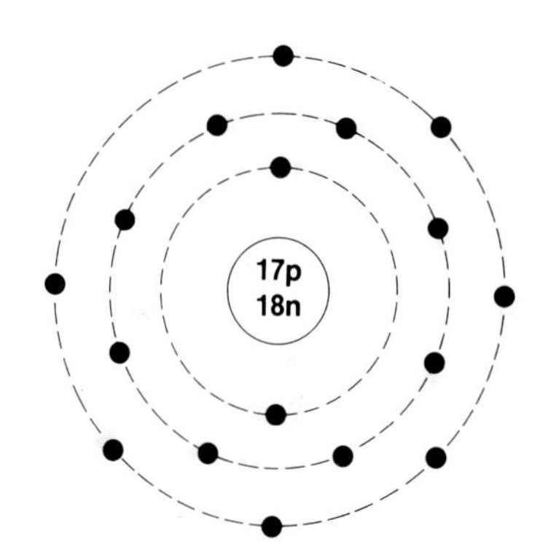

Figure 3-9 The atomic model of a chlorine atom.

The formation of compounds is linked to an atom's strong tendency to form completely filled energy levels. The outer energy level of a sodium atom contains only one electron, whereas the outer level of a chlorine atom lacks only one electron to complete it. A sodium atom will readily lose its one outer electron to achieve two complete energy levels. The loss of an electron means the sodium atom has lost a negative charge. This leaves the sodium atom with a charge of 1+. A chlorine atom will readily accept an electron to achieve three complete outer energy levels. The addition of one electron gives the chlorine atom a charge of 1−. Atoms that have lost or gained electrons such that they become electrically charged particles, like those formed from sodium and chlorine atoms, are called **ions.** The two oppositely charged ions attract each other to form the compound sodium chloride (NaCl).

Elements combine chemically with each other according to rules that have established an atom's oxidation number. The **oxidation number** of an atom is the charge that atom has in a particular molecule or ion. That charge results when an atom loses, acquires, or shares electrons during a chemical reaction. The electrons in the outer energy level are the ones that take part in the reactions.

In its elemental state (not combined chemically with other different atoms), the oxidation number of an element is zero. Likewise, in any compound, the sum of the oxidation numbers of all atoms in the compound must equal zero. In a combination of two elements, some elements tend to acquire electrons and have a negative oxidation number that equals the electrons acquired in the compound. Other elements tend to lose electrons and have a positive oxidation number that equals the number of electrons lost.

The sodium atom has an oxidation number of 1+ because it takes on a positive charge of one unit when it loses its outer electron and forms a sodium ion (Na^+). The oxidation number of chlorine is 1− because it takes on a negative charge of one unit when it acquires an electron and forms a chloride ion (Cl^-). Use Table 3-2 to find the oxidation numbers of some of the more common elements.

Notice that some of the elements listed in Table 3-2 have more than one oxidation number. Iron has an oxidation number of 2+ in some compounds and 3+ in others. This is because iron can lose either two or three electrons. Metals have positive oxidation numbers, since they tend to lose electrons in chemical reactions. Nonmetals usually have

TABLE 3-2

Oxidation Numbers of Some Common Elements

Name	Symbol	Oxidation Number
Aluminum	Al	3+
Barium	Ba	2+
Calcium	Ca	2+
Copper(II), (cupric)	Cu	2+
Copper(I), (cuprous)	Cu	1+
Iron(III), (ferric)	Fe	3+
Iron(II), (ferrous)	Fe	2+
Hydrogen	H	1+
Lead(II)	Pb	2+
Magnesium	Mg	2+
Potassium	K	1+
Silver	Ag	1+
Sodium	Na	1+
Zinc	Zn	2+
Bromine	Br	1−
Chlorine	Cl	1−
Fluorine	F	1−
Iodine	I	1−
Oxygen	O	2−
Sulfur	S	2−

negative oxidation numbers. They tend to acquire electrons during chemical changes. Which elements in Table 3-2 are nonmetals?

Notice that sodium appears in Group 1 of the periodic table (Figure 3-6). Recall that all elements in each group have many of the same properties. As with sodium, all elements in Group 1 have an oxidation number of 1+. Notice that chlorine appears in Group 17. What is the oxidation number of all elements in Group 17?

Many elements combine to form ionic bonds. The transfer or sharing of electrons between atoms forms chemical bonds. **Chemical bonds** are forces that hold two or more atoms together. As you have seen, many compounds are formed by the transfer of electrons from one element to another. The type of chemical bond formed when one atom transfers an electron to another atom is called an **ionic bond.**

Figure 3-10 shows the formation of sodium chloride by ionic bonding. The sodium atom readily gives up its outer electron, forming a sodium ion. The chlorine atom easily accepts an electron, forming a chloride ion. The outer energy levels of both elements are thus completed.

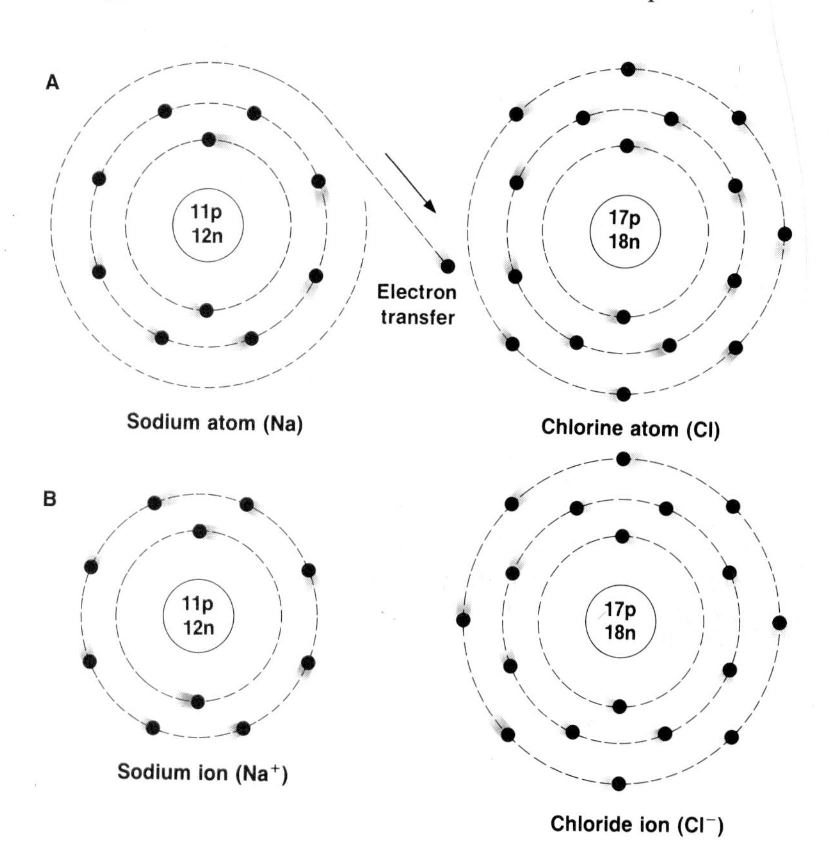

Figure 3-10 Sodium chloride is formed by the transfer of a single electron in the outer energy level of the sodium atom to the outer energy level of the chlorine atom (A). What kind of bond is formed in diagram B by this transfer?

The positive sodium ion and negative chloride ion are electrically attracted to each other, and form sodium chloride. This white solid compound is commonly called table salt. Its formula may be written as $Na^{1+}Cl^{1-}$. However, formulas are usually written without the oxidation numbers. Thus, sodium chloride is written as NaCl.

If you know the oxidation numbers of the elements, you can predict the formulas of the compounds they form. How would you predict the formula for calcium chloride? In Table 3-2, the oxidation number of calcium is listed as 2+. It is important to note that all compounds are electrically neutral. Thus, the total positive particles of a compound must equal the total negative particles. Since chlorine has an oxidation number of 1−, two chloride ions (Cl^{1-}) will combine with one calcium ion (Ca^{2+}). Therefore, the formula of calcium chloride is written as $CaCl_2$. In the same way, aluminum chloride is written as $AlCl_3$, since aluminum has an oxidation number of 3+. How would you write the chemical formula for calcium oxide?

Notice in Figure 3-10 that the nucleus of an atom is not changed as bonds form during a chemical reaction. Also notice that no atoms or parts of atoms are created or destroyed in this process. Energy, however, is always exchanged whenever chemical bonds are made or broken.

Some groups of atoms bond as a unit. The behavior of some groups of atoms resembles that of a single atom. That is, certain groups of atoms stay united during a chemical reaction. These groups are sometimes called radicals. Two common groups are the hydroxide group (OH) and the sulfate group (SO_4). Each group acts as though it were a single atom with a definite oxidation number. Table 3-3 lists the oxidation numbers of several groups. The oxidation number of a hydroxide group is 1− because this group forms a hydroxide ion (OH^-). It is important to note that the single negative charge does not belong to the hydrogen or to the oxygen, but to the group as a whole.

The symbol for a group is sometimes written in parentheses. This style is needed when a group is used more than once in writing a formula. For instance, sodium hydroxide is NaOH, calcium hydroxide is $Ca(OH)_2$, and aluminum hydroxide is $Al(OH)_3$.

Compounds are named from their formulas. Notice that in writing chemical formulas for compounds, the symbol for the element (or group) with the positive oxidation number is written first. This symbol is followed by that of the element (or group) with the negative oxidation number. In naming almost all compounds containing two elements, the

TABLE 3-3

Oxidation Numbers of Common Groups

Name	Formula	Oxidation Number
Ammonium	NH_4	1+
Acetate	$C_2H_3O_2$	1−
Hydrogen carbonate	HCO_3	1−
Carbonate	CO_3	2−
Chlorate	ClO_3	1−
Hydroxide	OH	1−
Nitrate	NO_3	1−
Phosphate	PO_4	3−
Sulfate	SO_4	2−
Sulfite	SO_3	2−

name of the last element ends in "-ide." For example, NaCl is sodium chloride, KI is potassium iodide, and CaO is calcium oxide.

The ending "-ate" indicates that the compound contains oxygen in addition to the other elements mentioned in the name. For example, $NaNO_3$ is sodium nitrate, $CaCO_3$ is calcium carbonate, and $KClO_3$ is potassium chlorate. How would you write the formulas for sodium hydrogen carbonate, ammonium chloride, and hydrogen sulfate?

Recall that some metals have more than one oxidation number. These metals form more than one positive ion. For example, when the oxidation number of iron is 2+, its compounds are called ferrous, or iron(II). When its oxidation number is 3+, its compounds are called ferric, or iron(III). Thus, there are two chlorides of iron. One is ferrous chloride, or iron(II) chloride ($FeCl_2$). The other is ferric chloride, or iron(III) chloride ($FeCl_3$). Copper is another element with two oxidation numbers: 1+ and 2+. What are the two chlorides of copper?

Some elements combine to form covalent bonds. Not all compounds are formed by the transfer of electrons from one atom to another in an ionic bond. A compound may also be formed when some atoms share common electrons. Nonmetals usually bond in this manner. When atoms share one or more pairs of electrons, a **covalent bond** is formed.

Water is an example of a molecule formed by the sharing of electrons. Refer back to the periodic table, and note that oxygen has six outer electrons. Thus, oxygen needs two

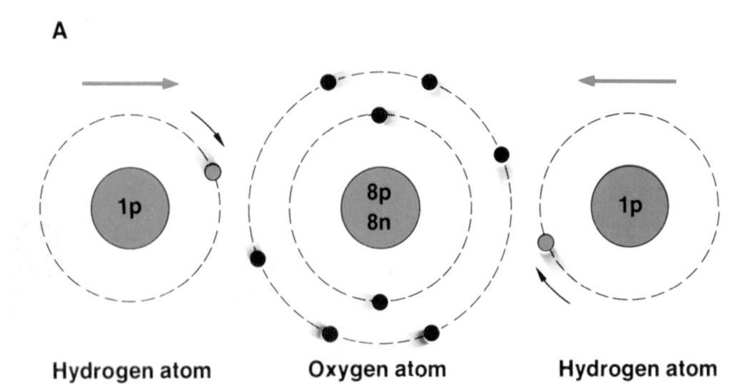

Figure 3-11 Diagram A represents two atoms of hydrogen and one atom of oxygen before bonding. Diagram B depicts the formation of a molecule of water from these three atoms. What kind of bonding has taken place?

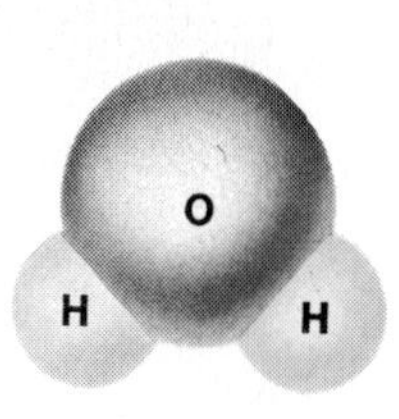

Figure 3-12 Explain why the oxygen (O) atom is larger than either hydrogen (H) atom in this laboratory model of a water molecule (H_2O).

electrons to complete its outer energy level. Hydrogen needs only one electron to complete its outer energy level. Figure 3-11 shows how a molecule of water (H_2O) is formed. Notice that pairs of electrons are being shared. In each bond, the oxygen atom and one hydrogen atom share the one electron from the hydrogen atom and one electron from the oxygen atom. Such sharing completes the outer energy level of each atom. A laboratory model of a water molecule is illustrated in Figure 3-12. Keep in mind that atoms are not really solid spheres. The sphere is merely a convenient way to show an atom in a model.

Molecules of common gases such as oxygen, hydrogen, chlorine, and nitrogen are also formed by the sharing of electrons. Since a hydrogen atom has only one electron, two hydrogen atoms share their electrons with each other. As a result, each atom has a complete energy level, and a hydrogen molecule is formed. The formula for this molecule is written as H_2. Molecules of other common gases contain covalent bonds. The formulas are written as O_2 for oxygen, Cl_2 for chlorine, and N_2 for nitrogen. These molecules containing two atoms are called diatomic molecules, because "di" means "two" (Figure 3-13).

As in ionic bonding, the formation of covalent bonds involves energy exchange. In most cases, when compounds are formed from elements, some form of energy is given off. This energy may be in the form of heat, sound, or light. When a compound is broken down, energy is required to break chemical bonds. The amount of energy absorbed in the breaking of bonds is equal to the amount of energy released during their formation.

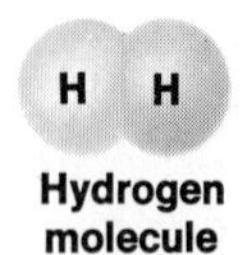

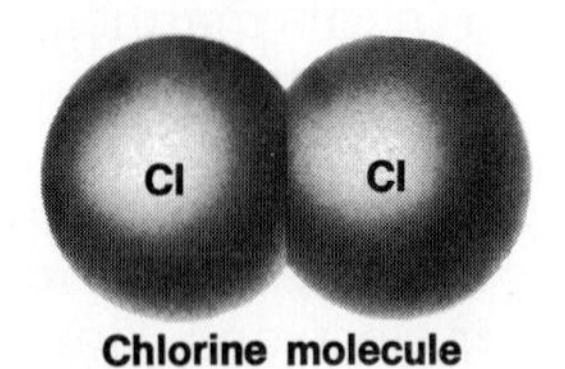

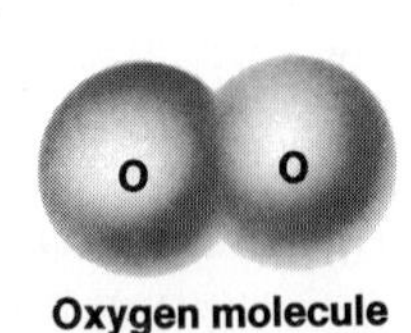

Figure 3-13 Laboratory models of the diatomic molecules of hydrogen (H_2), chlorine (Cl_2), and oxygen (O_2) gases. How are these molecules formed?

BIOGRAPHY

DMITRI MENDELEEV

father of the modern periodic table

Many attempts have been made to organize the known elements. In 1864, an Englishman, John Newlands, arranged all the known elements in the order of their atomic masses. Newlands noticed that the eighth element in this arrangement had chemical properties similar to the first element. For that reason, Newlands divided the elements into series of seven elements each, making the eighth element the first in a second series of seven elements. This arrangement was called the law of octaves. Later it was found that this system was limited in its usefulness since it only worked for the lighter elements.

The periodic table in use today is based largely on the pioneering work of a Russian chemist, Dmitri Mendeleev (MEN-*deh*-LAY-*eff*) (1834–1907). When Mendeleev first prepared a table of elements, there were only 66 known elements. He started his table by organizing the elements into vertical rows based on their atomic masses. When he did this, he realized that certain properties of the elements were being repeated. Some elements such as lithium, sodium, and potassium were very active. Others, like carbon and silicon, were excellent conductors of electricity and heat. Mendeleev concluded that there was a connection between the atomic masses of the elements and their properties. He later used this information to arrange the elements into horizontal rows.

Once the table had been completed, Mendeleev noticed that it contained gaps. He assumed that these gaps represented the elements that had not yet been discovered. He left the gaps in the table where the undiscovered elements seemed to fit. Mendeleev then predicted the properties that these undiscovered elements should have. When the elements were later discovered, they fit into the gaps just as Mendeleev had predicted. Mendeleev was a man of great genius, regarded as the most outstanding teacher of his time.

Some 45 years later, a brilliant English scientist, Henry Moseley, used X-ray patterns as a basis for listing the elements in order of their atomic numbers. Today, scientists know that atomic number, rather than atomic mass, is the proper basis for the order of the elements in the periodic table.

CHAPTER REVIEW

SUMMARY

1. All atoms are made of protons, neutrons, and electrons.
2. The atomic number of an element is equal to the number of protons in its nucleus. It also identifies an element in the periodic table.
3. In a neutral atom, the number of protons (+) in the nucleus equals the number of electrons (−) in the energy levels around the nucleus.
4. The mass number is equal to the total number of protons and neutrons in the nucleus of an isotope.
5. The charge an atom has in a particular molecule or ion is indicated by its oxidation number.
6. Ionic bonds are formed by the transfer of electrons from the outer energy level of one element (usually a metal) to the outer energy level of another element (usually a nonmetal).
7. Covalent bonds are formed when elements (usually nonmetals) share outer energy level electrons with each other.

VOCABULARY

Match the item in the left column with the best answer in the right column. Do not write in this book.

1. atomic mass	a. bond formed by sharing electrons
2. atomic number	b. number of neutrons plus protons
3. chemical bond	c. element on right side of periodic table
4. covalent bond	d. indicates the charge an atom has
5. electron	e. negatively charged particle in an atom
6. ionic bond	f. force holding atoms together
7. ions	g. positively charged particle in an atom
8. isotopes	h. number of protons
9. mass number	i. arrangement of elements according to their properties
10. metal	
11. neutron	j. element on left side of periodic table
12. nonmetal	k. bond formed by electron transfer
13. oxidation number	l. formed when electrons are lost or gained
14. periodic table	m. atomic particle with no charge
15. proton	n. atoms with the same number of protons and different numbers of neutrons
	o. average of the masses of an element's isotopes

REVIEW QUESTIONS

1. Name the three basic particles of an atom and the charge each has.
2. What is a scientific model?
3. Where in an atom is most of the mass found? Explain.
4. How does an atom of iron differ in structure from an atom of gold?

5. What is the maximum number of electrons in each of the first three energy levels of an atom?
6. Set up a table of 10 elements. List the following information about each: the number of protons, neutrons, and electrons.
7. Where in the periodic table are the most active elements located?
8. Where in the periodic table are the metals?
9. What is the oxidation number of Group 1 elements?
10. Name five elements that are included in the "nitrogen group."
11. What are the formulas for molecules of nitrogen and chlorine?
12. Write the formula for each of the following compounds: (a) calcium hydroxide, (b) ammonium sulfate, (c) hydrogen phosphate, and (d) copper(II) chloride.
13. Name the compound represented by each of the following formulas: (a) $KClO_3$, (b) NH_4OH, (c) Na_2CO_3, (d) $Pb(C_2H_3O_2)_2$, and (e) K_2S.
14. Describe two ways atoms combine to form compounds. Give examples.
15. Write a brief summary of the modern atomic theory of matter.
16. Explain the difference between the atomic mass of an element and the mass number of an atom.
17. Explain why some elements have positive oxidation numbers and other elements have negative oxidation numbers.
18. How has the periodic table been of value in advancing chemistry?
19. What type of chemical bonding would you expect to take place in the making of lithium fluoride? Explain.

CRITICAL THINKING

20. Sketch a diagram of the following atoms: (a) carbon-12, (b) magnesium-24, (c) nitrogen-14, and (d) sulfur-32.
21. The force of attraction holds the particles together within an oxygen atom. When the two oxygen atoms come near each other, both repellent and attractive forces act on the atoms. Describe the particles in each atom that produce these forces.
22. What evidence would you look for to tell that a chemical reaction has taken place?

PROBLEMS

1. Write the chemical formulas for each of the following compounds: (a) barium chloride, (b) calcium iodide, (c) copper(II) oxide, (d) lead(II) nitrate, and (e) aluminum hydroxide.
2. Write the names of each of the following compounds: (a) $MgSO_4$, (b) Na_2O, (c) $Ba(OH)_2$, (d) Li_2SO_4, (e) NH_4Cl, and (f) $PbCO_3$.
3. Give the oxidation number of the following underlined elements or groups: (a) $\underline{Ca}Cl_2$, (b) $\underline{Ag}_2S$, (c) $K_2\underline{SO_4}$, (d) $\underline{Cu}_2O$, and (e) $Al_2\underline{S}_3$.

FURTHER READING

Ardley, Neil. *Atoms and Energy*. New York: Franklin Watts, Inc., 1976.

Asimov, Isaac. *How Did We Find Out About Atoms?* New York: Walker, 1976.

Ley, Wily. *The Discovery of the Elements*. New York: Delacorte, 1983.

How the Sciences Work Together

The Problem in Physical Science: Can We Expand Our Use of Geothermal Energy?

Fossils fuels—coal, oil and natural gas—meet almost 90 percent of the world's energy needs. However, burning these, or any other kind of fuels such as wood or manure, is a major source of environmental pollution. They contribute to acid rain and smog. Carbon dioxide, produced by burning these fuels, may trap increasing amounts of solar energy, leading to unpredictable shifts in global climate.

Concern over the effects of burning fuels has led some researchers to seek energy alternatives that don't involve burning. Among these alternatives is geothermal energy which is produced within the earth. One estimate indicates that the total amount of useful geothermal energy available is greater than the energy stored in all known deposits of fossil fuels.

Conventional geothermal power plants harvest the earth's interior heat by tapping steam and hot water from underground streams that pass through hot rock formations. The steam is used to run electrical generators or provide heat and hot water for nearby homes.

Firehole River and hot springs in Yellowstone National Park.

Burning fossil fuels to generate electricity contributes to pollution.

Many areas around the world are located above hot rock formations. Most of these areas, however, have no underground streams. Without water to bring heat to the surface, these reservoirs of underground energy have gone untapped. Scientists are seeking ways

Pipes carry steam from underground at a geothermal power plant in New Zealand.

of expanding the use of geothermal energy by injecting water into the hot dry rock, creating artificial streams.

The Connection to Earth Science

The outer solid layer of the earth is called the *crust.* The temperature of the crust increases below the surface by an average of 30°C per kilometer. In some locations the temperature rises over 70°C per kilometer. Rock temperatures have to be above 360°C to make artificial techniques feasible. The source of this underground heat is deposits of unstable forms of uranium, thorium, and potassium below the earth's crust. When these elements break down they release energy which heats the rock layers in the crust above.

The Connection to Physics

For artificial geothermal wells to work, the injected water must circulate through a large area of hot rock in order to be raised to a useful temperature. This steam or hot water must then be returned to the surface without draining away through cracks in the rock.

At a test site in New Mexico, workers have drilled what they call an *injection well* over four kilometers deep. In order to create the large surface area needed for adequate heating, water was pumped into this well until the pressure at the bottom reached over 400 times that at the surface. This caused the rock at the bottom to fracture into hundreds of sheets. Water pumped into this well seeps through these layers of hot rock and picks up heat.

Artificial geothermal well located in Fenton Hill, New Mexico.

When the rock is fractured, shock waves are produced. These shock waves are picked up by detectors at the surface, and the location of the fractured area can be determined. A second well called a *production well* is then drilled to intersect the fractured area. This forms a complete loop so that cool water injected from the surface is forced back through the production well as steam or hot water.

The Connection to Chemistry

Water pumped down the injection well must be kept at pressures 10 times greater than normal atmospheric pressure. Without special measures, most of this water would gush back out of the injection well and never reach the hot rock. Chemists keep this back flow in check with a *polymer plug.* This is a plastic material that molds around the pipe in the injection well and forms a watertight seal.

The test well in New Mexico has produced 20 liters of water per second at 190°C. This is enough energy to provide electricity for about 50 average homes. So far, artificial geothermal wells are being investigated in England, Japan, and the Soviet Union.

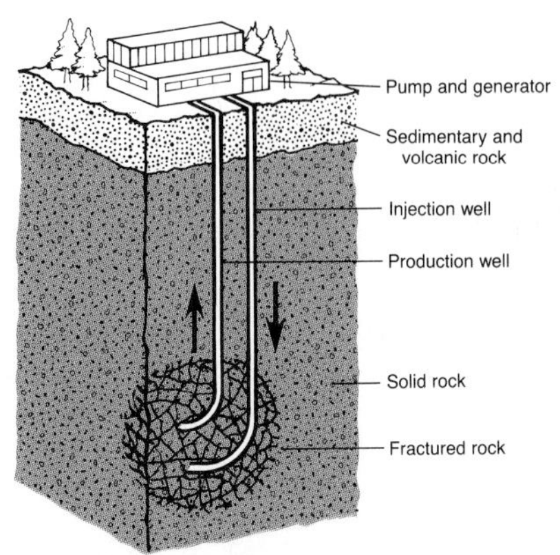

Cold water is pumped underground and recovered as hot water or steam at an artificial geothermal well.

MATTER INTERACTS

UNIT 2

Imagine a vast chamber deep below the surface of the earth: cool and clammy air, the sound of water falling drop by drop, and an overwhelming musty odor. The cavern seems quite still, yet it is actually the setting for countless chemical reactions and phase changes. Acidic groundwater seeps in through cracks, reacting with solid limestone to form a milky liquid. Icicle-like stalactites grow as solutions evaporate, leaving behind solid limestone. And calcite crystals solidify from solutions on the cavern walls.

Whether the place is an underground cavern, the earth's surface, the sun, or almost anywhere, matter is interacting and reacting. Reacting substances regroup to form new substances with new properties. Matter can also react in such a way that energy is released. In this unit, you will investigate different kinds of reactions, including chemical reactions and nuclear reactions. You will learn about the changes that take place in solids, liquids, and gases and you will be introduced to the various kinds of solutions produced when matter interacts.

CHAPTERS

THE KINETIC THEORY OF MATTER

As warm ocean air collides with cooler air above San Francisco Bay, tiny liquid drops form in the moving air mass. Suddenly, Golden Gate Bridge is enveloped in fog. What change in the state of matter has occurred? Is energy gained or lost with this change?

CHAPTER 4

SECTIONS

OBJECTIVES

- ☐ Explain the kinetic theory of matter.
- ☐ Identify the three phases of matter as they relate to the forces between particles and the energy these particles have.
- ☐ Identify kinds of solutions and their properties.
- ☐ Describe the nature of electrolytes and the formation and properties of ions.

4.1 MOLECULES IN MOTION

The kinetic theory is a model for the behavior of matter. If ammonia water was placed in an open dish in the front of your classroom, you would smell the odor in a very short time. You would smell this odor even if there were no air currents in the room. Molecules of ammonia leave the dish and travel in all directions. Eventually, they reach your nose. How fast the molecules of ammonia water move depends upon the energy they have. This energy is described as **kinetic energy**, or the energy of motion. The higher the temperature, the greater the speed of the molecules, and thus the greater their kinetic energy.

The statement that matter is made up of particles (atoms, ions, or molecules) that are in constant motion is part of the **kinetic theory** of matter. This theory has been used to explain how matter behaves and to predict new facts about matter.

MOTION OF MOLECULES

activity

OBJECTIVE: Perform an experiment to determine the effect of temperature on the solubility of a liquid in solution.

PROCESS SKILLS

In this activity, you will *observe* and *collect data* and then *analyze* the data concerning the relationship between temperature and solubility.

MATERIALS

2 100-mL beakers	food coloring
cold tap water	paper
hot tap water	pencil

PROCEDURE

1. Fill one of the beakers with cold water. Fill a second beaker with hot water. Wait several minutes.
2. Add one drop of food coloring to each beaker at the same time. Be careful not to disturb the beakers.
3. Observe what happens in each beaker. Prepare a diagram of each beaker to illustrate your observations made over time.

OBSERVATIONS AND CONCLUSIONS

1. Compare the behavior of the food coloring in the hot water and in the cold water. Explain any differences.
2. Why was it important to wait several minutes before adding food coloring to the beakers?

Diffusion indicates that molecules are in random motion. A few drops of hydrochloric acid (HCl) are placed on a wad of paper in the bottom of a cylinder. Then some strong ammonia water is placed on a wad of paper in the bottom of another cylinder. When the open ends of the two cylinders are brought together, as shown in Figure 4-1, a white smoke forms. This smoke is the compound ammonium chloride (NH_4Cl). Ammonium chloride forms as a result of a chemical reaction between the vapors of hydrochloric acid and ammonia. The movement of molecules from an area of higher concentration to an area of lower concentration is called **diffusion**. Diffusion takes place in liquids, solids, and gases.

Porosity is the quality of having very small holes, or openings, through which gases or liquids can pass. Solids having this quality are said to be **porous.** Gases diffuse easily through porous solids. For example, oxygen gas diffuses through the membranes that cover the cells of plants and animals and waste gases diffuse out of cells.

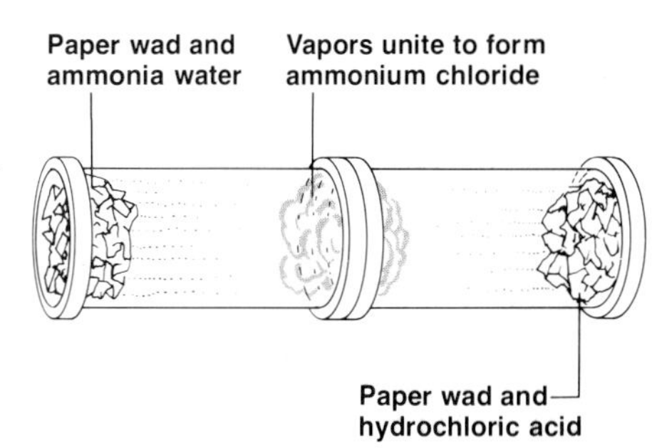

Figure 4-1 This diagram illustrates the diffusion of gas molecules of ammonia (NH_3) and ammonium chloride (NH_4Cl). How do you know that a chemical reaction has taken place?

There are three phases of matter. Most matter can be divided into three groups: solids, liquids, and gases. These groups are called phases, or states, of matter. Wood and steel are examples of matter in the solid phase. Water and gasoline are examples of matter in the liquid phase. The air you breathe is an example of matter in the gas phase.

In the solid phase, matter has a definite volume and a definite shape. The atoms of a solid move slowly. They are usually vibrating in place as shown in Figure 4-2A. Strong forces hold solid particles together very tightly. Solids cannot be compressed; they have a high density and their rate of diffusion is very slow.

Matter in the liquid phase has a definite volume but no definite shape. Liquids flow freely. As you can see in Figure 4-2B, the particles of a liquid move more freely than the particles of a solid. Liquid particles also move faster and have greater kinetic energy than solid particles. In addition, there are greater distances between the particles of a liquid. Like solids, liquids cannot be compressed.

Matter in the form of a gas does not have a definite volume or shape. A gas takes on the shape and size of its container. When you blow up a balloon, you can see how the air fills its container. Notice in Figure 4-2C that the spaces between gas particles are large. Unlike liquids, gases can be compressed. They have low densities (liquids are about 1,000 times denser than gases) and diffuse very rapidly.

Scientists recognize a fourth state of matter, known as plasma. This form of matter is similar to a gas. It is made up of charged particles and exists only at very high temperatures. Plasma is rare on earth. The northern lights, or *aurora borealis,* is an example of matter in a plasma state in the upper atmosphere. Plasma is the major state of matter in the stars. Scientists have estimated that over 99 percent of all matter in the universe is plasma.

Another unusual form of matter is that of neutron stars. This form of matter is so compressed that electrons and protons are squeezed together to form neutrons. The

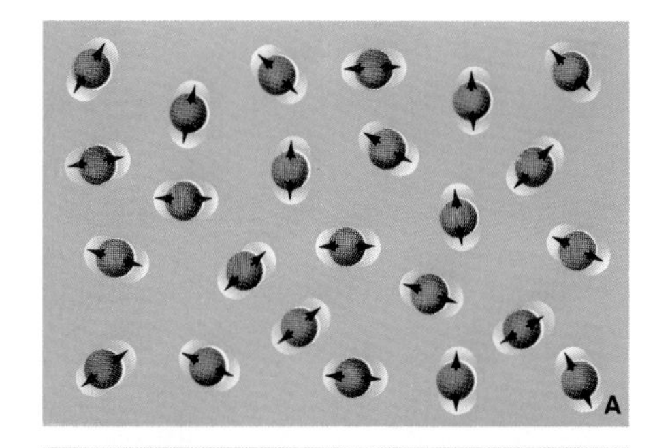

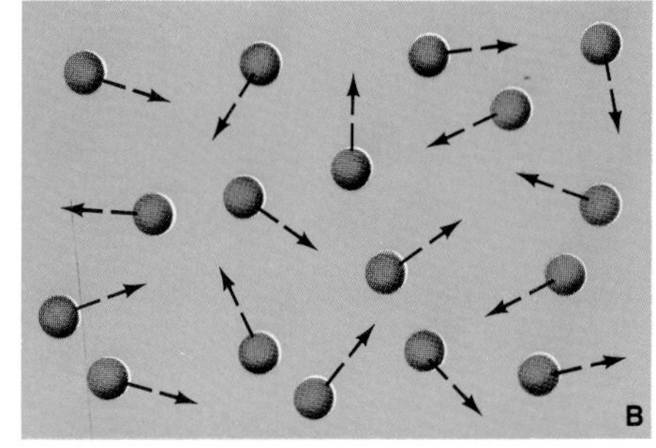

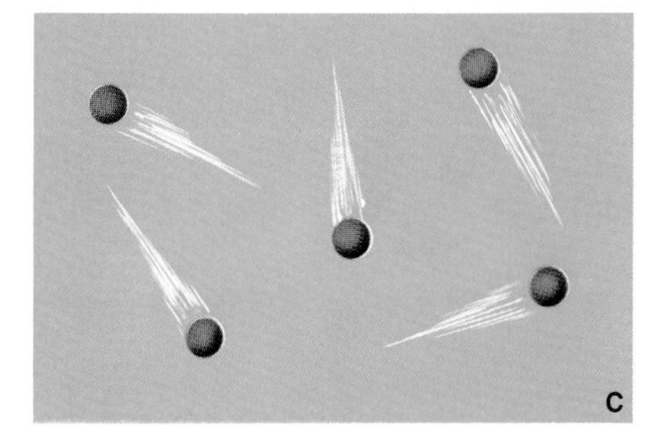

Figure 4-2 (A) Particles of a solid vibrate about fixed positions. (B) Weak forces of attraction between particles of a liquid allow for greater movement. (C) Particles of a gas are widely separated and move farther away.

density of a neutron star is many billions of kilograms per cubic centimeter. If the earth were compressed in a similar manner, it would measure only about 200 m across.

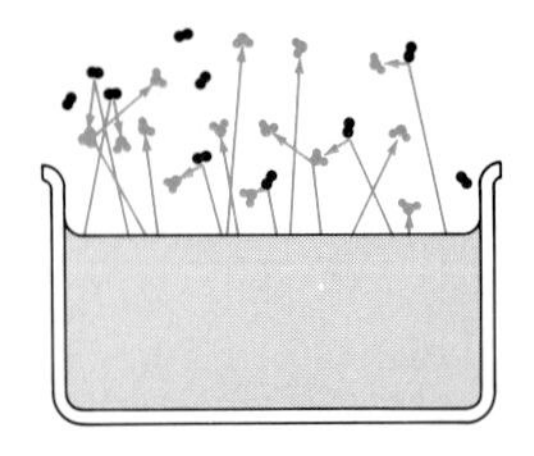

Figure 4-3 Water evaporates because some molecules receive enough energy to leave the surface. Some water molecules return to the surface after hitting molecules of gases in the air. What happens to the evaporation process if the temperature of the water is increased?

Changes in phase involve changes in energy. All forms of matter can be changed from one phase to another by a change in temperature, or pressure, or both. When a piece of ice (solid) is heated, it changes to water (liquid). Adding heat energy to ice increases the kinetic energy of the molecules and weakens the forces holding the ice molecules together. If enough heat is added to a solid, these forces will not be able to hold the solid in a definite shape. The solid will then melt into a liquid. Melting is the process of changing a substance from a solid to a liquid.

As the water continues to be heated, its molecules continue to gain energy. The forces will not be able to hold the rapidly moving molecules together in the liquid state. The water molecules will spread apart into a gas—water vapor. The gas molecules will have more energy than the liquid molecules and will rise to the surface. This process is known as boiling.

In melting or boiling, the kinetic energy of the molecules overcomes forces of attraction between molecules. The chemical bonds within a molecule do not break when a solid melts or when a liquid boils, however. Melting and boiling are not chemical changes. They are physical changes.

A substance can also change from a liquid to a gas through evaporation. Evaporation is the escape of some of the faster-moving molecules from the surface of a liquid. This change of phase can take place whether the liquid is hot or cold. Boiling, however, takes place only at a definite temperature. Water evaporates because some of its molecules at the surface have enough energy to break away from the rest of the molecules and become a gas (Figure 4-3). Energy is removed by these molecules, and the remaining liquid is cooled. Gently blow on a wet finger and on a dry one at the same time. Which finger is losing energy? How can you tell?

When heat is removed from molecules of steam, the molecules will slow down until the forces of attraction are once again strong enough to hold them together to form liquid water. The process of changing from a gas to a liquid is called **condensation.** As more heat is removed, the liquid will finally freeze into a solid the molecules of which are held together by very strong forces of attraction. The total amount of heat energy removed as a sample of matter freezes is equal to the total amount of heat energy absorbed as the same sample of matter melts.

Changes in phase can also be brought about by a change in pressure. For example, pressure applied to ice will cause it to melt. If you have ever walked in the snow, you may have noticed some snow sticking to the soles of your shoes. Your weight provides the pressure to melt the snow. As you lift your foot, the pressure is removed and the water refreezes on the sole of your shoe in the form of ice. The next step melts more snow and adds more ice to your shoe.

Sometimes matter changes from a solid to a gas without passing through the liquid phase. Dry ice (carbon dioxide in the solid phase) is an example of matter having no liquid phase at ordinary temperatures. At normal air pressure, it changes phase directly from the solid to the gaseous phase. A substance that does this is said to **sublimate,** or **sublime.** The process itself is called sublimation (SUB-*luh*-MAY-*shun*). Because of its low temperature, dry ice is often used to keep ice cream from melting.

4.2 STRUCTURE OF CRYSTALS

Many solids have a definite internal structure. If a salt solution is allowed to stand in an open dish, the water will evaporate and the salt will remain in the dish. If you look closely at the salt particles, you will see that each one has the same shape. All of the particles have flat surfaces and straight edges. A solid containing atoms that are arranged in a regular pattern is called a **crystal**. Crystals are regular geometric structures. Common table salt, for instance, is made up of tiny, cube-shaped crystals like those shown in Figure 4-4.

The size of crystals depends on the time it takes them to form. The longer it takes, the larger the crystals become.

Figure 4-4 Magnified crystals of common table salt (NaCl). Describe the shape of these crystals.

Figure 4-5 The garnet crystals on the top and the quartz crystals on the bottom illustrate that crystal growth follows a definite pattern.

Each crystal of a given substance has the same shape and distinct pattern as the other crystals of the same substance. (See Figure 4-5.) The atoms and molecules of which they are formed are arranged in regular, repeating patterns. Many other solids, such as plastic, gum, and even glass are not crystals. The molecules of these substances do not form a regular, repeating structure.

Crystals can be grown from seed. Crystals of copper(II) sulfate ($CuSO_4$) are blue, six-sided solids the opposite sides of which are parallel. These crystals can be prepared by allowing a hot concentrated solution of copper sulfate to cool. When the solution is poured off, crystals of copper sulfate remain at the bottom of the container.

Recall that the size of a crystal depends on the time it took to form. A large copper sulfate crystal can be made by allowing a solution of copper sulfate to evaporate slowly for two weeks. At this point, a small copper sulfate crystal dropped into the solution will serve as a seed. The copper sulfate will collect on the seed as the crystal grows. It is possible to grow crystals having a mass of as much as 1 kg by this method. CAUTION: *Since copper sulfate is poisonous, care should be taken in handling the crystals.*

Crystal patterns can be used to help identify a substance. The shape of a crystal depends on the pattern in which the atoms or molecules of the solid are arranged. Scientists use X-rays to study crystal patterns. Each crystal creates a unique pattern on an X-ray picture. By measuring such patterns, scientists can tell how far apart the particles in a crystal are and how they are arranged.

Crystals are grouped into six basic crystal systems. The simplest is the cubic system. This shape has axes at right angles to each other, as in the cube shown in Figure 4-6. All of the axes in this system are of equal length. Lead sulfate (PbS), sodium chloride (NaCl), and potassium chloride (KCl) are examples of compounds in the cubic system. Diagrams of other crystal systems also appear in Figure 4-6.

Figure 4-6 These diagrams represent the six basic crystal systems. The differences among crystal systems are due to the position of the axes in each one. How would you describe the position of the axes in the tetragonal system? the monoclinic system?

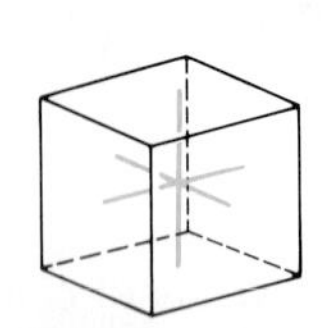

1. Cubic (sodium chloride)

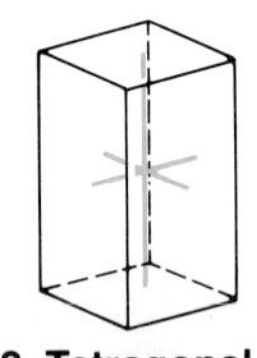

2. Tetragonal (cassiterite)

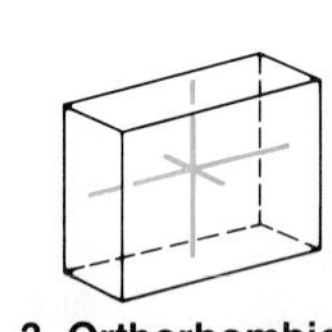

3. Orthorhombic (sulfur)

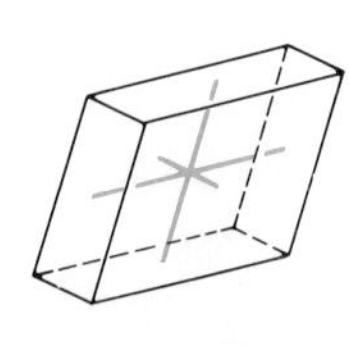

4. Monoclinic (gypsum)

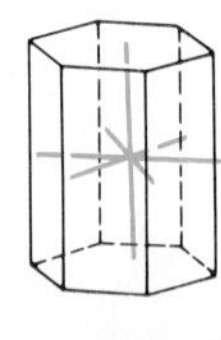

5. Hexagonal (quartz)

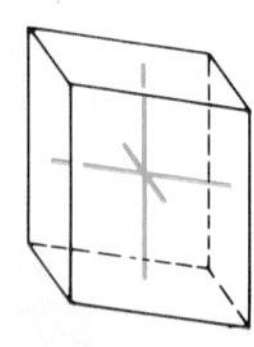

6. Triclinic (copper sulfate)

CRYSTAL FORMATION

OBJECTIVE: Perform an experiment to witness crystal growth.

PROCESS SKILLS

In this activity, you will *observe* and *collect data* about crystal growth and *communicate* information about crystal shape.

MATERIALS

table salt (NaCl)	microscope slide
distilled water	microscope
stirring rod	hand lens or magnifier
50-mL beaker	paper
medicine dropper	pencil

PROCEDURE

1. Fill a 50-mL beaker with distilled water. Add several pinches of salt and stir until dissolved.
2. Use a medicine dropper to place one drop of the solution on a microscope slide. Place the slide under a microscope. View the slide under low power.
3. As the water evaporates, watch for the growth of salt crystals.
4. Once the water has evaporated, place the slide on your laboratory table. Examine the crystals with a hand lens.

OBSERVATIONS AND CONCLUSIONS

1. Describe the growth of the salt crystals as the water evaporated.
2. Do all your crystals and those of your classmates have the same shape? Describe the shape(s) you see.

4.3 SOLUTIONS

Solutions are uniform mixtures. If you place a cube of sugar in a liter of water and stir, the sugar will seem to disappear. In reality, the molecules of sugar have simply spread evenly through the water. The sugar has dissolved in the water.

In this example, the liquid (water) in which the solid (sugar) dissolves is called a **solvent**. The solid (sugar) that is dissolved is called the **solute** (*sahl*-YOOT). The liquid formed by dissolving a solute in a solvent is called a **solution**. A solution is a uniform mixture of a solute and a solvent. The solute in a solution does not settle out over time, nor can it be removed by filtration. It can, however, be removed by evaporating the liquid solvent. A solution resulting from the mixing of a solid with a liquid is the most common type of solution. It is possible, however, for solvents, solutes, or solutions to be solids, liquids, or gases.

Water is sometimes called a "universal solvent" because it can dissolve so many solutes. It can dissolve gases, such as the mixture of gases in air, or a liquid, such as alcohol.

There are some substances that water cannot dissolve. These substances are said to be insoluble in water. Examples include oils, gums, and resins. These substances will dissolve in other liquids such as alcohol, gasoline, or turpentine. Different solvents are used to meet various needs. For instance, turpentine is used to dissolve certain paints.

There are several kinds of solutions. Solutions can contain varying amounts of solute. A spoonful of sugar dissolved in a liter of water is an example of a dilute solution. A cupful of sugar dissolved in the same amount of water is said to be concentrated.

In the example above, the terms concentrated and dilute are not exact measurements. The term **concentrated** means that a large amount of solute has been added to the solvent, and **dilute** means that a small amount has been added. In the chemistry laboratory, you may see bottles labeled "Concentrated Sulfuric Acid" or "Concentrated H_2SO_4." These labels indicate that there is not much water (solvent) present in the acid.

If you were to keep adding sugar to a concentrated sugar solution, you would finally reach a point at which no more sugar would dissolve at that temperature. The solid sugar would begin to settle to the bottom of the container. This solution would now be described as saturated. A solution is said to be **saturated** when additional solute will not dissolve. Such a solution contains a definite amount of solute. This amount is always the same for a given temperature and solute. As you can see in Figure 4-7, the amount of solute that will dissolve in a given quantity of solvent at a given temperature varies with different solutes.

If you heat a saturated solution of sugar and water, you will find that more sugar can be dissolved. In fact, a hot,

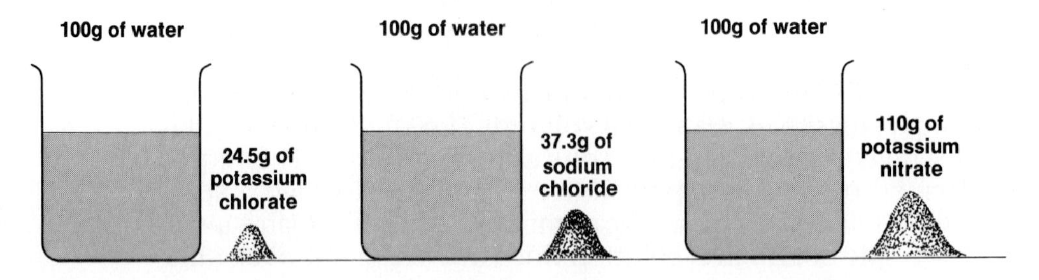

Figure 4-7 The amount of solute that will dissolve to produce a saturated solution varies according to the solute used. In the diagram, that amount is shown (in grams) for three solutes. The water temperature is 60° C. Which solute is most soluble?

saturated solution contains a much greater amount of solute than the same solution at a lower temperature. Most solutes are more soluble in hot solvents than in cold ones.

The kinetic theory explains the speed at which a solute dissolves. Because a solid is generally more soluble in a hot solvent than in a cold one, the time taken for a solid to dissolve is shorter in a hot solvent than in a cold one.

When heat is applied, the particles of the solute have extra kinetic energy and tend to move apart faster. The particles of the solvent move faster, also. The speed at which the solute goes into solution generally increases.

The speed of dissolving can also be increased by stirring or shaking. After adding sugar to a glass of lemonade, you stir it vigorously with a spoon. This stirring action brings fresh parts of the liquid in contact with the undissolved portions of the solute. The solute dissolves more quickly.

Grinding solid solutes into a fine powder will also increase the speed of dissolving. Grinding greatly increases the surface area of the solid that is exposed to the liquid, as illustrated in Figure 4-8. Notice that the 2-cm cube in the diagram on the left has a surface area of 24 cm^2. Now imagine that the 2-cm cube is cut into 1-cm cubes, as illustrated in the diagram on the right by the dashed lines. The eight new cubes now have a total surface area of 48 cm^2. The surface area exposed to a liquid has been doubled. Finely powdered solutes will dissolve more rapidly in a solvent than large lumps of the same substance.

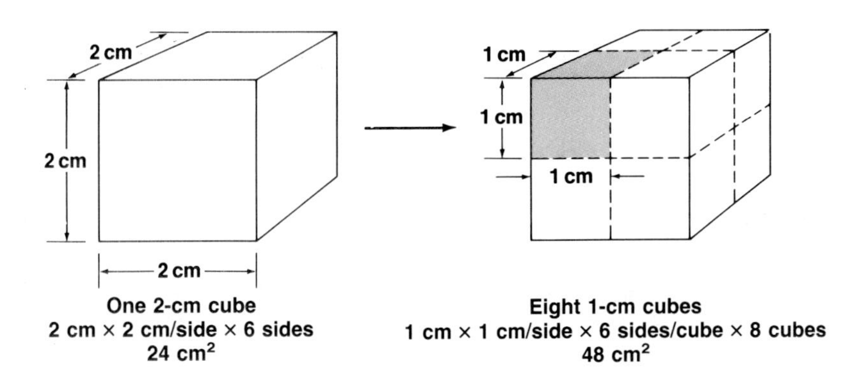

Figure 4-8 Dividing a cube into smaller parts increases total surface area. Does the volume change? Does the mass change?

Gases are soluble in liquids. Recall that a solid is generally more soluble in hot liquids than in cold ones. With gases, the reverse is true. Gases are more soluble in cold liquids than in hot ones. Very cold water contains a small amount of dissolved air. If a glass of cold water is left standing in a warm room, gas bubbles will form on the inside of the glass.

Figure 4-9 As cold water becomes warm, bubbles of air collect on the sides of the glass. What general statement can you make about the solubility of gases in liquids?

Look at Figure 4-9. As the water warms up, it loses some of its ability to hold air in solution.

Larger amounts of gas can be dissolved in a liquid if pressure is applied. Soda, for example, is water in which carbon dioxide gas has been dissolved under pressure. When the cap is removed, the pressure within the bottle decreases and the gas escapes from the liquid.

Some liquids are soluble in other liquids. Liquids such as alcohol and water mix completely with each other. Two such liquids that are soluble in each other are said to be **miscible** (MIS-*uh-bull*). A different condition exists when oil and water are combined. When oil and water are mixed, they tend to separate on standing. A definite boundary can be seen between the oil and water. Liquids that are insoluble in each other are said to be **immiscible.**

When shaken together, immiscible liquids can form a temporary mixture, called an **emulsion** (*ih*-MUL-*shun*). An emulsion consists of tiny droplets of one liquid scattered throughout another. When oil and water are shaken together in a container, an emulsion is formed. The suspended oil particles will come to the surface as soon as you stop shaking the container. Why should you shake a mixture of oil and vinegar salad dressing before using it?

Whole milk is another example of an emulsion. Upon standing, the tiny fat particles in the milk may come together and rise to the top as cream. To keep this from happening, milk is homogenized. **Homogenization** (huh-MAHJ-*uh-nih*-ZAY-*shun*) is the process of breaking up the particles in an emulsion so they are small enough not to separate. The fat particles in milk are broken down into still smaller particles by machines that thoroughly mix the milk. Compare the sizes of the fat particles shown in Figure 4-10. Homogenization forms a more lasting emulsion.

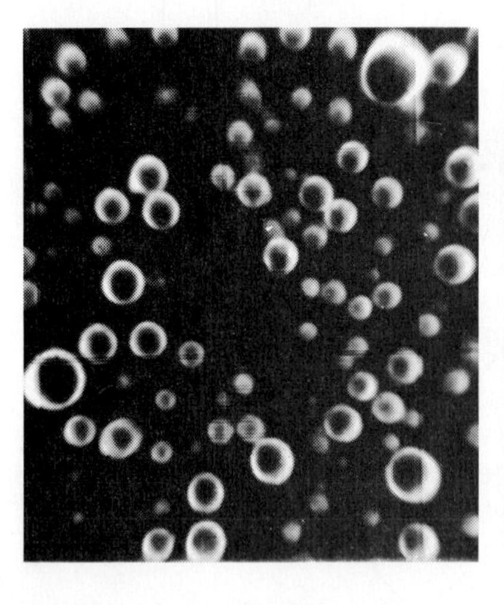

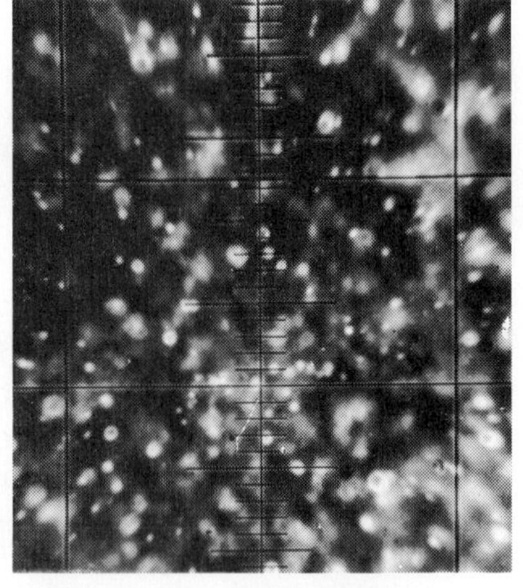

Figure 4-10 Compare the size of fat particles in homogenized milk (right) and in milk that has not been homogenized (left). Which particles would take longer to settle out?

Another way to make an emulsion last longer is to add a special substance to the liquids themselves. Such a substance is called an emulsifier (*ih*-MUL-*sih*-FIE-*ur*). Usually, a substance that is partly soluble in both liquids is used as an emulsifier. For example, kerosene and water, which do not mix well, will form an emulsion that stays mixed for a long time if a soap solution is added to the mixture. The soap acts as an emulsifier. Mayonnaise is an emulsion formed by adding egg yolk to a mixture of olive oil and vinegar. Egg yolk is partly soluble in both olive oil and vinegar.

4.4 ELECTROLYTES

Some solutions conduct an electric current. Substances that conduct an electric current when dissolved or melted are called **electrolytes** (IH-LEK-*truh-lites*). You can determine which substances are electrolytes and which are not by performing a simple experiment. Prepare a setup like the one shown in Figure 4-11. To begin the test, fill the beaker with pure water and close the switch of the electric circuit. Note that the lamp does not light. A dark lamp means that pure water does not provide a path to complete the electric circuit. Now add a few drops of sulfuric acid to the water. The lamp now glows brightly, and the solution completes the electric path, or circuit. This solution (sulfuric acid and water) conducts an electric current.

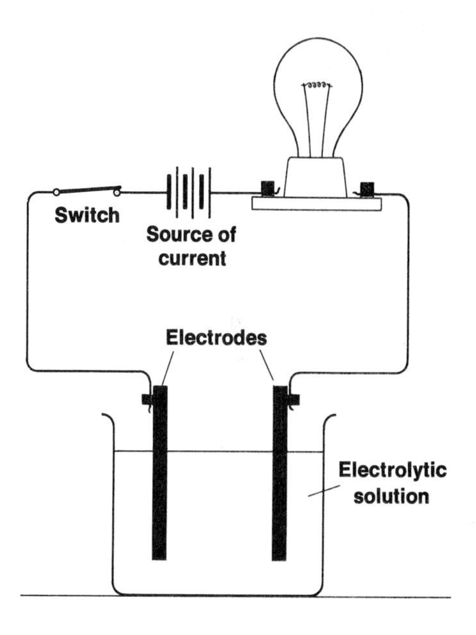

Figure 4-11 What causes the bulb to light when the switch is closed?

Try other liquids, such as solutions of common table salt (NaCl), hydrochloric acid (HCl), and sodium hydroxide (NaOH). You will see that the lamp also glows brightly when these solutions are used.

Now try a solution of sugar, alcohol, or glycerin. You will see that these substances do not conduct an electric current. Substances that do not conduct an electric current are called nonelectrolytes. CAUTION: *Do not experiment with the suggested solutions without your teacher's permission and supervision.*

Electrolytic solutions contain ions. Recall from Chapter 3 that sodium chloride is made up of sodium ions and chloride ions. The internal structure of a sodium chloride crystal resembles the diagram in Figure 4-12. When salt is dissolved in water, it separates into its ions. The Na^+ and Cl^- ions are released from their fixed places in the crystal pattern and become free ions in the water. This breaking down of sodium chloride can be written as follows:

$$NaCl(s) \rightarrow Na^+(aq) + Cl^-(aq)$$

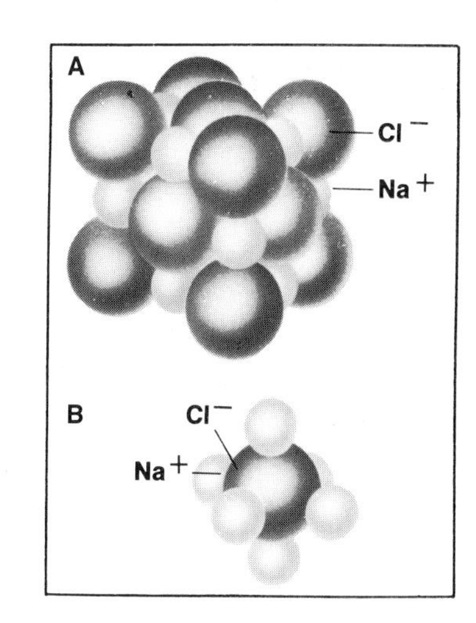

Figure 4-12 (A) Sodium chloride crystals are composed of equal numbers of sodium ions (Na^+) and chloride ions (Cl^-). (B) Focusing in on a single chloride ion in sodium chloride, how many sodium ions surround a chloride ion?

When substances react with water to form ions, they are said to be **ionized.** Ionization forms a complete path for electric current flow. Initially, the positive ends of the H_2O molecules atttract the Cl^- ions, and the negative ends of the H_2O molecules attract the Na^+ ions (Figure 4-13). If an electric current is applied, the Na^+ ions will then be attracted to the negative pole of a battery, while the Cl^- ions will be attracted to the positive pole.

A few covalent compounds, such as HCl gas, also react with water to form ions. The reaction is

$$HCl(g) \rightarrow H^+(aq) + Cl^-(aq)$$

Most covalent compounds do not form ions in water. For instance, sugar remains in solution as individual sugar molecules ($C_{12}H_{22}O_{11}$). Sugar is a nonelectrolyte. The particles of a nonelectrolyte do not ionize. Nonelectrolytes cannot conduct an electric current.

Acids, bases, and soluble salts are an important group of electrolytes. You will study these compounds in Chapter 5. The study of electrolytes is very important in chemistry. The following summary contains key points about them.

1. Electrolytes in water exist in the form of ions.
2. An ion is an atom or group of atoms that carries an electric charge.
3. The ions are free to move in solution and serve to conduct an electric current.
4. The total charge of the positive ions equals the total charge of the negative ions.
5. Water plays an important part in the formation of ions.

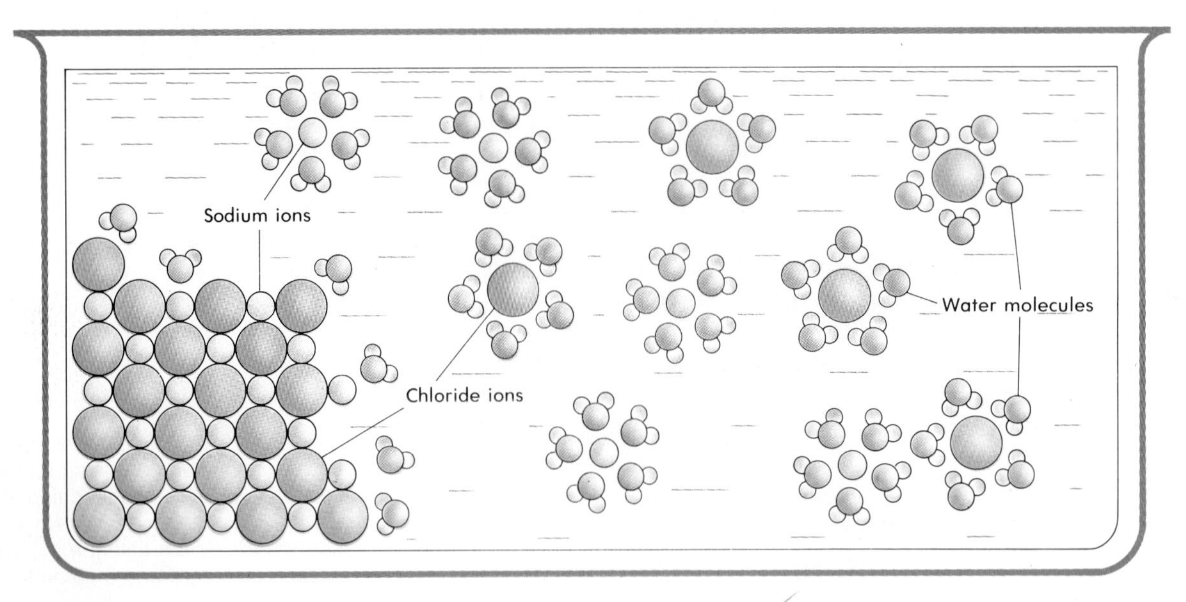

Figure 4-13 When sodium chloride is added to a beaker of water, the attractive force of the water molecules acts on the salt, eventually breaking the bonds that hold the sodium and chloride ions together.

CAREERS

SCIENCE TEACHING

combining education and science

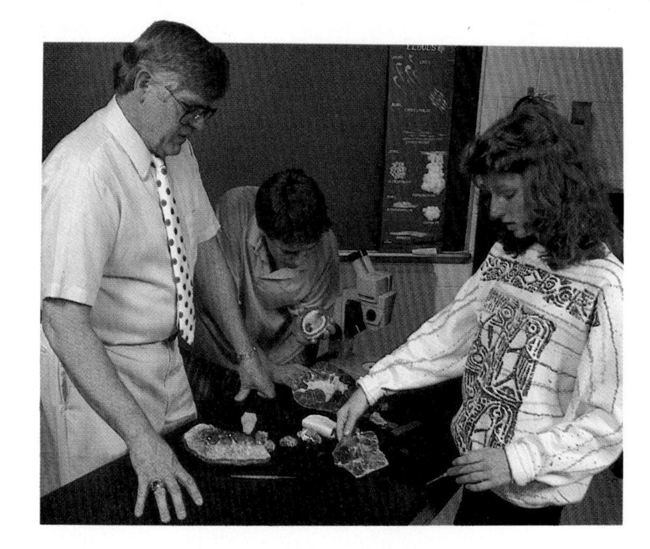

Science teachers work at three levels: elementary, secondary, and college. Teachers in the elementary grades may teach science as part of their work in reading, writing, and arithmetic. In some elementary schools, science teachers are specialists who teach science only. Elementary school teachers work with their students on basic skills in order to further their students' mental, physical, social, and cultural development.

At the secondary level, science teachers perform a number of tasks. They help the student learn about the general areas of life science, earth science, and physical science. Teachers prepare lesson plans and keep records of student progress. They also hold conferences with parents, students, and other teachers. Science teachers also set up demonstrations, order laboratory supplies, conduct field trips, and act as advisors to science clubs.

Science teachers at the college level usually specialize in one branch of science such as physics, biology, meteorology, or geology. These teachers, or professors as they are often called, also conduct original scientific research. They may be called upon to give lectures to large or small student groups, and teach several specialized subjects.

Teachers in public schools are required to be certified by their state education department. Most states require at least a bachelor's degree from a college or university. In most cases, practice teaching is also necessary. For secondary teachers, a major field of study must be completed in biology, chemistry, physics, or general science. To become a college science teacher, a Ph.D. is almost always necessary. Earning a Ph.D. degree requires several years of study beyond a four-year undergraduate program.

Science teachers must act as both scientists and educators. As a result, they are required to do a great deal of reading and studying after getting their degrees. Reading helps them keep up with new developments in science as well as with improved teaching methods.

For more information on a career as a science teacher, contact

National Science Teachers Association
1742 Connecticut Avenue, NW
Washington, DC 20009

CHAPTER REVIEW

SUMMARY

1. The kinetic theory of matter describes the rapid motion of particles of matter.
2. The higher the temperature of a substance, the greater the speed of the molecules, and the greater the kinetic energy.
3. Matter exists in three ordinary phases: solid, liquid, and gas.
4. Crystals have an orderly arrangement of atoms.
5. A solution is a uniform mixture of a solute and a solvent.
6. Solutions can be dilute, concentrated, or saturated.
7. Generally, an increase in temperature will increase the amount of solute that will dissolve in a solvent.
8. The speed at which a solute dissolves depends on the temperature, the size of the particles, and the amount of stirring of the mixture.
9. Substances that conduct an electric current when dissolved or melted are called electrolytes.

VOCABULARY

Match the item in the left column with the best answer in the right column. Do not write in this book.

1. concentrated
2. condensation
3. diffusion
4. dilute
5. electrolyte
6. emulsion
7. homogenization
8. immiscible
9. ionized
10. kinetic energy
11. kinetic theory
12. miscible
13. porous
14. saturated
15. solute
16. solution
17. solvent
18. sublimate, or sublime

a. phase change from solid to gas
b. statement that matter is made up of particles in motion
c. substance that conducts an electric charge when dissolved or melted
d. substance being dissolved
e. immiscible liquids forming a mixture
f. movement of molecules from an area of higher to lower concentration
g. substance in which a solute dissolves
h. energy of motion
i. substance formed when a solute is dissolved in a solvent
j. no additional solute can be added
k. substances that form ions in water
l. much solute added to the solvent
m. liquids that are soluble in each other
n. little solute added to the solvent
o. liquids that are not soluble in each other
p. change from a gas to a liquid
q. breaking up particles in an emulsion
r. solids through which gases or liquids can pass

REVIEW QUESTIONS

1. What is the kinetic theory of matter? Why is it an important idea?
2. Give several examples of diffusion. How does diffusion occur?
3. What are the three ordinary phases of matter? Give examples of each.
4. How can one phase of matter be changed into another?
5. How does the process of boiling differ from that of evaporation?
6. Describe a method for growing crystals.
7. What phase change is involved in the formation of crystals?
8. How are a solute, a solvent, and a solution related?
9. What two factors help to dissolve more gas in a liquid?
10. List several properties of a solution.
11. What is an emulsion? Give several examples.
12. How does temperature affect the solubility of most solids? of gases?
13. What is the difference between a dilute solution and a concentrated one? What is a saturated solution?
14. What is an electrolyte? Name three electrolytes and nonelectrolytes.
15. Why is it necessary for a substance to form ions in solution to be an electrolyte?
16. How is the speed of molecules in matter related to kinetic energy?
17. How can the rate at which a solid dissolves in a liquid be increased?

CRITICAL THINKING

18. Describe how the movement of atoms and molecules in solids, liquids, and gases would be affected by a temperature increase too small to produce a change in phase.
19. Using the kinetic theory of matter, describe how the melting of a solid and the boiling of a liquid take place.
20. Freshwater fish often die in large numbers during summer "heat waves." Explain why this happens.
21. What effect would an increase in temperature have on the rate of diffusion? Explain your answer.
22. Which form of ice would cool your drink faster—ice cubes or crushed ice? Why?

PROBLEM

The maximum amount of sodium chloride (NaCl) that can be dissolved in 100 mL of water at a given temperature is 35 g. At the same temperature, what is the minimum amount of water that will dissolve 175 g of NaCl? What is the name for this type of solution?

FURTHER READING

Farrell, J. J. "Molecules in Color." *Byte* 11 (December 1988): 329–330.

Holden, Alan, and Phyllis S. Morrison. *Crystals and Crystal Growing*. Cambridge, Mass.: MIT Press, 1982.

Watson, Philip. *Liquid Magic*. New York: Lothrop, Lee, and Shepard Books, 1983.

ACIDS, BASES, AND SALTS

Acidic spring waters, rich in calcium, seep into the waters of Mono Lake in California. With evaporation, the level of the lake fluctuates and columns of tufa (calcium carbonate) crystallize from the waters. What does the term acidic mean? What role do acids play in the crystallization of salts?

CHAPTER 5

SECTIONS

OBJECTIVES

- ☐ Identify the properties of acids, bases, and salts.
- ☐ Describe the making and uses of several common acids, bases, and salts.
- ☐ Distinguish among four major types of chemical reactions.

5.1 ACIDS

Acids have certain properties. Have you ever tasted pure lemon juice? If you have, you know that it has a sour taste. This sour taste is due to the presence of citric (SIT-*rik*) acid in the juice. All foods that contain acids have a sour or sharp taste. Vinegar tastes sour because of the acetic (*uh*-SEE-*tik*) acid that it contains. Many foods turn sour when they spoil because the starches and sugars they contain break down into acids. For instance, when milk turns sour some of the milk sugar is changed to lactic (LAK-*tik*) acid.

CAUTION: *You should never taste unknown liquids in the laboratory to find out if they are acids.* Some chemicals will burn your tongue and others are very poisonous. There is a better way of identifying acids. Figure 5-1 on page 78 illustrates some common properties of acids.

Sour taste like lemon

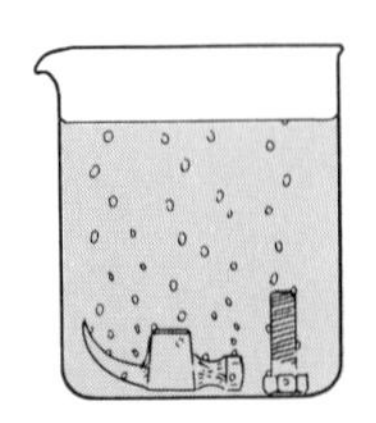
Corrode active metals

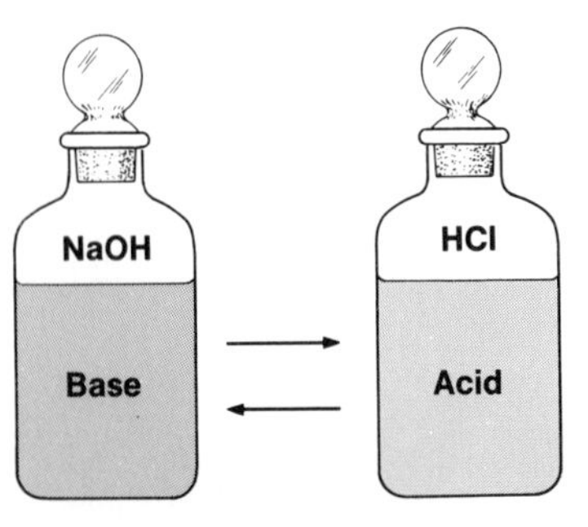

Acids neutralize bases

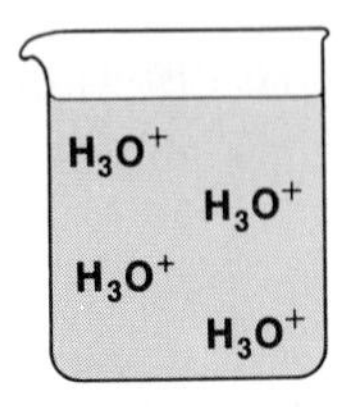

Form hydronium ions in solution

Turn litmus red

Figure 5-1 Common properties of acids. Why is it a good practice not to taste materials in the laboratory?

Acids can be identified by testing them with special paper called litmus paper. Litmus is a plant dye that changes color when it comes into contact with acids. Substances, such as litmus, that change color in the presence of certain ions are called **indicators**. Blue litmus paper is an indicator that turns red when dipped in acid.

Acids have properties that are opposite to those of another group of compounds, called bases. Acids and bases react with each other to neutralize each other's properties. Bases will be discussed later in the chapter.

Acids produce hydronium ions in aqueous solutions. If hydrogen chloride (HCl) gas is bubbled into water, a solution called hydrochloric acid (HCl) is obtained. The equation for this reaction is

$HCl(g)$	+	$H_2O(l)$	→	$H_2O \cdot H^+(aq)$	+	$Cl^-(aq)$
hydrogen chloride		water		water united with hydrogen ion		chloride ion

Notice that the hydrogen chloride gas reacts with water in a special way, forming $H_2O \cdot H^+$ ions and chloride ions. The dot (·) in the formula indicates that a water molecule is loosely connected to the hydrogen ion. The $H_2O \cdot H^+$ ion, called the hydronium ion, can be written as H_3O^+. All acids produce hydronium ions in aqueous solutions. An **acid** is defined as a substance capable of ionizing in water and giving hydrogen ions (H^+) to water molecules to form hydronium ions (H_3O^+). It is the hydronium ion that turns blue litmus paper red.

The equation for the ionization of an acid can also be shown in a simpler way:

$$HCl(g) \rightarrow H^+(aq) + Cl^-(aq)$$

Because the ions formed are in an aqueous solution, it is understood that the process takes place in water. The H^+ shown in the equation is the hydrogen released by the acid. It is understood that this ion will combine with a water molecule to form a hydronium ion. Therefore, the hydronium ion is not written in the equation. Following this simpler notation, the rest of the text will use the term hydrogen ion (H^+) in place of the hydronium ion.

The names and formulas of some common acids are listed in Table 5-1. Since you will be learning more about these acids in later chapters, it is important to become familiar with their formulas at this time.

Notice that all of the acids appearing in Table 5-1 contain hydrogen. The symbol for hydrogen is always written first in the formulas. In an aqueous solution, this hydrogen is the source of the hydrogen ions. Certain acids, such as acetic and tartaric, contain added hydrogen as part of a radical.

TABLE 5-1

Names and Formulas of Some Common Acids	
Acid	**Formula**
Sulfuric	H_2SO_4
Hydrochloric	HCl
Nitric	HNO_3
Carbonic	H_2CO_3
Boric	H_3BO_3
Phosphoric	H_3PO_4
Acetic	$HC_2H_3O_2$
Tartaric	$H_2C_4H_4O_6$

This hydrogen does not go into solution as ions.

Not all substances that contain hydrogen are acids. For instance, there is hydrogen in water (H_2O), methane (CH_4), cane sugar ($C_{12}H_{22}O_{11}$), and ammonia (NH_3). These compounds are not acids. They do not release hydrogen ions or produce hydronium ions.

Acids that produce large numbers of H^+ ions in water are called strong acids. The reason why so many hydrogen ions are produced is that strong acids ionize completely, or nearly so, in water solutions. Hydrochloric, nitric, and sulfuric acids are examples of strong acids. Acids that produce few hydrogen ions in water are called weak acids. Examples of weak acids include acetic, citric, and carbonic acids. These acids are only slightly ionized in water solutions. Do not confuse the terms strong and weak with the terms concentrated and dilute. It is possible to have a dilute solution of a strong acid.

Acids react with metals. Place a few clean iron nails in a beaker that contains dilute sulfuric acid (H_2SO_4). Notice that gas bubbles are given off. This indicates that a reaction is taking place. The gas given off is hydrogen (H_2). The nails become thinner as the iron atoms react with the acid. This experiment shows that acids are corrosive, meaning that they attack metals. After some time, if there is enough acid, the nails disappear, leaving a green solution behind. This solution is iron(II) sulfate ($FeSO_4$). The equation for this reaction is

$$Fe(s) + H_2SO_4(aq) \rightarrow FeSO_4(aq) + H_2(g)$$

Notice that Fe has replaced the H_2 in the sulfuric acid to form $FeSO_4$. This type of reaction is called single replacement. A **single replacement reaction** is one in which one element replaces another in a compound.

Acids react with some other metals, such as magnesium and zinc, in the same way they react with iron. Consider the single replacement reaction involving magnesium and hydrochloric acid. The equation for this reaction is

$$Mg(s) + 2HCl(aq) \rightarrow MgCl_2(aq) + H_2(g)$$

Notice that hydrogen gas is produced. What element has replaced hydrogen in this reaction?

Sulfuric acid is a vital chemical. The amount of sulfuric acid that a nation uses indicates the size of that nation's industry. Sulfuric acid is one of the most widely used industrial chemicals. The primary materials used to make this acid are water, oxygen, and sulfur. Sulfur can be found underground or derived from oil and natural gas.

The first step in making sulfuric acid is to burn sulfur in air. In this process, sulfur combines with oxygen to form sulfur dioxide (SO_2):

$$S(s) + O_2(g) \rightarrow SO_2(g)$$

Then the sulfur dioxide is changed to sulfur trioxide (SO_3):

$$2SO_2(g) + O_2(g) \rightarrow 2SO_3(g)$$

Finally, the sulfur trioxide is added to water to produce sulfuric acid (Figure 5-2):

$$SO_3(g) + H_2O(l) \rightarrow H_2SO_4(aq)$$

The three equations above are examples of another type of chemical reaction, called synthesis (SIN-*thuh-sus*). A **synthesis reaction** is one in which two or more substances combine to form a more complex substance.

Figure 5-2 The vertical towers at this sulfuric acid plant include a catalytic converter, which process SO_2 gas, and two absorbing towers in which the final conversion from SO_3 to H_2SO_3 takes place.

Figure 5-3 A steel strip is being drawn from an acid treatment tank. Why is it necessary to treat metals with sulfuric acid?

Sulfuric acid has many uses. One of the major uses of sulfuric acid is in the making of fertilizers. Minerals containing calcium phosphate do not dissolve readily in moist soil. By treating calcium phosphate with sulfuric acid, a soluble fertilizer is created that releases phosphates to the soil, thus enabling plants to readily absorb them.

Metals are treated with dilute sulfuric acid to remove any corrosion or oxide (rusty) coatings. This process, shown in Figure 5-3, cleans metals before they are plated or enameled. Sulfuric acid is used in making chromium-plated bumpers for cars and the enameled surfaces of refrigerators and other appliances.

Sulfuric acid is also used in making photographic film and in the making of many oil by-products, including gasoline. Sulfuric acid is needed to make paints, plastics, rayon, explosives, and many other products. You may be familiar with the use of sulfuric acid as the electrolyte in car batteries.

Sulfuric acid is widely used as a drying agent. A drying agent is a substance that removes water from other substances. When concentrated sulfuric acid is left in an open

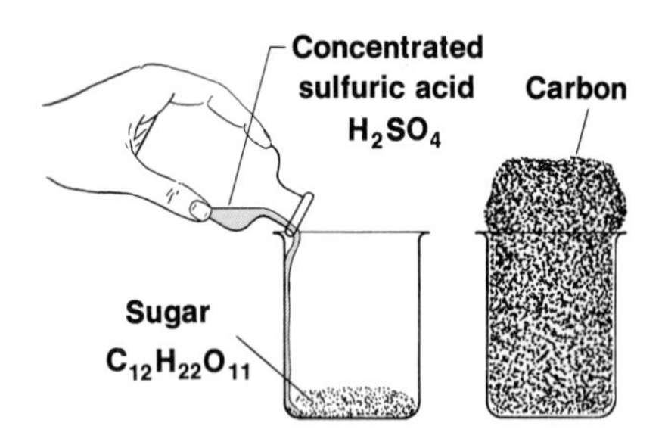

Figure 5-4 Concentrated sulfuric acid removes hydrogen and oxygen atoms from sugar, leaving a black carbon residue. What name is given to a material that removes water from other materials?

dish, its volume increases because the acid absorbs moisture from the air and dilutes itself. Sulfuric acid also removes hydrogen and oxygen from common table sugar ($C_{12}H_{22}O_{11}$) in the proportion needed to form water. After the acid reacts with the sugar, carbon is left behind as a residue (Figure 5-4). Sulfuric acid reacts in the same way with other materials containing carbon. For example, wood or paper turns black when placed in sulfuric acid.

Despite its wide range of uses, sulfuric acid can be quite dangerous. It, as well as all other acids, should be handled with care. For instance, if you add concentrated sulfuric acid to water, much heat is given off. When mixing an acid and water, it is important to use a container that can withstand the large amount of heat that is produced. To avoid danger, slowly add acid to the water while stirring. CAUTION: *Never add water to acid.* Be sure to wear protective clothing and to wear goggles to shield your eyes, as the mixture is likely to spatter. Concentrated sulfuric acid will burn your skin. Even dilute sulfuric acid can make holes in your clothes. If you spill acid on your skin or clothing, immediately wash the acid off with cold water.

Hydrochloric acid can be made in the laboratory. If you add concentrated sulfuric acid to sodium chloride and heat the mixture, a choking gas is produced. This gas is hydrogen chloride. Hydrogen chloride gas is soluble in water, forming hydrochloric acid. The laboratory setup for making small amounts of hydrochloric acid is illustrated in Figure 5-5. The equation for this reaction is

$$2NaCl(aq) + H_2SO_4(aq) \rightarrow Na_2SO_4(aq) + 2HCl(g)$$

This reaction is a good example of still another type of chemical change. It is called a double replacement reaction.

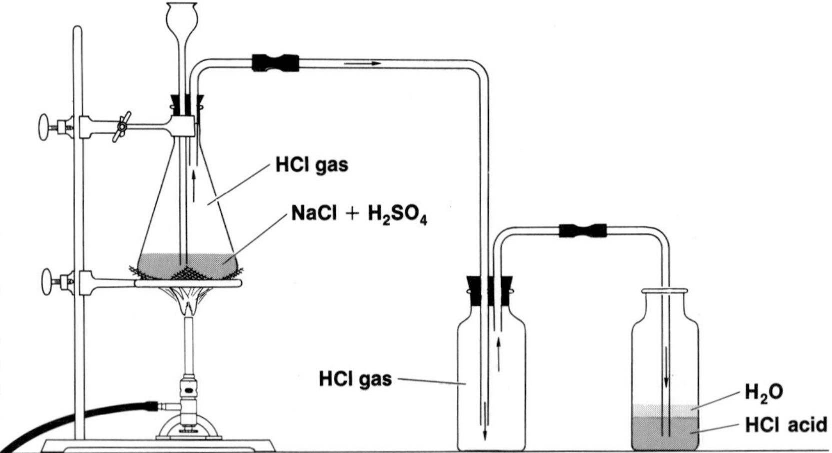

Figure 5-5 Heating sodium chloride with sulfuric acid produces hydrogen chloride gas. Describe what happens when the gas is bubbled into water. **CAUTION: Do not attempt this experiment without your teacher's permission and supervision. Remove the outlet tube from the water before the reaction stops.**

A **double replacement reaction** is one in which substances change places in a chemical reaction. It can be compared to two couples changing partners on a dance floor.

Hydrochloric acid has many uses. Galvanized iron is made by coating sheet iron with melted zinc. Before the metal is dipped in the zinc, hydrochloric and sulfuric acids are used to clean the metal. Hydrochloric acid is also used in cleaning excess mortar from stone and brick.

You may be surprised to learn that there is acid in your stomach. Gastric juices in your stomach contain concentrated hydrochloric acid that helps digest foods.

Nitric acid has many uses. Small amounts of nitric acid can be prepared in the laboratory by heating a mixture of sodium nitrate ($NaNO_3$) and concentrated sulfuric acid. Nitric acid gas (HNO_3) is produced and then condensed to a liquid. How would you write a balanced equation for this reaction?

Much of the nitric acid produced in the United States is used in making fertilizers. Nitric acid is also used in making dyes and plastics. A mixture of nitric and hydrochloric acids, called *aqua regia*, is used to dissolve gold.

Nitric acid is used in making all common explosives. Among these are smokeless gunpowder and dynamite. A particularly powerful explosive made from nitric acid is TNT, which is used in artillery shells. The molecules of all these explosives are unstable; that is, they break down quickly, forming gases that expand with great force.

Care must be used in handling nitric acid to avoid staining the skin yellow. Nitric acid produces this yellow stain on any protein and so is used as an indicator for protein. For example, a drop of nitric acid added to a slice of hard-boiled egg white will turn the egg white yellow. This color change indicates the presence of protein in egg white. The yellow color deepens to a bright orange when ammonia water is placed on the sample.

5.2 BASES

Bases can be identified by certain properties. You are probably familiar with many common bases. These bases are usually considered as household items. For example, household ammonia, a common cleaning agent, is a base. So is milk of magnesia, a mixture of magnesium hydroxide and water that is used as a medicine. Lye, a commercial grade of sodium hydroxide, is a base used for cleaning clogged sink drains.

Base solutions feel slippery

Just as acids in dilute solutions have a sour taste, bases have a bitter taste. CAUTION: *Never taste any chemicals in the laboratory.* Solutions of bases change red litmus paper to blue. Here is an easy way to remember these facts about bases:

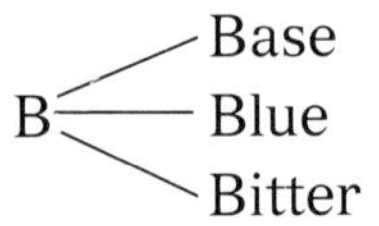

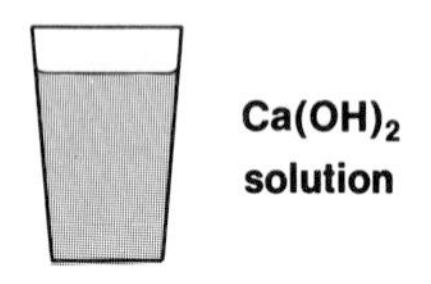

Bases taste bitter

If you were to rub a drop of a dilute base between your fingers, it would feel slippery. It is, however, an unwise practice to test for a base in this way. Most bases are caustic, which means they attack the skin and other tissues. In addition, strong bases destroy hair and wool. Bases are very harmful to the eyes. Several properties of bases are illustrated in Figure 5-6.

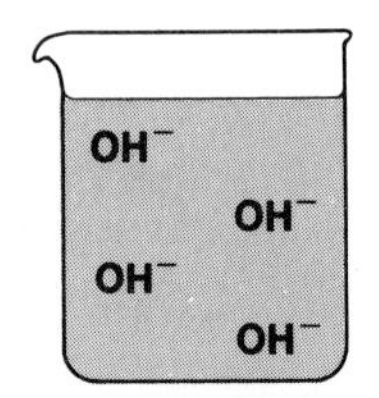

Form hydroxide ions in solution

Bases contain hydroxide ions. A base is any soluble substance that can neutralize an acid when the two are mixed. The most common bases are soluble hydroxides. The word alkaline (AL-*kuh-line*) is often used to describe substances with basic properties.

A **base** is defined as a substance capable of ionizing in water and accepting hydrogen ions from water to form hydroxide ions (OH^-). Each base contains one or more hydroxide groups. In water, soluble bases will ionize as in the following examples:

$$NaOH(s) \rightarrow Na^+(aq) + OH^-(aq)$$

$$Ca(OH)_2(s) \rightarrow Ca^{2+}(aq) + 2OH^-(aq)$$

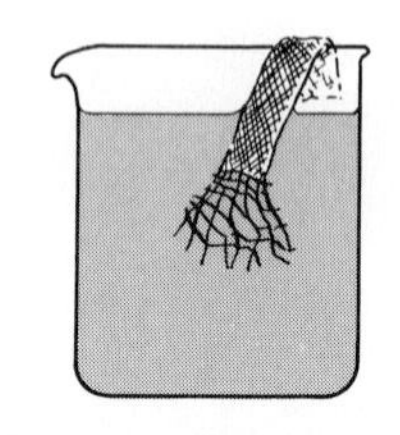

Caustic action on wool

Turn litmus blue

Figure 5-6 Common properties of bases. Why is it a good practice not to feel a substance to determine whether or not it is a base?

Table 5-2 shows the names and formulas of several bases. In naming bases, the metal ion, such as sodium (Na^+) or calcium (Ca^{2+}), is named first, followed by the word hydroxide. Thus, NaOH is called sodium hydroxide and $Ca(OH)_2$ is called calcium hydroxide. The symbol for the

TABLE 5-2

Some Common Bases and Their Formulas

Base	Formula
Sodium hydroxide	NaOH
Potassium hydroxide	KOH
Calcium hydroxide	$Ca(OH)_2$
Magnesium hydroxide	$Mg(OH)_2$
Aluminum hydroxide	$Al(OH)_3$
Ammonia water	NH_4OH
Iron(III) hydroxide	$Fe(OH)_3$

metal is written first in the formula. Ammonia water, a common base often called ammonium hydroxide (NH_4OH), does not contain a metal. In this base, the ammonium group (NH_4^+) acts like a metal.

Not all substances that have OH^- groups are bases. For example, the formula for methyl alcohol, or wood alcohol, is written CH_3OH. Since this compound does not ionize in water to form OH^- ions, it is not a base.

Sodium hydroxide and potassium hydroxide are very soluble in water. Their solutions are considered to be strong bases because of the high numbers of OH^- ions they contain. The word alkali (AL-*kuh-lie*) is often used in reference to strong bases.

Ammonia water is an example of a base that is weak because it is only slightly ionized. Calcium hydroxide, even though it ionizes completely in water, is a moderately strong base because it is only slightly soluble in water.

Sodium hydroxide can be made from salt and water. One way to make sodium hydroxide is to pass an electric current through a water solution of sodium chloride. The equation for the chemical change that takes place is

$$2NaCl(aq) + 2H_2O(l) \rightarrow 2NaOH(s) + H_2(g) + Cl_2(g)$$

Sodium hydroxide is a white crystal solid that is often sold under the name of lye. It reacts with grease to form soap. For this reason, sodium hydroxide is used to clean drains clogged with grease. Large amounts of sodium hydroxide are used in making rayon and cellophane, and in refining gasoline and other petroleum products.

Ammonia water is a weak base. Ammonia (NH_3) is a colorless gas with a sharp, penetrating odor. It is soluble in water. When ammonia gas reacts with water, ammonium ions (NH_4^+) and hydroxide ions are formed as shown by the following equation:

$$NH_3(g) + HOH(l) \leftrightharpoons NH_4^+(aq) + OH^-(aq)$$

The double arrow ($\leftrightharpoons$) in the equation indicates that this reaction is reversible. In other words, ammonia and water react to form ammonium and hydroxide ions. At the same time, these ions react with each other to form ammonia and water. The uneven size of the two arrows indicates that the reaction takes place predominantly in one direction. The shorter arrow pointing to the right indicates that only a small percentage of ammonia molecules are reacting with water at any given time. As a result, few hydroxide ions are in solution and ammonia water is a weak base. Nearly all particles in ammonia water solution are covalent molecules of NH_3.

CAUSTIC EFFECT OF STRONG BASES

activity

OBJECTIVE: Perform an experiment to determine the effect of a strong base on different fabrics.

PROCESS SKILLS
In this activity, you will *observe* the effect of a strong base on different fabrics and *infer* properties of bases.

MATERIALS

2 200-mL beakers
solid NaOH
metric balance
hot tap water
stirring rod
safety goggles
laboratory apron
gas burner
wire gauze
ringstand
ring
striker
tongs
piece of woolen cloth (5 cm × 5 cm)
piece of cotton cloth (5 cm × 5 cm)
paper
pencil

PROCEDURE
1. Prepare a 5 percent NaOH solution. To do this, first determine and record the mass of a 200-mL beaker. Add 5 g to this mass value of the beaker. Move the riders on the balance up to the new mass (mass of beaker + 5 g). Pour enough NaOH into the beaker to return the pointer to the zero (balanced) mark. CAUTION: *Do not touch the NaOH; it is highly corrosive. Wear goggles, gloves, and an apron, and use a fume hood.*
2. Remove the beaker from the balance and place it on your laboratory table. Pour enough hot tap water into the beaker to make 100 mL of solution. Mix the solution with a stirring rod. CAUTION: *Add the water slowly while stirring continuously. A great deal of heat will be liberated.*
3. Set up the ringstand, ring, and wire gauze. Place the beaker on the gauze and use a gas burner to heat the solution gently until it boils.
4. Use tongs to remove the beaker from the ringstand. Place it securely on your laboratory table.
5. Place both pieces of cloth in the solution. Stir the cloth with the stirring rod. Observe.
6. Once the reaction has stopped, use tongs to remove any cloth remaining in the beaker. Discard the cloth in a trash can. CAUTION: *Do not touch the cloth with your fingers.* Before discarding the solution, turn on the sink faucet. Pour the solution down the drain.

OBSERVATIONS AND CONCLUSIONS
1. What happened to the woolen cloth? to the cotton cloth?
2. Why should tongs be used when removing the cloth from the beaker?
3. Suggest a practical use for this experiment.

Ammonia water is one of the most useful laboratory bases. It also has a special use as a household cleaning agent because it removes grease and dirt.

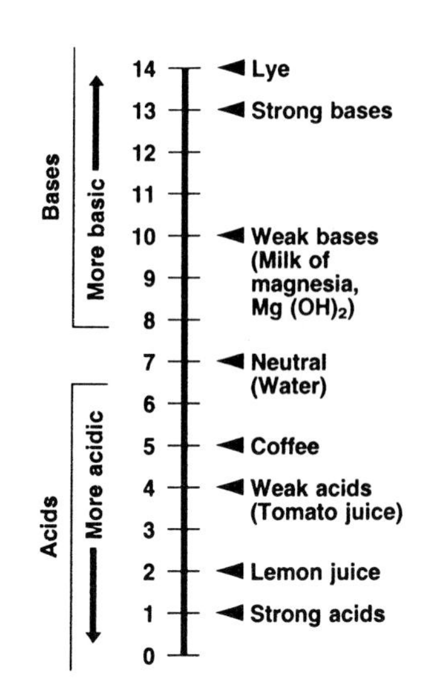

Figure 5-7 The pH scale ranges from 0 to 14. It indicates the strength of an acid or base. What does a pH of 7 indicate about the numbers of hydrogen ions and hydroxide ions in a solution?

Calcium hydroxide is a moderately strong base. The making of calcium hydroxide [$Ca(OH)_2$] is a two-step process. Calcium hydroxide is made from limestone, or calcium carbonate ($CaCO_3$). In the first step of the process, calcium carbonate is heated in a large furnace. The heat converts the $CaCO_3$ into calcium oxide (CaO) and carbon dioxide (CO_2) as shown by the following equation:

$$CaCO_3(s) \xrightarrow{\text{heat}} CaO(s) + CO_2(g)$$

This type of chemical change is called a decomposition reaction, the reverse of the synthesis reaction. A **decomposition reaction** is one in which a complex substance, like $CaCO_3$, is broken down into two or more simpler substances, like CaO and CO_2.

In the next step, water is added to the calcium oxide. Calcium oxide, usually called lime, is a white solid that reacts violently with water. A great deal of heat is given off during this step. The reaction of CaO with water in making calcium hydroxide can be written as

$$CaO(s) + H_2O(l) \rightarrow Ca(OH)_2(aq)$$

Calcium hydroxide is used in large amounts for making mortar and plaster. It is also used to treat acid soils. A water solution of calcium hydroxide, called limewater, is used to detect the presence of carbon dioxide gas. As a moderately strong base, calcium hydroxide is much less caustic than sodium hydroxide or potassium hydroxide.

The pH of a solution indicates how acidic or basic the solution is. Determining the strength of an acid or a base can be done using a pH scale. The pH scale is a series of numbers ranging from 0 to 14. Any solution with a pH less than 7 is acidic and has an excess of hydrogen ions. A solution with a pH greater than 7 is basic and has an excess of hydroxide ions. A solution with a pH of 2 is more acidic than one having a pH of 3. A solution with a pH of 12 is more basic than one having a pH of 10. Pure water is neutral and has a pH of 7. The pH values of some other common liquids appear in Figure 5-7.

The pH of a solution can be determined by using Hydrion paper such as that shown in Figure 5-8. When this paper is moistened with a solution, the paper changes color. The color of the test paper is then compared with the color scale mounted on the case. When the colors match, the pH number can be read directly from the scale.

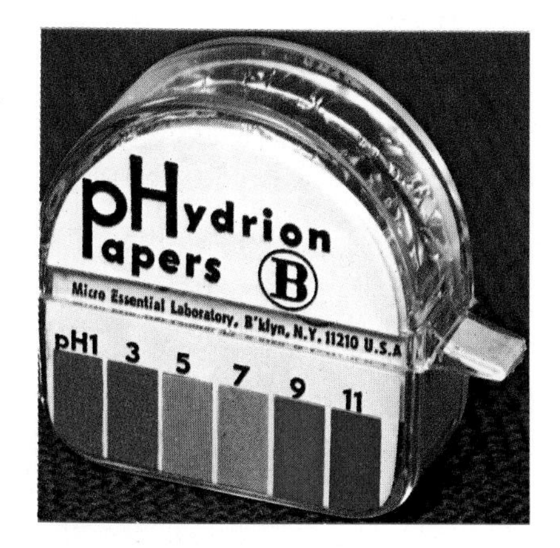

Figure 5-8 Hydrion paper provides a convenient method for determining the pH of a solution. What are some limitations on the accuracy of this method?

5.3 SALTS

A salt is one of the products of an acid-base reaction. When proper amounts of an acid and a base are mixed, the properties of the two solutions cancel each other. The products of such a reaction are a salt and water. A **salt** is defined as a compound made up of the positive ions of a base and the negative ions of an acid. The following is a word equation for the reaction that takes place:

$$\text{acid} + \text{base} \rightarrow \text{salt} + \text{water}$$

When solutions of hydrochloric acid and sodium hydroxide are mixed, the reaction can be written as

$$\underset{\text{acid}}{HCl(aq)} + \underset{\text{base}}{NaOH(aq)} \rightarrow \underset{\text{salt}}{NaCl(aq)} + \underset{\text{water}}{HOH(l)}$$

The reaction of any acid with any base produces a salt. In this case, sodium chloride is the salt produced by the reaction of hydrochloric acid and sodium hydroxide. The sodium chloride consists of the positive ions from the base (Na^+) and the negative ions from the acids (Cl^-). Since the acid, base, and salt are highly ionized in a water solution, the reactants and products can be written using their ionic forms:

$$\underset{\text{hydrochloric acid}}{H^+(aq) + Cl^-(aq)} + \underset{\text{sodium hydroxide}}{Na^+(aq) + OH^-(aq)} \rightarrow$$

$$\underset{\text{sodium chloride}}{Na^+(aq) + Cl^-(aq)} + \underset{\text{water}}{HOH(l)}$$

If the solution is heated to evaporate the water, crystals of sodium chloride (table salt) remain in the container.

Notice that in the above ionic equation sodium ions (Na^+) and chloride ions (Cl^-) appear on both sides of the equation. These ions take no part in the reaction. They are called spectator ions, because they do not participate in the reaction. Thus, the previous equation can be written more simply as this ionic equation

$$H^+(aq) + OH^-(aq) \rightarrow HOH(l)$$

This equation shows that the hydrogen ions of the acid and the hydroxide ions of the base join together to form water. The H^+ ions and the OH^- ions lose their effect in the solution by forming a nonionized compound, water. This reaction between the hydrogen ions of an acid and the hydroxide ions of a base is called a **neutralization reaction.**

In some neutralization reactions, an equal number of H^+ ions and OH^- ions are present and the resulting solution is neutral. If there are more of the H^+ ions than OH^- ions, some H^+ ions will be left at the end of the reaction. The solution will be acidic. If, on the other hand, there are more OH^- ions than H^+ ions, some OH^- ions will be left after the reaction. The solution will be basic. Therefore, an acid and a base completely neutralize each other only if the same number of H^+ ions and OH^- ions are present. No matter how small the volumes of acids and bases used, many billions of ions will be involved in the reaction. You cannot count billions of ions. However, you can use indicators, like litmus, to tell you when proper amounts of the acid or base have been added to give a neutral solution.

The properties of salts are quite varied. Salts as a group do not have any special properties. Some taste salty, some taste bitter. Some salts form crystals and some do not. Salts vary in color, although most are white.

In Chapter 4, you learned that certain substances that form ions in water have the ability to conduct an electric current. Those salts that are soluble in water break down freely as ions and conduct an electric current. Many dissolve well in water. However, others dissolve slightly, and still others do not dissolve at all. Table 5-3 lists the solubility of common salts in a water solution. Knowing the solubility of salts helps to predict reactions between ions. Salts known to be insoluble can be formed to separate an ion from solution. Some common salts and their uses appear in Table 5-4 on page 90.

TABLE 5-3

Solubility of Salts in Water
1. Common sodium, potassium, and ammonium compounds are soluble.
2. Common nitrates, acetates, and chlorates are soluble.
3. Common chlorides are soluble except silver, mercury(I), and lead. Lead(II) chloride is soluble in hot water.
4. Common sulfates are soluble except calcium, barium, strontium, and lead.
5. Common carbonates, phosphates, and silicates are insoluble except sodium, potassium, and ammonium.
6. Common sulfides are insoluble except calcium, barium, strontium, magnesium, sodium, potassium, and ammonium.

TABLE 5-4

Some Important Salts and Their Uses		
Name	**Formula**	**Uses**
Ammonium chloride (sal ammoniac)	NH_4Cl	dry cell batteries, medicine
Potassium nitrate (saltpeter)	KNO_3	making of fertilizers, fireworks
Silver nitrate	$AgNO_3$	medicine, photography
Potassium carbonate (potash)	K_2CO_3	making of glass and soap
Sodium hydrogen carbonate	$NaHCO_3$	making of baking powder, source of carbon dioxide gas for fire extinguishers
Copper(II) sulfate	$CuSO_4$	copper plating, making of fungicides
Aluminum sulfate	$Al_2(SO_4)_3$	water purification, deodorant
Calcium sulfate (plaster of paris)	$CaSO_4$	molds and casts
Sodium hypochlorite	$NaClO$	bleaching solution

Sodium chloride exists in abundance on earth. Sodium chloride, table salt, is found all over the world. Much of it is dissolved in the oceans. Salt from sea water can be obtained by allowing water to dry up in evaporating ponds. This method has been used since people first noticed salt crystals appearing in trapped pools of sea water.

Most salt comes from salt deposits below the earth's surface. The salt is recovered by room-and-pillar mining or by solution mining. In room-and-pillar mining, the method used in mining coal, miners remove blocks of salt to create huge underground rooms. Solution mining involves drilling a well to reach the salt. A large pipe is driven down to the center of an underground salt deposit. Water is then pumped through the pipe to dissolve the salt. The salt solution is brought to the surface through a smaller pipe placed inside the larger one. This technique is illustrated in Figure 5-9. The salt solution, or brine, that comes from the salt wells is then filtered, purified, and sold as table salt.

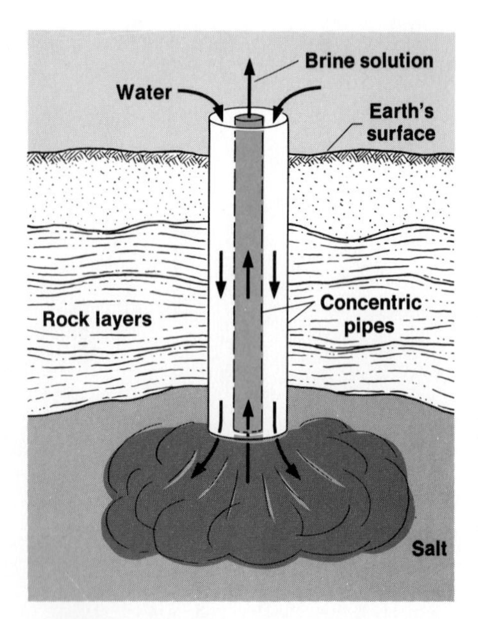

Figure 5-9 Water is forced down a salt well to dissolve underground salt. The salt solution, or brine, is then brought to the surface. How is the salt recovered from the crude salt solution?

Sodium chloride is important to us. Sodium chloride is vital to our diet and is found in solution in body fluids. It is important in the making of many chemicals such as lye, chlorine, and baking soda. Sodium chloride is also used to prevent decay in foods.

PURIFYING SALT

OBJECTIVE: Perform an experiment to separate pure salt from the impurities in rock salt.

PROCESS SKILLS

In this activity, you will *observe* the difference between rock salt and pure salt and *experiment* in order to remove the impurities from rock salt.

MATERIALS

metric balance
rock salt
2 200-mL beakers
hot tap water
safety goggles
filter paper
funnel
stirring rod
ringstand
ring
gas burner
tripod
wire gauze
evaporating dish

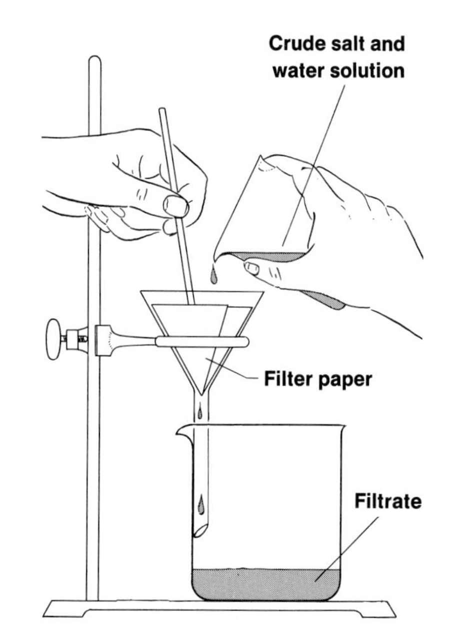

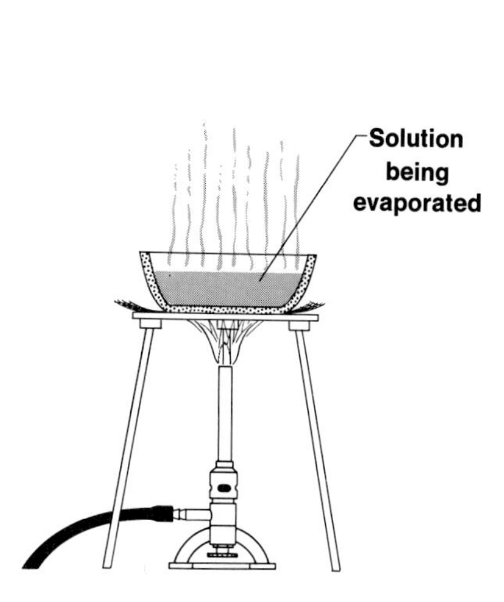

PROCEDURE

1. Use a metric balance to obtain 100 g of rock salt. Dissolve the rock salt in a beaker of hot tap water. Use a stirring rod to mix the solution.
2. Set up the filtering apparatus as shown at left above. Filter the solution using the technique illustrated in the diagram.
3. Carefully pour the filtrate from the beaker into an evaporating dish. Heat the dish as shown at right above until the water boils and the salt begins to separate out.
4. Let the solution in the dish cool to room temperature. Pour off the liquid and observe the salt.

OBSERVATIONS AND CONCLUSIONS

1. Why was it necessary to filter the rock salt solution?
2. Describe and identify the crystals remaining in the evaporating dish. Compare these crystals to the rock salt.
3. Where are the impurities originally contained in the rock salt?

BIOGRAPHY

SVANTE ARRHENIUS

a scientific pioneer

Svante Arrhenius (SVAHN-*teh ah*-RAY-*nee-oos*) was a remarkable person. Born in 1859 in Vik, Sweden, Arrhenius could read and write by the time he was three years old. During childhood, he became interested in mathematics and science. His love of these two subjects continued through his high school and university years and led him to study physics and chemistry while attending college. Arrhenius's interests eventually narrowed to the study of electrolysis and he attended graduate school to learn more about this topic.

It was during his years at graduate school that Arrhenius proposed a theory regarding ion formation. In 1883, he published his first paper on the subject. Arrhenius's theory stated that when an acid or a base was mixed with water, molecules of these substances would break apart to form ions. He thought that an acidic solution contained large numbers of hydrogen ions (H^+) and a basic solution contained large numbers of hydroxide ions (OH^-). Arrhenius also believed that these ions were electrically charged and could therefore conduct a current.

Like other pioneering scientists who have proposed new concepts, Arrhenius's work was not initially accepted. In 1884, Arrhenius presented his ideas to the university's faculty. The professors listened to his ionization theory in disbelief. Arrhenius received the lowest grade for his work. His revolutionary theory almost cost him his degree. Over time, however, other scientists studied his work and realized that Arrhenius's theory was correct. From that point on, Arrhenius was treated with a great deal of respect. He was offered teaching positions at many of the most prestigious universities in Europe.

Arrhenius's ionization theory has been modified only slightly since its proposal as a result of additional research. For his work regarding ion formation, Arrhenius was awarded the Nobel Prize in chemistry in 1903. He died at age 68 in Stockholm, Sweden.

CHAPTER REVIEW

SUMMARY

1. A substance that ionizes and gives up hydrogen ions in water to form hydronium ions is called an acid.
2. Acids have a sour or sharp taste, turn blue litmus paper red, are corrosive, and neutralize bases.
3. Substances that change color in the presence of certain ions are called indicators.
4. A substance that ionizes and accepts hydrogen ions to form hydroxide ions when dissolved in water is called a base.
5. Bases have a bitter taste, turn red litmus paper blue, are caustic, and feel slippery.
6. A salt is a compound made up of the positive ions of a base and the negative ions of an acid.
7. Neutralization is the reaction between the hydronium (hydrogen) ions of an acid and the hydroxide ions of a base to form water.
8. Four general types of chemical reactions are synthesis, single replacement, double replacement, and decomposition.

VOCABULARY

Match the item in the left column with the best answer in the right column. Do not write in this book.

1. acid
2. base
3. decomposition reaction
4. double replacement reaction
5. indicator
6. neutralization reaction
7. salt
8. single replacement reaction
9. synthesis reaction

a. changes color in the presence of certain ions
b. reaction in which one element replaces another in a compound
c. forms hydroxide ions when dissolved in water
d. product of an acid-base reaction
e. reaction in which substances combine to form a more complex substance
f. reverse of a synthesis reaction
g. reaction in which substances change places in a chemical reaction
h. forms hydronium ions when dissolved in water
i. reaction between an acid and a base, forming water

REVIEW QUESTIONS

1. What is an acid? List four properties of acids.
2. How does the hydronium ion differ from the hydrogen ion?
3. What two substances are formed when dilute sulfuric acid reacts with iron? Write the balanced equation for this reaction.
4. Describe a safe method of diluting concentrated sulfuric acid.
5. Describe what happens when concentrated sulfuric acid is used as a drying agent.
6. How is sulfuric acid prepared for commercial use?
7. Describe how you would test a food for protein.
8. How is hydrochloric acid prepared in the laboratory?
9. What is a base? List four properties of strong bases.
10. What is the difference between a strong base and a weak base?
11. What substances can be used to make sodium hydroxide? Name two by-products of the reaction.
12. How could you prove that a piece of cloth is all wool?
13. What important uses do calcium hydroxide and sodium hydroxide have?
14. The pH of a solution is found to be 4.7. What does this mean? Compare this solution with a solution having a pH of 3.2.
15. Describe one method for finding a solution's pH in the laboratory.
16. What are salts?
17. What is meant by neutralization?
18. Do salts have general properties like acids and bases? Explain.
19. Describe how table salt is obtained from salt wells.

CRITICAL THINKING

20. Explain the statement, "All acids contain hydrogen, but not all compounds containing hydrogen are acids."
21. A dish of gasoline and a dish of concentrated sulfuric acid are weighed and left uncovered for several hours. Then they are weighed again. What change in mass, if any, would you expect to occur in each substance? Explain.
22. Write a balanced equation for the reaction between zinc and hydrochloric acid. Identify any spectator ions. Describe what occurs during the reaction.
23. Why is a solution of ammonia gas in water more accurately called ammonia water rather than ammonium hydroxide, a name commonly used?
24. If you spill acid on yourself, you should wash it off with water. Why should you NOT try to neutralize the acid with a base?
25. Write the chemical formulas for the following salts. Use Table 5-3 to predict the solubility of these salts. (Refer to Tables 3-4 and 3-5 for oxidation numbers of elements and radicals.)
    a. lead(II) nitrate
    b. barium carbonate
    c. sodium sulfide
    d. calcium chloride
    e. lead(II) sulfate
    f. ammonium chloride

26. Write balanced chemical equations for the following reactions. Using Table 5-3, predict if a soluble or an insoluble product (precipitate) forms.
    a. sodium sulfate + barium chloride
    b. magnesium sulfate + ammonium chloride
    c. zinc sulfate + barium sulfide
27. Copy the chemical equations below on a separate sheet of paper and balance the equations. Do not change the formulas. Which general type of chemical reaction is shown by each equation?
    a. $Na(s) + HOH(l) \rightarrow NaOH(aq) + H_2(g)$
    b. $H_2O(l) \rightarrow O_2(g) + H_2(g)$
    c. $H_2(g) + Cl_2(g) \rightarrow HCl(g)$
    d. $Mg(s) + HOH(l) \rightarrow Mg(OH)_2(s) + H_2(g)$
    e. $Na(s) + Cl_2(g) \rightarrow NaCl(s)$
    f. $NaCl(aq) + AgNO_3(aq) \rightarrow AgCl(s) + NaNO_3(aq)$
28. In each equation in Question 27, identify any acids or bases.

PROBLEMS

1. A sample of sea water contains 4 percent mineral salts. If you evaporated 5 kg of the sample, how much mineral salts would you have?
2. The mass of an empty beaker is 100 g. Salt solution is added until the total mass reaches 500 g. The water is evaporated and the salt is left. The beaker and salt now have a mass of 116 g. (a) What is the mass of the salt solution? (b) What is the mass of salt obtained? (c) What percent of the solution was salt?

FURTHER READING

Cecil, George. *Salt.* New York: Franklin Watts, Inc., 1976.

Jensen, W. B. *The Lewis Acid-Base Concepts: An Overview.* New York: John Wiley and Sons, 1980.

Multhauf, R. P. *Neptune's Gift: A History of Common Salt.* Baltimore: Johns Hopkins University Press, 1978.

Pearson, R. G., ed. *Hard and Soft Acids and Bases.* New York: Van Nostrand Reinhold, 1973.

Robinson, G., and J. Boyle. "Acid-Base Balance: An Educational Computer Game." *Bio Science* 37 (July/August 1987): 511–13.

NUCLEAR REACTIONS

The sun (seen here in total eclipse) uses fusion, the fusing of light atomic nuclei, to generate enormous amounts of power. Today's nuclear power plants use fission, the splitting of heavy atomic nuclei, in order to generate power. How are nuclear fusion and fission different from other sources of power we rely on?

CHAPTER 6

SECTIONS

OBJECTIVES

- ☐ Describe the nature, properties, types, and uses of nuclear radiation.
- ☐ Identify and use nuclear equations.
- ☐ Describe how nuclear fission and fusion may be used as a source of energy.
- ☐ Discuss the uses and safety rules for radiation produced in nuclear reactors.
- ☐ Define the terms half-life, radioisotope, critical mass, and moderator.

6.1 NATURAL RADIOACTIVITY

Radioactivity was discovered by accident. In 1896, a French scientist named Henri Becquerel (*Bek*-REL) made a startling discovery. At the time, he was studying the effects of sunlight on fluorescent minerals. At the end of a long day, he wrapped one of the minerals, a compound of uranium, in paper and placed it in a desk drawer next to photographic plates. Later, when Becquerel picked up the photographic plates, he discovered that they had been exposed. Becquerel knew that the plates had not been exposed to sunlight.

Figure 6-1 This photograph was produced by radiation from a tiny piece of uranium ore. The radiation exposed the film even though the film was wrapped in thick, black paper.

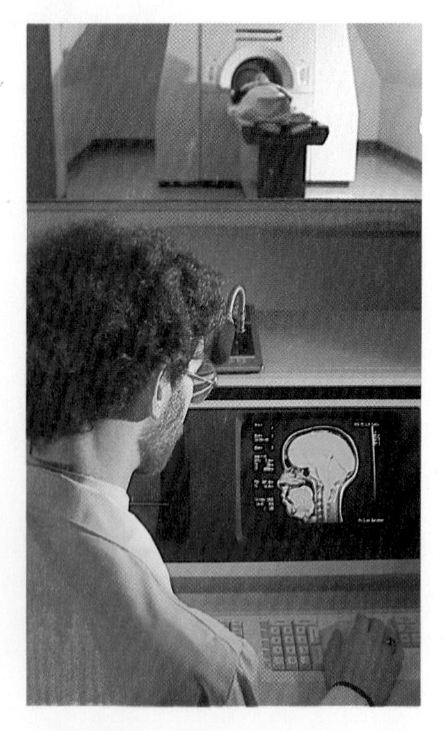

Figure 6-2 Magnetic Resonance Imaging (MRI) is a technique used to diagnose and study disease of internal organs. With this technique, the large machine in the background is used to create a magnetic field that affects the nuclei of certain isotopes in the body. The computer in the foreground translates the effect on the isotopes into images that can be analyzed for disease.

Therefore, he reasoned that they had been affected by the uranium. After further study, Becquerel found that uranium releases high-energy rays and atomic particles. He attributed the release of energy to changes in the nucleus of the uranium atom. The process by which energy is given off as a result of changes in the nucleus is called **radioactivity**.

Since Becquerel's discovery, much has been learned about radioactive elements. All radioactive elements and their compounds release radiation (energy rays and atomic particles) at a constant rate. Neither extremes of hot nor of cold temperatures affect the rate at which nuclear changes occur and energy is released. The radioactivity of an element is not even affected when the element is chemically combined with other elements.

Radioactive elements have common properties. All of the elements with atomic numbers greater than 83 are radioactive. A few elements with atomic numbers smaller than 83 are also radioactive. The unstable isotopes of radioactive elements are known as **radioisotopes.** Radioactive elements are alike in the following ways:

1. They penetrate certain kinds of materials. Radiation darkens photographic film even though the film is wrapped in heavy black paper and kept in the dark. The effect of radiation on photographic film is shown in Figure 6-1. The radiation also passes through paper, wood, flesh, and even thin sheets of metal.
2. They break down into different atoms. The atoms of all radioactive elements continually decay into atoms of either a new isotope of the same element or an isotope of a different element.
3. They produce ions in gas molecules. Radiation knocks outer electrons from gas molecules in the air. The gas molecules then become positive ions.
4. They damage living tissue. Large doses of radiation are fatal to all living things. Small doses of radiation can produce flesh wounds that heal slowly. Symptoms of radiation sickness include nausea, diarrhea, vomiting, internal bleeding, and a feeling of weakness. However, if the amount of radiation is controlled, it can be used for the detection and destruction of diseased tissue in the treatment of cancer and certain skin diseases (see Figure 6-2).

Radioactive elements may emit three kinds of radiation. The nucleus of an atom is made up of particles held together by powerful binding forces. When these forces are upset, energy and matter escape from the nucleus. This process is described as radioactive decay, and it produces radiation in several forms. Radiation may be in the form of alpha particles, beta particles, or gamma radiation. These three kinds of radiation can be separated by means of an electrically charged set of plates as illustrated in Figure 6-3.

1. Alpha (α) particles are positively charged helium nuclei, made up of two protons and two neutrons. Alpha particles lack great penetrating power. They can be stopped by a thin sheet of aluminum foil, by a few sheets of paper, or by a few centimeters of air.
2. Beta (β) particles are high-speed electrons traveling at nearly the speed of light. Beta particles have about 100 times the penetrating power of alpha particles. They can pass through wood nearly 3 cm thick.
3. Gamma radiation (γ) is similar to visible light but has much more energy. Gamma radiation is the most penetrating form of radiation. Even thick layers of lead or concrete will not completely stop all gamma radiation. Gamma radiation is similar to X-rays, which are discussed in Chapter 28.

Alpha and beta particles are seldom given off at the same time from the same nucleus. Gamma radiation, however, is frequently produced when alpha or beta particles are given off.

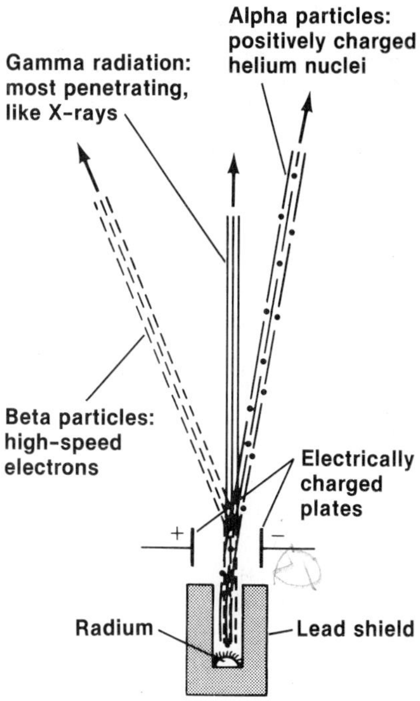

Figure 6-3 Radium, a radioactive element, continuously releases three forms of radiation from its nucleus. Why is the degree of deflection for beta particles greater than the degree of deflection for alpha particles?

Each radioactive element has a half-life. The rate at which the nucleus of an atom breaks down varies with different elements. This rate is called an element's half-life. The **half-life** of a radioactive element is the time needed for half of the atoms of a given sample to break down. For example, the half-life of radium is about 1,600 years. No matter how many atoms of radium there are to begin with, after 1,600 years half of them will remain unchanged. During the next 1,600 years, half of the remainder will break down, and so on. Another radioactive element called carbon-14 has a half-life of approximately 5,730 years. The breakdown, or radioactive decay, of carbon-14 is illustrated in Figure 6-4. For some radioactive elements, the half-life is less than 0.0000001 sec. For others, it is billions of years.

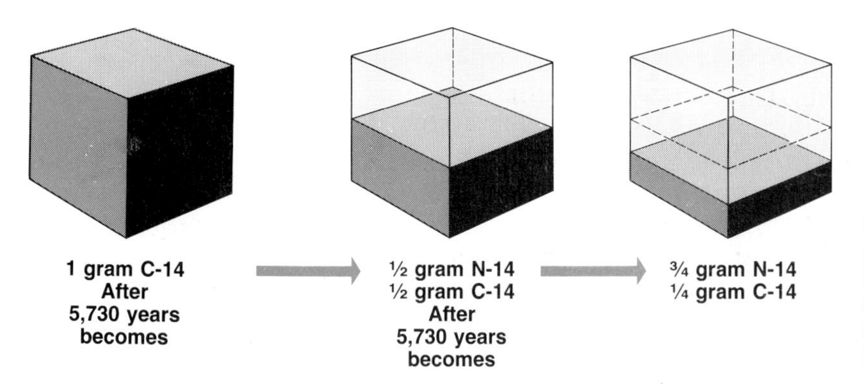

Figure 6-4 The half-life of radioactive carbon-14 is 5,730 years. With time, the amount of carbon-14 decreases at a constant rate and the amount of its daughter product, nitrogen-14, increases.

Atoms do not disappear during the breakdown of the nucleus. As stated earlier, the radioactive element decays into a different isotope of the same element or an isotope of a completely different element. The product of radioactive decay is called a **daughter product.** A daughter product can itself be radioactive, in which case further decay occurs until a stable daughter product is formed. Note that in Figure 6-4, the stable daughter product of carbon-14 is nitrogen-14.

Radioactive dating provides a good time clock. When radioactive uranium-238 breaks down, it decays into several intermediate daughter products and eventually becomes the stable element lead-206. By finding the amount of lead mixed in with a sample of uranium-bearing ore, scientists can compute how long ago the rock was formed. Knowledge of the half-life of radioactive minerals helps scientists estimate the age of the earth and its fossil remains.

One method used to date early human history is through the breakdown of radioactive carbon-14. Carbon-14 forms when high-energy particles (cosmic rays) from outer space strike nitrogen-14 atoms in the atmosphere. The stable nitrogen atoms are transformed into unstable carbon-14 atoms when the nitrogen atoms absorb a neutron and lose a proton. As stated earlier, C-14 has a half-life of about 5,730 years and decays back to N-14.

Plants and animals contain carbon dioxide made up of both unstable carbon-14 atoms and stable carbon-12 atoms in constant proportions. Plants absorb carbon-14 from carbon dioxide in the atmosphere. Humans and other animals take in carbon-14 mainly from the plants they eat. The intake of carbon-14 stops when an organism dies. However, the breakdown of carbon-14 continues.

The approximate age of early plant or animal life can be found by determining the amount of carbon-14 left in the test sample, since the amount of the daughter product (N-14) cannot be accurately measured. Because the half-life of carbon-14 is 5,730 years, a plant fossil with only one-quarter as much carbon-14 as a living plant must have been alive about 11,460 years ago (2 half-lives $\times$ 5,730 years/half-life). Carbon-14 is useful in making age estimates of organisms up to approximately 50,000 years old because scientists believe the ratio of C-14 to C-12 in living organisms has been constant over the last 50,000 years. The process of measuring carbon-14 content in order to determine the age of an ancient object is referred to as radiocarbon dating. Archaeologists and geologists have used radiocarbon dating since the late 1940s to learn about prehistoric humans, animals, and plants.

6.2 ARTIFICIAL RADIOACTIVITY

Rutherford was the first to change an atom. The study of radioactive elements in their natural state led scientists to believe that they could make other elements in the laboratory. Before different elements could be made, however, scientists had to develop a method to add particles to the nucleus of an atom. In 1919, Ernest Rutherford, a British scientist, found a way to penetrate the electric barrier surrounding a nucleus. Rutherford bombarded atoms with high-speed alpha particles and was able to hit a small number of nuclei. These direct hits caused changes in the nuclei; that is, a change from one element to another.

Rutherford's first successful nuclear reaction was changing nitrogen atoms into oxygen atoms. The nucleus of a nitrogen atom was struck by alpha particles given off by radium. One proton was released from the nucleus and an isotope of oxygen was formed.

Remember that an alpha particle (α) is the nucleus of a helium atom (He) and that a proton is a hydrogen nucleus. Therefore, the word equation for Rutherford's nuclear reaction can be written as

$$\text{helium} + \text{nitrogen} \rightarrow \text{oxygen} + \text{hydrogen}$$

During the reaction, a helium nucleus, traveling at high speed, strikes the nitrogen nucleus and is absorbed by it. The newly formed nucleus is unstable and breaks apart, giving off a high-speed hydrogen nucleus (proton). The result of the breakdown is the formation of an oxygen nucleus. This process is illustrated in Figure 6-5.

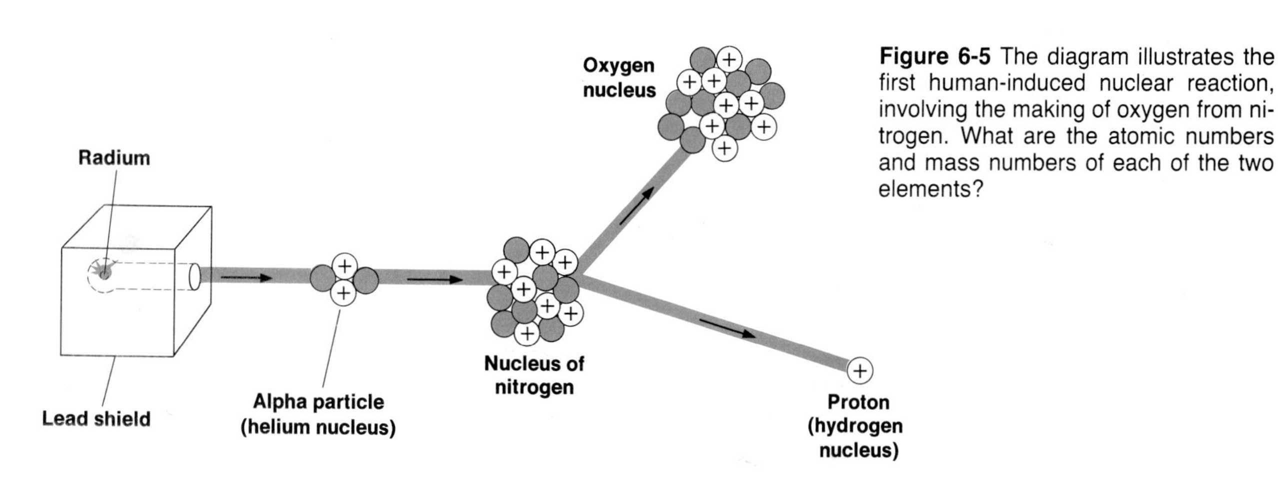

Figure 6-5 The diagram illustrates the first human-induced nuclear reaction, involving the making of oxygen from nitrogen. What are the atomic numbers and mass numbers of each of the two elements?

The equations used to describe nuclear reactions look much like chemical equations. The word equation for the reaction just described can also be written as

$$\begin{array}{ccccccc} {}^{4}_{2}He & + & {}^{14}_{7}N & \rightarrow & {}^{17}_{8}O & + & {}^{1}_{1}H \\ \text{helium nucleus} & & \text{nitrogen nucleus} & & \text{oxygen nucleus} & & \text{hydrogen nucleus} \end{array}$$

The number to the upper left of each symbol indicates the mass number of the element. The mass number represents the total number of protons and neutrons in the nucleus. The sum of the mass numbers on each side of the equation (in this case, 18) must be equal.

The number to the lower left of each symbol is the atomic number of the element, that is, the number of protons in its nucleus. Notice that the total number of protons (in this case, nine) is the same on each side of the equation.

During a nuclear reaction, alpha and beta particles can be given off. The loss of an alpha particle (${}^{4}_{2}He$) from the nucleus of an atom means the loss of two protons and two neutrons. This decreases the atom's atomic number by two and its mass number by four. When a beta particle (electron, ${}^{0}_{-1}e$) is given off, an electron is lost and a neutron in the nucleus turns into a proton. Thus, the loss of a beta particle increases the atomic number by one but does not change the mass number. Table 6-1 lists some additional properties of these and other subatomic particles and radiation.

TABLE 6-1

Subatomic Particles and Radiation

Symbol	Name	Charge	Mass Number
${}^{1}_{0}n$	Neutron (n)	0	1
${}^{1}_{1}H$	Proton (p) (hydrogen nucleus)	1+	1
${}^{0}_{-1}e$	Electron (beta particle, β)	1−	0
${}^{0}_{+1}e$	Positron (positive electron)	1+	0
${}^{4}_{2}He$	Alpha particle (α)	2+	4
γ	Gamma radiation	0	0

Rutherford's experiments were important because they proved that the nucleus of an atom could be changed. However, the final key to unlocking nuclear energy had not yet been discovered. Although the resulting hydrogen nuclei had more energy than the helium "bullet," there were so few hits on nitrogen nuclei that energy was actually lost in the total process. Not until scientists split the uranium atom were great amounts of energy released.

SAMPLE PROBLEM

The radioactive element thorium ($^{234}_{90}Th$) loses a beta particle ($^{0}_{-1}e$). (a) What is the new isotope formed and what is its symbol? (b) The new isotope formed is also radioactive and loses an alpha particle ($^{4}_{2}He$). What isotope is then formed and what is its symbol?

SOLUTION TO (a)

Step 1: Analyze

You know that a beta particle is lost from the element thorium, which has an atomic number of 90 and a mass number of 234. The new isotope formed by this process is the unknown.

Step 2: Plan

Since you know that the loss of a beta particle increases the atomic number by one and does not change the mass number, you know the mass number of the new element is 234. You can calculate the new atomic number by adding one. Then, you can identify the new substance by finding its mass number on the periodic table (Figure 3-6).

Step 3: Compute and Check

$$\text{new atomic number} = \text{old atomic number} + 1$$
$$= 90 + 1 = 91$$

The new isotope is protactinium ($^{234}_{91}Pa$). Since this isotope has the same mass number and a greater atomic number than $^{234}_{90}Th$, your answer is correct.

SOLUTION TO (b)

Step 1: Analyze

You know that an alpha particle is lost from protactinium, which has an atomic number of 91 and a mass number of 234. The new isotope formed by this process is the unknown.

Step 2: Plan

Since you know that the loss of an alpha particle decreases the atomic number by two and the mass number by four, you can calculate the new atomic and mass numbers. Then, you can identify the new isotope by finding its mass number on the periodic table.

Step 3: Compute and Check

$$\text{new atomic number} = \text{old atomic number} - 2$$
$$= 91 - 2 = 89$$
$$\text{new mass number} = \text{old mass number} - 4$$
$$= 234 - 4 = 230$$

The new isotope is actinium ($^{230}_{89}Ac$). Since this isotope has smaller atomic and mass numbers than $^{234}_{91}Pa$, your answer is correct.

***S**ynthetic radioisotopes are useful in scientific research.* Radioisotopes of all elements can be made by using neutrons and other particles as bullets. Radioisotopes are quite useful in scientific research. With the aid of radioisotopes, more can be learned about the laws of nuclear structure. Radioisotopes can also be used as "tagged atoms" or tracers. By detecting the radioactivity of a few "tagged atoms," the behavior of the rest of the atoms can be inferred. For example, chemists can follow the path of tagged atoms during an experiment and thus learn what is happening in complex chemical reactions. Radioactive tracers are used in biological and medical research as well (see Figure 6-6).

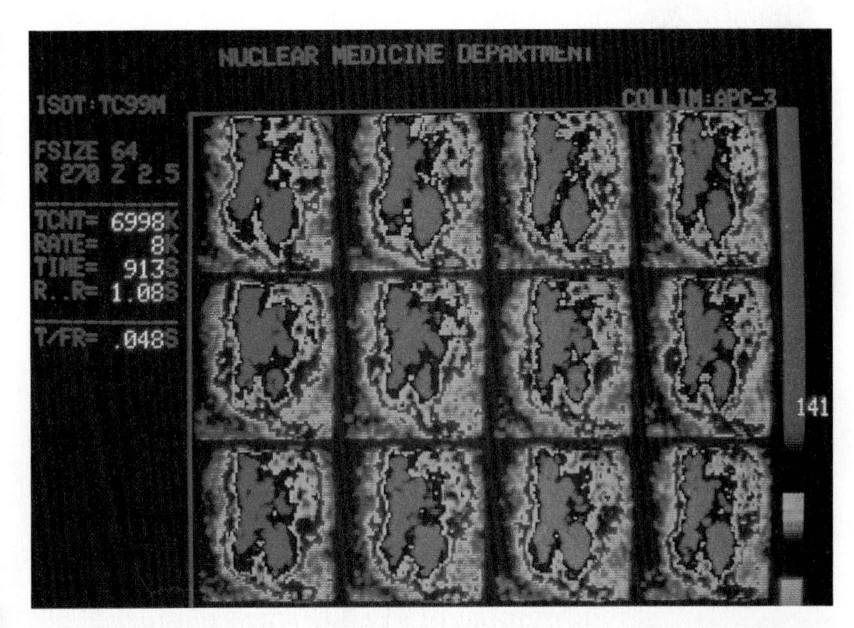

Figure 6-6 Images of the human heart are produced using the radioactive tracer thallium-201. After the tracer is administered into the body and circulates to the heart, a scanner is used to detect the resulting pattern of radioactivity, which then can be examined for abnormalities.

Radioisotopes are often used in the diagnosis and the treatment of cancer. Radioactive iodine is used in the diagnosis of cancer of the thyroid gland. Radiophosphorus is used to treat certain types of bone cancer. Some examples of radioisotopes used in medicine and other areas are listed in Table 6-2.

The isotope cobalt-60 ($^{60}_{27}Co$) is formed when a neutron enters the nucleus of an atom of ordinary cobalt ($^{59}_{27}Co$). The powerful gamma radiation given off by cobalt-60 is often used in the fields of medicine and industry as a substitute for X-rays. Cobalt-60, as well as other radioisotopes, is also used to test large metal castings for flaws (Figure 6-7) and welds in oil pipelines for weak points. The gamma rays are directed to pass through the metal and they darken photographic film at locations opposite the flaws and weak points.

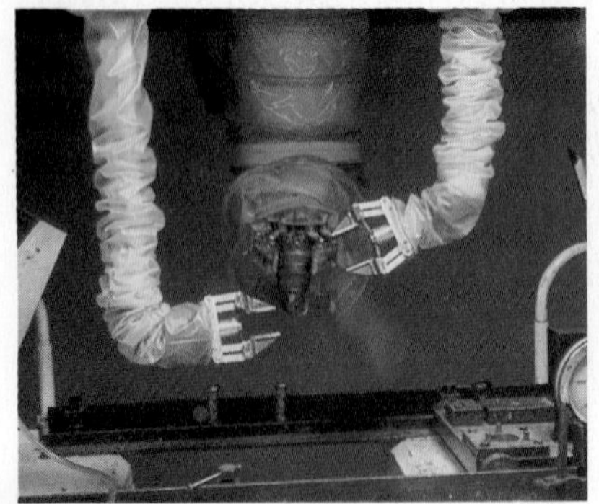

Figure 6-7 Radioisotopes are used to test metal casting for flaws. Highly radioactive materials must be handled by mechanical hands. These hands are operated by a person standing behind a glass shield that is about 1 m thick.

TABLE 6-2

Some Common Radioisotopes			
Radioisotope	**Type of Radiation**	**Half-life**	**Use**
Carbon-14	β	5,730 years	treating tumors, measuring age of fossils
Iron-59	β, γ	46 days	blood studies
Phosphorus-32	β	14 days	studying use of fertilizers by plants, treating bone cancer
Sodium-24	β, γ	15 hours	study of circulatory diseases
Strontium-90	β	27 years	treating lesions

6.3 NUCLEAR FISSION

Accelerators shoot particles into target atoms. Once Rutherford succeeded in penetrating the nucleus of the atom, scientists began to improve on his technique by building powerful machines to provide bullets with more penetration power. These machines apply oscillating voltages that increase the speed of charged nuclear particles. Charged particles subjected to these electric fields move more and more rapidly until their speed is sufficient for bombardment of the target atoms.

One such machine in use today is the linear accelerator. This machine applies successive bursts of alternating voltage to tubes inside the accelerator. These pulses pull the charged particles through a long path at ever increasing speeds. The paths of some accelerators similar to the one in Figure 6-8 are over 1.5 km long. A variety of particles is used inside linear accelerators to bombard target atoms. The nuclei of the lighter elements, such as hydrogen, deuterium, and lithium, are preferred.

Figure 6-8 This linear accelerator is used in nuclear research. How is the speed of charged particles increased in this device?

Neutrons make good atomic bullets. The speed of beams of neutrons cannot be increased in high-voltage machines. The reason is that neutrons, having no net charge, are not affected by an electric field. Even so, neutrons do have good penetrating power for atom smashing. Since neutrons are not charged, they are neither attracted nor repelled by

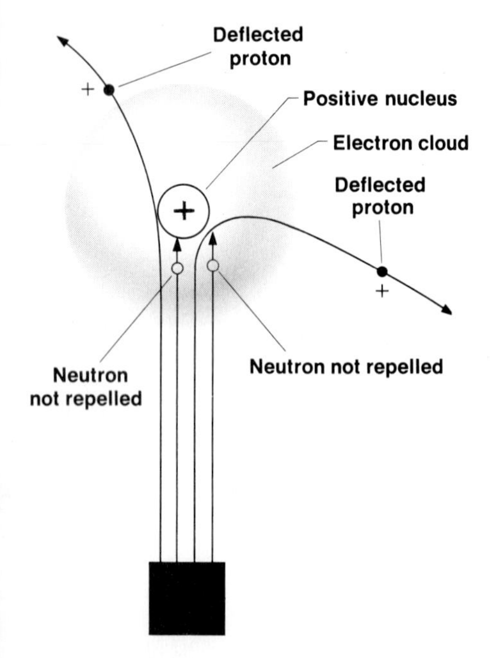

Figure 6-9 Atomic particles are shot like bullets toward target atoms in an accelerator. Why do neutrons make good bullets for striking target atoms?

the positive charge on the target nucleus (Figure 6-9). Thus, the neutron bullets are more likely to hit and be captured by the target atom.

It is not easy to obtain neutrons for use as bullets. Neutrons are tightly bound in atomic nuclei and can be freed only by using other bullets. In addition, freed neutrons cannot be stored for several reasons. For example, neutrons would pass through the walls of the container in which they were stored. Another reason is that neutrons, when removed from the nuclei of atoms, break down into protons and electrons. The average life of a neutron before it breaks apart is only about 17 minutes. Thus, neutrons must be collected when and where they are needed. One way to collect neutrons is to shoot helium nuclei ($^{4}_{2}He$) at a beryllium ($^{9}_{4}Be$) target. The nuclear equation for such a reaction is

$$\underset{\text{beryllium}}{^{9}_{4}Be} + \underset{\text{alpha particles}}{^{4}_{2}He} \rightarrow \underset{\text{carbon-12}}{^{12}_{6}C} + \underset{\text{neutron}}{^{1}_{0}n} + \text{energy}$$

Notice that energy, in the form of heat, is given off in the above reaction. When these experiments were first performed, scientists knew that the nucleus was a great storehouse of energy. Nuclear reactions seemed to support Einstein's theory that energy and mass are related. That is, mass can be changed into energy.

During nuclear reactions, some mass disappears and an equivalent amount of energy is given off in its place. This energy is mostly in the form of heat. The relation between mass and energy is given by Einstein's equation

$$E = mc^2$$

In this equation, E stands for the energy obtained, m for the mass that disappears, and c^2 for the square of the speed of light.

Using Einstein's equation, we find that when 1 g of a given mass is completely destroyed, an amount of energy equal to the burning of about 3,000 metric tons of coal is obtained. At the present time, no way has been found to change all of a given mass into energy. Only a small portion of the mass can be changed to energy.

Splitting the atomic nucleus gives off a large amount of energy. By 1938, several scientists, including Enrico Fermi, Lise Meitner, and Otto Hahn, were using neutrons to hit different kinds of atomic targets. When uranium-235 ($^{235}_{92}U$) was used as a target, the nucleus split into two smaller nuclei and two or three fast neutrons. In addition, large amounts of energy were released. This process, called nuclear fission, is illustrated in Figure 6-10. **Fission** is the splitting of a heavy nucleus into two or more lighter nuclei.

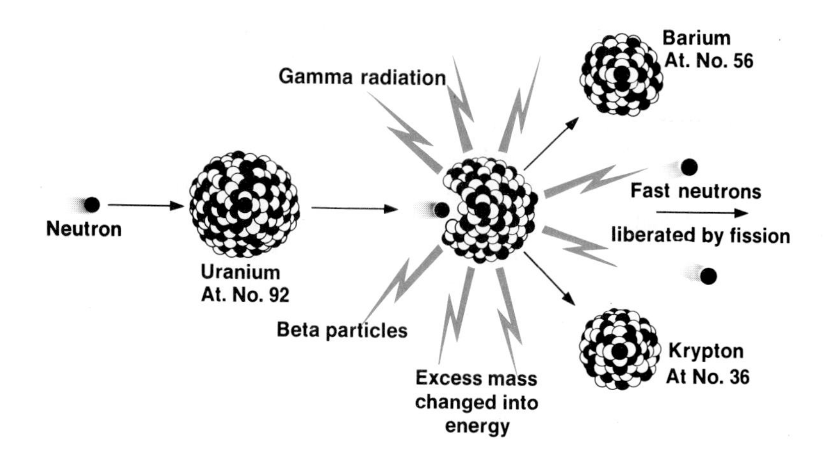

Figure 6-10 When a neutron bullet hits a $^{235}_{92}U$ nucleus, fission occurs. Many kinds of radioactive fission products result, depending on how the nucleus divides. What products are produced in the reaction shown in the diagram?

After fission has taken place, the total mass of all the pieces is slightly less than that of the original uranium nucleus. The mass that disappears is changed to energy as predicted by Einstein's equation, $E = mc^2$.

There are two main reasons why the discovery of nuclear fission excited the scientific world. First, large amounts of energy were set free. Second, it was believed that the fission process could lead to a chain reaction. If this were true, the neutrons set free by fission of one uranium nucleus might, in turn, cause fission in nearby nuclei. Thus, the reaction would quickly spread and huge amounts of energy could be set free. The fission process was discovered just as World War II was starting. The possibility of using fission energy in a bomb gave great urgency to nuclear research.

Uranium-235 will produce a chain reaction only if the amount available is large enough and pure enough. If the mass is too small, many neutrons that could produce fission escape before they can strike other nuclei. When the amount of uranium is large enough, most of the neutrons set free by one fission will strike other nuclei before they can escape. The smallest mass of material needed for a chain reaction of nuclear fission to occur is called the **critical mass**. The effect of the energy released when a critical mass has been achieved is shown in Figure 6-11.

Figure 6-11 The change of mass into energy is demonstrated by the explosion of this nuclear bomb. What is the relationship between energy and mass?

Nuclear fission can be controlled. Most uses of nuclear energy require a controlled chain reaction. Enrico Fermi and other scientists were the first to build a nuclear reactor capable of controlling the chain reaction. It was built at the University of Chicago where the first controlled nuclear fission reaction was set off on December 2, 1942.

Only slow neutrons can start and maintain a chain reaction in uranium. The neutrons released by the fission of $^{235}_{92}U$ move too fast to cause more fissions. The material used to slow the movement of neutrons in a fission reaction is called a **moderator**. The atoms of a moderator work by slowing neutrons in much the same way that a billiard ball slows down each time it strikes another ball. Water and graphite, a form of carbon, are effective moderators.

The first nuclear reactor was a room-sized pile of graphite blocks and small rods of uranium. This construction led to the name "atomic pile" for a nuclear reactor. Cadmium and boron steel are good absorbers of neutrons. Therefore, rods of these materials are used to control the reaction. When the control rods are lowered into the pile, the chain reaction stops. More neutrons are absorbed by the control rods than are set free by the fission of $^{235}_{92}U$. Thus, the reactions taking place in the pile can be slowed down or speeded up merely by moving the control rods in or out. Modern nuclear reactors are designed in much the same way as the one in Figure 6-12.

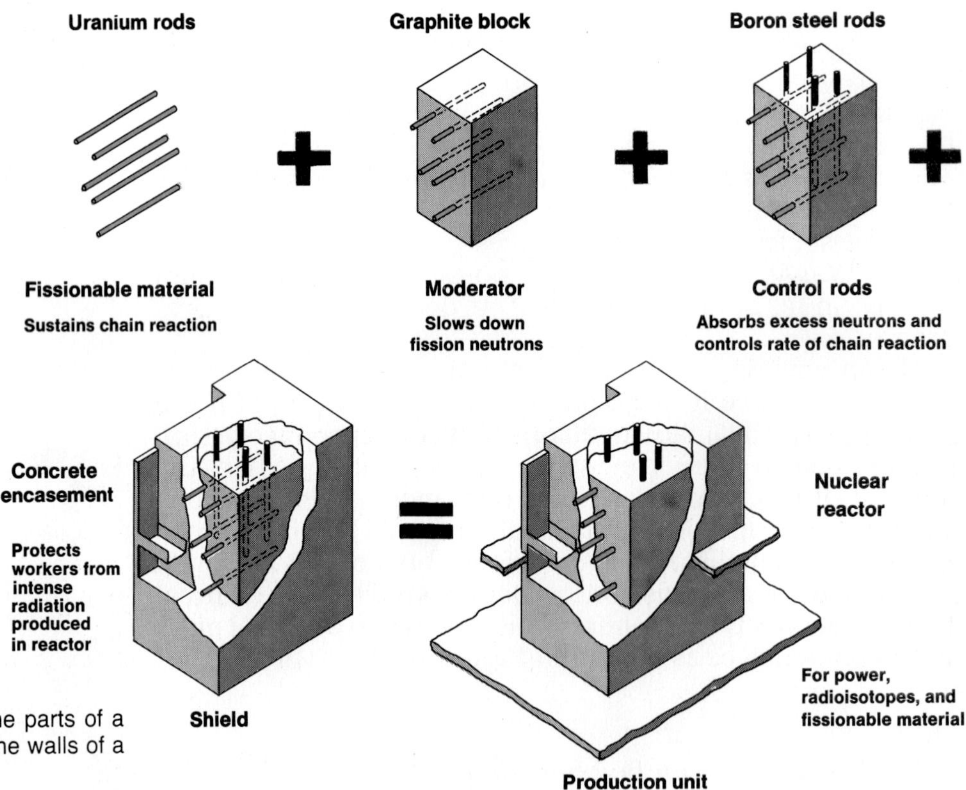

Figure 6-12 Diagram of the parts of a nuclear reactor. Why are the walls of a nuclear reactor so thick?

Since the first nuclear reactor was built, many other types of reactors have been developed. A breeder reactor, as in Figure 6-13, produces new fissionable material at a greater rate than the fuel can be used up. In a breeder reactor, one of the more abundant natural isotopes, uranium-238 ($^{238}_{92}U$), is converted into fissionable plutonium-239 ($^{239}_{94}Pu$).

Nuclear reactors have special uses. There are several basic uses for nuclear reactors. First, reactors produce large amounts of heat that can be used as a source of energy. The equipment needed to convert nuclear heat and coal heat into electricity is compared in Figure 6-14. Second, the radiation that is set free by nuclear reactors can be used to produce useful radioisotopes. Third, the nuclear reactor is an excellent source of neutrons, which can be used in research.

The United States was the first country to use nuclear energy to power ships and submarines. When a standard submarine is underwater, it cannot run on its diesel engines because the engines quickly use up all the oxygen. Thus, electric motors, drawing on energy from storage batteries, are used when the submarine is underwater. However, the batteries soon run down and the submarine must surface and use diesel power to recharge its batteries.

Because nuclear reactors do not use oxygen, a nuclear-powered submarine can stay underwater for a long time. In early tests, one submarine cruised underwater for over two months without surfacing. Unlike conventional submarines that must surface occasionally, nuclear submarines have been able to make scientific studies under the ice of the Arctic Ocean. The fissionable material in the submarines' reactors must be replaced every few years.

Nuclear energy is also being used to run surface ships. The first of these, the *Savannah*, sailed over 430,000 km after launching in 1961 before being refueled in 1968. This is equal to a distance of 12 times around the world.

Figure 6-13 Bundles of uranium fuel rods are being inspected before they are placed in a breeder reactor. Which isotope of uranium is most commonly used in a breeder reactor?

Nuclear reactors have many safeguards. Due to the intense heat and radiation given off during a nuclear reaction, nuclear reactors must be surrounded by thick metal, concrete, or water, or be buried underground. For example, the core of a typical nuclear power reactor near New York City is located inside a vessel with steel walls that are 18 cm thick. The vessel itself is in a water-filled steel tank, surrounded by a thick concrete wall. The whole unit, in turn, is enclosed in a steel sphere.

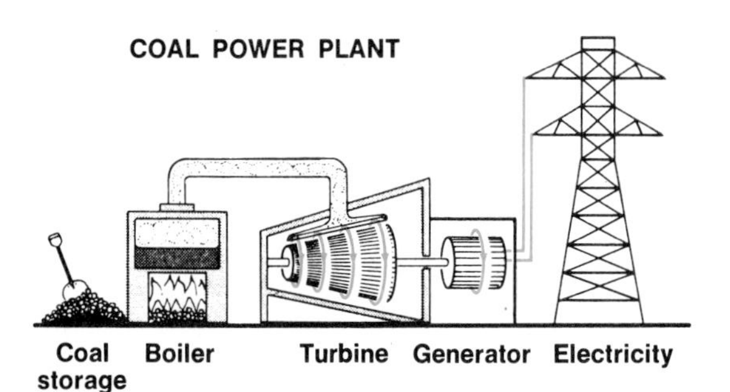

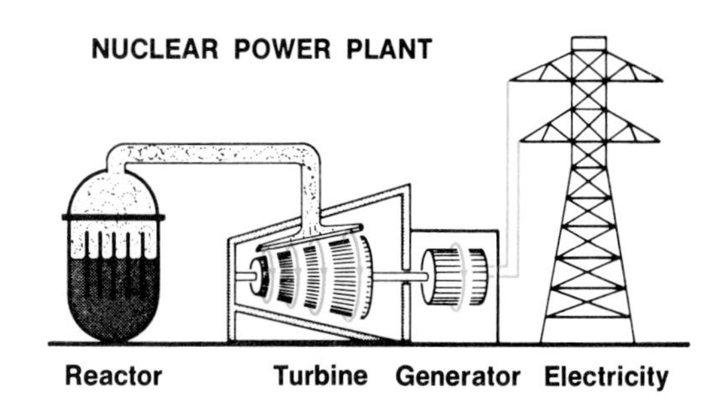

Figure 6-14 The difference between a nuclear power plant and other power plants, such as a coal power plant, is the way in which the steam to run the turbines is made. What safety measures are needed in a nuclear power plant?

Many people believe that nuclear reactors can be used safely, even in populated areas. However, there is a possibility that radioactive pollution of the environment may occur. Because of this possibility, some people are unwilling to have nuclear power plants in their communities.

Fossil-fuel power plants cannot continue to supply growing demands for energy. More nuclear power plants may be built unless other sources of energy can be developed to meet the growing energy needs of our society.

6.4 NUCLEAR FUSION

An atomic nucleus can be built up. For many years, people did not understand how the sun has been able to pour out huge amounts of energy for billions of years, without cooling off. We now know that nuclear reactions are the source of the sun's energy. Just as large nuclei can be split into smaller ones (fission), it is also possible to combine lighter nuclei to form heavier ones. The merging of two or more atomic nuclei to form a single heavier nucleus is called **fusion.** One important example of a nuclear fusion reaction is the fusion of four hydrogen nuclei to form one helium nucleus. The energy of the sun and certain other stars comes from this process.

Four hydrogen nuclei ($^{1}_{1}H$) react to form a helium nucleus ($^{4}_{2}He$) and two positrons ($_{+1}^{0}e$). A **positron** is an electron with a positive instead of a negative charge. This fusion reaction is indicated by the following equation:

$$\underset{\text{4 hydrogen nuclei}}{4\,^{1}_{1}H} \rightarrow \underset{\text{1 helium nucleus}}{^{4}_{2}He} + \underset{\text{2 positrons}}{2\,_{+1}^{0}e} + \text{energy}$$

Compare the total mass on the left side of the equation with that on the right side. In atomic mass units (u), a helium nucleus has a mass of 4.0015, a hydrogen nucleus has a mass of 1.0073, and a positron has a mass of 0.0005. Calculations for the $^{1}_{1}H$ fusion reaction are shown in Table 6-3.

TABLE 6-3

Mass Calculations (in u) for Fusion Reaction

Left Side:	Right Side:
$4\,^{1}_{1}H$ at 1.0073 = 4.0292	$^{4}_{2}He$ at 4.0015 = 4.0015
Total = 4.0292	$2\,_{+1}^{0}e$ at 0.0005 = 0.0010
	Total = 4.0025
Difference = 0.0267	

Notice that when four hydrogen atoms react to form one helium atom, a mass of 0.0267 u is lost. The difference in mass shown in Table 6-3 is called the nuclear mass defect. It is the mass that is changed to energy according to the equation $E = mc^2$, when smaller nuclei join to form a larger one.

The above nuclear fusion reaction will take place only at temperatures in the millions of degrees. For this reason, it is known as a thermonuclear reaction.

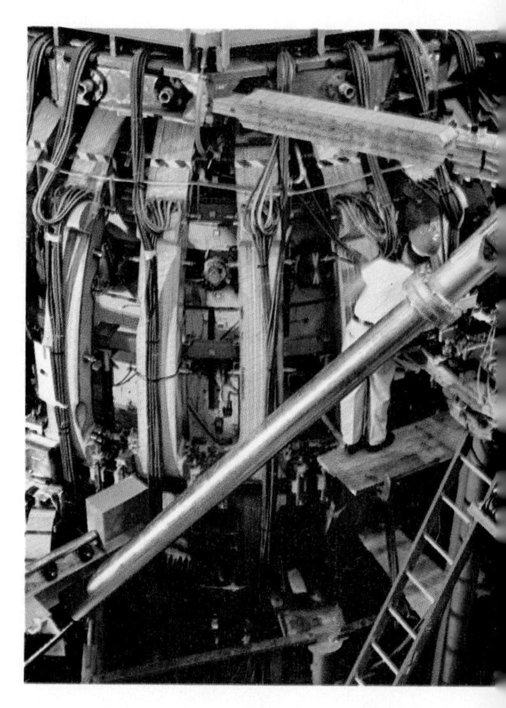

Figure 6-15 The high temperature (60,000,000° C) required for nuclear fusion can be produced by a reactor called the Princeton Large Torus. What is another requirement for obtaining power from nuclear fusion?

Controlled fusion may provide for future energy needs. One way of making a large amount of energy by the fusion process is by heating a compound of lithium and heavy hydrogen called lithium hydride ($^{6}_{3}Li^{2}_{1}H$) to a very high temperature. The nuclear reaction is

$$^{6}_{3}Li^{2}_{1}H \rightarrow 2^{4}_{2}He + \text{energy}$$

For a given mass of material, the energy given off by a fusion reaction is much greater than that for a fission reaction. However, as stated, fusion reactions can take place only at extremely high temperatures. This high temperature is hard to produce and maintain in the laboratory. At such a high temperature, all materials become completely ionized gases, called plasma. This plasma cannot be contained in any known material, since the container would also turn into plasma. One way to hold the plasma together is by means of a very strong magnetic field outside the container. This field also prevents the plasma from reaching the container walls. Despite many problems, fusion reactors like the one pictured in Figure 6-15 show promise for providing energy in the future.

The fusion process is preferable to the fission process for the following reasons:

1. Fuel sources are more plentiful since the oceans of the world can be used as a source of heavy hydrogen ($^{2}_{1}H$), the isotope used as a fusion fuel. Another fusion fuel can be obtained from lithium ore.
2. A smaller amount of radioactive waste per unit of energy is produced.
3. There is less heat lost to the environment.
4. Fusion energy is expected to cost less than present-day fossil and nuclear-fission fuels.

Antiparticles are a form of matter. For every kind of nuclear particle, there exists a "left-handed" twin or opposite counterpart, called an antiparticle. Many kinds of antiparticles have already been observed. When a particle meets its antiparticle, they both disappear, leaving only energy. Antiprotons are now being made routinely in large

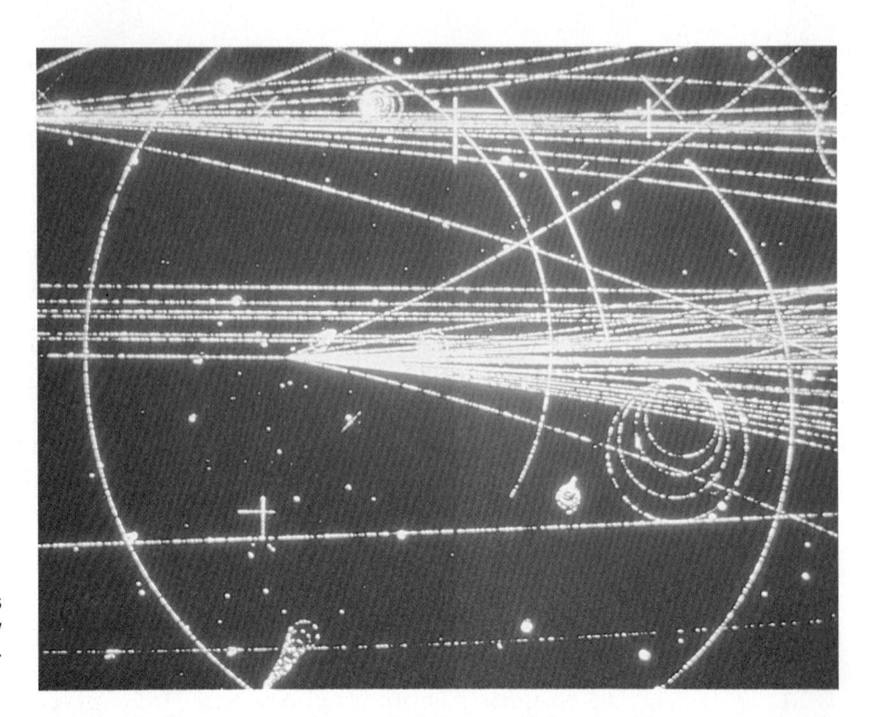

Figure 6-16 These bubble patterns formed in liquid hydrogen actually record the movements of charged particles and antiparticles.

amounts from atomic nuclei in particle accelerators. Antiprotons are the same as ordinary protons except that they have a negative charge instead of a positive one.

The opposite, or antiparticle, of an electron is a positron. When electrons meet positrons, they destroy each other to form gamma radiation. The inverse of this destruction can also occur, where energy is converted directly to matter. If a gamma ray of sufficient energy passes close to a nucleus, an electron and a positron can be generated. Matter and antimatter particles are always produced as pairs.

One device used to detect and study positrons is a bubble chamber. The chamber is a pressurized area containing liquid hydrogen that is very close to its boiling point. As charged particles or antiparticles pass through the chamber, ions are formed. Some of the hydrogen vaporizes, forming tiny bubbles around the ions. The paths of the particles become visible as a series of bubbles that can be recorded on photographic film (Figure 6-16). The bubble paths can then be analyzed to determine the energy and momentum of the antiparticles.

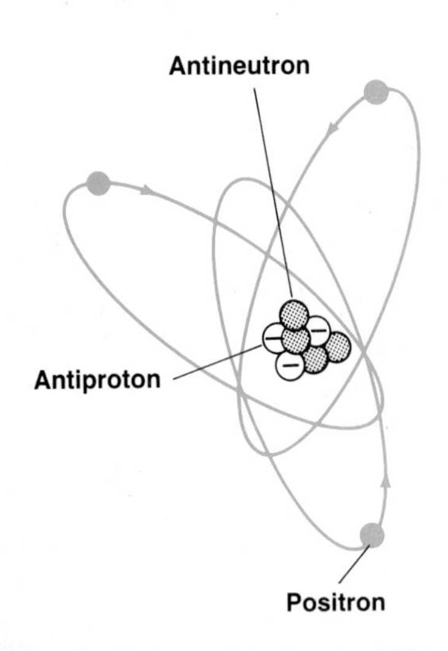

Figure 6-17 A model of an antilithium atom. What is the charge on the positrons outside the nucleus?

Most scientists agree that antimatter exists elsewhere in the universe. High-altitude balloons have detected antiprotons in the upper atmosphere. Atoms of antimatter would have a nucleus of antiprotons and antineutrons, with positrons instead of electrons outside the nucleus. Antimatter atoms would resemble the model of antilithium in Figure 6-17. What would happen if a small amount of antimatter hit the same amount of ordinary matter? The total mass would instantly change into a huge amount of energy as predicted by Einstein's energy-mass equation, $E = mc^2$.

MARIE CURIE

contributor to the modern atomic theory

Marie Sklodowska Curie was born in Warsaw, Poland, in 1867. Curie knew when she first started school that she wanted to become a scientist. After finishing high school, Curie worked for a family, teaching and training their children. After she had earned enough money, she went to Paris at the age of 24 to enter the university.

Curie had very little money and suffered from hunger and cold. She faced these hardships willingly in order to continue her studies. After much hard work, she graduated with the highest grade in her class and was awarded her degree in physics.

In 1895, the physicist married a French chemist named Pierre Curie. Within a few months, Curie and her husband began working with another French scientist, Henri Becquerel, on radioactivity.

In her experiments, Curie found that pitchblende, the ore from which uranium is removed, was far more radioactive than the uranium it contained. This suggested to her that a new element might be present. The Curies worked together for four years on this phenomenon. Their laboratory was a wooden shed with a dirt floor and a leaky roof. The work they were doing required a tremendous amount of patience. They had to sift through large amounts of pitchblende to obtain enough of a pure sample of the new element to test. Finally, in 1899, the Curies announced the discovery of not one, but two radioactive elements—polonium and radium. For this work, the Curies and Henri Becquerel shared the 1903 Nobel Prize in physics.

After the accidental death of her husband in 1906, Curie continued her research alone. She also took over Pierre's teaching position at the Sorbonne (University of Paris). In 1911, she was awarded a second Nobel Prize in chemistry.

Marie Curie was the first person to receive two Nobel prizes. Her work provided the foundation for the modern atomic theory. It also inspired practical applications of nuclear physics and chemistry. She died in 1934 due to overexposure to nuclear radiation.

CHAPTER REVIEW

SUMMARY

1. Radioactive elements give off alpha particles (helium nuclei), beta particles (high-speed electrons), and gamma radiation (high-energy light).
2. Radioactive elements darken film, ionize molecules, break down into simpler atoms, and have a damaging effect on living things.
3. The half-life of a radioactive element is the time needed for half the atoms of a given sample to decay or break down.
4. One element can be changed into another by bombarding a target nucleus with fast-moving charged particles.
5. Einstein's equation states that energy is equal to the mass destroyed times the square of the speed of light ($E = mc^2$).
6. Nuclear fission is the splitting of a heavy nucleus into two or more lighter nuclei with the release of energy and neutrons.
7. Nuclear fusion is the joining of lighter nuclei to form heavier ones, with the release of large amounts of energy.

VOCABULARY

Match the item in the left column with the best answer in the right column. Do not write in this book.

1. critical mass
2. daughter product
3. fission
4. fusion
5. half-life
6. moderator
7. positron
8. radioactivity
9. radioisotope

a. energy released as a result of changes in the nucleus
b. time needed for half of the atoms of a given sample to break down
c. smallest mass of a material needed to start a chain reaction
d. splitting a heavy nucleus into two or more lighter nuclei
e. electron with a positive charge
f. combining lighter nuclei to form heavier nuclei
g. material used to slow down neutrons
h. radioactive isotope of an element
i. product of radioactive decay

REVIEW QUESTIONS

1. Which elements are naturally radioactive? List four common properties of these elements.
2. Describe the three kinds of radiation emitted naturally by the nucleus of a radioactive atom.
3. What is meant by the half-life of a radioactive element?
4. Describe the characteristics of a daughter product.
5. How do the nuclei of oxygen-16, oxygen-17, and oxygen-18 differ?

6. What happens to an atom when a beta particle is given off?
7. What happens to the mass number and atomic number of an atom of a radioactive element when an alpha particle leaves the nucleus?
8. What safeguards are necessary in handling radioisotopes?
9. What is Einstein's energy-mass equation? Define the symbols used.
10. Why is it impossible to speed up neutrons in a high-voltage accelerator?
11. Why are neutrons so effective as bullets for smashing atoms?
12. What causes nuclear fission?
13. What is the purpose of a moderator in a nuclear reactor? What two materials are widely used as moderators?
14. Describe the use of control rods in nuclear reactors.
15. What is the chief hazard in the operation of nuclear reactors?
16. Compare the penetrating power of the three types of radiation from a radioactive element.
17. Outline the chain reaction of the $^{235}_{92}U$ fission process.
18. In what way is the breakdown of radioactive carbon-14 an aid to the study of early human history?
19. Discuss the advantages of nuclear submarines over other submarines.
20. Where does the energy given off by nuclear fusion come from?
21. Describe an antihelium atom in terms of antimatter particles.

CRITICAL THINKING

22. What two reasons can you think of to explain the need to use different radioisotopes to date different rocks and fossils? What types of isotopes would you use to date a very old and a very young rock?
23. Suggest a means of storing a radioisotope that gives off only beta particles.

PROBLEMS

1. How long would it take for 1 g of radium to break down so that only 1/8 of the sample is still radium? (Use 1,600 years as the half-life of radium.)
2. Thorium ($^{230}_{90}Th$) gives off an alpha particle and becomes radium ($^{226}_{88}Ra$). Write the nuclear equation for this reaction.
3. When bismuth ($^{210}_{83}Bi$) breaks down, it loses a beta particle. (a) What new radioactive element is formed? The new element gives off an alpha particle. (b) What is the symbol for the final nucleus? (c) Write the nuclear equation for the formation of the final daughter product.
4. When neutrons are fired at $^{238}_{92}U$ target atoms, the atoms capture some of the neutrons and temporarily become $^{239}_{92}U$. Write the nuclear equation for this reaction.

FURTHER READING

Asimov, Isaac. *How Did We Find Out About Nuclear Power?* New York: Walker, 1976.

Clark, Ronald William. *Einstein: The Life and Times.* New York: Avon Books, 1984.

McGowen, Tom. *Radioactivity: From The Curies to the Atomic Age.* New York: Franklin Watts, Inc., 1986.

How the Sciences Work Together

The Process in Physical Science: Looking Inside the Body

You may have heard the old saying, "a picture is worth a thousand words." The saying may be overused but modern medicine has come to rely on it. New techniques and instruments that produce "pictures" of the inside of the human body are providing powerful tools for diagnosing disease. They are also helping us learn about the basic functions of the human body.

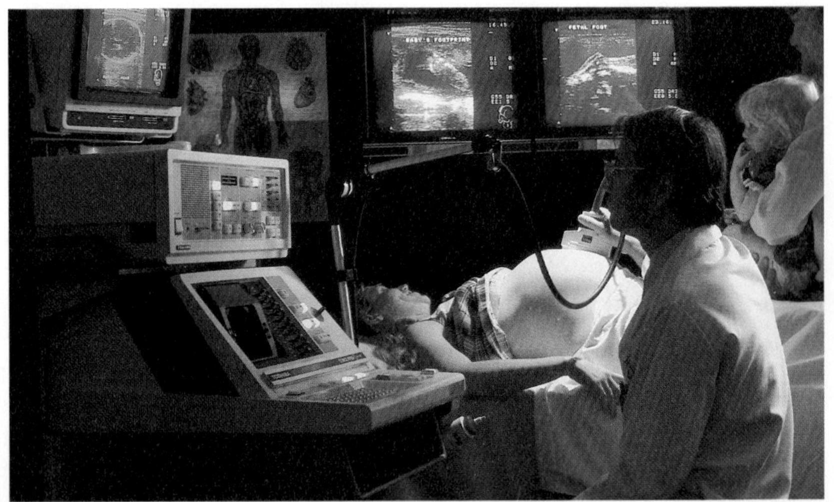

Ultrasound images help determine the health and development of a human fetus.

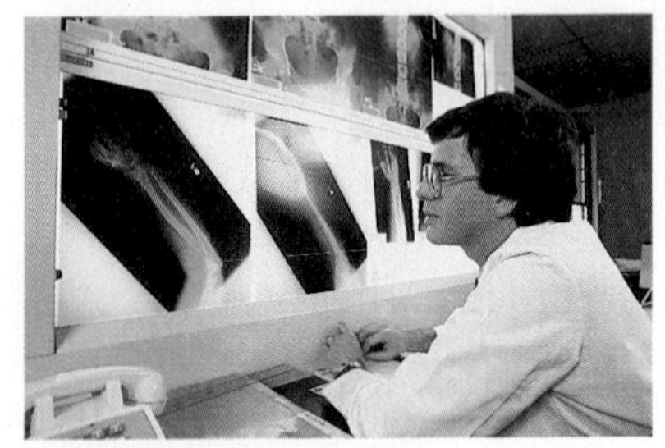

X-rays are used to diagnose broken bones and other medical problems.

The Connection to Health Sciences

One of these techniques is already familiar to most of us. The *X-ray* gave us our first glimpse at the inside of objects. The images formed by X-ray machines show bones and many of the organs within the body. However, new imaging techniques have been developed that, with the help of computers, can be used to study objects and processes that go undetected by X-rays.

Ultrasound techniques, for instance, are commonly used to provide moving images of the human fetus. Ultrasound uses sound waves which are over 20,000 cycles per second and beyond the range of human hearing. These sound waves are projected against parts of the body and bounce back at levels which depend on the density of tissue they encounter. The returned waves are analyzed by computer and an image is formed.

Another device, called *computerized axial tomography* (CAT), uses conventional X-rays to produce three-dimensional images. In a CAT scan, X-rays are beamed through the body at various angles. The information obtained is then converted by a computer into cross-sectional images. CAT scans are used in diagnosing brain disorders, fractures, and tumors of the liver and pancreas.

Another technique, *positron emission tomography*

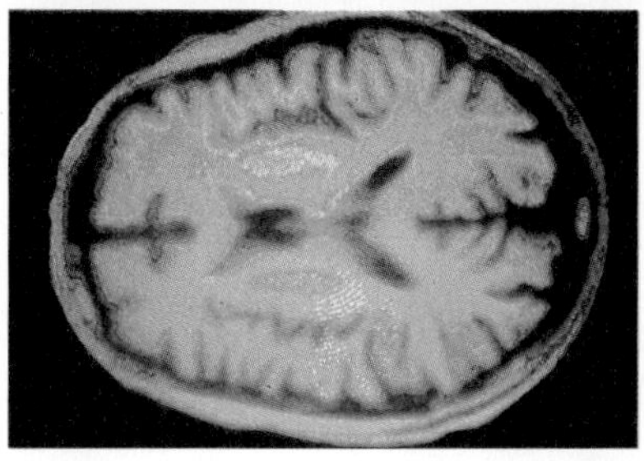
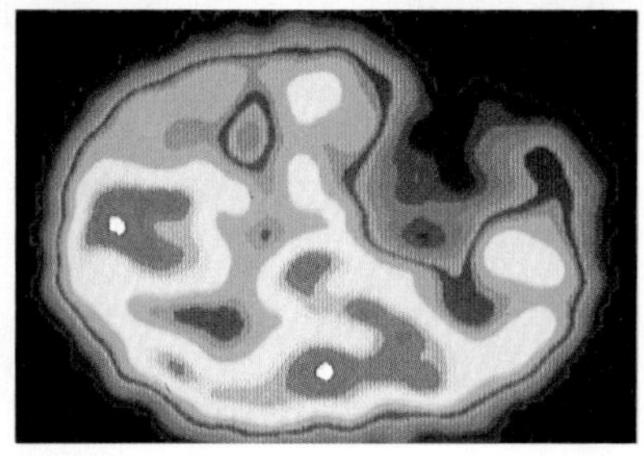
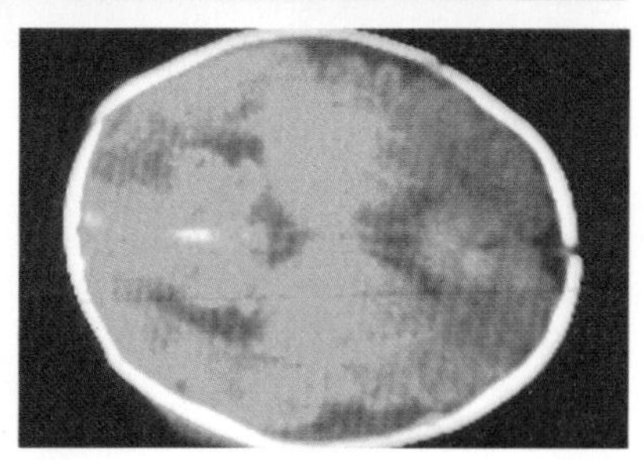

Cross-sectional images of the human head using (left to right) magnetic resonance imaging (MRI), positron emission tomography (PET), and computerized axial tomography (CAT).

(PET) scans radioisotopes that are inhaled or injected into the body. These specially-prepared substances contain radioactive atoms that emit positively charged particles called *positrons*. When these positrons meet inside the body with their negatively charged counterparts, the electrons, they annihilate each other to produce gamma rays that can be detected and scanned by a computer.

PET not only shows the shape and size of internal organs, as can X-rays, it also indicates the sites of energy use within the body. Though it is still a costly and not widely used procedure, PET has been successful in studying brain functioning and blood flow, lung functions, and heart problems.

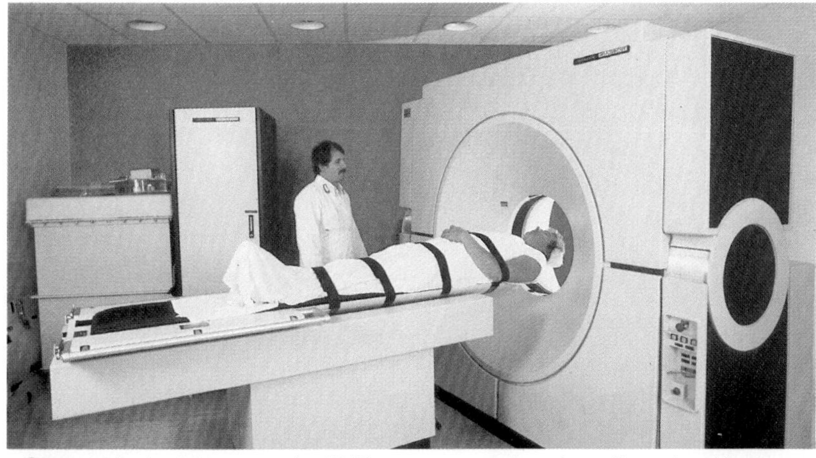

Computerized axial tomography (CAT) scans can produce three dimensional images.

The Connection to Physics and Chemistry

A diagnostic technique called *magnetic resonance imaging* (MRI), invented by a physicist, employs radio waves and strong magnets. MRI avoids the risk of radiation exposure from CAT and PET scans. Bones do not interfere with the MRI signal; therefore it has enabled us to get clear images of the brain and spinal cord for the first time.

When subjected to a powerful magnet, hydrogen atoms within the body align themselves with the magnetic forces. When a radio wave is aimed at the "magnetized" atoms, they absorb energy. When the magnetic field is switched off, the atoms return to a random orientation and, while doing so, emit a weak radio signal. A computer translates these signals into an image.

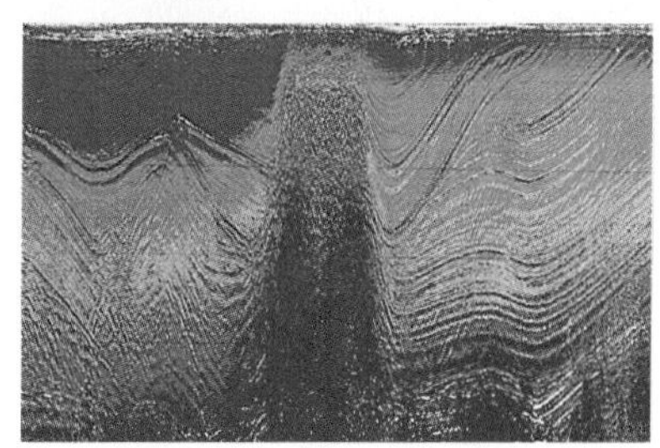

Computer enhanced images of rock below the earth's surface, called seismograms, are used to explore for oil.

The Connection to Earth Sciences

Imaging devices similar to those used in medicine can probe below the surface of the earth to study the earth's structure. Using these scanning techniques, coal deposits have been discovered without breaking ground.

CHEMISTRY IN OUR WORLD

UNIT 3

Inside the Silverdome in Pontiac, Michigan, thousands of fans roar as the Detroit Lions drive for the goal line. Outside, the wind roars too, driving sheets of icy rain down on the stadium. But the fans could not care less. No umbrellas are opened, no raincoats buttoned up. Not a drop of rain splatters on an unprotected head. The stadium, it turns out, has a very special roof. It is made of a fabric originally designed for the spacesuits of astronauts: a Teflon-coated Fiberglas invented by chemists.

As you will discover in this unit, chemists produce many materials that make our lives more comfortable. They also explore the mysteries of chemicals that are part of our natural environment—our atmosphere and the water and land all around us—and what can go wrong when we mistreat it.

CHAPTERS

COMMON GASES OF THE ATMOSPHERE

Billions of years ago, the earth looked like this. How might you describe its surface? its atmosphere? In what ways has the earth's surface and atmosphere changed?

CHAPTER 7

SECTIONS

OBJECTIVES

- ☐ Outline some of the physical and chemical properties of the common gases of the atmosphere.
- ☐ Describe the laboratory and industrial preparations of oxygen, carbon dioxide, ammonia, and hydrogen.
- ☐ Explain some of the uses of the gases of the atmosphere.
- ☐ Differentiate between respiration and photosynthesis.

7.1 OXYGEN: THE BREATH OF LIFE

Air consists of many gases. The mixture of gases that make up the earth's atmosphere is referred to as **air.** Clear air is a colorless, tasteless, and odorless gas. It is a mixture mostly of nitrogen and oxygen. Other gases such as argon, carbon dioxide, water vapor, and helium are present in very small amounts.

TABLE 7-1

Gases of the Air	Percent by Volume
Nitrogen (N_2)	78
Oxygen (O_2)	21
Argon (Ar)	0.94
Carbon dioxide (CO_2)	0.04
Other gases	0.02

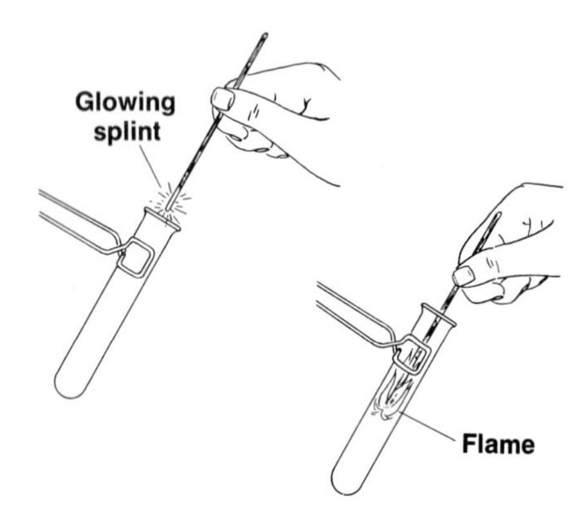

Figure 7-1 If a glowing wooden splint is thrust into a test tube of oxygen, the splint will burst into flame. What does this observation indicate?

Look at Table 7-1. Notice that there is more nitrogen in air than oxygen. However, oxygen is more plentiful on the earth as a whole. Combined with other elements, it is found in water, in many compounds occurring in the earth's crust, and in plant and animal tissues.

Oxygen can be identified by a simple test. A wood splint burns in air, but it burns more brightly in pure oxygen. Therefore, if a glowing splint is lowered into a bottle of pure oxygen, the splint will burst into flame at once (Figure 7-1). This reaction is frequently used as a test to indicate the presence of oxygen. Oxygen is the only common, odorless gas that promotes the burning of a wood splint.

Priestley discovered oxygen. Joseph Priestley, an English chemist, is given credit for the discovery of oxygen. He was the first scientist to publish the results of his experiments with this gas.

The nature of burning, rusting, and other processes involving oxygen was a mystery that challenged many scientists of the eighteenth century, including Priestley. In 1774, while studying the behavior of mercury(II) oxide, Priestley heated the powdered oxide by focusing the rays of the sun on it with a lens.

He saw that the red oxide of mercury turned black as it was heated. Then a small drop of mercury metal formed. Priestley stirred the hot mixture with a wooden stick. Much to his surprise, instead of smoldering, the stick burst into flame. Next, he heated some of the oxide in a container and collected the gas that was given off (Figure 7-2). He found that a lighted candle burned much more brightly in this gas than in air.

In another experiment, Priestley exposed a mouse to the gas and observed that the animal became very active. Priestley breathed some of the gas himself and found that he felt strangely "light" and "easy" for some time afterward. He described the gas as "good air." He did not realize that this gas (oxygen) is present in the air.

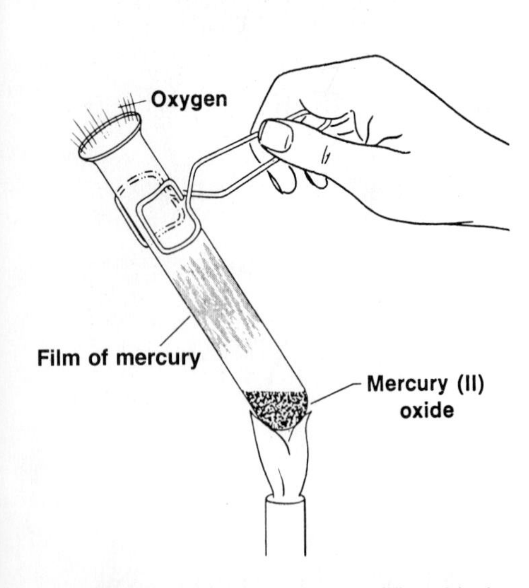

Figure 7-2 When mercury(II) oxide is heated, it breaks down into mercury (Hg) and oxygen (O_2). **CAUTION: Mercury vapors are poisonous. Do not perform this experiment yourself.**

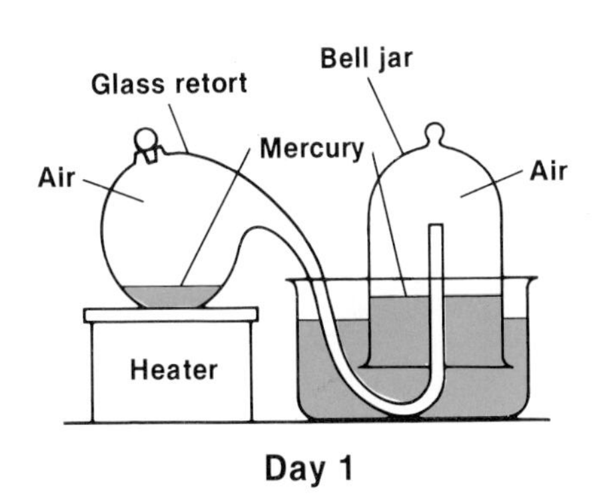

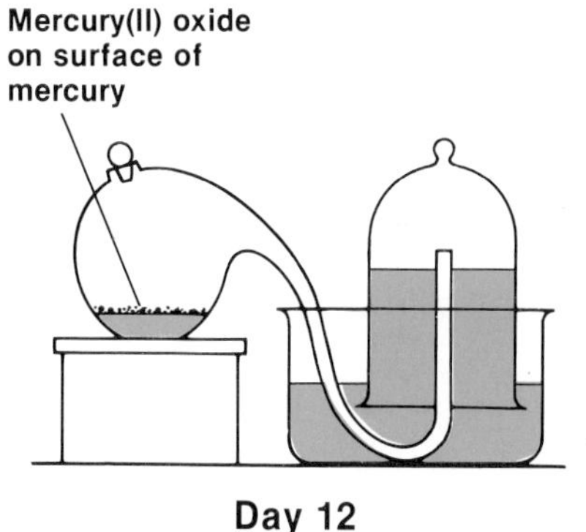

Figure 7-3 Lavoisier heated mercury (Hg) in a glass retort, producing mercury(II) oxide (HgO). Upon heating, the level of mercury in the bell jar rises. Why does this happen?

Lavoisier advanced the knowledge of oxygen. In October, 1774, Antoine Lavoisier (*la*-VWAH-*zee*-AY), a French chemist, learned of Priestley's experiments involving gases. Lavoisier thought that Priestley's new gas might also be present in the air, so he conducted a 12-day experiment to test this theory.

Using a glass retort and bell jar, Lavoisier heated some mercury in the presence of air as shown in Figure 7-3. A reddish powder began to gather on the surface of the mercury in the retort. After 12 days, no more red powder formed. Moreover, the volume of the air in the apparatus had decreased by one fifth. Mercury rose in the bell jar to take the place of that portion of the air that had reacted with the mercury. Today we know that about one fifth of air is oxygen. The oxygen of the air had combined with the heated mercury, forming the reddish powder mercury(II) oxide (HgO). See Figure 7-4.

Figure 7-4 Antoine Lavoisier, a French scientist, is shown performing his historic experiment in his laboratory. Describe his famous 12-day experiment.

In another experiment, Lavoisier heated the oxide to a higher temperature in a smaller retort. The mercury(II) oxide changed back to mercury and a colorless gas was given off. When he tested this gas, Lavoisier found that substances burned brightly in it, just as Priestley had seen. With these two experiments, Lavoisier provided proof that approximately one fifth of the air is an active gas that unites chemically with other substances. Lavoisier named the gas oxygen.

Oxygen gas can be made in several ways. When mercury(II) oxide (HgO) is heated, it decomposes (breaks down) into mercury (Hg) and oxygen (O_2), as shown by the following equation:

$$2HgO(s) \xrightarrow{\text{heat}} 2Hg(l) + O_2(g)$$

DISCOVERING THE OXYGEN CONCENTRATION IN AIR

activity

OBJECTIVE: Perform an experiment to find the percent of oxygen in air.

PROCESS SKILLS
In this activity, you will *measure* and *collect data* and then *analyze* the data in order to determine the concentration of oxygen in the air.

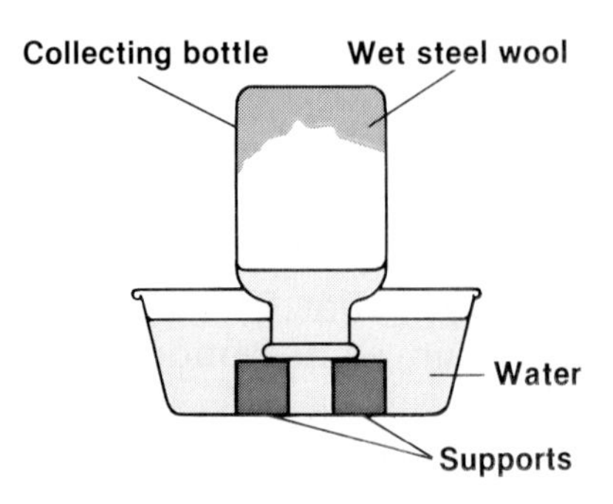

MATERIALS
steel wool
detergent
large-mouth jar
large pan
glass plate
graduated cylinder

PROCEDURE
1. Wash a piece of new steel wool with a detergent and rinse.
2. Place the wet steel wool in the jar so that it will remain firmly in place. Invert the jar in a pan of water, as shown in the diagram.
3. After a day or two, carefully put a glass plate under the mouth of the jar while it is still underwater without letting any water leak out.
4. Holding the glass plate tightly against the jar's mouth, lift the jar from the water and quickly turn it upright.
5. With a graduated cylinder, measure the volume of water in the jar. This volume is equal to the volume of oxygen from the air that combined with the steel wool.
6. Fill the jar with water and again measure its volume.
7. From these two measured volumes, find the percent of oxygen that is in the air sample. Compare your value with that given in Table 7-1.

OBSERVATIONS AND CONCLUSIONS
1. What evidence have you uncovered to lead you to think a chemical reaction has or has not taken place?
2. What was the reason for washing the steel wool in a detergent?
3. What was the reason for allowing the steel wool to be wet when it was placed in the jar?
4. What facts did you assume to find the percent of oxygen?
5. How might your results be affected if any of your assumed facts were wrong?

Oxygen also can be made by mixing and heating equal parts of potassium chlorate ($KClO_3$) and manganese dioxide (MnO_2), using the apparatus shown in Figure 7-5. The heat breaks down the potassium chlorate into potassium chloride (KCl) and oxygen. The equation for this reaction is

$$2KClO_3(s) \xrightarrow{\text{heat}} 2KCl(s) + 3O_2(g)$$

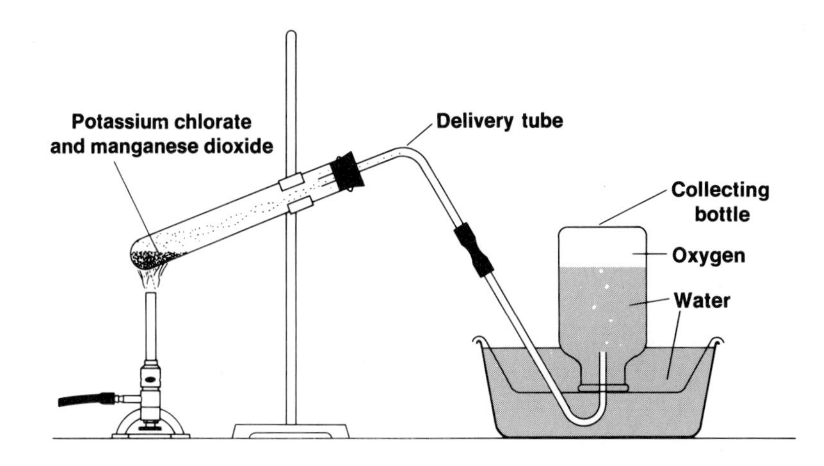

Figure 7-5 The apparatus for making and collecting oxygen. Oxygen is made in the test tube, moves through the delivery tube, and displaces the water in the collecting bottle. Why must you remove the delivery tube before removing the burner? **CAUTION: Do not attempt this experiment without your teacher's permission and supervision.**

The manganese dioxide is not included in this equation because it is not permanently changed in this reaction. Its addition to potassium chlorate enables $KClO_3$ to give up its oxygen more quickly and at a lower temperature than when heated alone. Manganese dioxide acts as a **catalyst** (KAT-*uh-list*), a substance that changes the rate of a chemical reaction but is not itself changed. Catalysts are widely used in chemistry. In this case, a catalyst (MnO_2) is used to speed up a chemical change. Sometimes catalysts are used to slow down chemical reactions.

Heating a mixture of potassium chlorate and manganese dioxide produces oxygen. Does the oxygen come from $KClO_3$ or MnO_2? Heating these compounds one at a time in a test tube will give the answer. When only the catalyst, manganese dioxide, is heated, oxygen is not given off. However, when heated to high temperatures, potassium chlorate melts and gives off a gas that tests positively as oxygen. When potassium chlorate is heated without the catalyst, it must be heated to a higher temperature before oxygen is given off. Again, the manganese dioxide catalyst allows the breakdown of potassium chlorate to take place sooner, at a lower temperature.

CAUTION: *Because of the potentially volatile reaction of potassium chlorate with manganese dioxide, making oxygen in the laboratory must be done with extreme care. The heated mixture can explode. Therefore, this method of making oxygen is not recommended for use in school laboratories.*

The described methods of making oxygen are not suitable for producing the large amounts needed for commercial uses. Large amounts of oxygen are usually made either by electrolysis of water or by evaporation of liquid air.

In Chapter 2 (Figure 2-4), you learned how electrolysis can be used to break down water into its elements, oxygen and hydrogen. Air can be made into a liquid if it is highly compressed and, at the same time, cooled to a very low temperature (−200° C). ("C" refers to Celsius, a scale used to measure temperature. You will learn about this scale in Chapter 19.) Liquid air is made up of mostly nitrogen and oxygen. Liquid nitrogen has a boiling point (−196° C) that is about 13° C lower than that of liquid oxygen (−183° C). Therefore, as the liquid air warms on standing, the nitrogen will boil away first. Mostly liquid oxygen is left behind, which can then be collected. The nitrogen gas produced is also recovered for industrial use.

Oxygen has many uses. Most living things use oxygen in the air for respiration. **Respiration** involves the use of oxygen to produce energy from nutrients. This energy is used to carry out the processes of life. See Figure 7-6.

Figure 7-6 The oxygen we breathe helps us to produce the energy our bodies need to be physically active. Why might this climber need additional oxygen?

Some people require pure oxygen, especially those who have breathing problems. Such people suffer from a deficiency of oxygen supply to their body tissues and require more oxygen than is normally contained in the air. As a result, hospitals and ambulances always have a supply of pure oxygen.

Additional oxygen is also necessary for people at high altitudes, such as those in airplanes, where the air holds less oxygen. To solve this problem, modern planes have pressurized cabins. Air from outside the plane is compressed inside the cabin, which increases the density of the air and, thus, the amount of oxygen it holds.

Firefighters and rescue workers often need gas masks and tanks of oxygen when entering smoke-filled buildings. Divers use tanks of compressed oxygen so that they will be able to breathe while underwater.

Oxygen has uses other than those involved in living processes. For example, oxygen is used with another gas, acetylene (*uh*-SET-*uh*-LEEN), in a torch that makes a very hot flame. An oxyacetylene torch is hot enough to cut through metals or to weld pieces of metal together (Figure 7-7).

Figure 7-7 The oxyacetylene flame cuts through slabs of steel quickly and efficiently. Why do the workers wear helmets over their heads?

Today, large amounts of oxygen are needed in industry. Modern steel mills, for example, use many metric tons of oxygen daily to produce iron. Blast furnaces use oxygen-enriched air to make iron from iron ore. By using this oxygen-air mixture, the ore can be changed to iron faster and impurities can be removed more efficiently. Rockets use great amounts of oxygen. Huge liquid-fueled rockets use lox (liquid oxygen) to burn fuel more efficiently.

Oxidation is a common chemical change. Modern chemistry really began when scientists discovered the important part oxygen plays in the chemical changes taking place all around us. When a substance such as coal, wood, or gasoline burns, it undergoes oxidation. The atoms of carbon and hydrogen that make up these substances join with oxygen in the air to form new compounds. **Oxidation** is a chemical reaction in which an element loses one or more electrons. It can also be defined as the chemical combination of oxygen with other substances.

Oxidation that takes place quickly and gives off heat and light is called **rapid oxidation** or **combustion**. Most often, the oxygen needed for burning is supplied by air. What examples of rapid oxidation (combustion) can you give?

Many serious explosions have taken place in grain elevators, coal mines, and other places where the air was filled with burnable dust. These explosions were caused when a spark or flame set off the very rapid oxidation of the dust particles.

What causes dust to explode whereas larger pieces of combustible material oxidize more slowly? Consider that a log burns more quickly when split into smaller pieces. These pieces will burn even more quickly when made into smaller particles and scattered over a fire. The increased speed of burning is due to an increase in the surface area of the wood in contact with the oxygen of the air. How can explosive combinations of dust, oxygen, and sparks be avoided? One precaution taken in coal mines and many industrial plants is to keep the air clear of burnable dust (Figure 7-8).

Figure 7-8 This coal miner is operating a device known as the EIMCO 2810 Miner by radio remote control. The device not only mines the coal but cleans the air of coal dust by means of an attachment called a scrubber.

The decay of dead plants and animals is another example of oxidation. Unlike burning, the process of decay is a very slow chemical change. Decay is also referred to as slow oxidation. Slow oxidation, like rapid oxidation, gives off heat. The decay process takes place over such a long time that the heat is not usually noticeable.

The breakdown of dead plant and animal material by slow oxidation actually contributes to the food chain. The decay process produces substances such as nitrogen compounds that nourish growing plants. However, slow oxidation may also cause undesirable changes. For example, when an iron barbecue grill is left outdoors exposed to air and moisture, the iron unites with the oxygen in the air to form an iron oxide, commonly called rust.

Because rusting of iron causes millions of dollars worth of damage every year, a strong effort is made to prevent it. Steel (processed iron) tools are coated with a thin film of oil to keep the oxygen from coming into direct contact with the metal. Steel bridges and water tanks are usually painted with a mixture of red lead and linseed oil, and often with a coat of aluminum paint. Some steel products are plated with metals such as tin, nickel, zinc, or chromium. These metals also oxidize, but the resulting oxide forms a thin coating that protects the iron or steel below it from oxidation.

If you leave a pile of oily rags in a closed place, you may see another example of the effects of slow oxidation. The rags may heat up and then burst into flame. Such a fire is caused by spontaneous combustion. The word spontaneous means by itself.

The fire starts because the oily rags slowly oxidize, building up heat. If the heat cannot escape, the temperature of the rags will rise. Finally, the rags get so hot that they burst into flame. The lowest temperature at which a material begins to burn, whether by spontaneous combustion or another cause, is known as the material's **kindling temperature** (Figure 7-9).

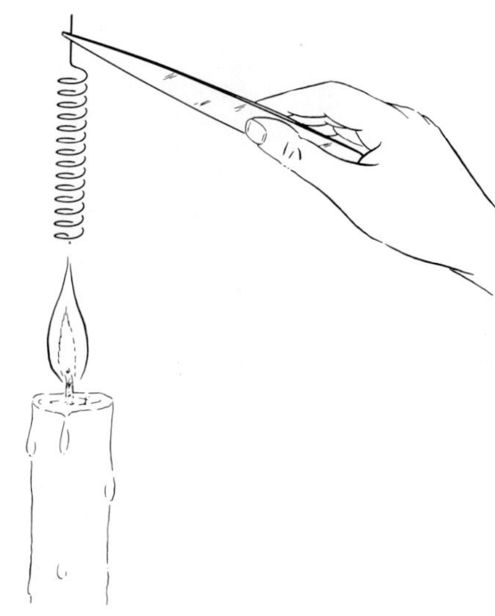

Figure 7-9 A copper coil lowered into a candle flame conducts heat away from the candle so quickly that the wick is cooled below its kindling temperature. What happens to the flame?

To reduce the chances of spontaneous combustion of oily rags, most industries have a strict rule that such rags must be placed in covered metal containers. People should follow the same practice at home. For instance, rags that have been used to polish furniture or to wipe oil and paint should be properly stored in an airtight container, or dried outside and disposed of.

7.2 OTHER GASES OF THE AIR

Nitrogen is the most abundant gas of the air. In contrast to oxygen, nitrogen (N_2) is a rather inactive element. It does not combine readily with other elements and it does not support burning. Like oxygen, however, nitrogen has no odor, no taste, and no color.

Nitrogen is found in many compounds. For example, common explosives such as dynamite, gunpowder, TNT (tri-nitro-toluene), and nitroglycerin each contain nitrogen. The destructive power of these explosives comes from the rapid breakdown of their molecules, which can be triggered by a flame, electricity, or vibrations. This breakdown gives off heat and causes gases to expand quickly, producing very rapid combustion.

Figure 7-10 The nodules on these pea roots contain nitrogen-fixing bacteria. Describe the functions of these special bacteria.

All living things need nitrogen to live and grow. Most plants cannot use nitrogen directly from the air and take in nitrogen compounds from the soil. However, with the help of bacteria that live on their roots, other plants such as peas, beans, clover, and alfalfa can use nitrogen in the air. The bacteria change the nitrogen of the air into usable nitrogen compounds that the plants take in. This process is called **nitrogen fixation**. See Figure 7-10. Through the bacteria on the plant roots, some of the nitrogen compounds are returned to the soil for use by other plants. Another way of adding nitrogen compounds to the soil is through the use of fertilizers.

Argon accounts for less than one percent of the air. The word argon comes from the Greek term for lazy, a name that correctly describes the activity of this element. It is one of the **noble gases**, found in Group 18 of the periodic table (Figure 3-6). The noble gases are sometimes called the inert or rare gases.

Argon is used with nitrogen in electric light bulbs. Normally, the hot tungsten filament in the bulb evaporates, blackening the inner surface of the bulb. Argon slows the rate of evaporation so that the bulb stays bright longer.

Carbon dioxide (CO_2) is a compound found in the air. Although the amount of carbon dioxide in the air is very small, this gas is needed for life on earth. Green plants must have carbon dioxide to make their own food.

The food-making process in green plants is called **photosynthesis** (FOE-*toe*-SIN-*thuh-sis*). Plants use sunlight, water from the soil, and carbon dioxide from the air to make sugars and starches. These sugars and starches are used by the plants for growth and development. Oxygen is given off as a waste product.

Animals, including humans, depend on plants for their food and oxygen. Animals breathe the oxygen that plants put into the air and give off carbon dioxide as one of the products of food oxidation. Much of the carbon dioxide in the air today comes not from animals, however, but from the burning of fuels in homes, cars, and factories. Plants remove some of the carbon dioxide from the air. Most of it, however, is absorbed by the oceans where a mixture of water and carbon dioxide is formed. Despite this absorption, the amount of carbon dioxide in the atmosphere is increasing, causing an increase in the amount of the sun's energy trapped by our atmosphere. This increase in trapped energy is known as the **greenhouse effect**. Many scientists believe that this will lead to increased tempera-

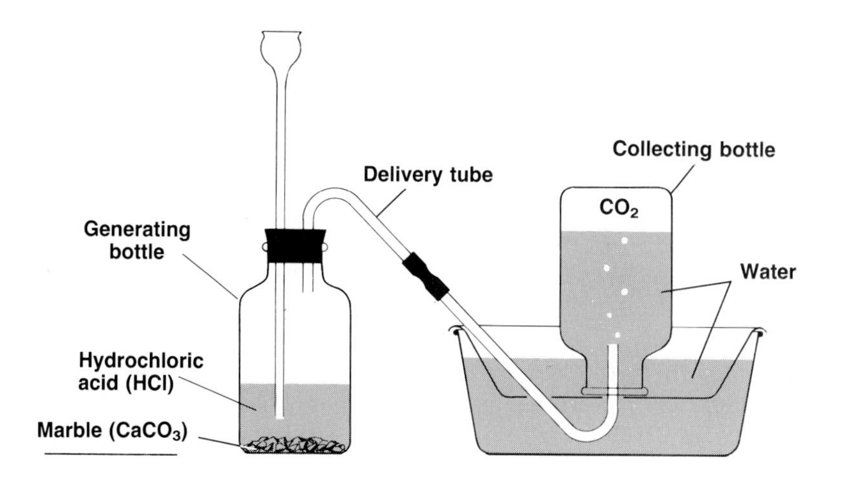

Figure 7-11 A common laboratory setup for making carbon dioxide. Which of the reacting substances produces the carbon dioxide? **CAUTION: Do not attempt this experiment without your teacher's permission and supervision.**

tures around the world, which will affect weather, climate, and ocean levels. Scientists speculate that the impact on farming and living conditions could be great.

In the laboratory, carbon dioxide can be made by adding hydrochloric acid (HCl) to chips of marble. Marble is impure calcium carbonate ($CaCO_3$). The setup for this procedure is shown in Figure 7-11. The equation is

$$2HCl(aq) + CaCO_3(s) \rightarrow CaCl_2(aq) + H_2O(l) + CO_2(g)$$

Carbon dioxide can be detected by performing a test based on the following facts. If limewater, a solution of $Ca(OH)_2$, is added to a bottle of CO_2 gas and shaken, the limewater turns white. The white color is due to the formation of an insoluble product, or precipitate, in this case calcium carbonate ($CaCO_3$). The equation for the reaction is as follows:

$$Ca(OH)_2(aq) + CO_2(g) \rightarrow CaCO_3(s) + H_2O(l)$$

The formation of the white precipitate is a common test used to detect CO_2.

Carbon dioxide does not burn or support the burning process. If you collect a bottle of CO_2 and thrust a burning splint into the bottle, the splint will be extinguished.

Other trace gases of the air are present in variable amounts. They include water vapor (H_2O), noble gases such as helium (He) and neon (Ne), ammonia (NH_3), and hydrogen (H_2). Although the concentration of most of these substances is generally constant in the air, the amount of water vapor varies according to the air temperature, weather conditions, and location on earth.

Let us examine some of the air's trace elements. The element helium gets its name from *helios*, the Greek word for the sun. Helium was detected in the sun before it was discovered on earth. Helium is obtained from inside the earth, for example, from natural gas wells in Kansas, Oklahoma, and Texas. Helium is used for filling balloons because it is less dense than air and so makes balloons rise. Moreover, because helium will not burn it is safer to use in balloons than hydrogen, which is also less dense than air but burns. Helium is also used in welding metals.

The element neon derives its name from the Greek word *neos*, meaning new. Neon is widely used in illuminated signs and for airplane beacon lights. A high-voltage charge applied across a neon-filled glass tube causes the gas to give off a familiar, glowing, orange-red light. Various mixtures of gases are used to produce a variety of colored lights (Figure 7-12). For instance, argon mixed with mercury will produce a bright blue color; argon mixed with helium produces a yellow color.

Noble gases such as helium and neon were once thought to be completely inert, or chemically inactive. This lack of chemical activity can be explained by their atomic structure. The outer shell of electrons of the noble gases is filled. Scientists thought that the noble gases would not combine with other elements because the gases do not tend to give, receive, or share electrons. However, scientists now know that it is possible to make a few compounds of the heavier noble gases (krypton, xenon, and radon) and active substances such as fluorine, chlorine, and oxygen.

Figure 7-12 So-called neon lights are used in illuminated signs and can also be used to create works of art, as in this abstract sculpture. What produces the various colors of light in these tubes?

Ammonia results from protein decay. Ammonia (NH_3) is one of the most important of all chemical compounds. A very small amount of ammonia gas is found in the air. Ammonia is formed when proteins in dead plants and animals are broken down in the decay process.

Ammonia is a colorless gas, less dense than air, with a sharp, stinging odor. It is poisonous and extremely soluble in water. One liter of water can dissolve about 700 L of ammonia gas at room temperature.

In the laboratory, ammonia (NH_3) can be made by heating a mixture of calcium hydroxide [$Ca(OH)_2$] and ammonium chloride (NH_4Cl) as shown in Figure 7-13. The chemical change that takes place with heating is described by the following equation:

$$Ca(OH)_2(s) + 2NH_4Cl(s) \xrightarrow{\text{heat}} CaCl_2(s) + 2H_2O(g) + 2NH_3(g)$$

Large amounts of ammonia are made for the commercial market by reacting nitrogen (N_2) with hydrogen (H_2), using a catalyst to speed up the process. This reaction takes place at about 600° C and at a pressure of about 1,000 times normal air pressure. The equation for this reaction is:

$$N_2(g) + 3H_2(g) \rightarrow 2NH_3(g)$$

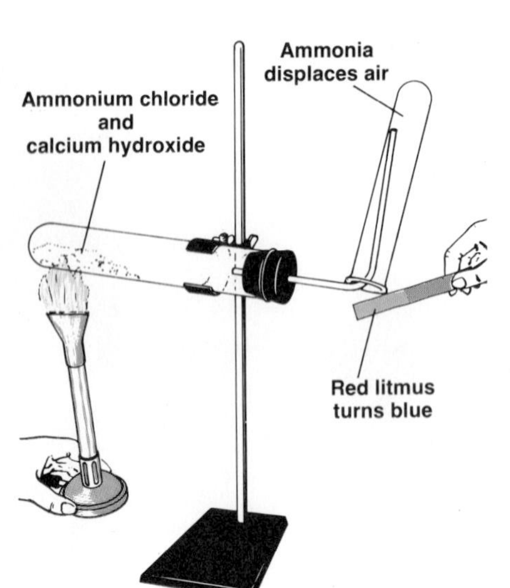

Figure 7-13 A typical setup for making small amounts of ammonia gas (NH_3). Why does the wet litmus paper change color as shown above? **CAUTION: Do not attempt this experiment without your teacher's permission and supervision.**

Because ammonia contains about 82 percent nitrogen (by mass), large amounts of ammonia are used in fertilizers (Figure 7-14). Ammonia is also used in making nitric acid for explosives and ammonia water for cleaning fluids.

Hydrogen is the least dense gas. Hydrogen gas (H_2) is very scarce in our atmosphere. However, analysis of light given off by stars shows that the stars are mostly composed of hydrogen. Scientists estimate that hydrogen atoms make up about 90 percent of all atoms in the universe. (The universe is mostly empty space.) Hydrogen compounds are common on earth and are found, for example, in all living things, in fuels, and in water.

Most active metals will replace the hydrogen in an acid, thereby setting the hydrogen free. In the laboratory, hydrogen can be made by the reaction of zinc (Zn) with dilute sulfuric acid (H_2SO_4). The reaction is described by the following equation:

$$Zn(s) + H_2SO_4(aq) \rightarrow ZnSO_4(aq) + H_2(g)$$

The zinc sulfate ($ZnSO_4$) that is formed remains in the generating bottle shown in Figure 7-15. Hydrogen gas travels through the delivery tube and drives the water out of the collecting bottle.

Hydrogen has the lowest density of all the elements. This property is one of the reasons it was once used to fill balloons. However, hydrogen can explode, which explains why it is no longer used for this purpose.

Hydrogen is not very active at room temperature, but it reacts with many substances when heated. For example, suppose a lighted candle is held carefully to the mouth of a small open (plastic) container in which hydrogen has just been generated. There will be a "pop" as the hydrogen burns rapidly. However, if the bottle is allowed to stand for a minute or two and a burning candle is brought near the

Figure 7-14 Liquid ammonia is applied directly to the soil as a fertilizer. What are some other uses for ammonia?

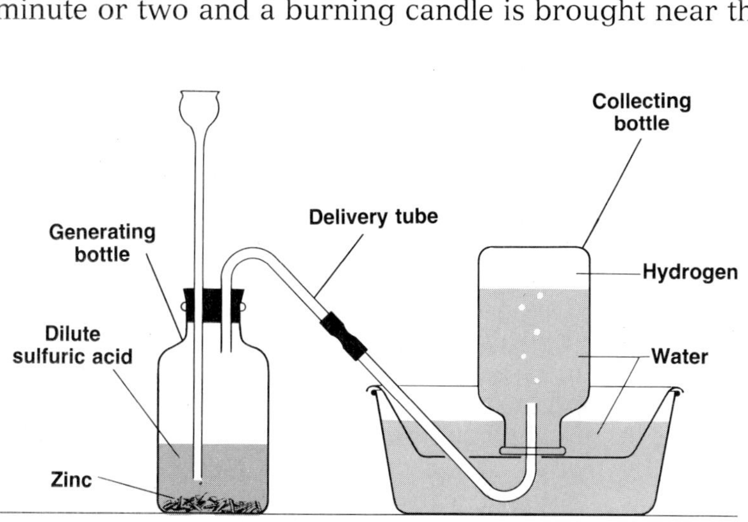

Figure 7-15 A typical setup for the laboratory preparation of hydrogen gas (H_2). **CAUTION: Do not attempt this experiment without your teacher's permission and supervision.**

MAGNESIUM AND AN ACID

activity

OBJECTIVE: Determine how magnesium reacts with an acid.

PROCESS SKILLS
In this activity, you will *observe* the reaction of magnesium with acid and *classify* the products formed.

MATERIALS
magnesium ribbon   test tube   dilute hydrochloric acid

PROCEDURE
1. Place a 5-cm length of magnesium (Mg) ribbon in a test tube containing a small amount of dilute hydrochloric acid (HCl). CAUTION: *Magnesium ribbon is highly flammable. Wear safety goggles and an apron.*
2. Observe what happens.

OBSERVATIONS AND CONCLUSIONS
1. Describe your observations.
2. Identify the gas that evolves. What method did you use?
3. Write the balanced chemical equation for this reaction.

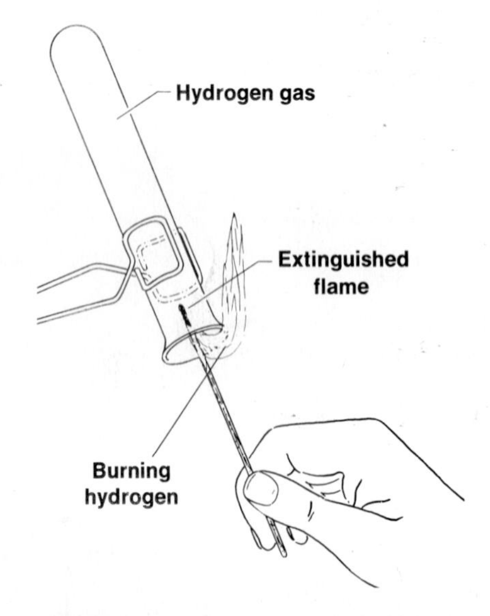

Figure 7-16 Burning a mixture of hydrogen and air. **CAUTION: Do not try this test for hydrogen without your teacher's supervision.**

bottle, nothing will happen because the hydrogen, which is less dense than air, has had time to escape from the open bottle.

If a flaming splint is placed inside an inverted bottle of hydrogen, the flame will go out, but the hydrogen will continue to burn at the mouth of the bottle (Figure 7-16). This is a test for the presence of hydrogen gas. While the hydrogen gas itself burns, it cannot support the burning of other substances such as the splint.

When hydrogen is added to cottonseed, coconut, or other food oils, it combines chemically to form solid cooking fats. This process is called **hydrogenation** (HIE-*drah-juh*-NAY-*shun*). Fine grains of nickel (Ni) are used as a catalyst to speed up the process. Peanut butter is hydrogenated to prevent the separation of the oils from the pulp.

Special welding and cutting devices use hydrogen to produce extremely hot flames. Hydrogen is also used to make high-grade gasoline from petroleum. Liquid hydrogen is used as fuel for rockets.

BIOGRAPHY

NEIL BARTLETT

Everyone said it was impossible to make a compound of a noble gas and another substance. As a matter of fact, the noble gases—helium, neon, argon, krypton, xenon, and radon—got their name because they did not react with other more "common" substances. The noble gases, like the royalty after which they were named, kept to themselves.

Then, in 1962, along came an English chemist by the name of Neil Bartlett. At the time, Bartlett was 30 and doing experimental chemistry in British Columbia, Canada. Bartlett's pet project involved the study of extremely reactive substances including one called platinum hexafluoride (PtF_6).

Based on theoretical calculations, Bartlett was convinced that PtF_6 was so reactive that it could get the noble gas xenon (Xe) to combine with it to form a substance never before seen. But was he correct? To find out, Bartlett built a special apparatus in which colorless xenon gas under pressure was suddenly mixed with deep red PtF_6 vapor. As Bartlett watched, the gases became transformed into something altogether new—a yellow solid! But what was this material? After careful chemical analysis, Bartlett uncovered the answer. The yellow solid was xenon hexafluoroplatinate ($XePtF_6$)—the "impossible" compound!

Bartlett's dramatic experiment revealed that what appears to be science "fact" one day may be science "fiction" the next. Soon other compounds of xenon, as well as of krypton and radon, were made. Perhaps Bartlett's most lasting contribution was to make scientists—and the rest of us—be more careful when using the word "impossible."

CHAPTER REVIEW 7

SUMMARY

1. Air is a mixture of many gases, mostly nitrogen and oxygen.
2. Oxygen can be made in the laboratory by breaking down mercury(II) oxide or potassium chlorate.
3. Oxygen can be made commercially by electrolysis of water and by evaporation of liquid air.
4. Oxygen is the only common odorless gas that supports burning.
5. Oxidation is a chemical reaction in which an element loses one or more electrons, also defined as the chemical combination of oxygen with other substances.
6. Catalysts are substances that change the speed of a chemical reaction but are not themselves changed by the reaction.
7. Nitrogen compounds are needed for the proper growth of plants.
8. Carbon dioxide, water, and sunlight are needed in photosynthesis.
9. Gases that occur as small traces in the air include water vapor, helium, neon, ammonia, and hydrogen.
10. Ammonia is a colorless gas that is less dense than air, has a strong stinging odor, is poisonous, and is very soluble in water.
11. Hydrogen, the element with the lowest density, is commonly found in combined form on earth.

VOCABULARY

Match the item in the left column with the best answer in the right column. Do not write in this book.

1. air
2. catalyst
3. greenhouse effect
4. hydrogenation
5. kindling temperature
6. nitrogen fixation
7. noble gases
8. oxidation
9. photosynthesis
10. rapid oxidation
11. respiration

a. relatively inactive elements
b. process by which hydrogen combines with oils to form solid fats
c. when substances join with oxygen to form new compounds
d. mixture of gases in earth's atmosphere
e. not changed as it changes the rate of a chemical reaction
f. use of oxygen to provide energy in living things
g. occurs quickly, and gives off heat and light
h. lowest temperature at which a material begins to burn
i. process whereby bacteria produce nitrogen compounds from nitrogen in the air
j. food-making process in green plants
k. trapping of sun's energy by earth's atmosphere

REVIEW QUESTIONS

1. Describe a convenient way of making oxygen in the laboratory.
2. Is the result of the activity shown on p. 124 an example of slow oxidation? Justify your answer.
3. How can oily rags cause a fire by spontaneous combustion?
4. What are some uses of nitrogen compounds?
5. How is carbon dioxide made in the laboratory? How is the gas identified?
6. List three conditions that are necessary for photosynthesis to take place.
7. How is ammonia gas made in the laboratory? How is it made commercially? Of what use is it?
8. How is hydrogen made in large quantities for commercial use?
9. What are four uses for hydrogen?
10. What is the most common use of helium?
11. Describe how dust explosions take place.
12. Where in the periodic table (Figure 3-6) are the noble gases found?
13. Why are the noble gases inactive compared to other elements?
14. Why is argon used in making electric light bulbs?

CRITICAL THINKING

15. What evidence can you give to support the idea that air is a mixture?
16. The gases of nitric oxide (NO), nitrogen dioxide (NO_2), and ozone (O_3) are all present in the atmosphere in varying amounts. The decomposition of ozone takes place in two stages represented by the following equations:

$$O_3 + NO \rightarrow NO_2 + O_2$$
$$NO_2 + O \rightarrow NO + O_2$$

    Which of the three gases mentioned above acts as the catalyst in this chemical process? Explain your answer.
17. Describe how the evolution of green sea plants (algae) and green land plants affected the evolution of life on earth.
18. One of the major products of photosynthesis is a carbohydrate ($C_6H_{12}O_6$), otherwise known as glucose. Write the word equation for the chemical reaction that takes place during photosynthesis, and then write the balanced chemical equation.

FURTHER READING

Allen, Oliver. *Atmosphere.* (From the *Planet Earth Series*). Chicago: Time-Life Books, 1983.

Gay, Kathlyn. *The Greenhouse Effect.* New York: Franklin Watts, Inc., 1986.

Gedzelman, Stanley D. *The Science and Wonders of the Atmosphere.* New York: John Wiley and Sons, 1980.

Kerr, R. A. "Stratospheric Ozone is Decreasing." *Science* 239 (March 25, 1988): 1489–91.

Tangley, L. "Ozone Update." *Bio Science* 38 (February 1988): 87.

THE CHEMISTRY OF WATER

All living things on our planet rely on water for their survival. We often take water for granted because it seems to be all around us. But what is water and where does it come from? What process does it undergo in order to become the drinkable tap water in your home?

CHAPTER 8

SECTIONS

OBJECTIVES

- ☐ Explain why water is a good solvent and why it exists in a liquid phase at room temperature.
- ☐ Describe and explain the reason for each step in the process of purifying water.
- ☐ Explain what causes hard water and describe the processes used to make it soft.
- ☐ Describe the treatment and disposal of sewage.
- ☐ Define the following terms: hydrogen bond, suspension, potable water, distillation, freeze separation, reverse osmosis, and ion exchange.

8.1 WATER: AN EVER-PRESENT COMPOUND

Water has some remarkable properties. Pure water is a liquid that has no color, odor, or taste. Water exists in all three phases: solid, liquid, and gas. Under normal conditions, water freezes at 0° C and boils at 100° C. (See Figure 8-1 on the next page.)

Figure 8-1 The change of phase from water to ice (freezing) or from ice to water (melting) takes place at a temperature of 0° C. At what temperature does water boil under normal conditions?

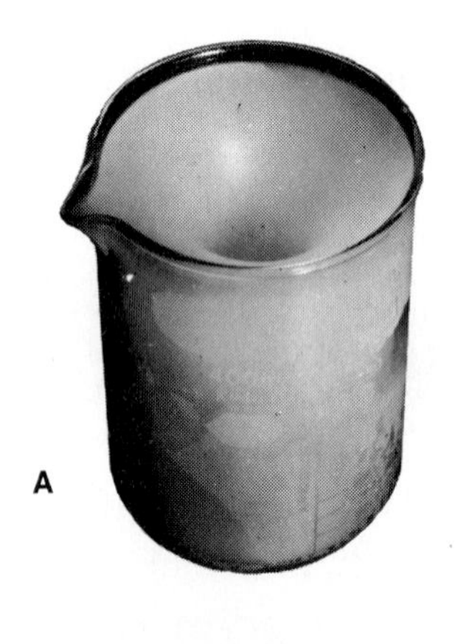

A

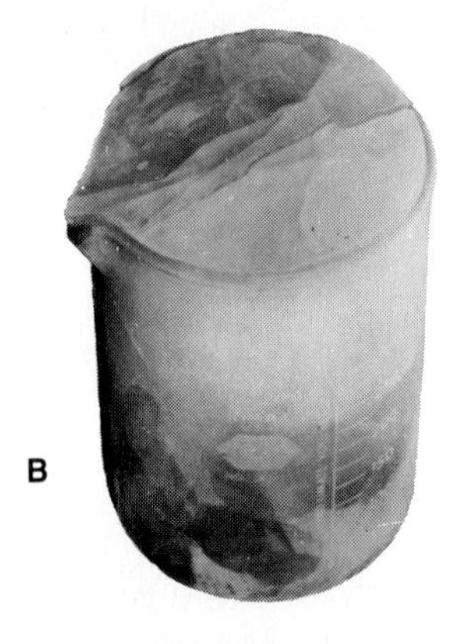

B

Figure 8-2 Most substances, like wax, contract when they change from a liquid to a solid (A). However, water expands when it freezes (B).

In many ways, water is a most remarkable compound. Almost all liquids shrink when they freeze. For instance, when a beaker full of melted wax is allowed to cool, a hollow will form in the center where the freezing liquid shrinks, as illustrated in Figure 8-2. Unlike wax, water expands when it freezes. When a tankful of water freezes, it exerts about 20,000 newtons (N) of force on each square centimeter of the tank. A tightly sealed container filled with water may burst when the water freezes, even if it is quite strong. Water pipes and car radiators are likely to burst in winter if the water in them freezes.

When ice at 0° C is warmed, it shrinks until the temperature reaches 4° C. At 4° C, water reaches its greatest density. As it is warmed further, the water slowly expands.

Water is a very stable compound. Water molecules do not break apart until heated to about 2,700° C.

Water is a good solvent. Water is a covalent molecule composed of an oxygen atom sharing a pair of electrons with each of two hydrogen atoms. This arrangement is shown in Figure 3-11 of Chapter 3. A laboratory model reveals that a water molecule has a bent structure (Figure 8-3). The angle of bonding, or bond angle, is 105°.

Like all molecules, water molecules are electrically neutral. However, you can see from Figure 8-3 that the hydrogen ends are weakly positive. The oxygen atom at the opposite end is weakly negative. A molecule with an uneven distribution of electrical charge due to its bent structure is called a **polar** molecule. The polar, covalent property of the water molecule is one of the main reasons why water is such a good solvent. Because of its polar nature, the water molecule can attract other molecules or ions of a solute, surround them, and pull them into solution. A molecule with a straight-line structure, such as CO_2 (O=C=O), is a symmetrically charged, or **nonpolar** covalent molecule.

The fact that water is a good solvent accounts for the saltiness of the oceans. For billions of years, rain has been falling on land areas of the earth and running off into streams, then into rivers, and finally into oceans. Rainwater dissolves solid matter from the land, mostly in the form of salts, and carries it to the sea. Thus, for billions of years, salts have been added to the oceans. The water that evaporates from the oceans returns to the land as rain, leaving behind the salts. Therefore, the ocean water that remains becomes more salty as time goes on.

Ocean water is a storehouse of raw materials. The waters that run from the land into the ocean carry with them not only common salt, but many other dissolved solids as well. A wide range of chemical resources is found within this vast storehouse. From a single cubic meter of sea water, 21,000 g of sodium chloride, 100 g of magnesium, 400 g of potassium, and 60 g of bromine can be obtained. Other substances, such as sulfur and calcium, are also present in smaller amounts.

The resources of the ocean are, for the most part, unused. It is difficult and expensive to remove these substances from the water. Only sodium chloride, magnesium, bromine, and iodine are removed in large amounts.

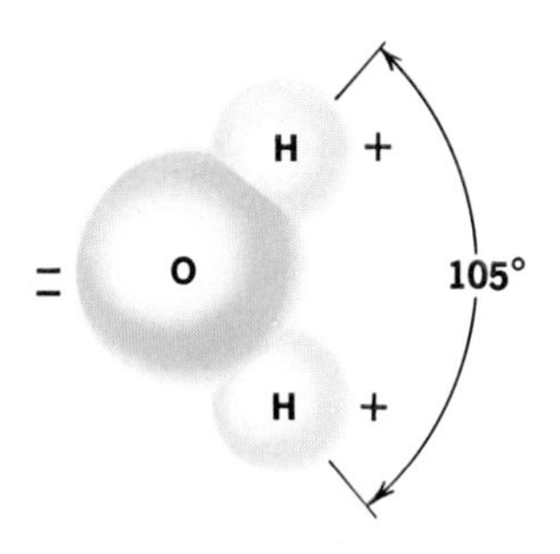

Figure 8-3 The diagram shows the polar nature of the water molecule. Why do you think the hydrogen atoms are partially positive and the oxygen atom is partially negative?

Water molecules are attracted to each other by weak forces. Water molecules are held together by the attraction of the positive hydrogen ends of each water molecule to the negative oxygen ends of other water molecules. This attraction forms a weak, but effective, **hydrogen bond** that is illustrated in Figure 8-4. This diagram shows that water is not simply a group of separate H_2O molecules, but rather a large number of H_2O molecules linked together by hydrogen bonds. Water is a liquid at room temperature because molecular groups are formed by hydrogen bonding. If hydrogen bonding did not exist, water would be a gas at room temperature.

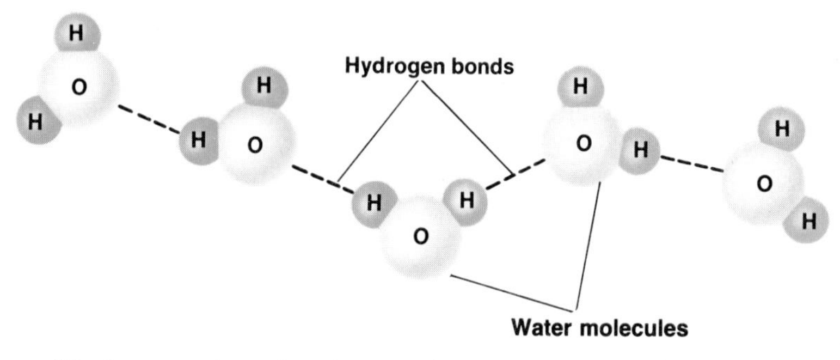

Figure 8-4 Hydrogen bonds are formed when the hydrogen atoms of the water molecule are weakly attracted to oxygen atoms of other water molecules.

Hydrogen bonds play an important part in determining the melting and boiling points of many substances. These bonds give ice an arrangement of crystals with many open spaces. This open structure shown in Figure 8-5 accounts for the fact that ice has a lower density than water.

Figure 8-5 A comparison of molecules in (A) water vapor (gas), (B) liquid water, and (C) ice (solid). As the water molecules freeze, they become bonded together in a structured order.

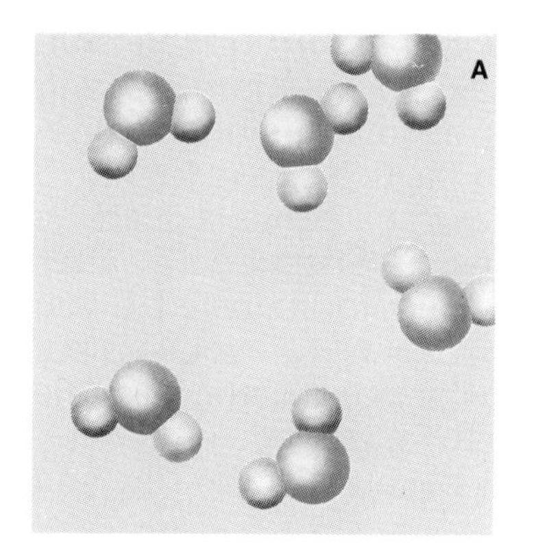

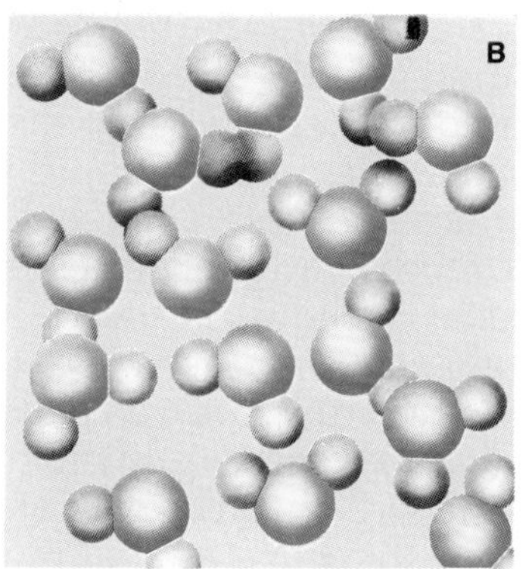

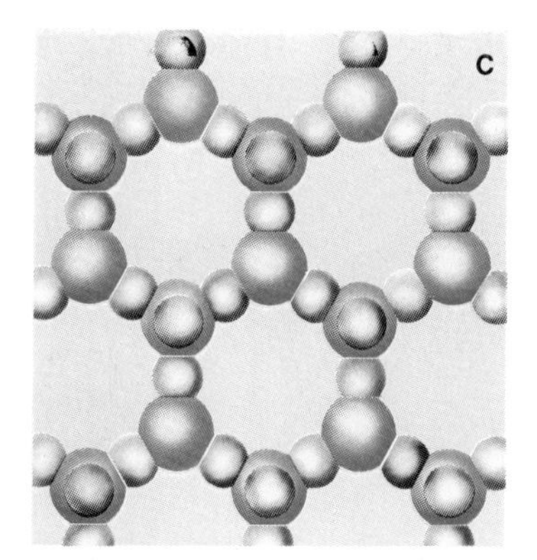

Pure water is not common. Although there is much water on the earth, pure water is not common because water readily dissolves many substances. The purest water in nature is rainwater, but it too contains impurities. As rainwater falls, it dissolves oxygen and other gases of the air, along with floating bits of dust and bacteria. Even so, rainwater is fairly free of dissolved solids, except in heavily polluted areas.

As soon as rainwater touches the ground, it dissolves minerals and impurities from plant and animal sources. These impurities often make water unfit for drinking because of the presence of harmful bacteria. However, water containing impurities that are harmless to the human body may be suitable for drinking.

Water that is fit to drink is called **potable water.** Potable water is clear, colorless, pleasant-tasting, free of harmful bacteria, and fairly free of dissolved solids. A small amount of dissolved air and minerals gives the water a better taste. Some types of dissolved minerals may make water unsuitable for laundry purposes.

Acid rain is one type of impure water. Any precipitation, such as rain or snow, that is acidic is called **acid rain.** Acids are produced when water vapor reacts with oxides of sulfur and nitrogen occurring as pollutants in the air. These pollutants come from chemical compounds given off by automobiles, factories, power plants, and other sources that burn coal or oil. When the water vapor in the air reacts with these compounds, sulfuric and nitric acids are formed and acid rain results. Acid rain has a pH below 5.6.

Many areas in eastern North America and in northwestern and central Europe in particular are adversely affected by acid rain. It pollutes lakes, rivers, and streams, can cause damage to statues and buildings, and is harmful to crops, forests, and soil. Highly acidic rain causes death to plant and animal life.

Water contains matter in suspension and in solution. In addition to dissolved solids held in solution, flowing water holds material in suspension. A **suspension** is a mixture formed by mixing a liquid with particles that are much larger than ions and molecules. For example, chocolate powder mixed with milk is a suspension. Suspended matter will settle if the liquid is allowed to stand undisturbed. If you look at a glass container filled with river water, you can see suspended matter, such as clay and organic materials. The river may have carried this matter for hundreds of kilometers. Some streams are very muddy because of the clay and silt held in suspension.

Inorganic, or nonliving, matter found in rivers and streams is most likely to consist of mineral compounds. The kinds of inorganic matter depend on the composition of the soils and rocks through which the water flows. Some of these materials either become dissolved in the water or are carried along by the river in suspension. A stream flowing through a limestone region will contain calcium compounds either in suspension or in solution. These materials affect the taste of the water. "Mineral water" is rainwater that has seeped underground through rocks, dissolving minerals and gases as it travels. Containing large quantities of mineral matter such as lime, salt, and iron, mineral water is sometimes recommended as part of a healthy diet.

SOLIDS IN POTABLE WATER

OBJECTIVE: Recover solids in water samples.

PROCESS SKILLS

In this activity, you will *observe* the results of boiling two water samples and *infer* any differences in their suspended solids.

MATERIALS

watch glass	distilled water
beaker	tap water
gas burner	safety goggles

PROCEDURE

1. Put on your safety goggles.
   Place a small amount of tap water in a watch glass.
2. Place the glass over a beaker of slowly boiling water as shown above. Let the water in the watch glass evaporate.
3. Observe the watch glass after the water has evaporated. Record your observations.
4. Repeat this procedure using a small sample of distilled water in a clear watch glass. Record your observations.

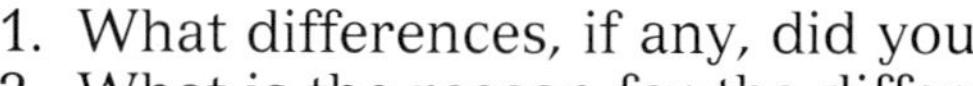
OBSERVATIONS AND CONCLUSIONS

1. What differences, if any, did you note in your observations?
2. What is the reason for the differences you found in the water samples?

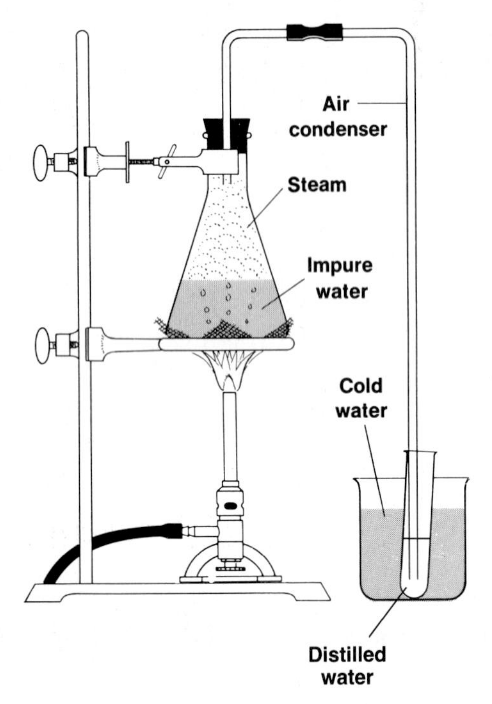

Figure 8-6 The liquid in the flask vaporizes, leaving behind solid impurities. Explain why the test tube is placed in cold water.

Substances derived from animals or plants, or organic matter, are often found mixed with water. These substances may be very harmful depending on their nature and the amount contained in the water. What appear to be harmless bits of leaves or straw can be dangerous because bacteria grow rapidly on these materials.

Boiling and distillation purify water. Water that has been boiled for about 10 minutes is probably safe to drink. While boiling the water does not remove dissolved or suspended matter, it does kill harmful bacteria, making the water potable.

Chemists and pharmacists use distilled water in making solutions or suspensions. Distilled water is also used in car batteries and steam irons. **Distillation** is the process of evaporating a liquid and then condensing the vapors in a separate container.

Water is distilled by boiling it in a container until steam is formed, as illustrated in Figure 8-6. The steam is then led through a tube to a collecting container that is cooled, usually by another container filled with cold water. This cold container causes the steam to condense back into liquid water. Dissolved solids that are not changed into gases remain in the container. Water that is free of dissolved solids can be obtained by this process. Impurities that evaporate easily, however, are not removed. Why do you think this is so? Most often, the first sample of distilled water is thrown away.

The distillation process purifies water. However, this method of making pure water is too costly when large amounts of pure water are needed.

Drinking water comes from many sources. The source of your drinking water depends on where you live. In rural areas, water is pumped from wells, springs, or streams.

Wells should be dug as far away as possible from places where organic materials might make the water unfit for human use. Many farm wells are a health hazard because they are poorly located or poorly built. Impure surface water may seep into the well.

Rivers and lakes are the major sources of drinking water in this country. They supply about 75 percent of the water used by communities for drinking, and by farmers for irrigation of crops.

DISTILLATION OF WATER

activity

OBJECTIVE: Determine the effect of distillation on a sample of impure water.

PROCESS SKILLS
In this activity, you will *observe* the product of distillation and *communicate* what you observe.

MATERIALS
- gas burner
- ringstand with ring
- clamp
- test tube
- 250-mL beaker
- rubber tubing (about 3 cm)
- Erlenmeyer flask
- one-hole rubber stopper
- wire gauze
- glass tubing bent at 90° angles (see Figure 8-6)
- safety goggles
- powdered copper(II) sulfate
- warm and cold water

PROCEDURE
1. Put on your safety goggles. Dissolve about 5 g of copper(II) sulfate in 100 mL of warm water. Add this solution to the flask. The blue color indicates that the water is not pure.
2. Set up the apparatus as shown in Figure 8-6.
3. Heat the solution by boiling and distill the liquid until a small amount of distilled water has been collected.

OBSERVATIONS AND CONCLUSIONS
1. Study the distilled sample and describe what you observe.
2. Explain this result.

Most of the water used by New York City comes from the Catskill Mountains, about 160 km away. The water flows through large, underground pipes. The water needs of the rapidly growing city of Los Angeles are so great that its water is brought from several sources, including the rivers of northern California more than 800 km away. In addition, part of its water supply is brought through long pipes from the Sierra Nevada Mountains (400 km away) and from the more distant Colorado River.

St. Louis and other large cities get their water from the Mississippi River. This water must be filtered and treated with chemicals before it is used.

Cities along the Great Lakes have a supply of fresh water nearby. The water commonly enters an intake pipe about 15 m below the surface of the lake and about 5 km from the shore. The water is then treated to make it safe to drink.

Large cities use many millions of liters of potable water daily. In the United States, an average family uses over a million liters of water a year. Unless water conservation practices are encouraged, it is believed that this country will need to double the present amount of available fresh water by the end of the century. To meet these needs, the United States government is experimenting with methods of getting fresh water from the ocean.

Cities and towns prepare water for drinking in many ways. Water obtained from a river or lake is seldom safe to drink. Such water, as you have learned, often contains organisms that may cause disease. Therefore, cities and towns treat water before it is used. To make water potable, sediment or suspended material is removed and harmful bacteria are killed. This process is usually carried out through the following five basic steps:

1. Water is taken from its source and allowed to stand in a large pool called a settling basin. The basin may cover an area of 5,000 m^2, perhaps the size of a small pond. Here, most of the heavy suspended matter and debris settle to the bottom.

Figure 8-7 Chemicals such as alum, chlorine, lime, and fluoride are often added to water as it goes to the second settling basin. Particles in the water stick to the alum, which sinks to the bottom of the basin.

2. The water is then fed into a second settling basin (see Figure 8-7). Since small particles suspended in water settle slowly, chemicals are used to speed up the settling process. Small amounts of aluminum sulfate [$Al_2(SO_4)_3$] and calcium hydroxide [$Ca(OH)_2$] are added to the water. These two chemicals react with each other to form sticky precipitates of aluminum hydroxide [$Al(OH)_3$] and calcium sulfate ($CaSO_4$). The double replacement reaction is

$$Al_2(SO_4)_3(aq) + 3Ca(OH)_2(aq) \rightarrow 2Al(OH)_3(s) + 3CaSO_4(s)$$

The fine particles present in the water stick to the precipitates. These small particles coagulate, or come together, to form heavier particles, and settle to the bottom. Bacteria also become caught in the precipitates. Much more of the original suspended material and bacteria is removed in the second basin.

3. The water is next passed through filters that may be larger than your classroom. These filters are composed of a layer of sand about 2 m thick that is placed above a layer of gravel. A layer of charcoal is often used between the sand and gravel to take out colored matter and foul-tasting substances. Figure 8-8 illustrates the layers in a sand filter.

4. The filtered water is often sprayed into the air, much like a large fountain. This process is called **aeration** and is depicted in Figure 8-9. Oxygen of the air dissolves in the water and kills some types of bacteria. Aeration also improves the taste of water.

5. In the United States, chlorine gas is often used to make sure that all harmful bacteria are killed. The amount of chlorine used depends on the condition of the water and the season of the year. In the fall, more chlorine is used because there is more organic matter in the water.

Figure 8-10 summarizes the basic steps usually undertaken in water purification in towns and cities.

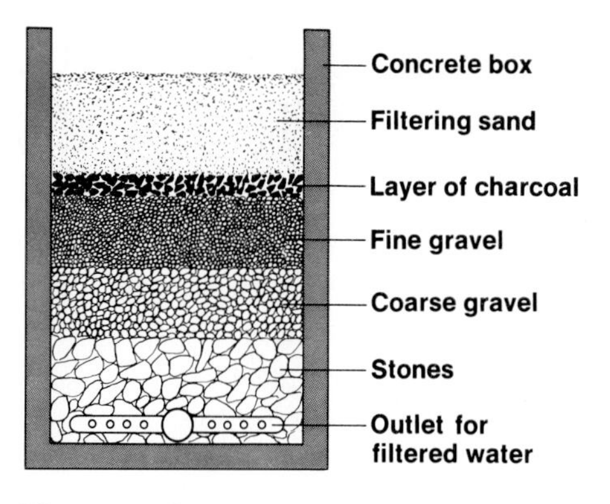

Figure 8-8 Cross section of a sand filter used in water purification.

Figure 8-9 In the aeration process, oxygen of the air dissolves in water.

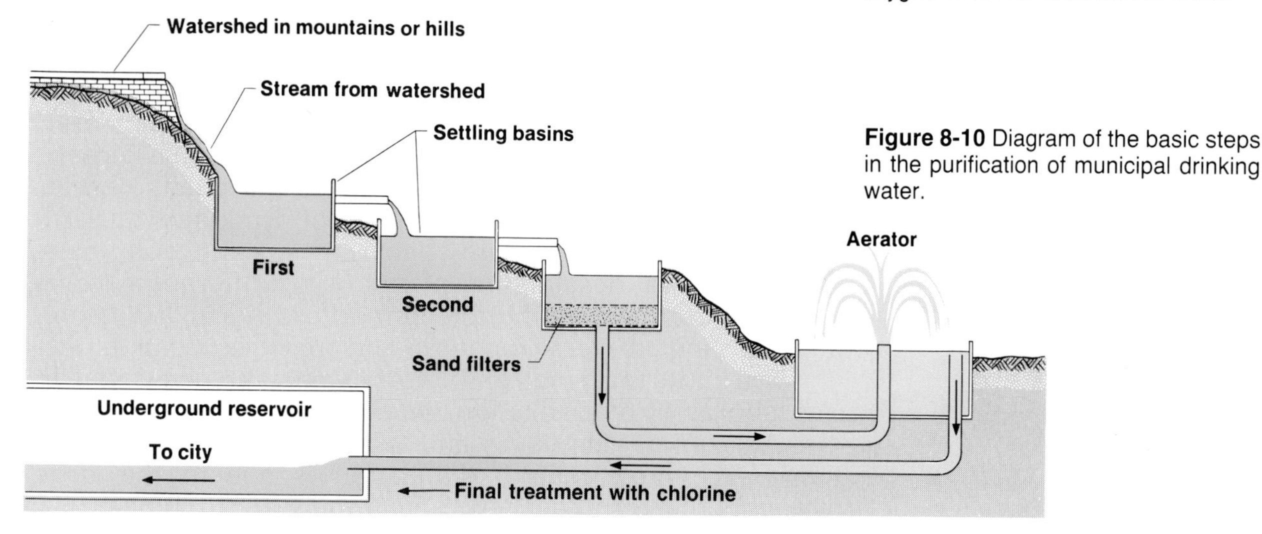

Figure 8-10 Diagram of the basic steps in the purification of municipal drinking water.

REMOVING COLORED MATTER FROM WATER

activity

OBJECTIVE: Perform a procedure that removes colored matter from water.

PROCESS SKILLS
In this activity, you will *observe* how colored matter can be removed from water and *analyze* your observations.

MATERIALS
10 g brown sugar
3 100-mL beakers
filter paper
funnel
10 g boneblack (animal charcoal)

PROCEDURE
1. Dissolve about 10 g of brown sugar in a 100-mL beaker of warm water.
2. Filter a portion of this solution into a second beaker. Observe the color of the liquid that passes through the filter paper.
3. Add 10 g of boneblack to the remaining solution.
4. Stir and filter this solution into a third beaker. Observe the color of the liquid.

OBSERVATIONS AND CONCLUSIONS
1. Compare the color of the two liquids that passed through the filter paper.
2. Explain what you observe.

Potable water can be prepared by desalting sea water. The oceans provide resources other than minerals. Perhaps the most abundant of these resources is potable water derived from salt water. The making of drinking water is a very old process that dates back more than 3,000 years. Today, several desalination plants using different systems are in operation. Among the methods used are distillation, ion exchange, reverse osmosis, and freeze separation.

You know that small amounts of water can be purified in the laboratory by distillation. On a large scale, however, this process can be very costly and is used only in special situations. The high cost of distillation is due to the large amounts of energy needed to heat the water (Figure 8-11).

Figure 8-11 This water plant makes millions of liters of potable water a day from brackish (salty) water by the distillation process. What is a disadvantage of this method?

Ion exchange is another method of making potable water. In this method, two ionic compounds in solution are mixed. The ions of the same charge exchange places, resulting in new compounds. One of these new compounds may precipitate and can be separated from the solution. Ion exchange is also used to soften hard water. You will learn more about hard water in the next section.

Osmosis is a process that occurs in nature. Water naturally passes through a thin membrane, or skin, from an area of greater concentration of water (in this case, the fresh water) to an area of lesser concentration of water (the salt water). **Reverse osmosis** is a simple way of reversing nature's normal process and is illustrated in Figure 8-12. Salt water under pressure flows to a membrane with pores just large enough for water molecules. Water is able to pass through the membrane and the salt (larger sodium and chloride ions) is left behind. This process also removes bacteria from the salt water. Reverse osmosis uses a small amount of energy and shows great promise for making large amounts of potable water from sea water.

Freeze separation is another method by which fresh water can be obtained from ocean water. When water freezes it forms ice crystals that are almost free of salt. The ice is then melted to obtain fresh water. It requires less energy to freeze sea water than to distill it.

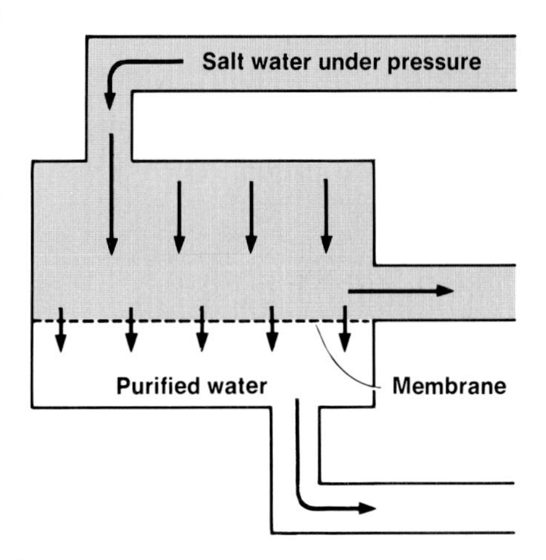

Figure 8-12 With reverse osmosis, pressure is applied to salt water to overcome natural osmotic pressure and force fresh water through a thin cellulose membrane.

8.2 HARD AND SOFT WATER

Water is either hard or soft. Making water potable is only one problem. Many communities face still another major task. If you live in a section of the country where the water is hard, then you are well aware of the trouble it causes. **Hard water** is water that precipitates, or curdles, soap due to a high mineral content.

The word hard means it is hard to make suds with. Clothes cannot be properly cleaned with ordinary soap in hard water. Furthermore, the curd that forms leaves a greasy film on the fabric and gives it a dingy look. Distilled water is free of hardness and is called **soft water.** Soft water has a low mineral content and easily makes suds with soap.

Hard water contains certain minerals. As rainwater falls, it dissolves a small amount of the carbon dioxide in the air. When the water soaks through the ground, it dissolves more carbon dioxide from dead roots and other plant matter. If this carbon dioxide solution flows over limestone ($CaCO_3$), a reaction takes place forming calcium hydrogen carbonate [$Ca(HCO_3)_2$]. The calcium hydrogen carbonate dissolves in the water, forming calcium ions (Ca^{2+}) and hydrogen carbonate ions (HCO_3^-). The chemical equation for this reaction is

$$CO_2(g) + H_2O(l) + CaCO_3(s) \rightarrow Ca(HCO_3)_2(aq) \rightarrow Ca^{2+}(aq) + 2HCO_3^-(aq)$$

Calcium ions in solution are the most common cause of hard water. When soap is added to hard water, a curd forms. This sticky material shown in Figure 8-13 forms as a product of the reaction of the calcium ions in the water with the soap. As a result, an insoluble calcium compound is formed. You may have seen this curd as the greasy ring that forms around the washbowl or bathtub. Soluble magnesium and iron salts also produce ions that cause hardness in water.

To get a lasting lather with soap and hard water, all of the curd-forming ions must first be precipitated out of solution. More soap is required to form the precipitate. Obviously, this method of taking out the unwanted ions is wasteful, and better means of softening water exist.

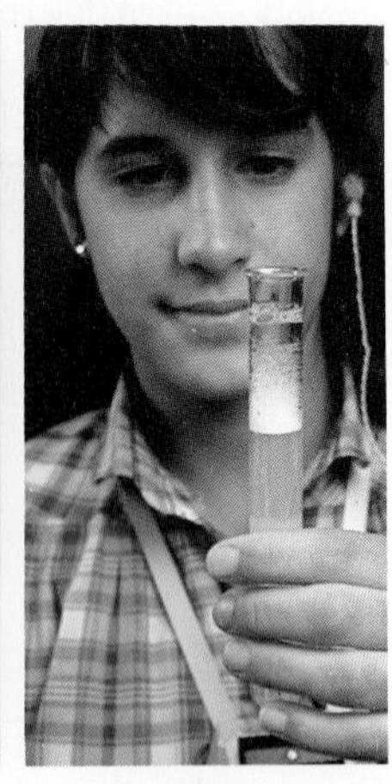

Figure 8-13 Equal amounts of soap powder were added to hard water (left) and to soft water (right). Notice the difference in results.

Some water can be softened by boiling. Hard water containing calcium hydrogen carbonate can be softened by boiling. Boiling causes the minerals to settle out when the calcium hydrogen carbonate decomposes. This breakdown is shown by the following equation:

$$Ca(HCO_3)_2(aq) \xrightarrow{heat} H_2O(l) + CO_2(g) + CaCO_3(s)$$

The carbon dioxide is given off into the air and the calcium carbonate precipitate settles out. The action of the soap is not affected by the calcium carbonate because calcium carbonate is insoluble and does not break down into calcium ions. However, softening large amounts of water containing calcium hydrogen carbonate by boiling is costly because it uses too much energy.

Most soft water is prepared chemically. The calcium, magnesium, or iron ions that cause hard water can be removed by adding soda ash, slaked lime, or borax. A typical reaction for the removal of the calcium ions using slaked lime [$Ca(OH)_2$] is

$$Ca(HCO_3)_2(aq) + Ca(OH)_2(aq) \rightarrow 2H_2O(l) + 2CaCO_3(s)$$

The calcium ions in solution are precipitated as insoluble calcium carbonate and settle to the bottom of the container. Many cities and industries use slaked lime to soften the water as well as to make it potable. This method is an effective way of softening large amounts of hard water. Soda ash and borax work just as well for softening water. They are convenient when small amounts of soft water are needed for home use.

Certain sodium compounds exchange their sodium ions for calcium ions. In this ion-exchange process, hard water containing calcium ions flows through a tank containing particles of a compound known as sodium zeolite. The calcium ions in the hard water replace the sodium ions in the zeolite, as described by the following word equation:

$$\text{calcium ion} + \text{sodium zeolite} \leftrightharpoons \text{calcium zeolite} + \text{sodium ion}$$

The calcium ions form insoluble calcium zeolite, which remains in the water softening tank. The sodium ions that are set free have no effect on the hardness of water.

When all of the sodium zeolite in the tank is used up, a concentrated solution of sodium chloride is added to the tank. The chemical action is then reversed, changing the calcium zeolite back into sodium zeolite. The zeolite can be used repeatedly. After salt is added, the tank is again ready to soften the water. The zeolite-type softener is used in both homes and industries.

Industries need to use soft water. Hard water may be no more than a nuisance at home, but it is a major problem in many industries. It is most troublesome in steam boilers and

Figure 8-14 Mineral deposits of magnesium and calcium salts on this heating element reduce the efficiency of the hot-water heater.

hot water pipes. Minerals, which are deposited due to heating of the hard water, clog the pipes. The deposits shown in Figure 8-14 insulate the boiler from the source of heat, thus reducing its efficiency. For this reason, water-softening methods are vital to industry. Steam-electric power plants must use carefully treated water in their boilers.

Paper mills need a supply of water that is free of iron compounds, which may stain the paper. Other compounds may keep the paper from having a smooth, glossy surface. Soft water is also needed in textile mills to produce exact shades of color evenly spread through the yarn and cloth.

In homes, synthetic detergents have almost totally replaced soap for laundry use. These detergents work well in both hard and soft water. The major advantage of these cleaning agents is that they do not form a scum with the calcium ions in hard water.

8.3 SEWAGE: TREATMENT AND DISPOSAL

Waste water must be treated. When a large volume of rain or snow falls on an urban area, the water is drained from roofs, yards, and streets into storm sewer lines. These lines normally carry the water to nearby rivers or streams, or into other large bodies of water. Sewer water carries trash, organic matter, and chemicals. Pieces of rock, soil, metal, oil and grease, fertilizers, pesticides, and microorganisms may also be carried by the water.

To dispose of human waste and other organic and inorganic matter, as well as detergents, another system of pipes, called sanitary sewer lines, is used. These wastes are treated in sewage disposal plants before the waste water is discharged into large bodies of water. In some cities, the sanitary and storm sewer lines are combined. This combination of lines causes many problems. When rain falls or snow melts, the amount of water in the single sewer system increases greatly. This combined waste water and runoff can easily overload the sewage treatment plant. When such an overload occurs, the excess bypasses the plant and takes the raw sewage with it. Thus, the river or lake that receives the overflow becomes polluted.

The waste water includes a rich source of phosphate ions (PO_4^{3-}) from detergents and fertilizers, and nitrate ions (NO_3^-) from organic matter. These nutrient ions, when dumped into bodies of water, speed the growth of algae and other water plants. When these plants die and decompose, the dissolved oxygen in the water is used up. Fish and other animal life may die from the lack of oxygen. Plant growth and decay take place at a much slower rate. Increased amounts of nutrient ions from untreated sewage speed the "death" of many lakes and streams. This process is called **eutrophication** (YOO-*troe-fuh*-KAY-*shun*) (Figure 8-15).

Figure 8-15 University of Minnesota researchers are shown testing samples of water from a peat bog that is threatened by mining activities. Nutrient ions added to the water ecosystem could cause eutrophication.

In recent years, ocean water has also become polluted from raw sewage. Many communities along the Eastern seaboard, particularly the New Jersey shore, have suffered because of ocean dumping. Beaches and waters became polluted and were often closed to the public, causing local communities to lose millions of tourist dollars.

Since untreated sewage contains bacteria and other material harmful to health and safety, it must be treated before it is discharged into rivers, lakes, and oceans. The goal in sewage treatment is to make the waste water as much like potable water as possible.

In the past, most sewage was dumped into the waterways without treatment. As long as the amount of sewage was fairly low, the natural action of water movement purified the water. As the volume of sewage grew, however, nature was no longer able to do the job effectively. In recent years, people have insisted on full sewage treatment before discharge into the waterways.

There are three basic steps in sewage treatment. These steps are called primary, secondary, and tertiary.

Primary treatment of sewage is the first step. In primary treatment, the waste is first fed through a grit chamber, where huge grates screen out the large, bulky matter. Then the liquid goes to settling tanks. There, suspended material is allowed to settle. The solid material that settles as sludge on the bottom of the tanks is then pumped into large tanks, called digesters. In the digesters, the bacteria in the sludge break down any existing organic matter. This solid material is then dried and used as fertilizer. After primary treatment, the liquid may either be dumped into the waterways or passed on for secondary treatment. Chlorine may be added

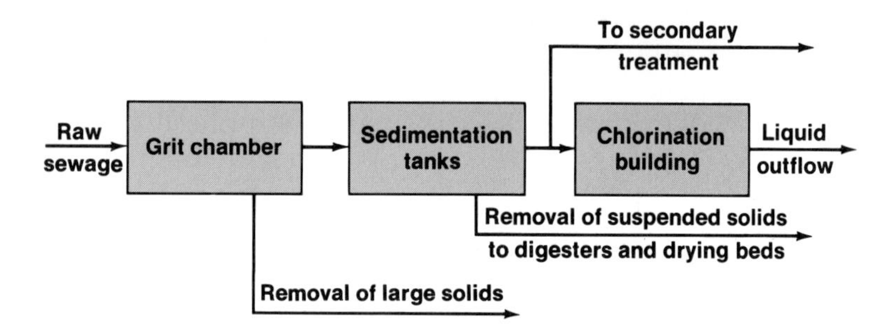

Figure 8-16 A flow chart showing solid waste removal in primary sewage treatment. What is the value of the chlorine treatment?

before discharge in order to kill most of the disease-causing bacteria and to reduce odors. Figure 8-16 shows a simplified flow chart of this filtering process.

Secondary treatment is the next step. Primary treatment of sewage by itself is not adequate. In secondary treatment, up to 90 percent of the organic material is removed from the sewage after it comes from the settling tanks.

One method used to remove the organic material is to allow the waste water to pass through gravel filters to an aeration tank, where it is mixed with oxygen of the air. This aeration allows bacteria to break down more organic material. After passing through another settling tank where more solids and suspended materials are removed, the liquid is treated with chlorine to complete the secondary treatment. The liquid is then discharged into the waterway or passed to the third stage of treatment.

The third step is called tertiary treatment. Tertiary is the last stage of sewage treatment and the most advanced. This process is complex and costly. In this step, the phosphate and nitrate ions are taken out of the water. Reverse osmosis, ion exchange, and distillation are among the methods used in tertiary treatment.

Figure 8-17 A sewage treatment plant where tertiary treatment removes almost all pollutants. Describe what is done to the sewage in this stage of treatment.

In the ion-exchange method, the phosphate and nitrate ions are replaced with chloride (Cl^-) and carbonate (CO_3^{2-}) ions, which do not promote increases in plant growth. When water is finally discharged from tertiary treatment, it is almost free of pollutants and very close to drinking-water quality. Figure 8-17 shows one type of tertiary treatment plant. Scientists are now studying methods of turning raw sewage into potable water.

Sewage treatment produces useful by-products. The dried sludge from the sewage that is taken from the secondary and tertiary steps contains many nutrients that can be sold as fertilizers when properly treated. Much of the solid, rocky material can be used in road building, for making building blocks, or for land fill.

ACID RAIN

The high concentration of sulfuric and nitric acids in the atmosphere is responsible for the deterioration of this statue in Chicago.

a problem of our own making

Item: A once crystal-clear lake in Ohio becomes a soggy swamp choked with stagnant clumps of green algae. At the bottom of the lake, a carpet of decaying algae consumes most of the oxygen in the water. Robbed of this life-sustaining gas, the fish that live in the lake begin to die. Researchers estimate that 3,000 lakes in the eastern United States are suffering similar fates.

Item: In a major city in the northeastern United States, air pollution is causing more and more cases of lung disease. Levels of toxic metals, such as mercury, aluminum, and lead, are increasing in the city water supplies.

Item: Approximately 16,000 km^2 of a once lush forest in Germany lay barren as if swept by some deadly plague.

What is the cause of all this devastation? Acid rain! As you have learned in this chapter, acid rain is the general term for precipitation—rain, snow, hail, sleet—that contains high levels of sulfuric and nitric acids. These acids can eat away at buildings, leach metals from soil or water pipes and carry the metals into our drinking water, irritate human lungs, kill trees, and raise the acidity of lakes, making them unfit for life.

What can be done to stop acid rain? We could stop burning high-sulfur coal to produce energy, since this activity is a primary source of the sulfur dioxide in acid rain. However, completely eliminating the use of coal means jobs lost for coal miners. Another option is to reduce the sulfur content in coal before it is burned. The disadvantages of this process are higher manufacturing costs and reduction rather than elimination of sulfur pollution. Another option is to use special equipment known as scrubbers to remove the sulfur dioxide from exhaust gases before those gases are released into the air. Unfortunately, the added expense of installing and maintaining the scrubbers would be passed on to the consumer. Yet another option might be to rely more on alternative sources of energy such as water, wind, solar, and nuclear power. But these sources of energy are not yet widely available and, even if they were, they would be more costly than coal.

FOR FURTHER RESEARCH AND DISCUSSION

How would you try to solve the acid rain problem without increasing fuel costs? Is it possible to reverse the damage?

CHAPTER REVIEW

SUMMARY

1. A hydrogen bond is a weak attraction between the negative oxygen ends of a water molecule and the positive hydrogen ends of other water molecules.
2. Potable water is water that is fit to drink.
3. Distillation is a process of evaporation of a liquid followed by condensation of its vapors in a separate container.
4. Community water supplies are made potable by settling, filtration, aeration, and by the use of chlorine and other chemicals.
5. Hard water is water containing ions such as calcium and magnesium, which form precipitates with soap.
6. Among the methods used in desalting sea water are distillation, ion exchange, reverse osmosis, and freeze separation.
7. The three basic stages of sewage treatment are (1) screening and settling, (2) removal of organic material by filtration and oxidation, and (3) removal of nutrient ions.

VOCABULARY

Match the item in the left column with the best answer in the right column. Do not write in this book.

1. aeration	a. water that is safe to drink
2. distillation	b. attraction among water molecules
3. freeze separation	c. mixture of liquid and solid particles
4. hydrogen bond	d. process of evaporating liquid and then condensing its vapors
5. ion exchange	e. removing salt from water by freezing
6. nonpolar	f. filtered water sprayed into the air
7. polar	g. molecule with a bent structure
8. potable water	h. salt water is forced through a membrane
9. reverse osmosis	i. replacing an ion with another of like charge
10. suspension	j. molecule with a straight-line structure

REVIEW QUESTIONS

1. What is usually meant by "pure" water? Why is rainwater not entirely pure?
2. What makes the oceans salty?
3. What is the difference between organic and inorganic matter?
4. What is acid rain? Why is it harmful?
5. Why is organic matter dissolved in water harmful to health?
6. How is water distilled?
7. List five steps commonly used for making city water potable. Describe the purpose of each step.

8. What chemicals are used to settle suspended particles in water? Describe their action.
9. Describe a sand filter. How is it used to purify water?
10. What is meant by hard water? Describe the difference between hard water and potable water.
11. What causes a deposit to form in steam boilers and hot water pipes? Why are these deposits harmful?
12. Why is soft water needed in paper and textile plants?
13. What are some methods used to remove salt from ocean water?
14. Describe reverse osmosis as a method used to make water potable.
15. What does sewage consist of? Why should it be treated?
16. Distinguish between a storm sewer and a sanitary sewer system.
17. What nutrient ions that cause plant growth are contained in sewage? Where do they come from? Why should these ions be removed?
18. What does primary treatment of sewage usually consist of? Is this treatment sufficient for most purposes? Explain.
19. Explain why water is such a good solvent.
20. What is the difference between a solution and a suspension? Which could you separate by filtering? Explain.
21. How do calcium ions get into the water supply? Do these calcium ions make water harmful for drinking?
22. How can water that contains calcium hydrogen carbonate be softened by boiling? Write the equation for this reaction.
23. Explain how slaked lime softens water containing calcium hydrogen carbonate. Write the equation for this reaction.
24. How does sodium zeolite soften water by the ion exchange method? How is a zeolite unit made reusable?
25. Describe how potable water may be obtained by freezing salt water.
26. Describe in order the three steps of sewage treatment.
27. What can be made of the dried sewage sludge from secondary and tertiary treatment? What can be made of the rocky, insoluble material?
28. Why is sewage treated with chlorine before it is discharged into rivers and streams?
29. Why is water a liquid at room temperature?

CRITICAL THINKING

30. How would you set up an experiment to find the percent by mass of solid material in a sample of river water?
31. You are given a sample of liquid that has no color, odor, or taste. How could you prove the liquid is water?

FURTHER READING

Arnov, B. *Water: Experiments to Understand It.* New York: Lothrop, Lee, and Shepard Books, 1980.

Gunston, Bill. *Water.* Morristown, N.J.: Silver Burdett, 1982.

Keough, Carol. *Water Fit to Drink.* Emmaus, Pa.: Rodale Press, 1980.

Lehr, J. H., et al. *Domestic Water Conditioning.* New York: McGraw-Hill, Inc., 1979.

McCormick, John. *Acid Rain.* New York: Gloucester Press, 1986.

METALLURGY

This gold jaguar was hand crafted in Peru sometime between the ninth and fifth centuries B.C. Ancient cultures used gold and other metals to create ornaments, tools, and weapons because of the metals' attractive luster and ability to be worked into many forms. How are metals used today?

CHAPTER 9

SECTIONS

OBJECTIVES

- ☐ Describe the processes of making pig iron and steel.
- ☐ Summarize the chemical and heat treatments used to produce desired properties in refined metals.
- ☐ Explain how electrolysis is used to refine copper and aluminum.
- ☐ Relate the properties of metals and alloys to the uses of these materials.

9.1 IRON

Iron is the fourth most abundant element in the earth's crust. You learned in Chapter 2 (Table 2-1) that iron accounts for 4.7 percent of the earth's crust. This element is rarely found in its pure, or free, state. Instead, it is found in a variety of ores (compounds). Iron ore is important because steel, which is made from iron, has many industrial uses.

Iron and other metals have been in use for thousands of years. During this time people have found many sources of metals and many ways to process them. **Metallurgy** (MET-*ul*-UR-*jee*) is the science of taking useful metals from their ores, refining the metals, and preparing them for use. Not every piece of rock that contains a useful metal is an ore. An **ore** is a rock or mineral from which a metal can be obtained profitably.

There are several common ores of iron. The most important iron ore is a reddish-brown mineral called hematite (HEM-*muh*-TITE) (Fe_2O_3). This iron ore may contain up to 70 percent iron. For many years, the main source of hematite in the United States was the area around Lake Superior.

Figure 9-1 An open-pit mine near Lake Superior. What are some environmental problems of open-pit mining?

One such mine is shown in Figure 9-1. Most of the Lake Superior hematite has been extracted. The ore is now imported from Quebec, Labrador, and South America.

The second most important iron ore is magnetite (MAG-*nuh*-TITE) (Fe_3O_4). This ore may contain up to 72 percent iron. Magnetite is so named because it is a natural magnet. Other less important ores of iron are limonite (LIE-*muh*-NITE), siderite (SID-*uh*-RITE), and pyrite. Limonite ($Fe_2O_3 \cdot 3H_2O$) is about 63 percent iron, and siderite ($FeCO_3$) is only 48 percent iron. Pyrite (FeS_2) is about 50 percent iron. It has a shiny, metallic appearance and is also referred to as "fool's gold."

A low-grade iron ore called taconite (TAK-*uh*-NITE) is a hard rock that contains only 25 percent to 50 percent iron. Taconite contains layers of iron oxide and silica (SiO_2). The rock must be crushed before it can be refined. This ore is becoming more important as other ores are used up.

Iron is refined in a blast furnace. The process of separating a metal from the other materials in its ore is known as **refining.** Iron ores contain oxygen combined with iron. To get rid of this oxygen, the ore must be refined by heating it with a reducing agent. A **reducing agent** is a substance that removes oxygen from a compound. This process of refining iron ore takes place in a blast furnace, as illustrated in Figure 9-2.

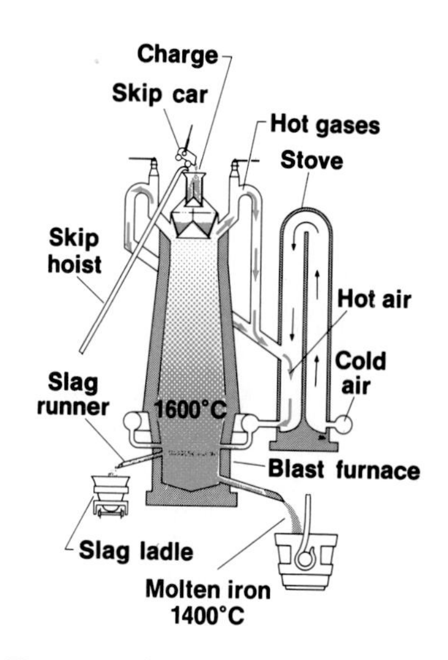

Figure 9-2 Section of a blast furnace. A blast furnace is about 10 stories high.

Coke, which is a fuel made from soft coal, produces the reducing agent used to refine iron ore. The coke is burned in a blast furnace. The carbon in coke combines with oxygen, first producing carbon dioxide (CO_2) and then carbon monoxide (CO). Carbon monoxide acts as the reducing agent, removing the oxygen from the ore. The process of refining hematite is shown by the following reactions:

Step 1. Burning of coke, which is mostly carbon (C):

$$C(s) + O_2(g) \xrightarrow{\text{heat}} CO_2(g)$$

Step 2. Formation of carbon monoxide from carbon dioxide and coke:

$$CO_2(g) + C(s) \xrightarrow{\text{heat}} 2CO(g)$$

Step 3. Carbon monoxide acts as the reducing agent and removes the oxygen from the hematite ore:

$$Fe_2O_3(s) + 3CO(g) \xrightarrow{\text{heat}} 2Fe(l) + 3CO_2(g)$$

The carbon dioxide formed as one of the products in Step 3 reacts with coke to form more carbon monoxide.

Burning coke releases a large amount of heat. This heat melts the iron, but it does not melt impurities in the iron ore. The kind and amount of impurities vary. Limestone ($CaCO_3$) is added to the blast furnace to remove the most common impurity, sand. Sand is mostly silicon dioxide (SiO_2) or quartz. When heated, silicon dioxide reacts with limestone to form calcium silicate ($CaSiO_3$) and carbon dioxide. The easily melted calcium silicate is a glassy waste product called slag. The reaction is:

$$SiO_2(s) + CaCO_3(s) \xrightarrow{\text{heat}} CaSiO_3(l) + CO_2(g)$$

The chemical reactions in a blast furnace take place continuously. Iron ore, coke, and limestone are poured into the top of the huge blast furnace using a skip hoist and skip car. This mixture of raw materials is called a charge. As materials react and melt, they sink to the bottom of the furnace. Because it is less dense, the slag floats on top of the liquid iron. Both of these liquids are drawn off at the bottom of the blast furnace, as shown in Figure 9-2.

The iron drained from the blast furnace is called pig iron. It contains about 92 percent to 94 percent iron. The rest is comprised of impurities such as carbon, manganese, phosphorus, silicon, and sulfur. Pig iron can be remelted and poured into molds. Iron used in this way is called cast iron. Cast iron is used to make heavy machinery and some kinds of pots and pans.

SAMPLE PROBLEM

What percent of iron is contained in a sample of pure magnetite (Fe_3O_4)?

SOLUTION

Step 1: Analyze

You are given the chemical formula of the substance magnetite. The percent of iron found in a pure sample of this substance is the unknown.

Step 2: Plan

The law of constant proportions (Chapter 2) states that every compound always contains the same proportion by mass of the elements of which it is formed. Since you know magnetite is Fe_3O_4 and can find the mass numbers of Fe and O on the periodic table (Figure 3-6), you can calculate the total mass of a molecule of magnetite and the proportion by mass of iron (Fe) in that molecule.

Step 3: Compute and Check

mass numbers: Fe = 56 u, O = 16 u

$$\text{mass of } Fe_3O_4 = 3(\text{Fe}) + 4(\text{O})$$
$$= 3(56 \text{ u}) + 4(16 \text{ u})$$
$$= 232 \text{ u}$$

$$\text{percent iron} = \frac{3(\text{Fe})}{Fe_3O_4} \times 100$$
$$= \frac{3(56 \text{ u})}{232 \text{ u}} \times 100$$
$$= 72 \text{ percent}$$

Since 72 percent agrees with the maximum percent of iron given in the text for magnetite, your solution is complete.

Iron from the blast furnace is used in making steel. An important difference between steel and iron is that steel contains fewer impurities. Steel always contains a small but measured amount of carbon. Steel may also contain small but controlled amounts of other metals added to give certain desired properties.

Figure 9-3 Iron from a blast furnace is being poured into a basic-oxygen furnace.

The basic-oxygen furnace turns pig iron into steel. As shown in Figure 9-3, molten pig iron from the blast furnace is poured into the basic-oxygen furnace. The pig iron is added through the pouring outlet shown in Figure 9-4. Limestone is added, and then oxygen is blown into the molten material through a lance to burn away the unwanted matter. No fuel is needed because the reaction between oxygen and the impurities releases so much heat that the mixture stays in a molten state.

After about 40 minutes to 60 minutes, the huge furnace is tilted and about 80 metric tons of finished steel are poured into large blocks. The steel can also be made directly into other products.

Because the basic-oxygen furnace maintains such high temperatures, it can melt pieces of used steel such as parts of car bodies and engines. As much as 40 percent of the metal processed in a basic-oxygen furnace is used steel. Thus, reusing steel decreases the amount of raw iron ore resources used. It also saves energy.

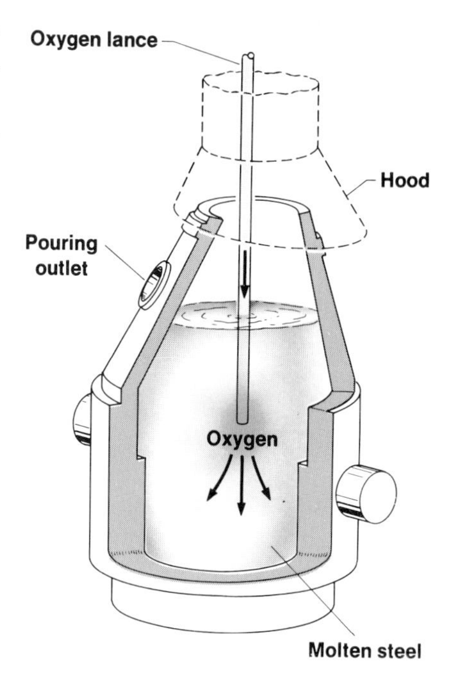

Figure 9-4 The basic-oxygen furnace turns molten pig iron into steel. What is the purpose of the oxygen?

Carbon and heat treatment give steel certain desired properties. The amount of carbon in steel varies from about 0.05 percent to 2 percent. If the percent of carbon is low, the product is very soft steel, such as that used in making paper clips. Steel with a high amount of carbon is hard and very brittle. The carbon crystals make the steel hard because

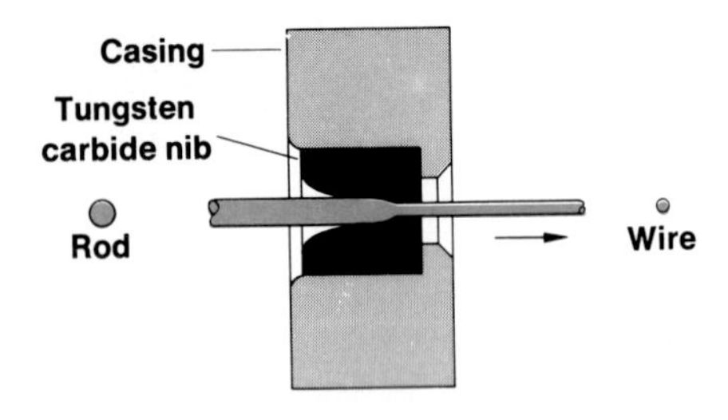

Figure 9-5 Cross section of a steel rod being drawn through a die to produce wire. Note the different cross sections of the rod and the wire. What kind of steel is needed to make wire?

they keep the atoms of iron from sliding past each other. The hardness of steel depends not only on the amount of carbon but also on how the carbon is joined with the iron, which is determined by the crystal structure after heat treatment.

Almost every type of steel must be given some form of heat treatment. The three main methods used in heat treatment of steel are quenching, tempering, and annealing.

Quenching is a heat treatment in which steel is heated to a given temperature and then plunged into water or oil for sudden cooling. This process makes the steel strong, hard, and brittle.

Although quenched steel is very hard, it is too brittle for many needs. The quenched steel can be reheated to a point below the quenching temperature and then recooled. The higher the reheating temperature, the softer the finished steel will be. The process of reheating and recooling quenched steel is called tempering. This process makes very strong steel. Tempered steel is used for many cutting devices, such as knives and scissors.

Annealing is a process in which heated steel is cooled slowly, first in a furnace and then in air. This process softens the steel so that it can be shaped into many objects, such as wrenches and other tools. The shaped object can then be hardened by heat. Annealed steel can also be made into wire. Figure 9-5 illustrates a rod of annealed steel being drawn, or pulled, through a device called a die. Compare the diameter of the wire with the diameter of the rod from which it was made.

Iron and steel can be oxidized. Unprotected iron and steel objects corrode, or rust, in moist air. The rust that forms is brittle so it flakes off. This exposes the metal below it to further rusting until the iron has rusted throughout.

The rusting of iron is an oxidation process. Iron combines with oxygen in the air, forming iron(III) oxide (Fe_2O_3), the same chemical compound as hematite. This reaction takes place slowly, although rainwater containing dissolved carbon dioxide can make the reaction proceed somewhat faster. The reaction is shown by this equation:

$$4Fe(s) + 3O_2(g) \rightarrow 2Fe_2O_3(s)$$

If iron can be protected from oxygen, the surface will not rust. The simplest way to protect iron surfaces is to cover them with paint. Keeping iron objects dry will also help stop rusting.

Iron can be oxidized by many kinds of foods. The containers that are commonly called "tin cans" are actually made of steel. However, they are lined with a thin layer of tin, which keeps the food from reacting with the steel.

activity

HEAT TREATMENT OF STEEL

OBJECTIVE: Compare the effect of two different heat treatments on steel.

PROCESS SKILLS

In this activity, you will *observe* the behavior of heat-treated steel and *infer* the effects of heat treatment on steel.

MATERIALS

safety goggles
gas burner
striker
2 bobby pins
tongs
beaker
water

PROCEDURE

1. Put on your safety goggles. Using tongs to hold a bobby pin, heat the bent portion of the pin to redness in the flame of a gas burner.
2. Cool the pin quickly by dipping it into a beaker of cold water.
3. Heat the bent portion of the second bobby pin to redness in the flame of a gas burner.
4. Remove the pin slowly from the hottest part of the flame and then allow the pin to cool slowly in the air.
5. Once cooled, test the flexibility of each bobby pin by pulling the points apart.

OBSERVATIONS AND CONCLUSIONS

1. Compare the flexibility and brittleness of the two heat-treated pins.
2. How does heating followed by rapid cooling affect steel? What is this treatment called?
3. How does heating followed by slow cooling affect steel? What is this treatment called?

9.2 COMMON METALS

Aluminum is obtained by the Hall process. Like iron, aluminum is not found in its free state. Aluminum is obtained from an ore called bauxite (BAWKS-*ite*). Bauxite contains 45 percent to 60 percent aluminum oxide (Al_2O_3).

Bauxite ore is often found in shallow deposits close to the earth's surface. The richest deposits are found in a wide band along either side of the equator. At present, large quantities of bauxite come from Jamaica, Surinam, and Guyana. In the United States, bauxite is mined in Arkansas, Georgia, and Alabama.

In 1886, Charles Martin Hall, a 23-year-old American, found an easy way of separating aluminum from its oxide ore. He mixed aluminum oxide with another aluminum compound, cryolite (KRIE-*uh*-LITE) (Na_3AlF_3). He found that when aluminum oxide and cryolite are heated to about 1,000° C, the molten mixture conducts an electric current. The current causes the aluminum metal to separate from the oxygen in the ore. The aluminum sinks below the mixture, and the oxygen is released. This reaction is shown by the following equation:

$$2Al_2O_3(l) \rightarrow 4Al(l) + 3O_2(g)$$

In Chapter 5 you learned that this kind of reaction is called a decomposition reaction. When an electric current is used to cause a decomposition reaction, the process is called electrolysis. The electrolysis of aluminum is diagrammed in Figure 9-6.

Part of the oxygen released from the ore combines with the positively charged carbon rods. Carbon monoxide, which burns on contact with the air, is formed. As electrolysis continues, the carbon rods are used up and must be

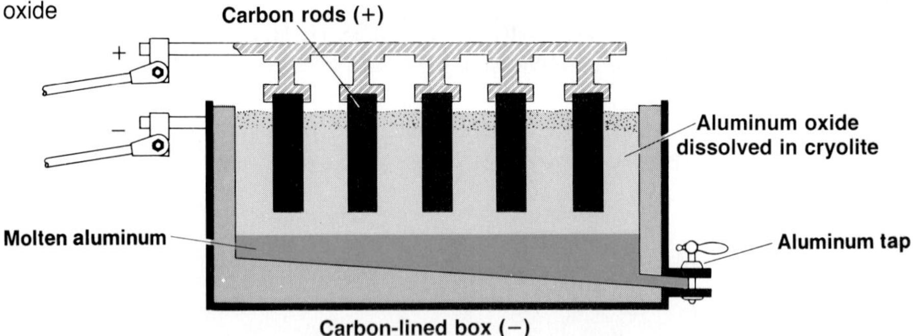

Figure 9-6 Aluminum is obtained by electrolysis of purified aluminum oxide and cryolite.

replaced. The molten aluminum collects at the negatively charged bottom of the container and is removed from there. This electrolysis process uses very large amounts of electric energy.

Aluminum is nonpoisonous and lightweight. It heats up and cools down quickly. Because of these properties, aluminum is commonly used in food packaging. Most beverage cans are aluminum and many other foods are wrapped in aluminum foil. Aluminum foil is also used in the walls and ceilings of buildings to keep heat out in the summer and hold heat in during the winter. Aluminum foil is used in this way because it reflects much of the heat energy that hits it. Aluminum is also used to make cooking utensils, automobile parts, and building materials such as siding, screens, window frames, and roofs.

Like iron, aluminum can be oxidized. But unlike iron oxide, aluminum oxide does not flake off the surface of the metal. Instead, it forms a thin layer that protects the metal below it. This makes aluminum very durable.

Aluminum is the most commonly recycled metal. Used aluminum cans are melted and formed into new products. Why does this save energy?

Copper occurs in a free state or in copper ores. Copper is rarely found as a free (native) metal, meaning a metal that is not chemically combined with other substances. Most copper comes from several kinds of ores. The most important copper ores are the sulfur compounds chalcopyrite (KAL-*koe*-PIE-*rite*) ($CuFeS_2$) and chalcocite (KAL-*kuh-site*) (Cu_2S). Another ore, cuprite (KYOO-*prite*) (Cu_2O), is an oxide of copper.

Deposits containing copper are removed from what are called open-pit mines at the surface and from mines deep beneath the surface. In the United States, most of the free copper and copper ore is found in Michigan, Arizona, Utah, Nevada, and New Mexico.

Free copper is found in igneous and sedimentary rocks, so it must be processed to be used. The rock is first crushed to a coarse powder. The lighter rock is then removed by a stream of water and the heavier copper metal remains. This step concentrates the copper.

The second step mixes the concentrated copper with coke and limestone in a small blast furnace. In the furnace, limestone unites with silica impurities to form slag, in the same way slag is formed when iron ore is reduced in a blast furnace. The burning coke melts the copper, which collects at the bottom of the furnace. The molten copper is then drawn off and cast into large plates. The copper is further refined by electrolysis.

Figure 9-7 In the ore-flotation process, the particles of Cu_2S stick to the air bubbles and are skimmed off. What settles to the bottom of the tank?

The following steps describe how copper is obtained from chalcocite, which is copper(I) sulfide (Cu_2S):

1. The ore is concentrated. The ore-flotation process is used to separate copper(I) sulfide from the rock (Figure 9-7). In this process, crushed ore, water, and clean oil are mixed in huge tanks. When air is blown through the mixture, bubbles form on top. The bits of copper(I) sulfide cling to the air bubbles on top, while the wet waste material, called tailings, settles to the bottom. Then the copper(I) sulfide is skimmed off into other tanks.

Figure 9-8 Blister copper being poured into 2,750-kg slabs from a blast furnace. What is the next step in making refined copper?

2. The Cu_2S is roasted. **Roasting** is a process in which a sulfide ore is heated in oxygen-enriched air to change it into an oxide. First, the Cu_2S is heated with oxygen to change most of it into copper(I) oxide (Cu_2O). Second, as in the process of refining iron, limestone is added to remove silicon dioxide. The equations for these two reactions are shown below.

$$2Cu_2S(s) + 3O_2(g) \rightarrow 2Cu_2O(l) + 2SO_2(g)$$

$$SiO_2(s) + CaCO_3(s) \rightarrow CaSiO_3(l) + CO_2(g)$$

A mixture of mostly copper oxide and some copper sulfide results.

3. The mixture is reduced. In a special blast furnace, the reduction of the mixture of the sulfide and oxide produces copper metal. The equation for the reaction is

$$Cu_2S(l) + 2Cu_2O(l) \rightarrow 6Cu(l) + SO_2(g)$$

As the molten copper cools into slabs, the escaping gases cause blisters on the surface of the metal. The slabs are therefore called blister copper (Figure 9-8).

4. The metal is refined. Copper is refined by electrolysis. Impurities in crude copper make it a very poor conductor of an electric current, so it must be purified. All crude copper is refined by electrolysis in a tank similar to the one in Figure 9-9. A tank of copper(II) sulfate solution contains a small amount of sulfuric acid to help conduct an electric current. The tank contains slabs of positively charged crude copper and thin sheets of negatively charged pure copper, referred to as starter sheets.

The electric current causes the copper ions (Cu^{2+}) dissolved in the solution to move to the negatively charged pure copper sheets. There, the ions are deposited in layers of pure copper (Cu). As the process continues, the positively charged impure copper slabs continue to supply copper ions for the solution.

During electrolysis, the slabs of crude copper slowly get smaller as the atoms of copper go into solution as copper ions. Meanwhile, the copper sheets increase in size as copper ions from the solution are deposited as pure copper. The copper obtained by this process is over 99.9 percent pure, which is the level of purity required for use in electrical conductors. Impurities from the crude copper drop to the bottom of the tank as a muddy deposit. This mud often contains gold and silver, valuable by-products of copper refining.

Copper is a reddish metal that is often covered with a brownish coat of tarnish. When you rub off the tarnish with sandpaper, you see the red color of pure copper underneath. Copper exposed to the weather becomes covered with a greenish coating of several copper compounds. These coatings do not flake off, thus they protect the copper underneath from further corrosion.

Copper is most widely used for making electric wire due to its ability to conduct electricity. The electrical industry uses approximately six-tenths of all copper produced. Because it resists corrosion, it is used for water pipes and roofing. Copper is also very malleable, or easy to shape, so it is suited for use in making jewelry and coins.

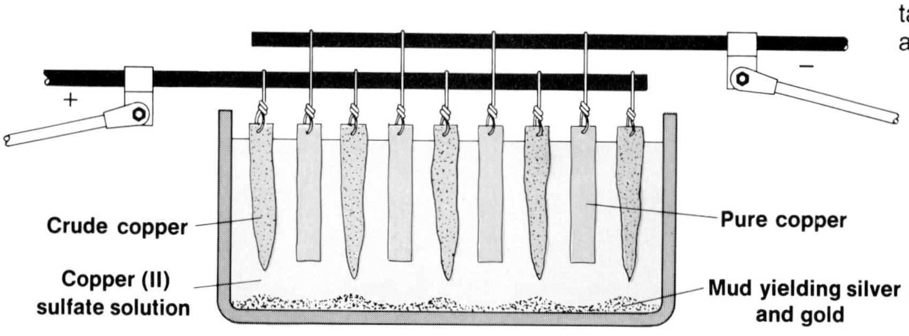

Figure 9-9 Crude copper is refined by electrolysis occurring in a tank that contains negatively charged pure copper and positively charged impure copper.

ELECTROPLATING COPPER

activity

OBJECTIVE: Determine the effect of an electric current on a solution that contains copper ions.

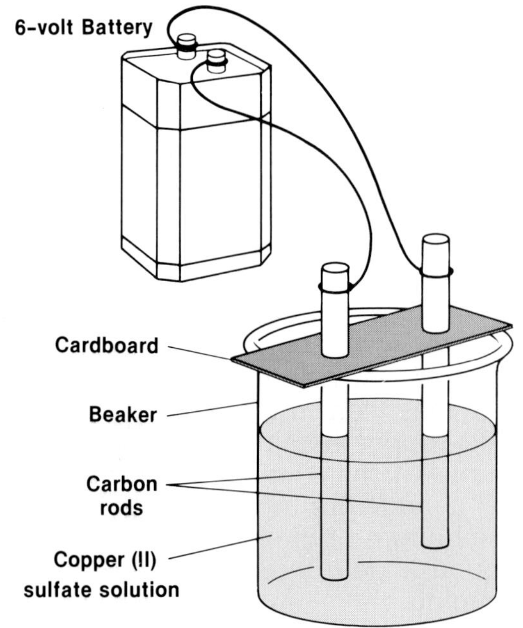

PROCESS SKILLS

In this activity, you will *observe* the process of electroplating and *hypothesize* about the action of metal ions in this process.

MATERIALS

safety goggles
gloves
beaker
copper(II) sulfate solution
sulfuric acid
dropper
2 carbon rods
cardboard
scissors
6-volt battery
2 pieces of wire

PROCEDURE

1. Put on your goggles and gloves. Pour about 50 mL of dilute copper(II) sulfate solution into a beaker. Add several drops of sulfuric acid. CAUTION: *These substances are toxic. Extreme caution should be exercised when handling them.*
2. Using the scissors, make two holes in the cardboard about 4 cm apart. Clean two carbon rods and push them through the holes in the cardboard. Use wire to connect the rods to the battery.
3. Place the rods in the beaker, as shown in the diagram. Observe what happens at each rod.
4. After several minutes, remove the rods from the beaker and disconnect the battery.

OBSERVATIONS AND CONCLUSIONS

1. What did you observe at the rods connected to the positive and negative terminals of the battery?
2. What change did you observe in the solution in the beaker?
3. What happened to the copper ions in the solution?
4. Electroplating involves two reactions, one at each rod. The combined equation for these is given below. Balance the equation, and identify the substance produced at each rod.

$H_2O(l) + Cu^{2+}(aq) \rightarrow O_2(g) + H^+(aq) + Cu(s)$

Silver is a precious metal. Metals that are useful but scarce and costly are called **precious metals.** You probably have heard of silver, gold, and platinum. Other precious metals that are not so well known are palladium, osmium, and iridium.

Silver was used as early as 2,400 B.C., and it was considered to be more valuable than gold by many ancient cultures because it was rarer in the native state. In addition to occurring as a free metal, it is commonly found in an ore called argentite, which is silver sulfide (Ag_2S), and may also occur in lead ores such as galena (PbS). Most commercial silver, however, is a by-product of processing copper, lead, and zinc ores. The best source of silver comes from the muddy deposits that sink to the bottom of the tank in the refining of copper.

When this mud is treated with sulfuric acid, silver sulfate (Ag_2SO_4) is formed. Strips of copper are dipped into the silver sulfate solution and pure silver collects on the strips. The single replacement reaction is

$$Ag_2SO_4(aq) + Cu(s) \rightarrow CuSO_4(aq) + 2Ag(s)$$

Figure 9-10 Silver bullion cast in this factory will be stored in huge vaults for safekeeping. Name two other precious metals.

Figure 9-10 shows silver that has been removed from the strips, melted, and formed into bars called silver bullion, or ingots.

Silver is a soft metal that shines with a beautiful luster. Although silver is the best conductor of electric current, it is too costly for most electric wiring. Like copper, silver is very malleable. As a result, large amounts of silver are used in jewelry and other ornaments. Sterling silver, used in tableware and jewelry, is 92.5 percent silver and 7.5 percent copper. The addition of a small amount of copper gives the very soft metal more durability under constant use.

A less expensive option for tableware is silver plate. It is made by coating inexpensive base metals with pure silver or a silver alloy using electrolysis. Although much less costly, silver plate is not as durable as solid silver because the silver coating wears through with use.

The largest user of silver in the United States is the photography industry. Chemical changes in silver salts used in photographic emulsion produce the images on film. Due to declining production and increased cost, the use of silver in coins has largely been replaced by other metals such as nickel and copper.

You have probably seen how silver tarnishes and turns brown or even black. This tarnish is a coating of silver sulfide (Ag_2S), produced by the reaction of silver with sulfur compounds commonly found in such foods as mustard and eggs or in polluted air.

Other metals are obtained from their ores. You have studied the methods of obtaining iron, aluminum, copper, and silver from their ores. Other common metals are also obtained in similar ways.

Table 9-1 provides information about other common metals, one of which is magnesium. Magnesium is a very reactive, silver-white metal. Although it is abundant in nature, it usually occurs in compound form due to its reactive character. Magnesium occurs dissolved in sea water and some groundwater, and in various minerals including dolomite and talc. The most commonly used method of obtaining pure magnesium is by the electrolysis of sea water. Magnesium is often mixed with aluminum and zinc for added strength and ease in shaping. Magnesium is best known for its strength and lightness, making it particularly suited for use in automobile and aircraft parts. Because it burns brightly, it is also used in making fireworks and emergency flares.

Tin is another common metal, particularly important in the production of tin plate. Tin plate consists of steel sandwiched between thin layers of tin and is used for making tin cans. The principal tin ore is cassiterite (SnO_2), which is processed by ore flotation and reduction. The leading producers of tin include Malaysia, Australia, and Bolivia.

TABLE 9-1

Other Common Metals and Their Uses

Metal (symbol)	Uses
Cadmium (Cd)	control rods for nuclear reactors, plating
Chromium (Cr)	plating, making stainless steel
Cobalt (Co)	mixed with other metals to make alloys, treatment of cancer
Gold (Au)	jewelry, ornaments, dentistry
Lead (Pb)	making car batteries, pipes
Magnesium (Mg)	car and plane parts, fireworks, flares
Mercury (Hg)	lamps, switches, thermometers, barometers
Nickel (Ni)	hardening steel, plating, catalyst
Platinum (Pt)	catalyst, electronics, lab-ware, jewelry
Tantalum (Ta)	surgery, making chemical equipment
Tin (Sn)	making tin plate for cans
Titanium (Ti)	combustion chambers for rockets and jet aircraft
Tungsten (W)	filaments for light bulbs, mixed with other metals to make alloys
Zinc (Zn)	making galvanized iron, brass articles

activity

CLEANING TARNISHED SILVER

OBJECTIVE: Observe the chemical reactions that take place as silver is cleaned.

PROCESS SKILLS

In this activity, you will *observe* the reaction of tarnished silver with a cleaning mixture and *hypothesize* about the role of metal ions in this process.

MATERIALS

safety goggles
beaker
water
sodium chloride
sodium hydrogen carbonate (baking soda)
aluminum pan
tarnished silver object (such as a spoon)

PROCEDURE

1. Put on your safety goggles. In a beaker, mix very hot water with small amounts of sodium chloride ($NaCl$) and sodium hydrogen carbonate ($NaHCO_3$).
2. Place a piece of tarnished silver in an aluminum pan. Pour the hot solution into the pan.
3. Allow the solution to stand for five minutes. Observe the silver object.

OBSERVATIONS AND CONCLUSIONS

1. What change did you observe in the silver?
2. What change did you observe in the aluminum pan?
3. What change did you observe in the solution?
4. Two chemical reactions took place in the cleaning mixture. Equations for these reactions are given below. Balance each equation.

   $NaHCO_3(aq) + Al^{3+}(aq) \rightarrow Na_3AlO_3(aq) + CO_2(aq) + H^+(aq)$
   $Ag_2S(s) + H^+(aq) \rightarrow Ag(s) + H_2S(aq)$

5. Do you think this process would work in a glass container? Explain your answer using the information in the equations.
6. Do you think any silver was lost from the object? Explain your answer.

9.3 ALLOYS

Most metal objects are made of mixtures of metals. Few metal objects in common use are made of a single, pure metal. One exception is the pure copper in electric wires.

A substance made up of two or more metals melted together is called an **alloy.** Steel and brass are two alloys that have many common uses. There are thousands of alloys. Different alloys are obtained from the same elements by mixing them in different proportions.

There are several ways metals can be blended to produce an alloy. One way is by making a solid solution. If one metal dissolves in another when they are melted together, and stays dissolved when cooled, the two metals form a solid solution. There are many different solutions of copper and zinc because these metals can be mixed in all proportions. On the other hand, only a limited amount of zinc can be mixed with lead to form an alloy. If more than the needed amount of zinc is added, the excess will simply form a separate layer. Most common alloys are solid solutions.

Sometimes metals form metallic compounds that have crystal structures. Although the elements involved form bonds, they do not follow the pattern of oxidation numbers discussed earlier in the text. These alloys have unusual formulas, such as Cu_5Zn_8, Na_4Pb, or Al_2Cu.

Other alloys are metallic mixtures. In these alloys, crystals of one substance are scattered throughout the mass, somewhat like raisins in a cake. Alloys in this class have a wide range of properties. Steel is an example of a metallic mixture. There are many different steel alloys. These alloys contain different substances and are formed under different conditions.

Some metals will not form alloys with certain other metals. For example, copper does not mix easily with iron. However, copper mixes well with tin and zinc, forming some of the most widely used alloys.

Alloys have desirable properties. The properties of alloys are usually quite different from those of the basic metals that they contain. For example, an alloy is often harder than the metals of which it is made. Brass, which is composed of zinc and copper, is harder than either of its component metals. The ability of brass to resist stretching is more than twice that of copper and more than four times that of zinc.

Alloys are usually poorer conductors of heat and electricity than the pure metals. Also, the melting point of an alloy is often lower than that of any of the metals of which it is made. Many other properties of metals are changed when they combine in alloys. These properties include color, elasticity, expansion, and magnetism.

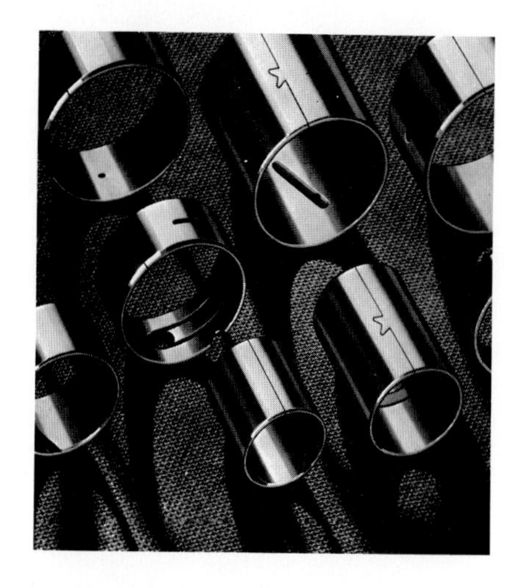

Figure 9-11 Bearings are made of an alloy that helps reduce friction from a rotating shaft.

Alloys have many uses. Bronze, which was the first alloy made in prehistoric times, consists of copper and tin. This alloy is harder and more durable than either copper or tin. You may have seen bronze memorial tablets and statues in parks and other public places. Bronze is used in the rotating parts of some electric motors.

Have you ever heard someone talk about burning out a bearing in a car? A bearing, like those in Figure 9-11, is a hollow metal tube that allows turning parts to rub against each other with minimum friction. An alloy called babbitt, which includes tin, copper, and antimony, is often used for such bearings. The friction produced between the alloy and the steel is much less than the friction produced between steel and steel. Babbitt bearings are used on the crankshaft of car engines.

Brass is used to make hinges, knobs, and other hardware around the home. Since foods react with it, brass cannot be used for cooking utensils. However, brass does make long-lasting water pipes, musical instruments, and jewelry. Naval brass is an alloy that contains lead in addition to copper and zinc. This alloy resists corrosion from salt water, so it is often used on ships.

As you have learned, a measured amount of carbon can be added to iron to make steel. Most steel contains a small amount of a number of other metals. The kind of metal added to the steel affects the properties of the finished product. For instance, nickel increases the heat and acid resistance of steel. Chromium is added to prevent rust and manganese is included to increase the hardness. Some steel used today contains as many as 15 different kinds of metals, depending on the properties needed.

Aluminum alloys are used to make objects that are light but strong. Alloys of aluminum, titanium, and vanadium are used in aircraft. Aluminum-titanium alloys are used in spacecraft. Small amounts of other metals are added to produce certain characteristics. For example, adding manganese will make an aluminum-titanium alloy stronger at high temperatures, and more resistant to corrosion. When would resistance to heat and corrosion be important to a space shuttle?

Another metal that is used in spacecraft alloys is beryllium. This metal is more elastic than steel, and stronger than other lightweight metals.

A relatively new use for alloys is in superconductors. A **superconductor** is a substance with little or no resistance to electric current. These materials can conduct electricity better than silver, copper, or aluminum. However, superconductors work well only at very low temperatures. For example, an alloy of niobium and tin becomes a superconductor at −250° C. Such alloys are not likely to replace copper or aluminum in electrical wiring, but they may be used to make computers work faster. You will learn more about superconductors in later chapters.

Table 9-2 lists some additional alloys with their properties and uses. The table also shows the most common makeup of each alloy. For example, stainless steel most commonly contains iron, chromium, and carbon, but small percentages of nickel, silicon, manganese, and titanium are sometimes added to modify the properties of the alloy for specific uses. Stainless steel is known for its ability to resist rust, tarnish, heat, and wear. It is commonly used in knives, flatware, pots, and pans, and in equipment for hospitals and restaurants.

TABLE 9-2

Alloys and Their Uses

Alloy	Composition	Property	Uses
Alnico	aluminum, nickel, cobalt, iron, copper	permanent magnet	magnets for telephones, loudspeakers, hearing aids
Duralumin (*doo*-RAL-*yoo*-MIN)	aluminum, magnesium, copper, manganese	strong, lightweight	tools, ladders, building materials, frames
Monel	copper, nickel	does not tarnish easily	sinks, drainboards, ice cream cabinets
Nichrome	nickel, chromium, iron, manganese	high electrical resistance	heating elements in electric irons, toasters, and ranges
Solder	lead, tin	low melting point	to join two pieces of metal together
Stainless steel	iron, chromium, carbon	resists rust	building materials, cabinets, sinks, kitchen utensils
Wood's metal	bismuth, cadmium, tin, lead	low melting point	plugs for sprinklers of automatic fire extinguishers

METAL FATIGUE

destroyer of metal structures

This bridge in Connecticut collapsed due to metal fatigue. When the structure of the bridge failed, lives were lost, vehicles were destroyed, and property was damaged.

On April 28, 1988, a 19-year-old Aloha Airlines Boeing 737 jet flying from Hilo, Hawaii, to Honolulu was forced to make an emergency landing on Maui when a large section of its fuselage was ripped away. One person was killed and six others were injured. What was the cause of the accident? Metal fatigue.

Metal fatigue is the gradual weakening of a metal caused by continued stress on that metal. In the case of the Aloha Airlines accident, metal fatigue was caused by the continual pressurization and depressurization of the jet's cabin during thousands of flights. These changes in cabin pressure cause the plane itself to expand and contract, putting stress on its metal shell. In the old 737 jet models, this change in size is particularly dangerous because of the way rivets were used to hold the metal shell together. The rivets were sunk into the metal so that they would lie flush with the outer surface of the plane. Unfortunately, this also meant that these rivets were more likely to cause surface cracks as more and more stress was put on the rivets.

In order to prevent future disasters, the Federal Aviation Administration (FAA) called for the replacement of these "flush" rivets in other 737s with another type of rivet that should prevent the plane's metal shell from cracking. Besides initiating this crucial mandate on plane construction, the Aloha accident also helped remind people of the need for more intensive inspection and repair of airplanes.

Airplanes are not the only structures that can suffer from metal fatigue. Stresses created by vehicles rolling over a bridge, for example, can cause cracks in the bridge's metal surface. This is especially true when there are scratches or defects existing in the metal before its use on the bridge.

FOR FURTHER RESEARCH AND DISCUSSION

What sorts of things should engineers consider when designing buildings, bridges, and airplanes to help eliminate problems caused by metal fatigue? How often should planes and structures be inspected? When should repairs be started: before or after cracks become obviously large?

CHAPTER REVIEW

SUMMARY

1. Metallurgy is the science of producing useful metals from their ores.
2. An ore is a rock or mineral from which a metal can be obtained in profitable amounts.
3. In making iron, a charge of iron ore, coke, and limestone is heated.
4. In the basic-oxygen furnace, iron is made into steel.
5. Steel contains iron, carbon, and sometimes traces of other metals.
6. Steel hardness depends on carbon content and heat treatment.
7. Ways to obtain metals from their ores include ore-flotation, roasting, reduction, and electrolysis.
8. Roasting is a process in which sulfide minerals are heated in air, or oxygen-enriched air, to form metal oxides.
9. Reduction is a process in which oxygen is removed from a metal oxide ore producing a free metal.
10. An alloy is made up of two or more metals melted together.

VOCABULARY

Match the item in the left column with the best answer in the right column. Do not write in this book.

1. alloy	a. substance with little or no electrical resistance
2. metallurgy	b. changing sulfide ores to oxides
3. ore	c. scarce, costly metal
4. precious metal	d. rock from which a useful metal is obtained profitably
5. reducing agent	e. mixture of metals
6. refining	f. separating a metal from its ore
7. roasting	g. science of metal production
8. superconductor	h. removes oxygen from a compound

REVIEW QUESTIONS

1. What is metallurgy?
2. Can all metal-bearing rocks be called ores? Explain your answer.
3. Name three important ores of iron.
4. What substances are contained in a blast furnace charge?
5. What compound is used to remove silicon dioxide in refining iron?
6. What function does coke serve in the process of refining iron?
7. What is the difference between iron and steel?
8. What substances are mixed in the basic-oxygen furnace?
9. Why is no fuel needed in a basic-oxygen furnace?
10. What effect does sudden cooling have on hot steel?
11. When iron rusts, what chemical change takes place?

12. What can be done to stop iron from rusting?
13. Name the ore of aluminum.
14. What two compounds are mixed in the Hall process?
15. List five uses for aluminum.
16. Name three ores of copper.
17. What is meant by roasting an ore? Why is this done?
18. How is blister copper refined? What is the purpose of this process?
19. Name two valuable metals that are by-products of copper refining.
20. What is a precious metal? List five precious metals.
21. List three sources of silver.
22. What causes silver to tarnish?
23. List three ways of making alloys.
24. List three ways alloys differ from pure metals of which they are made.
25. What metals are used to make bronze?
26. What effect does chromium have when added to steel?
27. What metals are used to make solder?
28. Explain by means of chemical equations how pig iron is produced in a blast furnace.
29. Compare the treatment of free copper with the refining of iron.
30. How is the refining of copper ore like the refining of aluminum ore?
31. How would you show that aluminum tarnishes?

CRITICAL THINKING

32. Coke ovens, where coke is derived from soft coal, are often located near steel mills. Why?
33. Why does reusing scrap iron save energy?
34. Compare the energy sources for the refining of iron and copper.
35. Why is the low melting point of Wood's metal a valuable property for its use as water plugs in the sprinkler systems of automatic fire extinguishers?

PROBLEMS

1. What is the minimum amount of iron you would expect to get from a metric ton of taconite? (1 metric ton = 1,000 kg)
2. If hematite were pure Fe_2O_3, what percent of iron would it contain? (Use the following mass numbers: Fe = 56; O = 16.)
3. Pure gold is marked 24K (carat). If a bracelet having a mass of 50 g is marked 14K, what percent of gold does it contain? Find the number of grams of gold in the bracelet.

FURTHER READING

Burt, O. W. *The First Book of Copper.* New York: Franklin Watts, Inc., 1982.

Coombs, Charles. *Gold and Other Precious Metals.* New York: William Morrow and Co., Inc., 1981.

Fisher, D. A. *Steel: From the Iron Age to the Space Age.* New York: Harper & Row Publishers, Inc., 1982.

ORGANIC COMPOUNDS

Trees, like this Baobab in Tanzania, are living organisms. All living things—whether they are plants, birds, insects, or humans—are made of organic compounds. How are these compounds different from those that make up some nonliving things, such as a quartz crystal or a glass of water?

CHAPTER 10

SECTIONS

OBJECTIVES

- ☐ Distinguish between organic compounds and inorganic compounds.
- ☐ Identify and give examples of five groups of hydrocarbons.
- ☐ List uses of organic compounds.
- ☐ Compare substituted hydrocarbons with their related alkanes.

10.1 ORGANIC CHEMISTRY

There are differences between organic compounds and inorganic compounds. Carbon compounds make up the structure of living things, or organisms. Thus, scientists classify carbon compounds as **organic compounds.** Other substances such as acids, bases, salts, and water do not contain carbon. These substances are classified as inorganic compounds. These two groups of compounds differ in several ways.

1. They contain different bonds. Organic compounds are formed as a result of covalent bonding, or the sharing of electrons. Most inorganic compounds are due to ionic bonding, or the transfer of electrons.
2. They dissolve differently. Most organic compounds do not dissolve in water. They do, however, dissolve in such organic liquids as alcohol or ether. Many inorganic compounds dissolve, more or less readily, in water.
3. They behave differently when heated. Organic compounds tend to decompose, or break down into simpler substances, when heated to a high temperature. Inorganic compounds melt and then vaporize without breaking down when heated to a high temperature.
4. They react at different rates. Reactions involving organic compounds take place relatively slowly. Organic reactions often require hours or even days for completion. However, most inorganic reactions take place quickly.

Organic compounds can be divided into a number of groups. Compounds in each group have similar properties and molecular structure. These groups include hydrocarbons (such as methane and propane gases), proteins, carbohydrates, fats, vitamins, alcohols, esters, and organic acids. Esters include the compounds that give fruits their odors and flavors. Organic acids include substances that provide the tartness of vinegar and citrus fruits. In this chapter you will study the hydrocarbon and alcohol groups.

10.2 HYDROCARBONS

The hydrocarbons include many important compounds. Compounds that are composed only of hydrogen and carbon are called **hydrocarbons** (HIE-*druh*-KAHR-*buns*). Hydrocarbons make up the simplest group of the organic compounds. There are thousands of different hydrocarbons, which are classified into subgroups called series.

The simplest and most abundant series of hydrocarbons is called the **alkane** (AL-*kane*) **series**. Alkanes differ from one another in the number of carbon and hydrogen atoms in the molecule. The first compound in this series is the gas methane (CH_4), the main component of natural gas.

The next compound in the alkane series is ethane (C_2H_6). Ethane is followed by propane (C_3H_8) and butane (C_4H_{10}).

Propane and butane are sometimes sold as "bottled gas," which can be used for cooking. Only these four smallest alkanes are gases at room temperature.

Table 10-1 lists some members of the alkane series. Notice in the table that the name of each compound ends in *-ane*. The beginning of each name tells how many carbon atoms there are in the molecule. As you will see, these prefixes are used in naming many organic compounds.

TABLE 10-1

The Alkane Series of Hydrocarbons

Name	Formula	Phase at Room Temperature
Methane	CH_4	gas
Ethane	C_2H_6	gas
Propane	C_3H_8	gas
Butane	C_4H_{10}	gas
Pentane	C_5H_{12}	liquid
Hexane	C_6H_{14}	liquid
Heptane	C_7H_{16}	liquid
Octane	C_8H_{18}	liquid
Nonane	C_9H_{20}	liquid
Decane	$C_{10}H_{22}$	liquid
Eicosane	$C_{20}H_{42}$	solid
Hexacontane	$C_{60}H_{122}$	solid

Notice that there is a pattern to the formulas of the alkanes. Each compound differs from the one before it by the addition of CH_2. To find the order of the compounds in this series, the general formula C_nH_{2n+2} is used. The first compound, methane, starts with $n = 1$.

$$C_1H_{2(1)+2} = CH_4$$

Some members of the alkane series contain more than 1,000 carbon atoms. What would be the formula of an alkane with 1,000 carbon atoms? Look again at Table 10-1. Predict the phase at room temperature of the alkane with this large a number of carbon atoms.

Structural formulas are used to represent organic compounds. Scientists often need to know how the atoms are arranged in an organic compound. A **structural formula**

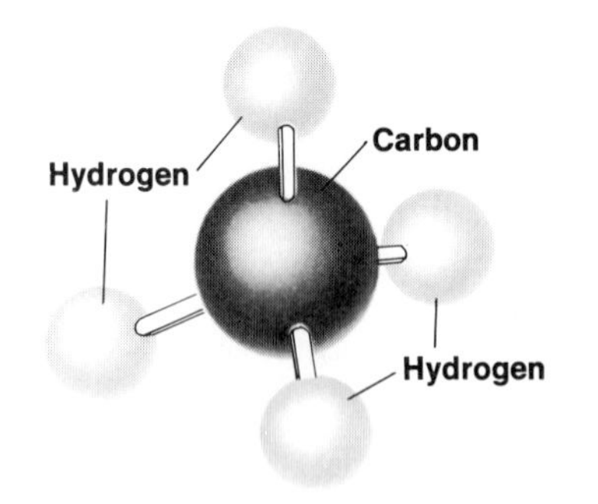

Figure 10-1 In the methane molecule, CH_4, represented above by a ball and stick model, a carbon atom forms covalent bonds with four hydrogen atoms. In the diagram, how are these covalent bonds represented?

shows the arrangement of atoms in a molecule. The structural formulas for each of the first four members of the alkane series are as follows:

```
     H                    H   H
     |                    |   |
  H—C—H                H—C—C—H
     |                    |   |
     H                    H   H
```

methane, CH_4 ethane, C_2H_6

```
   H  H  H                 H  H  H  H
   |  |  |                 |  |  |  |
H—C—C—C—H               H—C—C—C—C—H
   |  |  |                 |  |  |  |
   H  H  H                 H  H  H  H
```

propane, C_3H_8 butane, C_4H_{10}

The dashes (—) in the above formulas stand for covalent bonds. Thus, the structural formula for methane shows that each outer electron of the carbon atom is paired with the electron of one hydrogen atom. In the structural formulas above, notice that each carbon atom is surrounded by four bonds. Carbon forms four bonds because each carbon atom has four outer electrons that can be shared. A laboratory model of the methane molecule is shown in Figure 10-1. How many bonds does each hydrogen atom form?

Compounds can have different structural formulas. In some hydrocarbons, the carbon atoms are bonded together in a straight chain. The structural formulas for propane and butane already shown indicate that these compounds form straight chains. However, many hydrocarbon compounds also form a branched chain. The structural formulas of two forms of butane are shown below. Notice that one carbon atom forms a branched chain by bonding to the middle carbon atom. Compare these structural formulas with the models shown in Figure 10-2.

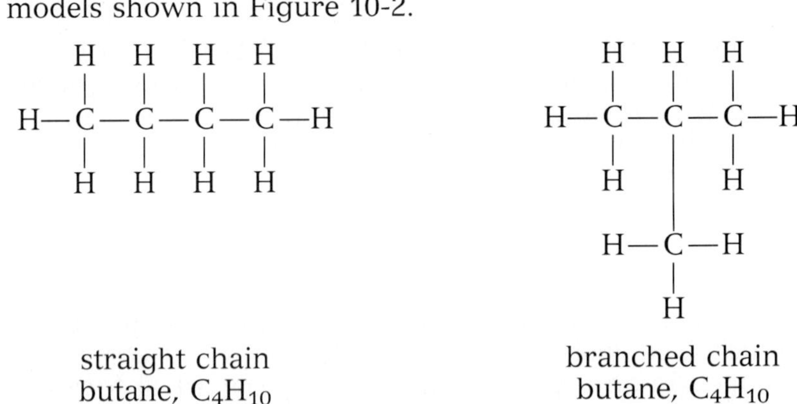

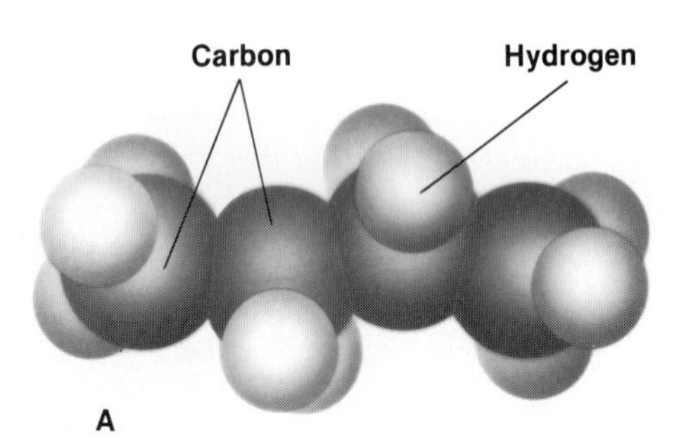

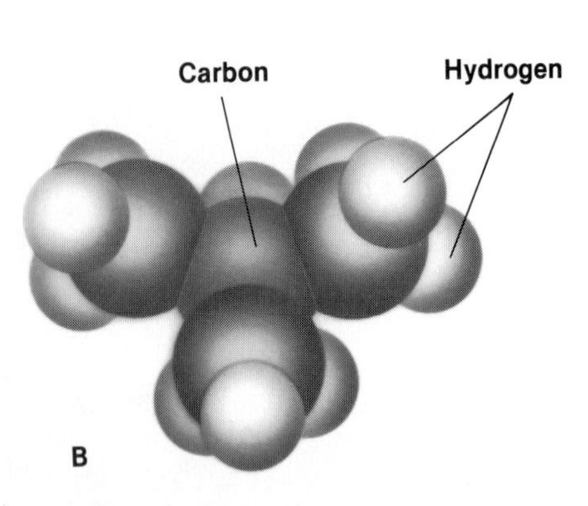

Figure 10-2 Models of straight-chain butane (A) and branched-chain butane (B). How many carbon atoms does each molecule contain?

These forms of butane are called isomers (IE-*suh-murz*). **Isomers** are compounds the molecules of which have the same number and kind of atoms but with a different arrangement. Although both isomers have the same chemical formula, they are, in fact, different compounds. The branched-chain compound has properties that differ from those of the straight-chain compound. For example, the branched-chain isomer of butane has a lower boiling point than does the straight-chain isomer.

The more carbon atoms in a hydrocarbon molecule, the more isomers it forms. Pentane, the next compound after butane in the alkane series, has three isomers. There are 75 possible isomers of decane ($C_{10}H_{22}$).

Gasoline contains many hydrocarbons. Gasoline is a complex mixture containing alkanes (including pentanes through nonanes) and other hydrocarbons. The amounts of the different isomers in gasoline affect its quality. For instance, gasoline with straight-chain hydrocarbons burns more quickly than does gasoline containing mostly branched-chain compounds. In Chapter 7 you learned that burning means combining with oxygen. Think of the shapes of different isomers. A branched-chain molecule is compact, but a straight-chain molecule is stretched out. Thus the straight-chain molecule is more easily reached by oxygen, and it burns more quickly.

Sometimes gasoline burns too quickly in a car engine. Instead of working smoothly, the pistons in the cylinder of the engine receive sharp, hammer-like blows, called knocking. This can result in lost power and possible damage.

Knocking can be prevented by using a slower-burning fuel. It can also be reduced by adding a catalyst to the gasoline. A catalyst that was widely used for this purpose is tetra-ethyl (TET-*ruh*-ETH-*uhl*) lead. This compound slows down the rate at which the gasoline burns, thus improving its antiknock qualities. However, this lead catalyst causes air pollution, so leaded gasoline is not widely used today. Most cars are now designed with engines that will work smoothly without the catalyst, using unleaded gasoline.

Gasoline is rated by octane numbers. Engineers have established a system of comparing the knocking property of gasoline. The ability of a gasoline to resist knocking is expressed by a rating called the **octane number**. The higher the octane number, the more knock-resistant the gasoline.

An isomer of octane with three branched chains has excellent antiknock properties. This compound, called isooctane, is given an octane number of 100. Straight-chain heptane, or normal heptane, knocks very badly. Therefore, normal heptane is given a rating of zero. Test mixtures of these two compounds are used as a standard by which gasoline can be compared.

For example, a mixture of 93 percent iso-octane and 7 percent normal heptane is assigned an octane number of 93. Any gasoline that has an antiknock quality as high as this standard test mixture is given the same octane number. Such a gasoline makes an excellent car engine fuel. Some small airplanes use 100-octane gasoline. Most unleaded gasolines are in the 89-92 octane range (Figure 10-3).

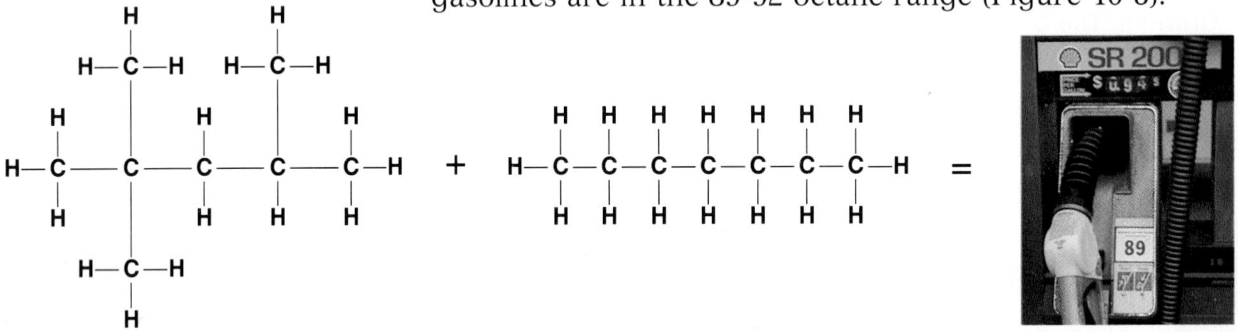

89% Iso-octane + 11% Normal heptane = 89 Octane

Figure 10-3 Gasoline is a complex mixture of hydrocarbons. To have a rating of 89, a gasoline mixture must have the same antiknock properties as the standard mixture illustrated.

Hydrocarbons can be saturated or unsaturated. Look back at the structural formulas of alkanes. Notice that every carbon atom in each molecule is bonded to another carbon atom or to a hydrogen atom. These bonds are called single bonds because one pair of electrons is shared in each bond. Each carbon atom has formed as many bonds as it can. No more hydrogen atoms can be added to the molecule. Such a molecule, containing only single bonds, is said to be **saturated**.

Other hydrocarbons are not saturated. Such molecules, which contain bonds other than single bonds, are said to be **unsaturated**. For example, the **alkene** (AL-*keen*) **series** is composed of compounds in which two carbon atoms in the molecule are connected by two bonds. These bonds are shown in the structural formulas by two dashes (═). In this type of bonding, two outer electrons of one carbon atom are paired with two outer electrons of another carbon atom, forming a double covalent bond. The structural formulas for two alkenes are shown below. Notice that there are still a total of four bonds around each carbon atom.

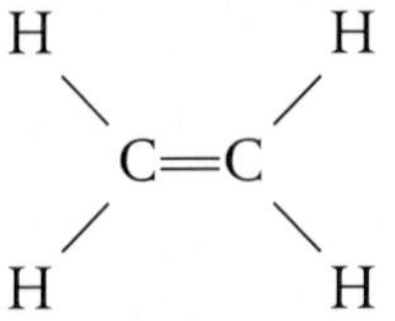

ethene, C_2H_4

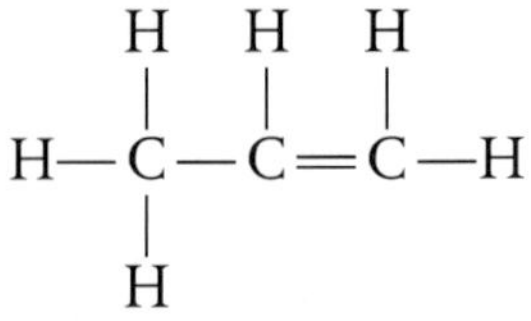

propene, C_3H_6

Notice that the names of these members of the alkene series are similar to the names of members of the alkane series having the same number of carbon atoms. The suffix *-ene* is substituted for the suffix *-ane*. For instance, ethane is the alkane with two carbon atoms. Thus, the alkene with two carbon atoms is named ethene. What is the name of the alkane with three carbon atoms?

```
     H   H   H
     |   |   |
H—C—C=C—H
     |
     H
```

Figure 10-4 Acetylene is used in welding metals together. What is the formula for this compound?

The third member of the alkene series has four carbon atoms and is called butene (C_4H_8). The fourth member of the series is called pentene (C_5H_{10}). The general formula for compounds in the alkene series is C_nH_{2n}, starting with $n = 2$ for ethene.

$$C_2H_{2(2)} = C_2H_4$$

Another group of hydrocarbons, the **alkyne** (AL-*kine*) **series**, is made up of compounds in which two carbon atoms share three pairs of electrons, forming a triple covalent bond. In a structural formula these bonds are shown by three dashes (≡). The structural formulas for two alkynes are shown below. How many bonds does each carbon atom form?

```
                        H
                        |
H—C≡C—H           H—C—C≡C—H
                        |
                        H
```

ethyne, C_2H_2 — propyne, C_3H_4

Ethyne (C_2H_2), commonly called acetylene, is the first compound of the alkyne series. Acetylene burns with a very hot flame. It is used as a fuel in welding and cutting metals (Figure 10-4).

The second compound in the alkyne series is propyne (C_3H_4). The general formula for compounds in the alkyne series of hydrocarbons is C_nH_{2n-2}, starting with $n = 2$ for ethyne.

$$C_2H_{2(2)-2} = C_2H_2$$

Organic compounds can form rings. Another group of hydrocarbons form ring-shaped molecules. The structural formula of benzene (C_6H_6), an important ring compound, is

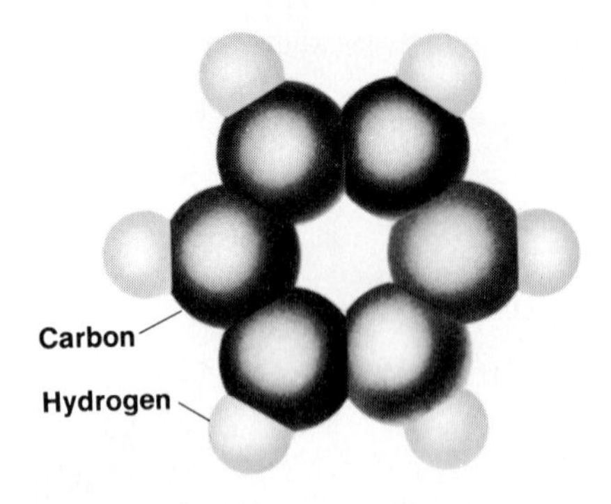

Figure 10-5 Laboratory model of the benzene (C_6H_6) molecule, the simplest of the ring compounds.

shown below. A model of the benzene molecule is shown in Figure 10-5.

```
        H
        |
        C
      //  \
  H—C      C—H
    |      ||
  H—C      C—H
      \\  /
        C
        |
        H
```

The **benzene series** is a group of hydrocarbons that contain rings similar to the benzene molecule. The general formula for compounds in the benzene series is C_nH_{2n-6}. The first member, benzene, starts with $n = 6$.

$$C_6H_{2(6)-6} = C_6H_6$$

Because ring compounds can also have branched chains, great numbers of these different hydrocarbons are possible. Benzene is the basic raw material for making thousands of organic compounds used in plastics, dyes, drugs, perfumes, and explosives. Benzene is an excellent solvent for many organic compounds, including rubber. Although benzene has many uses in industry, it must be handled very carefully because it can cause cancer.

10.3 ADDITION AND SUBSTITUTION REACTIONS

Atoms can be added to unsaturated compounds. Recall that alkanes are saturated hydrocarbons. No additional hydrogen atoms can be bonded to a saturated molecule. Thus, alkanes do not react with hydrogen.

```
    H  H
    |  |
 H—C—C—H (g)+ H2(g) → no reaction
    |  |
    H  H
```

saturated compound
(alkane)

Alkenes and alkynes are unsaturated compounds. These compounds contain a double or triple bond between carbon atoms. In certain reactions, the covalent bonds of an unsaturated hydrocarbon can be broken. Hydrogen atoms can then be added to the molecule. This change makes the molecule saturated. The following equations show **addition reactions**, which add hydrogen to unsaturated hydrocarbons:

$$\begin{array}{ccc} \mathrm{H} & & \mathrm{H} \\ & \diagdown \quad \diagup & \\ & \mathrm{C{=}C} & \\ & \diagup \quad \diagdown & \\ \mathrm{H} & & \mathrm{H} \end{array} (g) + H_2(g) \rightarrow \begin{array}{ccccccc} & & \mathrm{H} & & \mathrm{H} & & \\ & & | & & | & & \\ \mathrm{H} & - & \mathrm{C} & - & \mathrm{C} & - & \mathrm{H} \\ & & | & & | & & \\ & & \mathrm{H} & & \mathrm{H} & & \end{array} (g)$$

unsaturated compound (alkene) — saturated compound (alkane)

$$\mathrm{H{-}C{\equiv}C{-}H}\ (g) + 2H_2(g) \rightarrow \begin{array}{ccccccc} & & \mathrm{H} & & \mathrm{H} & & \\ & & | & & | & & \\ \mathrm{H} & - & \mathrm{C} & - & \mathrm{C} & - & \mathrm{H} \\ & & | & & | & & \\ & & \mathrm{H} & & \mathrm{H} & & \end{array} (g)$$

unsaturated compound (alkyne) — saturated compound (alkane)

Notice that in the alkene, one bond was broken and two atoms of hydrogen were added. In the alkyne, two bonds were broken and four hydrogen atoms were added. In both cases, alkanes are formed. Note that after an addition reaction, only a single bond remains between the carbon atoms where a double bond or a triple bond once was. The new hydrogen atoms become attached by single bonds to the carbon atoms.

Atoms in organic molecules can be replaced. As you have seen, double and triple bonds in unsaturated compounds can be broken, and hydrogen atoms can be added. Another type of reaction, called substitution, is also possible. In a **substitution reaction** one atom, or group of atoms, is replaced by another atom or group of atoms. This reaction can take place in both saturated and unsaturated compounds.

Under certain conditions, hydrocarbons react with fluorine, chlorine, bromine, or iodine. For example, methane reacts with chlorine to form monochloromethane (CH_3Cl).

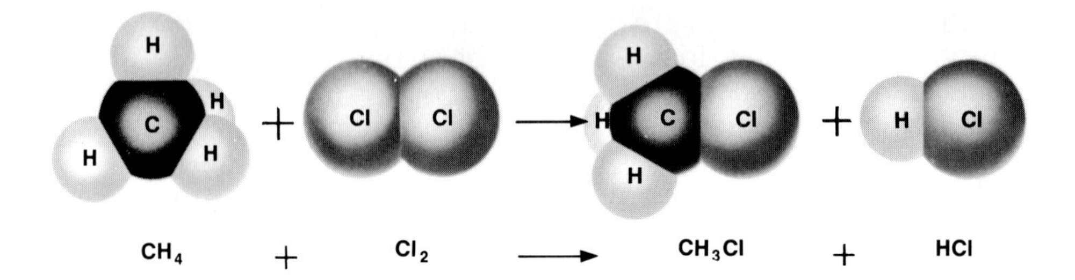

Figure 10-6 Methane and chlorine react to undergo a substitution, or replacement, reaction. What is another type of organic reaction?

Figure 10-6 shows models of this reaction. Compare the structural formulas of methane and monochloromethane shown below. Notice the chlorine atom that has replaced the hydrogen atom from methane.

$$\begin{array}{ccc} & H & \\ & | & \\ H- & C & -\boxed{H} \\ & | & \\ & H & \end{array} \qquad\qquad \begin{array}{ccc} & H & \\ & | & \\ H- & C & -\boxed{Cl} \\ & | & \\ & H & \end{array}$$

methane — monochloromethane

Addition and substitution reactions are among the most important reactions in organic chemistry. Many other organic compounds can be made from a hydrocarbon by replacing hydrogen atoms with radicals, groups of atoms, or individual atoms of other elements. For example, by substituting chlorine for hydrogen atoms in methane, a series of different compounds can be made:

$$\begin{array}{ccc} & H & \\ & | & \\ H- & C & -Cl \\ & | & \\ & H & \end{array} \qquad\qquad \begin{array}{ccc} & Cl & \\ & | & \\ H- & C & -Cl \\ & | & \\ & H & \end{array}$$

monochloromethane
CH_3Cl

dichloromethane
CH_2Cl_2

$$\begin{array}{ccc} & Cl & \\ & | & \\ H- & C & -Cl \\ & | & \\ & Cl & \end{array} \qquad\qquad \begin{array}{ccc} & Cl & \\ & | & \\ Cl- & C & -Cl \\ & | & \\ & Cl & \end{array}$$

trichloromethane
(chloroform)
$CHCl_3$

tetrachloromethane
(carbon tetrachloride)
CCl_4

Fluorine can also be substituted for one or more hydrogen atoms in methane. For example, two fluorine atoms and two chlorine atoms can be substituted for the hydrogen atoms in methane. The compound produced is dichlorodifluroromethane (CCl_2F_2). The common name for this compound is freon. Freon is used as a refrigerant for home refrigerators and air-conditioning systems. What is the structural formula for this compound? Freon is believed to be harmful to the earth's atmosphere. An international effort toward decreasing the use of freon as a coolant is now underway.

10.4 ALCOHOLS

Alcohols form a large group of organic compounds. An OH radical can be substituted for hydrogen atoms in a hydrocarbon. When one or more hydrogen atoms are replaced by OH radicals, an **alcohol** is formed. Although alcohol molecules contain OH groups, they are not bases. This is because alcohols do not ionize in water, unlike the inorganic bases you studied in Chapter 5.

In methyl alcohol (CH_3OH), the simplest alcohol, one OH radical replaces one hydrogen atom in methane (CH_4). Compare the structural formulas of methane and methyl alcohol.

```
     H                    H
     |                    |
 H — C — [H]          H — C — [OH]
     |                    |
     H                    H

  methane            methyl alcohol
```

Methyl alcohol is sometimes called wood alcohol because this alcohol can be produced from wood. Methyl alcohol is very poisonous. If swallowed, it can cause blindness or death. Methyl alcohol is used as a solvent for lacquers and shellac, and in the making of many organic compounds. Methyl alcohol is usually prepared by the reaction of carbon monoxide and hydrogen.

$$CO(g) + 2H_2(g) \rightarrow CH_3OH(l)$$

Ethyl alcohol (C_2H_5OH) contains one OH radical replacing one hydrogen atom of ethane (C_2H_6). The structural formulas of ethane and ethyl alcohol are shown below:

```
     H   H                    H   H
     |   |                    |   |
 H — C — C — [H]          H — C — C — [OH]
     |   |                    |   |
     H   H                    H   H

     ethane               ethyl alcohol
```

Figure 10-7 Denatured alcohol is not fit for human consumption.

Ethyl alcohol is sometimes referred to as ethanol. Also, you may have heard it referred to as grain alcohol. This is because it is most often made by fermenting grain or fruit juices. In this process, sugar is broken down by yeast, producing ethyl alcohol and carbon dioxide:

$$\underset{\text{sugar}}{C_6H_{12}O_6(aq)} \rightarrow \underset{\text{ethyl alcohol}}{2C_2H_5OH(aq)} + 2CO_2(g)$$

A fermented mixture may contain as much as 15 percent alcohol. At this point, the yeasts are killed as a result of alcohol poisoning and the process stops. The fermented mixture can be made more concentrated by distillation. Ethyl alcohol is the intoxicating substance in alcoholic drinks. In large amounts, it is poisonous.

Alcoholic beverages usually do not contain more than 45 percent alcohol, or 90 proof. The proof number is twice the percent number. Ethanol is also used in huge amounts in the chemical industry. For these uses, the mixture may be distilled until it contains up to 95 percent alcohol.

When prepared for industrial purposes, foul-tasting or poisonous substances are often added to ethyl alcohol to make it unfit to drink. The alcohol is then described as denatured. See Figure 10-7. The government requires that ethyl alcohol be denatured in order to protect its revenues gained from taxes imposed on alcoholic beverages. Methyl alcohol is often used as the denaturant.

Table 10-2 lists some of the members of the alkane series of hydrocarbons with the names and the formulas of their related alcohols. Notice that in each case one hydrogen atom has been replaced by one OH group.

TABLE 10-2

Some Alcohols from the Alkane Series

Hydrocarbon		Alcohol	
Methane	CH_4	methyl alcohol (methanol)	CH_3OH
Ethane	C_2H_6	ethyl alcohol (ethanol)	C_2H_5OH
Propane	C_3H_8	propyl alcohol (propanol)	C_3H_7OH
Butane	C_4H_{10}	butyl alcohol (butanol)	C_4H_9OH
Pentane	C_5H_{12}	amyl alcohol	$C_5H_{11}OH$
Hexane	C_6H_{14}	hexyl alcohol	$C_6H_{13}OH$

FERMENTATION OF SUGAR

activity

OBJECTIVE: Determine the products of fermentation.

PROCESS SKILLS

In this activity you will *observe* the process of fermentation and *analyze data* to *infer* the products of fermentation.

MATERIALS

10 percent sucrose solution
1-L bottle
dry yeast
one-hole stopper (to fit bottle)
glass tubing
rubber tubing
gas-collecting bottle and trough
wood splint

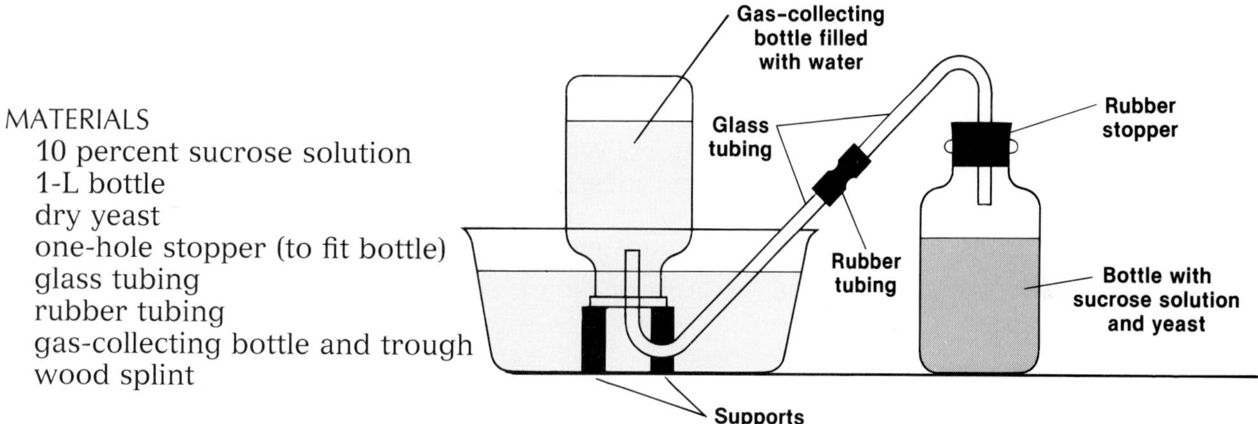

PROCEDURE

1. Fill half the 1-L bottle with sucrose solution. Add one package of dry yeast, and swirl to mix well.
2. Add sucrose solution until the bottle is nearly full.
3. Assemble the apparatus as shown in the diagram. CAUTION: *Ask your teacher to insert the glass tubing in the stopper.*
4. Set aside the apparatus until the gas-collecting bottle is full of gas. Make observations each day.
5. After the gas-collecting bottle is full, use a glowing splint to test the gas in the bottle.
6. Remove the stopper from the bottle and carefully smell the contents by fanning the fumes. CAUTION: *Never inhale fumes directly. Use your hand to fan the fumes toward your nose.*

OBSERVATIONS AND CONCLUSIONS

1. What happened when a glowing splint was placed in the collected gas?
2. Infer what gas was collected.
3. Did the liquid in the 1-L bottle have an odor? Describe it.
4. Before sucrose can be fermented, it must be broken down to a simple sugar. Balance the equation for this reaction:

   $C_{12}H_{22}O_{11}(s) + H_2O(l) \rightarrow C_6H_{12}O_6(aq)$

5. The simple sugar $C_6H_{12}O_6$ is then fermented by the yeast. Write a balanced equation for this reaction.

Some alcohols contain more than one OH radical. One such alcohol is ethylene glycol. It is made by allowing ethene to react with atmospheric oxygen to form ethylene oxide and then allowing the oxide to react with water as shown by the following equations:

$$2C_2H_4(l) + O_2(g) \rightarrow 2C_2H_4O(l)$$

$$C_2H_4O(l) + H_2O(l) \rightarrow C_2H_4(OH)_2(l)$$

The structural formula for ethylene glycol is shown below. Because of its two OH radicals, ethylene glycol is very soluble in water. This alcohol is widely used as an antifreeze in car radiators. Water freezes at 0° C. If four parts of water are mixed with six parts of ethylene glycol, the freezing temperature of the mixture is lowered to −49° C. If ethylene glycol is ingested, it can result in death.

$$\begin{array}{ccccc} & & H & & H & & \\ & & | & & | & & \\ H & — & C & — & C & — & H \\ & & | & & | & & \\ & & H & & H & & \end{array} \qquad \begin{array}{ccccccc} & & H & & H & & \\ & & | & & | & & \\ HO & — & C & — & C & — & OH \\ & & | & & | & & \\ & & H & & H & & \end{array}$$

ethane
C_2H_6

ethylene glycol
$C_2H_4(OH)_2$

Another important alcohol that contains more than one OH group is glycerol, commonly called glycerin. Glycerol has three OH radicals replacing three hydrogen atoms of propane. Compare the structural formulas shown below. Glycerol is a by-product of soap making from fats. It is a very thick liquid that is nontoxic and sweet tasting. Because of its ability to absorb moisture, it is used in skin creams and lotions. This compound is also used in the making of solvents, printer's ink, sweeteners, medicines and the explosive nitroglycerin.

$$\begin{array}{cccc} & & H & & \\ & & | & & \\ H & — & C & — & \boxed{H} \\ & & | & & \\ H & — & C & — & \boxed{H} \\ & & | & & \\ H & — & C & — & \boxed{H} \\ & & | & & \\ & & H & & \end{array} \qquad \begin{array}{ccccc} & & H & & \\ & & | & & \\ H & — & C & — & \boxed{OH} \\ & & | & & \\ H & — & C & — & \boxed{OH} \\ & & | & & \\ H & — & C & — & \boxed{OH} \\ & & | & & \\ & & H & & \end{array}$$

propane
C_3H_8

glycerol
$C_3H_5(OH)_3$

BIOGRAPHY

DOROTHY HODGKIN

nobel prize winner

Dorothy Crowfoot Hodgkin, a British chemist, is known for her ground-breaking work on the molecular structure of organic compounds. Born in Cairo, Egypt, in 1910, Hodgkin attended both Oxford and Cambridge Universities in England. In 1947, she was made a Fellow of the Royal Society. Hodgkin was awarded the Nobel Prize in chemistry in 1964.

Hodgkin was a pioneer in the use of a special technique called X-ray diffraction to study proteins and other organic compounds. She was also a member of the group of scientists who analyzed the structure of penicillin, an antibiotic.

Hodgkin did research on the structure of the insulin molecule. This protein is produced in the body and is necessary for the body's proper use of sugars. Many diabetics, people whose bodies do not produce enough insulin, take insulin every day.

Hodgkin's Nobel Prize was awarded for her research in the structure of the vitamin B_{12} molecule, which has the formula $C_{63}H_{90}O_{14}N_{14}PCo$. Vitamin B_{12} is found in meats, eggs, and fish. This vitamin is not found in plants. However, it is required in the diet of animals.

Vitamin B_{12}, when injected into the human body in an amount as small as 0.000006 g per day, can cure pernicious anemia. Pernicious anemia is a disease caused by failure of the liver to supply vitamin B_{12} to the bone marrow, which it needs to produce red blood cells in the body.

The exact ways in which vitamin B_{12} functions in living things are not yet known. Further research will continue to provide solutions to many problems related to complex biochemical reactions.

CHAPTER REVIEW

SUMMARY

1. Organic compounds include hydrocarbons, proteins, carbohydrates, fats, vitamins, alcohols, esters, and organic acids.
2. Hydrocarbons are grouped into the alkanes (single bond), the alkenes (double bond), and the alkynes (triple bond).
3. Carbon atoms are linked together in straight chains, branched chains, and ring compounds.
4. Single-bond hydrocarbon compounds are saturated. No other hydrogen atoms can be added.
5. Isomers are compounds with the same number and kind of atoms but with a different arrangement of atoms in the molecule.
6. The octane rating is a number that compares the antiknock properties of a gasoline with that of a standard test fuel.
7. Double- and triple-bond compounds are unsaturated. Bonds can be broken and other atoms added.
8. Two important types of organic reactions are addition and substitution (replacement) reactions.
9. Alcohols are obtained from a hydrocarbon by replacing one or more hydrogen atoms with one or more OH groups.

VOCABULARY

Match the item in the left column with the best answer in the right column. Do not write in this book.

1. alcohol
2. alkane series
3. alkyne series
4. benzene series
5. hydrocarbon
6. isomers
7. octane number
8. organic compound
9. substitution reaction
10. unsaturated

a. any compound containing carbon
b. compound composed of hydrogen and carbon
c. hydrocarbon that contains OH groups
d. one atom or group replaces another atom or group
e. hydrocarbons having single bonds
f. first member is called acetylene
g. compounds the molecules of which contain the same number and kind of atoms but with a different arrangement
h. describes ability of a gasoline to resist knocking
i. hydrocarbons containing ring compounds
j. molecules containing bonds other than single bonds

REVIEW QUESTIONS

1. What is an organic compound? Name four groups of organic compounds.
2. Contrast the properties of organic and inorganic compounds.
3. Draw the structural formula for straight-chain pentane. What does this formula tell you about the molecule?
4. Draw the structural formulas for straight-chain heptane and straight-chain octane. What are the differences between these compounds?
5. How are isomers alike? How are they different?
6. Name three alkanes in gasoline.
7. How is the octane rating of gasoline determined?
8. What is meant by a saturated compound? an unsaturated compound? Give an example of each.
9. Distinguish between the structure of the alkene and alkyne series.
10. Write the molecular formula for benzene.
11. List several uses of benzene.
12. Describe two important types of organic reactions.
13. How would you recognize the formula for an alcohol?
14. How are methyl and ethyl alcohols made commercially?
15. How does an inorganic base differ from an alcohol?
16. List three alcohols and one use for each.

CRITICAL THINKING

17. Each compound in the alkane series is saturated. What does this mean?
18. Write the molecular formula of the 50th member of the alkane series.
19. Draw the structural formulas for two branched-chain isomers of the hydrocarbon pentane.
20. High-octane gasoline has the best antiknock properties. However, it is more expensive than low-octane gasoline. How would you decide which gasoline to buy?
21. Draw the structural formulas for four substitution compounds that methane forms with iodine.

FURTHER READING

Baum, Stuart J. *Introduction to Organic and Biological Chemistry.* New York: Macmillan, 1987.

Morrison, Robert Thornton, and Robert Neilson Boyd. *Organic Chemistry,* 5th ed. Boston: Allyn & Bacon, 1987.

Williams, J. R., et al. *Ethanol, Methanol, and Gasohol.* Ann Arbor, Mich.: Ann Arbor Science, 1980.

CHEMISTRY AT WORK FOR YOU

Chemistry plays a major role in providing the basic human needs of shelter, food, and good health. Chemical processes produce the glass and plaster used to construct buildings, the substances added to soil to enhance crop growth, and drugs used to control disease. How are these and other products created by chemistry?

CHAPTER 11

SECTIONS

OBJECTIVES

- ☐ Compare the uses and properties of several kinds of building materials.
- ☐ Identify the materials used in making glass, and describe the glass-making process.
- ☐ Identify the materials used in the making of paints and varnishes, and describe the uses of each.
- ☐ Contrast the chemical reactions of several types of fire extinguishers.
- ☐ Explain how soaps and cleaning agents work, and describe the soap-making process.
- ☐ Describe how chemicals are used as fertilizers, insecticides, and fungicides.
- ☐ Compare uses of disinfectants, drugs, and antibiotics.

11.1 BUILDING MATERIALS

Portland cement is used to make concrete. Cement is made mostly from two abundant raw materials, limestone and clay. Both of these materials are low in cost and are widely distributed. However, large amounts of fuel, power, and heavy equipment are needed to produce cement.

Figure 11-1 Some rotating kilns are 150 m long and 4 m in diameter.

"Portland" is the name given to the cement-making process. (The name comes from Portland, England; the cement resembles the limestone found there.) A mixture of crushed limestone and clay is poured into the upper end of a slow-turning cement kiln as shown in Figure 11-1. The mixture slowly melts into lumps about the size of peas, which are called clinker. After the clinker cools, it is ground into a fine powder. The cement is then placed in bags, ready for use.

The most important use of cement is in the making of concrete. Concrete is made by mixing cement, sand, gravel, and water. The concrete is poured into forms that hold it in place until it hardens. The concrete sets, or becomes firm, in an hour or two but continues to gain strength over a period of days or weeks. This hardening is caused by the formation of crystals. The crystals lock together and make a very hard, artificial stone. To make a much stronger material, steel rods or steel mesh are placed in the form before the concrete is poured. The product of this process is called reinforced concrete. Reinforced concrete is used in the construction of such structures as skyscrapers and bridges.

Bricks and tiles are made from clay. Bricks have been used as building materials for thousands of years. Originally, bricks were made from blocks of clay and then baked in the sun. Today, much harder bricks are made by heating them in a very hot furnace.

In the brick-making process, dried clay and sand are sifted together. Then these materials are mixed with water to form a stiff paste. This paste is molded into bricks and then dried for a few days. Finally, the bricks are fired in a kiln. The temperature and length of firing depend upon the kind of clay used and the hardness wanted. Common bricks are fired at low temperatures, while glassy bricks are fired at higher temperatures. Different substances in the clay produce bricks and tiles of various colors, as in Figure 11-2. Iron compounds, for example, produce a red color.

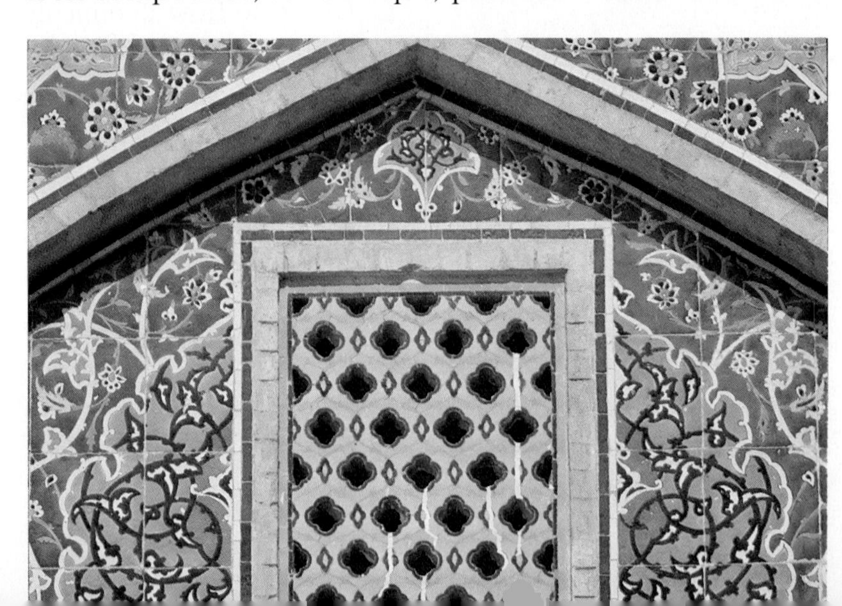

Figure 11-2 Clay is also used to make ornamental tiles for floors, walls, and ceilings. This photograph shows a section of tiles on a building in Isfahan, Iran.

Mortar holds together blocks, bricks, and stones. Mortar is a material used to bond brick, concrete blocks, and other similar building materials. It is a blend of Portland cement, sand, slaked lime [$Ca(OH)_2$], and water. Masons spread mortar on bricks or concrete blocks to build a wall (Figure 11-3). The mortar sets and becomes hard.

Figure 11-3 Bricks are usually laid in horizontal layers. The bricks are arranged in patterns called bonds. Why do you think bonding patterns are important?

The process by which mortar sets to a rock-like mass is not fully understood. Scientists believe that some of the slaked lime reacts slowly with the sand (SiO_2) to form calcium silicate ($CaSiO_3$) as follows:

$$Ca(OH)_2(s) + SiO_2(s) \rightarrow CaSiO_3(s) + H_2O(l)$$

In addition to the above reaction, carbon dioxide from the air slowly unites with some of the slaked lime to form limestone ($CaCO_3$) as shown below:

$$CO_2(g) + Ca(OH)_2(s) \rightarrow CaCO_3(s) + H_2O(l)$$

Plaster and dry wall are used on ceilings and walls. Putting plaster on the walls or ceiling of a building is a two-step process. First, an undercoat is applied, which consists of mortar containing hair or shredded fiber. Then a second, or finish, coat is applied over the undercoat. This coat is a mixture of slaked lime, water, and a powder called plaster of paris. The finish coat dries quickly with a smooth, hard surface.

Figure 11-4 Dry wall is used to panel the interior of homes instead of plaster. What are some advantages of using dry wall?

Plaster of paris is obtained from gypsum (JIP-*sum*). Gypsum, a mineral found in large deposits in some western states, is calcium sulfate ($CaSO_4 \cdot 2H_2O$). Recall that the dot ($\cdot$) in this formula indicates that water molecules are loosely connected to the rest of the compound. When gypsum is heated, it loses part of this water. A fine, white powder of plaster of paris [$(CaSO_4)_2 \cdot H_2O$] remains.

When water is added, plaster of paris forms a paste. This paste hardens, or sets, in a few minutes. As it sets, water molecules combine with the plaster of paris to form gypsum. When you use plaster of paris, you must work fast or the paste will set before you are finished. Obviously, plaster of paris must be stored in moisture-proof containers. Plaster of paris was used for mortar in building the pyramids in Egypt.

Plaster is no longer used in most buildings. Instead, large panels of gypsum, called dry wall, are used. Dry wall is very smooth and is much easier to use, faster to install, and much less costly than plaster (Figure 11-4).

Glass has been made for thousands of years. The Egyptians made glass containers nearly 3,500 years ago. One kind of natural glass is obsidian, produced by volcanic action. Obsidian was used by early humans, who shaped it into tools, arrowheads, jewelry, and money. In the past 80 years, scientists have developed many new and useful kinds of glass such as fiberglass, safety glass, and the photochromatic glass used in sunglasses.

Glass is a mixture, having no fixed boiling and freezing points. It is sometimes called an "undercooled" liquid. Glass can be made by melting together limestone ($CaCO_3$), soda ash (Na_2CO_3), and white sand (SiO_2). These materials are used to make about 90 percent of all glass used in the home, such as drinking glasses, bottles, and window panes. Broken scrap glass, called cullett, is often added to the glass mixture. The old glass, which would otherwise litter the land, provides a low-cost source of new glass.

The raw materials for making glass are mixed and poured into a furnace. The furnace looks like a small swimming pool, except that it is very hot and filled with melted glass. Some furnaces hold as many as 1,200 metric tons of glass.

The hottest part of the furnace, the shallow end, has a temperature greater than 1,500° C. At this temperature, glass is a pasty liquid, like molasses. It takes about a week for raw materials to diffuse from the shallow end of the furnace to the deep end. During this time, bubbles of carbon dioxide gas escape from the glass and some of the raw materials slowly change into a mixture of silicates (minerals containing silicon and oxygen groups). These reactions are shown in the following equations:

$$CaCO_3(s) + SiO_2(s) \xrightarrow{heat} CaSiO_3(l) + CO_2(g)$$

$$Na_2CO_3(s) + SiO_2(s) \xrightarrow{heat} Na_2SiO_3(l) + CO_2(g)$$

Many methods are used in making glass. The molten glass from the furnace is moved into various machines that form the glass into products. From the machines, the glass products are carried to annealing ovens where the products are slowly cooled. The slow cooling reduces strains in the glass that would otherwise be caused by uneven cooling.

The bubbles you see in some glass bottles are caused by trapped carbon dioxide gas. The bubbles become trapped if the temperature in the furnace is not high enough to drive off all of the gas.

To make plate glass, a machine dips a horizontal iron rod into the furnace. As the rod is lifted straight up, the melted glass clings to it, forming a large sheet. The thickness of the sheet is determined by controlling the speed of the rising rod and by keeping the glass in the furnace at the proper temperature. After the glass cools, the edges are trimmed.

A newer method of making plate glass is the float process. In this process, molten glass is floated on the surface of a bath of molten tin. Gravity keeps the tin very flat so that the glass layer also takes on this flat shape. See Figure 11-5.

Figure 11-5 Workers in a float glass plant in Chehalis, Washington, inspect newly processed sheets of plate glass. What other methods are used in making plate glass?

Most kinds of glass cannot withstand sudden changes in temperature without breaking or cracking. Pyrex brand glassware is one exception; it is often used to make laboratory equipment or in baking dishes for the home. When heated, Pyrex glass expands much less than ordinary glass. Boric oxide, added to the glass in small amounts, gives Pyrex this property.

If a glassmaker wants to produce a clear, colorless glass, pure ingredients must be used. Traces of iron oxide in the sand will give glass a pale green color. This type of glass is used in making some kinds of bottles.

Nearly all glass products are machine-made. Bottle-making machines imitate the process used by glassblowers (Figure 11-6). These machines have a great many pipes that dip into the melted glass. Compressed air is used to blow the glass. One machine can turn out thousands of bottles in one hour.

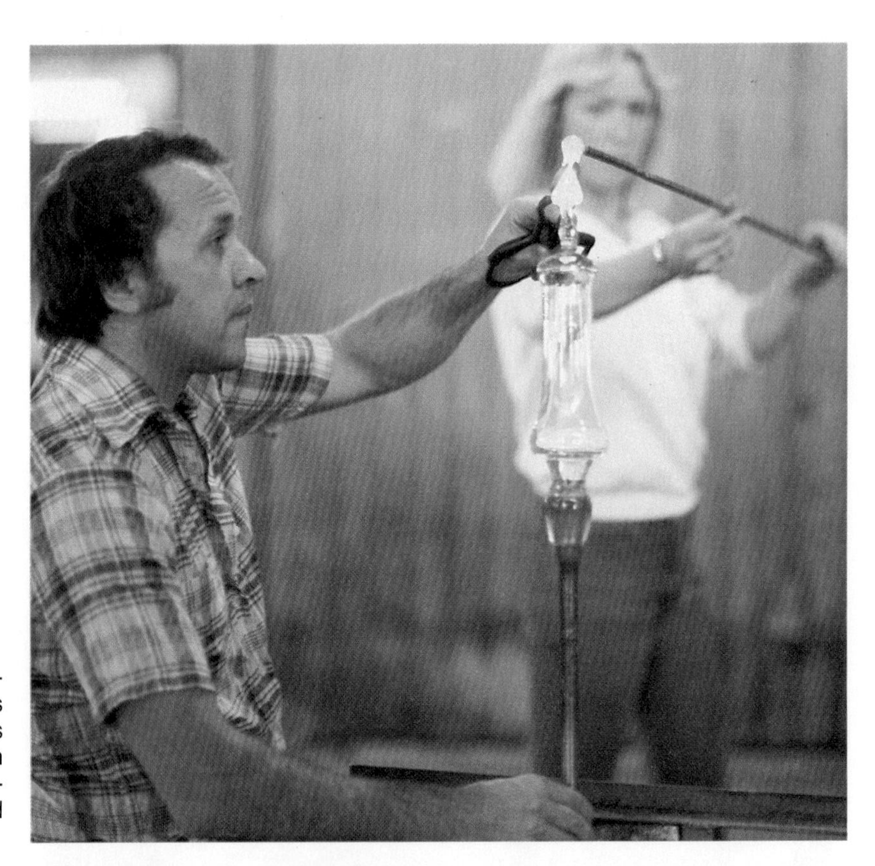

Figure 11-6 In glassblowing, the pear-shaped end of a hollow blowpipe is dipped in molten glass. The glass is shaped into a hollow bulb by blowing on the pipe. Periodically during this process, the glass is reheated until it is red hot so that it can be easily shaped.

Paint provides color. Paints were used by early humans to decorate their caves, to coat tools, and even to decorate their faces. Today, people use many kinds of paint, not only to protect surfaces, but also to make their surroundings more beautiful.

Figure 11-7 Paint is made by mixing colored pigment powder in a liquid that contains resins and a solvent.

Oil paint is mostly a mixture of the following types of substances: (1) a drying oil, such as linseed oil, to act as a binder and to absorb oxygen from the air as the paint dries; (2) a pigment such as zinc oxide, to give the paint body, covering power, and the desired color; (3) a thinner, such as turpentine, to make the paint spread easily; and (4) a drying agent, such as manganese dioxide, to act as a catalyst to speed up the drying process.

Drying oils in paints form elastic solids by reacting with oxygen in the air. The rubber-like coating that soon appears on the surface of an open can of paint is the result of this process.

Common drying oils include linseed oil, which is pressed from ripe flaxseeds; tung oil, from the nut of the China tung trees; and soybean oil. Other oils for this purpose come from seeds of castor, safflower, sunflower, and hemp.

Paint pigments are a mixture of several substances. Three of the best white pigments are white lead, zinc oxide, and titanium oxide, although leaded paints, which are toxic, are no longer used to paint the interiors of homes. For colored paints, small amounts of colored matter are added to the white pigments. See Figure 11-7.

Paints and varnishes protect surfaces. One of the most important functions of paint is to form a barrier between a material, like wood or metals, and air and moisture. Air and moisture contribute to the rotting of wood and the oxidation of metals such as iron. Oxidation, such as rust, weakens metals and the structures from which they are made. If you

TREATING METALS FOR RUST PROTECTION

activity

OBJECTIVE: Perform an experiment to find which materials will prevent rust.

PROCESS SKILLS

In this activity, you will *organize data* and *observe* the effects on the rusting process of coating nails with paint, oil, and copper sulfate.

MATERIALS

tongs
4 iron nails
sulfuric acid (dilute)
glass container
oil paint
oil
copper(II) sulfate solution

PROCEDURE

1. Using the tongs, carefully place the four nails into the container of dilute sulfuric acid for several minutes. CAUTION: *Sulfuric acid is corrosive. Do not touch the acid or get it on your clothes. Have eyewash available and wear goggles and a laboratory apron.*
2. Remove the nails with the tongs. Rinse with water and dry the nails.
3. Coat one nail with oil paint and one with oil. Dip the third nail into the copper(II) sulfate solution until the solution turns a copper color. Leave the last nail untreated as a control.
4. Hang all the nails on a support outdoors for at least a week. Observe what happens.

OBSERVATIONS AND CONCLUSIONS

1. Record your observations.
2. Which nails were protected from rust?
3. Explain how these nails were protected from rusting.

have ever wondered why the first coat of paint on a bridge is often a reddish-orange color, it is because red lead oxide paint is used to protect steel structures from rusting.

Emulsion paint products have largely replaced oil-based paints for most uses. Emulsion paints are water-thinned, which enables them to dry rapidly. When a resin emulsion is used as a binder in water-thinned paints, the paints are called latex paints. These paints have good covering properties and form easily cleaned, satin-like finishes. For most interior uses, latex paints are preferred over oil paints. When wet, latex paints can be washed away with water, which makes cleaning less of a chore.

Varnish is made by boiling certain gum-like materials, called resins, in oils. No pigment is added so that the grain of the wood will not be hidden. However, pigments can be added to varnish to make enamel. Enamel gives a higher shine and often a hard finish.

Lacquers are solutions of resins that dry to form a hard, shiny finish. Although oil paint hardens by taking in oxygen from the air, lacquers dry by evaporation of the solvent. The drying rate of lacquers depends upon the type of solvent used. Some lacquers dry so fast that you cannot use a brush to apply them. These lacquers must be sprayed on.

11.2 CHEMISTRY AND FIRES

Three conditions are needed to permit a fire to burn. A fire will start provided three basic requirements are met: (1) sufficient heat to bring a material to its kindling temperature; (2) a material, or fuel, that will burn; and (3) oxygen to support the burning. To put out a fire, it is necessary to remove only one of these ingredients.

Flooding a fire with enough water cuts off the oxygen supply and also cools the burning material to below its kindling temperature. Sand is another common material that can be used to smother or cut off oxygen from a fire.

Water should not be used on oil fires because it may spread the oil, and thus spread the fire. It is also dangerous to use water near electrical wiring because water conducts an electric current. Therefore, water is not an ideal substance to put out all fires.

Chemicals help to put out fires. The oldest chemical fire extinguisher still in common use today is the soda-acid type. This extinguisher consists of a strong copper tank nearly full of a solution of sodium hydrogen carbonate ($NaHCO_3$). A bottle of sulfuric acid (H_2SO_4) with a loose stopper hangs from the inside top of the tank, as shown in Figure 11-8.

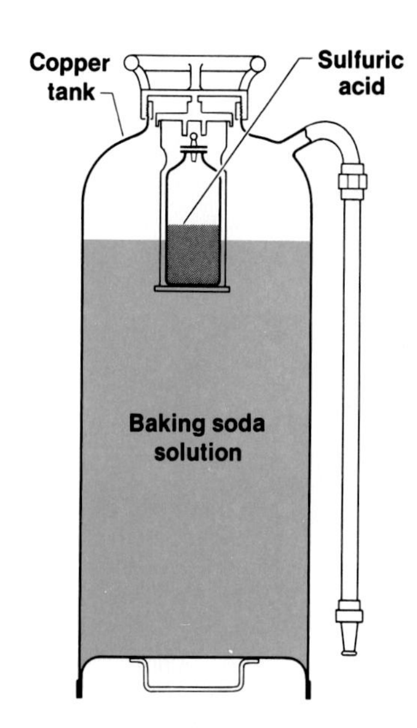

Figure 11-8 When a soda-acid fire extinguisher is turned upside down, the acid reacts with the baking soda solution to give off carbon dioxide.

When the tank is turned upside down, the stopper falls from the acid bottle. The acid spills into the baking soda solution. The reaction of the acid and baking soda forms sodium sulfate (Na_2SO_4), water, and carbon dioxide (CO_2) as shown in the following equation:

$$H_2SO_4(l) + 2NaHCO_3(aq) \rightarrow Na_2SO_4(s) + 2H_2O(l) + 2CO_2(g)$$

This reaction produces pressure inside the tank. The pressure forces a solution of water, carbon dioxide, and sodium sulfate out of the hose, which helps put out the fire by cooling and smothering the burning material. Why do you think this solution is more effective than plain water?

This type of extinguisher, however, has several drawbacks: (1) The solution always contains some acid that has not reacted with the soda. This acid could cause damage to clothing and furniture. (2) This type of extinguisher cannot be used around electric wiring because of the danger of an electric shock. (3) The soda-acid type of extinguisher is not effective for oil or gasoline fires since the stream of liquid tends to sink below the blazing fuel, thus causing the fire to spread. (4) The tank is large and heavy, and once started, there is no way of stopping it until it is empty.

A foam extinguisher, however, is very effective in fighting oil fires. This type of extinguisher produces a blanket of aluminum hydroxide and carbon dioxide. Foam extinguishers, however, are not often used around the house because the foam damages rugs and furniture.

A carbon dioxide extinguisher consists of a strong, steel cylinder that holds liquid carbon dioxide under great pressure. When the valve is opened, carbon dioxide gas rushes out through a cone-shaped nozzle. The cooling effect caused by the sudden expansion of the gas changes most of the escaping gas into carbon dioxide "snow."

The "snow" has a temperature of about −80° C. This lowers the temperature of the burning material below its kindling point. The heat from the fire causes the "snow" to change to dense CO_2 gas, which prevents oxygen from reaching the fire. The carbon dioxide does not harm home furnishings. The liquid CO_2 extinguisher works well against oil and electrical fires (Figure 11-9).

Figure 11-9 Carbon dioxide extinguishers are well suited for home use.

Fire retardants help prevent burning. Special substances can be added during the manufacture of textiles and building materials to make them less flammable. These substances, known as fire retardants, produce physical or chemical changes that interfere with combustion. When exposed to fire, some fire retardants cause the treated material to release gases that quench the flame. Other fire retardants swell up to form an insulating surface layer between the material and the fire. A material treated with fire retardants is said to be fire resistant. In the United States, fire-resistant materials must be used in the construction of houses, schools, and children's sleepwear.

11.3 CHEMISTRY AND CLEANING AGENTS

Many cleaning agents are used in the home. Cleanliness is essential to good health. Soap and other detergents are the most common cleaning agents found in the home. A **detergent** is a substance that helps remove dirt from various materials. Household chemicals used for cleaning also include scouring powders, solvents, and special purpose cleaners for metals, glass, and textiles.

Scouring powders loosen dirt by grinding or wearing it away. Powdered sand and pumice (a type of volcanic rock) are often used for scouring. Cleaning solvents dissolve grease and are often used for dry cleaning.

Household ammonia is used to clean metals, glass, and linoleum tiles. Ammonia is advantageous since it does not leave a film when it dries. This weak base is a good solvent for grease, although like all other bases, it is harmful to most painted surfaces.

Soap is a common detergent. Soap is made by boiling a fat, such as lard, with a solution of sodium hydroxide (lye), NaOH. Fats are compounds of carbon, hydrogen, and oxygen. The most common form of fat for making laundry soap is tallow, which is obtained from animal fats.

The word equation for the soap-making process is

$$\text{fat} + \text{lye} \rightarrow \text{glycerin} + \text{soap}$$

Glycerin remains in some soaps. In other soaps, however, the glycerin is recovered as a by-product. The glycerin then may be used in making cellophane, printer's ink, medicines, cosmetic lotions, and some explosives.

Soapmakers add oils to improve the quality of soap. Oils used to make toilet soap include coconut, palm, and olive. Soybean, corn, castor, and linseed oils are also used. Perfume is added to give the soap a pleasant aroma. Water softeners and dyes may also be added during the soapmaking process.

Soaps clean by physical action. In the cleaning process, soap and synthetic detergent molecules surround greasy dirt particles. A typical soap molecule has a rather long structure. The "sodium end" of the molecule is water soluble. The other end of the molecule is oil soluble and tends to

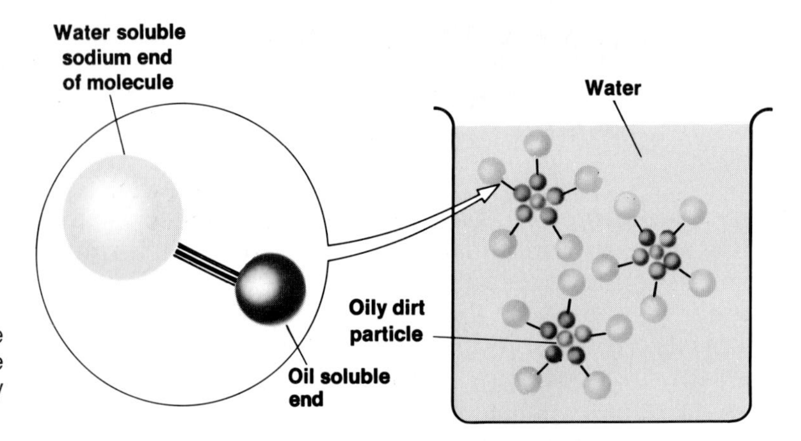

Figure 11-10 The oil-soluble end of the soap molecule prepares the way for the water-soluble sodium end to work away the dirt particles.

dissolve the oily covering of dirt particles. Figure 11-10 shows how the soapy layer acts on the oily covering of the dirt. After the oil-soluble end weakens the greasy film, the dirt particles are easily washed away by water.

Synthetic (soapless) detergents have largely replaced soap for laundry purposes. Like soap, all detergents on the market today are biodegradable, that is, they are broken down by living things such as bacteria. Of what significance is this?

The role of a soapless detergent is that of a wetting agent. Soapless detergents make water "wetter" by reducing the force of attraction between water molecules. One result is that detergents allow water to spread through cloth fibers more readily, loosening hard to reach dirt.

When a garment is dry-cleaned, it is rinsed in a solvent other than water. Among other things, the use of special dry-cleaning solvents reduces the shrinking of fabrics and keeps colors from running together, both of which tend to occur when certain fabrics are laundered in water.

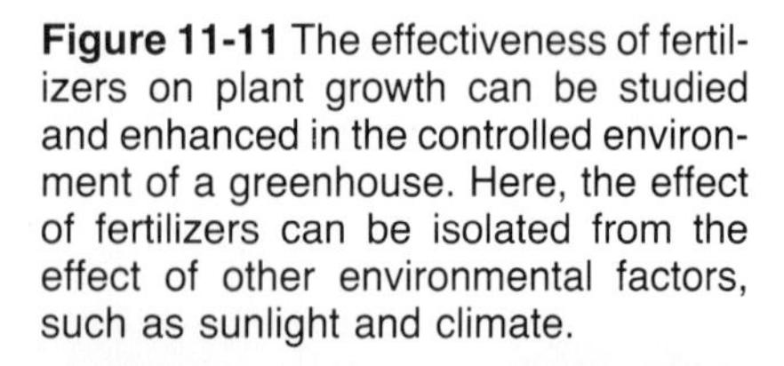

Figure 11-11 The effectiveness of fertilizers on plant growth can be studied and enhanced in the controlled environment of a greenhouse. Here, the effect of fertilizers can be isolated from the effect of other environmental factors, such as sunlight and climate.

11.4 CHEMISTRY AND FOOD

Fertilizers increase crop yields. Fertile soil contains minerals and living organisms such as bacteria, fungi, plants, and animals. Decayed living things account for the presence of plant nutrients, especially compounds of nitrogen, in the soil. If such nutrients are missing or scarce, soils must be fertilized in order to insure healthy crops (Figure 11-11).

About 16 elements are needed for the growth of healthy plants. The soil, or fertilizers, supply some of these, including nitrogen, phosphorus, and potassium which are needed in relatively large amounts. Nitrogen is necessary for the making of proteins. Phosphorus is needed for the formation of healthy roots. Potassium acts as a catalyst in the making of starches and sugars. These and other substances must be in an ionic and water soluble form before they can be used for the growth and development of plants. Therefore, most fertilizers contain soluble ionic compounds of the three major elements.

Many states have laws that require the components of commercial fertilizers to be stated on the label. For example, the label "4-6-8" means that the fertilizer contains 4 percent nitrogen, 6 percent phosphorus, and 8 percent potassium.

Organic fertilizers are also used to nourish crops. These fertilizers differ from commercial fertilizers in that they originate from plant or animal sources. Organic fertilizers promote growth of helpful bacteria in the soil. Also, they may increase the ability of soil to hold water and nutrients, and they do not pollute groundwater as can commercial fertilizers.

Chemicals protect crops from insects and diseases. To grow healthy plants, harmful insects must be controlled. Two common types of insect pests are the sucking insects and the chewing insects. These unwanted insects can be controlled in a number of ways, including the use of chemicals called **insecticides**.

For many years, DDT (dichloro-diphenyl-trichloroethane) was a widely used insecticide. This compound kills insects by destroying their nerve tissues. Even though DDT is very effective, its use is now banned in the United States for a number of reasons. For one thing, DDT remains active for years and is not insect-specific, that is, it kills useful insects as well as harmful ones. DDT is also stored in plants and in animals that eat plants, and has been shown to be poisonous to various animals. It can endanger the lives of birds and fish, and may contaminate food eaten by humans. DDT has moved through the world's food chains. It has been found in the bodies of polar bears and in the ice of the Antarctic. DDT has also been found in the fatty tissues of humans and is thought to be a cancer-causing agent.

Arsenic compounds of lead and calcium are often used in fruit sprays. Since these compounds and many other insecticides are poisonous to people and animals other than insects, fruits and vegetables must always be thoroughly washed before they are eaten. See Figure 11-12.

Figure 11-12 This photo shows how crops are sprayed with insecticides by airplanes. What are the advantages of "crop-dusting?"

Plants are also attacked by various kinds of fungi. The fungi include rusts, molds, mildew, and smut. All of these can injure plants. Chemicals that destroy fungi are called **fungicides**. Sulfur, applied as a fine dust, is used as a fungicide. Bordeaux mixture, containing copper(II) sulfate, lime, and water, is a standard spray for many plant fungus diseases.

Leavening agents are used in baking. When bakers make bread, they let the bread dough stand in the warm air for several hours before baking. During this period, the dough is kneaded several times to assure a more thorough mixing of the ingredients. This practice is needed for the dough to "rise." Years ago, bakers would save a lump of the dough to be mixed with the next batch of bread. Although people knew that these procedures were needed, they did not understand what caused the bread to rise.

Louis Pasteur, a French scientist, solved the mystery of this process. Pasteur found that while the dough was in warm air, tiny fungi cells, called yeasts, became embedded in the dough. The yeast cells produced chemicals called enzymes (EN-*zimes*), or organic catalysts. The enzymes set off a chemical reaction in the dough. This reaction, known as **fermentation**, produced bubbles of carbon dioxide gas. In baking, the heated bubbles increase in size, causing the dough to expand or rise. A substance that causes dough to rise is called a **leavening agent.**

Today, instead of adding a lump of old dough, yeast is added directly to the new dough. The enzymes released by the yeast cells start the fermentation in the dough. When a batch of dough is made, a little sugar is added to the flour. The yeast cells then provide the enzyme that changes the sugar to alcohol and carbon dioxide. The equation for this reaction using glucose, one type of sugar, is

$$\underset{\text{sugar}}{C_6H_{12}O_6(s)} \rightarrow \underset{\text{grain alcohol}}{2C_2H_5OH(l)} + \underset{\text{carbon dioxide}}{2CO_2(g)}$$

The heat of baking drives off the alcohol from the finished bread. Carbon dioxide bubbles cause air holes in the bread. These air holes make the bread lighter and more easily digested. One disadvantage of this method of baking is that it takes several hours for the dough to rise when yeast is used as a leavening agent. See Figure 11-13.

Baking soda and baking powders are also commonly used as leavening agents. Baking soda reacts quickly with acid to produce carbon dioxide. Therefore, baking soda is sometimes mixed with flour and sour milk. The lactic acid in the sour milk causes the release of carbon dioxide gas from the baking soda, and the dough rises.

Figure 11-13 In baking bread, yeast first causes the bread to rise. Then the heat of baking produces the "oven rise." What is one disadvantage of using yeast in baking?

A disadvantage of this process is that it is hard to control the reaction, since sour milk varies in lactic acid content. Specially prepared baking powders that do not require sour milk are now available for home baking.

Baking powders are leavening agents that contain a mixture of the following ingredients: (1) baking soda as a source of carbon dioxide gas; (2) a substance, such as cream of tartar, that produces an acid when water is added; and (3) corn starch to keep the mixture dry.

When water is added to the baking powder, it reacts with the acid substance to produce hydrogen ions. The hydrogen ions react with baking soda to produce carbon dioxide gas. Why are baking powders packed in moisture-proof containers?

The advantage of baking powder is that the reacting substances are present in just the right amounts. Today, you can find "ready mixes" for baking breads and cakes.

11.5 CHEMICALS THAT PROTECT YOUR HEALTH

Disinfectants stop the growth of germs. Chemicals that retard or stop the growth of bacteria on nonliving surfaces include **disinfectants**. When such chemicals are too strong to be used on the skin or other body tissues, they are called **germicides**. When the chemicals are safe to use on the skin but still strong enough to stop the growth of bacteria, they are called **antiseptics.** In many cases, antiseptics are dilute solutions of germicides. Soap and water is the most common antiseptic.

THE FERMENTATION PROCESS

activity

OBJECTIVE: Observe the process of fermentation

PROCESS SKILLS

In this activity, you will *observe, analyze data,* and *experiment* in order to find evidence for fermentation.

MATERIALS

10 percent unsulfured molasses solution
green peas
1/4 package of dry yeast
glass container
cloth
limewater

PROCEDURE

1. Prepare about 0.5 L of a 10 percent unsulfured molasses solution.
2. Add some green peas to the solution.
3. Add one-fourth of a package of yeast.
4. Cover the solution with a piece of cloth and keep it in a warm dark place for several days.
5. Describe the odor you detect after several days.
6. Devise a way to collect the gas given off. Bubble it through limewater. CAUTION: *Wear goggles, gloves, and a laboratory apron.*

OBSERVATIONS AND CONCLUSIONS

1. Record your observations.
2. Explain the odor you detected after several days.
3. What happened when you bubbled the gas through the limewater? Explain.

A three-percent solution of hydrogen peroxide (H_2O_2) is another common antiseptic. It produces oxygen, which inhibits the growth of some bacteria. Since strong light will break down this compound, hydrogen peroxide is stored in light-proof bottles. Even so, hydrogen peroxide slowly loses its strength by breaking down into water and oxygen as follows:

$$2H_2O_2(aq) \rightarrow 2H_2O(l) + O_2(g)$$

Germicides are too harsh to be used on living tissue. Sodium hypochlorite (NaOCl) is a germicide that is used as a household disinfectant and as a bleach. Chloride of lime ($Ca(OCl)_2$) is used as a swimming pool disinfectant. These compounds are poisonous and should be handled with great care. However, the amount of chlorine used in a pool is so diluted, it is not generally harmful to one's health.

Drugs, when used properly, can ease pain and fight disease. Primitive people found out by chance that some of the plants growing about them seemed to relieve pain, heal sores, and even cure diseases. These plants were the first drugs. **Drugs** are chemicals that affect living processes.

Drugs such as digitalis, which is used to regulate heartbeat, morphine, which is used as a pain-killer, and quinine, which is used to treat malaria, are still derived from plants. Other drugs are composed of minerals, such as Epsom salts ($MgSO_4$) and milk of magnesia [$Mg(OH)_2$]. Still other drugs—vaccines and gland extracts, for example—come from animal sources. However, most drugs today are prepared in large chemical laboratories or factories. Some drugs are now made through the techniques of genetic engineering. In this process, simple organisms such as bacteria and yeasts are transformed genetically so that they produce a desired drug. Human insulin, used to treat diabetes, is one such drug.

Drugs must be used wisely, under a doctor's care. Some drugs are habit-forming whereas others are poisonous. In fact, any drug can be poisonous if it is misused.

Drugs used to relieve pain are called **analgesics** (AN-*uhl*-JEE-*ziks*). Perhaps the most common analgesic drug is aspirin. Aspirin, which was first used in Germany over 80 years ago, is the most common painkiller available without a prescription. Americans consume millions of aspirin tablets every year.

Some pain-killing drugs, called **narcotics**, affect the nervous system. Although helpful, they are habit-forming. Narcotics are obtained from the opium poppy and other plants or are made from chemicals. Physicians use these drugs to relieve pain after operations or during a painful illness, or to cause sleep. However, continued and unsupervised use of narcotics can lead to drug addiction. Some well-known narcotics are codeine, morphine, and the much more potent drug heroin.

Other drugs besides narcotics can be habit-forming. Among these drugs are barbiturates and amphetamines. Barbiturates act as a sedative and are sometimes called "downers." Barbiturates also are often used to calm people or help them sleep. Regular use, however, may cause addiction, and large amounts can cause coma or death.

Doctors prescribe amphetamines to decrease appetite and to treat certain diseases. Amphetamines, sometimes called "uppers," are also used as stimulants to relieve fatigue. They cause an increase in heart rate and blood pressure. Large, repeated doses of amphetamines can cause tension, anxiety, and paranoia. An overdose can cause brain damage or death.

Since 1935, physicians have known that certain organic sulfur compounds, called sulfa drugs, produce outstanding results in controlling certain diseases. It was found that sulfa drugs prevent bacteria from multiplying. This gave the white blood cells in the body a chance to attack and destroy the bacteria.

The first sulfa drug, sulfanilamide (SUL-*fuh*-NIL-*uh-mide*), was introduced in time to save countless wounded soldiers from serious infections during World War II. In the past, sulfa drugs were used for treating certain types of pneumonia, strep throat, dysentery, venereal diseases, and blood poisoning. Today, sulfa drugs are used mainly to treat urinary tract infections.

Many diseases are controlled by antibiotics. Eventually, the use of sulfa drugs decreased because more powerful drugs were developed. Penicillin, the antibacterial properties of which were discovered in 1929 by British scientist Dr. Alexander Fleming, became the first successful antibiotic. **Antibiotics** (AN-*tie-bie*-AHT-*ihks*) are chemicals that are originally produced by living organisms and prevent the growth of bacteria, as shown in Figure 11-14. Penicillin is a crystalline substance taken from green molds similar to those that sometimes grow on bread.

Figure 11-14 Paper tabs containing antibiotics rest on a carpet of bacteria. What evidence is present to suggest that an antibiotic has inhibited the growth of the bacteria?

A serious problem in the use of sulfa drugs and antibiotics is that bacteria tend to build up a resistance to them. Sometimes, however, an infection that has become resistant to one drug can be treated by switching to another drug.

The proper use of toothpaste or tooth powder helps prevent tooth decay by keeping the teeth clean and by slowing bacterial growth. An effective low-cost tooth cleaner can be made by using one part of baking soda, two parts of table salt, and a little chalk ($CaCO_3$) as a mild abrasive. Some flavoring and a small amount of detergent may be added. Commercial toothpastes and powders, however, contain a small amount of chalk, calcium phosphate or calcium sulfate, flavoring, and a detergent. Fluoride toothpastes include very small amounts of tin(II) fluoride to harden tooth enamel.

Cosmetics have been used throughout history. The Egyptians of several thousand years ago were the earliest known users of cosmetics (Figure 11-15). Cosmetics are materials applied to make the body more attractive or alter

Figure 11-15 Cosmetics, such as the elaborate eye makeup shown in this photograph, were used by Egyptians thousands of years ago.

a person's appearance. The Food and Drug Administration regulates cosmetics in the United States, making sure they are safe and properly labeled for consumer use. Among the many cosmetics found on the market are face and bath powders, blush, facial creams, nail polish, and many kinds of lotions.

Face and bath powders contain calcium carbonate that helps absorb moisture. Kaolin (a basic material in clay) is added to help the powder stick to the skin. Titanium oxide and zinc oxide, finely ground and thoroughly mixed, are used as white pigments. Coloring matter and perfumes are then added.

Blush is a face powder to which some red iron(III) oxide has been added. Lipsticks are mixtures of fat or wax with dyes, to which an oil is added as a softener. About 50 percent of lipsticks are made using castor oil, perfume, and coloring matter.

Facial creams are emulsions of fats, waxes, and oils in water. Borax and other substances are used to make the oils and water blend together. The oils used include olive oil, almond oil, and mineral oil. Beeswax, mineral oil, and borax are contained in most cold creams. Lanolin, which is made from a greasy coating found on the wool of sheep, is an excellent skin softener and is often used in creams, lotions, and hair preparations. Some creams contain potassium soap for cleaning action.

Nail polish is made from lacquers, coloring matter, and a solvent. Nail polish remover contains organic solvents such as acetone.

BIOGRAPHY

LOUIS PASTEUR

discoverer of the germ theory

Louis Pasteur (1822–1895), a French scientist, was noted for his work on fermentation and decay. He developed the germ theory of disease and discovered that food could be made sterile by a heat treatment process, later called pasteurization.

One of the puzzling problems of Pasteur's day was the theory of "spontaneous generation," which stated that living things originated from nonliving matter. Pasteur observed that fermentation of certain liquids was hastened by their exposure to air. This caused him to wonder whether invisible organisms were always present in the atmosphere, or whether they were spontaneously generated. Pasteur performed a series of experiments. In one experiment, he filtered air and then exposed unfermented liquids to the clean air. Pasteur was able to show that the organisms causing fermentation were not spontaneously generated. Instead, they came from similar organisms present in ordinary air.

The wine makers of France found that their wine often became sour. The wine turned to vinegar when the fermentation process went too far. Pasteur found that certain kinds of bacteria caused the spoilage. When the bacteria were not present, no acids were formed and good wine was produced.

The harmful organisms spoiling the wine could have been killed by boiling, but this would have also destroyed the flavor of the wine. Pasteur found a solution by heating the fermented liquid high enough to kill the bacteria while keeping the temperature low enough to preserve the flavor of the wine. This discovery saved the French wine industry and introduced the process of pasteurization to the world.

Pasteur also reasoned that if bacteria could cause wine to become "sick," perhaps bacteria could also cause sickness in animals and people. He then performed experiments that produced evidence to support this idea, which later became known as the germ theory of disease.

CHAPTER REVIEW 11

SUMMARY

1. A mixture of clay and limestone is heated in a kiln to make Portland cement.
2. Concrete is made by mixing cement, sand, gravel, and water.
3. Common glass is made by mixing sand, soda ash, and limestone.
4. Paints, varnishes, enamels, and lacquers protect wood and metal surfaces.
5. Fires have three requirements to keep burning: (1) heat to raise the material to its kindling temperature, (2) something to burn (fuel), and (3) oxygen.
6. The action of detergents depends mainly on the structure of their molecules, one end of which is water soluble whereas the other end is oil soluble.
7. Nitrogen, phosphorus, and potassium are elements needed in relatively large amounts for the proper growth of plants.
8. Leavening agents in dough cause it to rise by forming bubbles of carbon dioxide gas.
9. Drugs are substances that effect living processes and many are used to fight disease or to ease pain.
10. Antibiotics prevent the growth of bacteria and are substances originally produced by living organisms.

VOCABULARY

Match the item in the left column with the best answer in the right column. Do not write in this book.

1. analgesics	a. substances that remove dirt
2. antibiotics	b. chemicals that destroy unwanted insects
3. antiseptics	c. chemicals that cure or prevent diseases, or ease pain
4. detergents	d. drugs used on skin to stop growth of bacteria
5. disinfectants	e. chemicals that destroy fungi
6. drugs	f. chemicals used on nonliving surfaces that kill or stop growth of bacteria
7. fermentation	g. substance that causes dough to rise
8. fungicides	h. chemical reaction that produces carbon dioxide gas
9. germicides	i. chemicals too strong to use on the skin or body tissues
10. insecticides	j. drugs used to relieve pain
11. leavening agent	k. pain-killing drugs that affect the nervous system
12. narcotics	l. chemicals originally produced by living organisms that prevent growth of bacteria

REVIEW QUESTIONS

1. Describe the making of Portland cement.
2. What is the difference between cement and concrete?
3. What is mortar made of?
4. Why is hair or shredded fiber added to plaster?
5. How are bricks made?
6. What are the raw materials used in making glass?
7. Describe how glass is made.
8. Name four items commonly found in oil paint.
9. What is latex paint? What are the advantages of using it?
10. Describe the difference between the drying action of ordinary oil paint and that of lacquer.
11. Describe how the soda acid fire extinguisher works.
12. List some advantages and some disadvantages of the soda-acid type fire extinguisher and of the foam-type extinguisher.
13. What is a detergent? List several.
14. Describe how synthetic detergents clean clothes.
15. Write the word equation for making soap.
16. What useful by-product is generated by soap making? How is this by-product used?
17. Describe the key reactions of baking powder.
18. Why may baking powder become useless after standing in an open container for a long time?
19. What elements are needed in large amounts for the proper growth of plants?
20. A fertilizer is labeled "5-10-5." What does this mean?
21. Explain the difference between insecticides and fungicides.
22. If you cut your finger, should you use an antiseptic or a germicide? Why?
23. Describe the action of hydrogen peroxide as an antiseptic. Why must it be stored in dark bottles?
24. Describe the reaction that takes place when yeast acts on sugar.
25. What effects do each of the following classes of drugs have upon the body: (a) narcotics, (b) barbiturates, and (c) amphetamines?
26. Describe the difference between the source of sulfa drugs and that of most antibiotics.

CRITICAL THINKING

27. What might account for bubbles appearing in finished glass?
28. Which type of paint is best used to cover the walls of a kitchen or bathroom? Explain your answer.
29. Which type of fire extinguisher is best used in the home? Explain your answer.
30. DDT has been found in the bodies of polar bears and in ice in the Antarctic. Suggest how this might occur.
31. Why might a doctor hesitate to prescribe long-term use of barbiturates and amphetamines?

FURTHER READING

Bansal, N. P., and R. H. Doremus. *Handbook of Glass Properties*. San Diego, Calif.: Academic Press, Inc., 1986.

Bates, Robert L. *Stone, Clay, Glass: How Building Materials Are Found and Used*. Hillside, N.J.: Enslow Publishers, 1987.

Meyer, Carolyn. *Being Beautiful: Story of Cosmetics from Ancient Art to Modern Science*. New York: William Morrow and Co., Inc., 1977.

Weis, Malcolm E. *Why Glass Breaks, Rubber Bends, and Glue Sticks*. San Diego, Calif.: Harcourt, Brace, Jovanovich, Inc., 1979.

Woods, Geraldine. *Drug Use and Drug Abuse*, revised ed. New York: Franklin Watts, Inc., 1986.

FOSSIL FUELS

A drilling rig off the coast of Alaska labors to satisfy the world's ever-increasing hunger for fossil fuels. What types of fuels are extracted from the ground in this way? How must they be refined before they can provide the energy to heat our homes and run our automobiles?

CHAPTER 12

SECTIONS

OBJECTIVES

- ☐ Describe the various stages in the formation of coal.
- ☐ Describe the making of charcoal and coke, and the formation of water gas and natural gas.
- ☐ Compare the products formed by complete and incomplete combustion of a hydrocarbon fuel.
- ☐ Describe the destructive distillation of soft coal and the fractional distillation of petroleum.

12.1 COAL, COKE, AND CHARCOAL

Coal is a fossil fuel. Coal gives off heat when it is burned. However, this trait alone does not make it a good fuel. A **fuel** is a substance that can be burned to produce heat at a reasonable cost. To be particularly useful, a fuel should be easy to store. It should burn cleanly, producing no unwanted by-products.

The greatest period of coal formation dates back approximately 300 million years. Some types of coal, however, began to form as early as 440 million years ago. Scientists believe that coal is mostly the buried remains of tropical plants. In the past, swampy sections of the earth were covered by very dense plant life. The huge tree ferns and mosses formed many layers as they died. With time, these layers of dead plants were covered with sediment and rock. Heat from the earth's interior and pressure from the weight of the earth turned the plant remains into coal. Some samples of coal contain fossils. Coal and other fuels that formed from organic matter are called **fossil fuels**.

Coal forms in several stages. **Peat** is the first stage in the formation of coal. Peat is a soft, brown, spongy material composed of plant matter (carbon) that has been changed through the action of heat and pressure. The fibrous structure of peat clearly shows its plant origin (Figure 12-1).

Figure 12-1 Four stages in the formation of coal include (from left to right) peat, lignite, bituminous, and anthracite. Note that the peat appears brown and fibrous.

Since peat contains a large amount of moisture, it produces a great deal of smoke as it burns, making it undesirable as a fuel. Despite this, peat has been used as a fuel in Europe for thousands of years because of its abundance.

As peat continues to decompose over a long period of time, it loses most of its fiber. This second stage in coal formation is known as **lignite**, brown coal. Lignite is harder than peat, contains less moisture, and has a higher percent of carbon. The amount of moisture in lignite is about 50 percent.

Heat and pressure slowly change some lignite into **bituminous** (*buh*-T(Y)OO-*muh*-NUS) **coal**, or soft coal. Bituminous coal contains even less moisture than lignite. However, it contains much less volatile matter. When heated, volatile matter forms gases that burn easily.

In some places, such as eastern Pennsylvania, the folding of the earth's crust has produced still greater pressure. This increase in pressure changes some bituminous coal into **anthracite** (AN-*thruh*-SITE), or hard coal. Anthracite is often found deeper in the ground, where pressures are greater, than bituminous coal. In addition, it contains a higher percent of carbon and a lower percent of volatile matter.

AMOUNT OF ASH IN COAL

activity

OBJECTIVE: Perform an experiment to find the percent of ash in coal.

PROCESS SKILLS
In this activity, you will *measure* and *collect data* and then *analyze data* in order to determine the concentration of ash in coal.

MATERIALS
- sample of coal
- mortar and pestle
- porcelain crucible
- ringstand with ring
- gas burner
- metric balance
- safety goggles

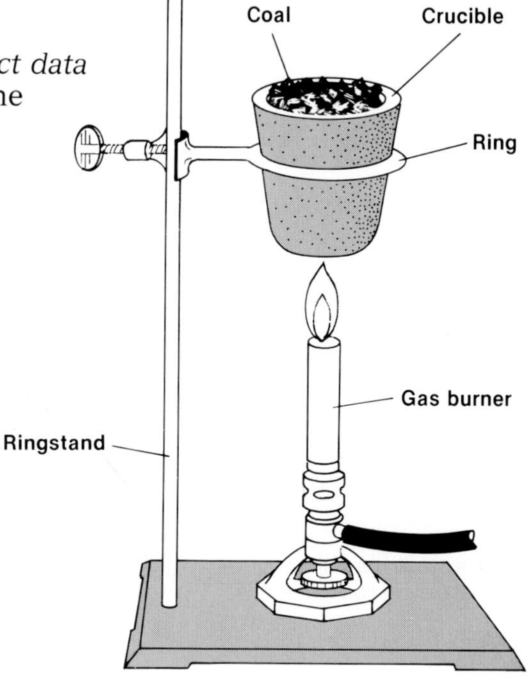

PROCEDURE
1. Put on your safety goggles. Using a mortar and pestle, crush a sample of coal and determine the mass of the sample by using a metric balance.
2. Determine the mass of an empty crucible. Set up the crucible on a ringstand and place the coal in the crucible, as shown in the diagram.
3. Heat the sample until it is completely burned. CAUTION: *This procedure should be done under a fume hood.*
4. Determine the mass of the ash residue. Calculate the percent of ash in the coal sample.

OBSERVATIONS AND CONCLUSIONS
1. What is the average value for the percent of ash for your class?
2. How does your value compare to the class average?

The content of coal varies. Coal varies in terms of the amount of sulfur, minerals, moisture, and carbon (organic matter) it contains. Most of the coal used in this country contains less than five percent sulfur. The burning of coal that contains any amount of sulfur is harmful, in that it contributes to air pollution and the formation of acid rain (see Chapter 8). Like sulfur, minerals are considered impurities in coal. Some minerals do not burn but turn to ash, thus lowering the amount of heat generated by the coal. Ash particles can also contribute to air pollution.

The carbon content of coal is highest in anthracite and lowest in peat. Conversely, the moisture content of coal is highest in peat and lowest in anthracite. A high moisture content limits coal's usefulness as a fuel and lowers its heating value. **Heating value** refers to the amount of heat produced by a given amount of fuel when burned. It is usually expressed in units of calories per gram (cal/g). Peat and lignite contain more moisture than bituminous and anthracite, so their heating values are lower. Bituminous has the highest heating value of any coal. It is the most plentiful form of coal and is the chief fuel used in steam-electric generating plants.

Known deposits of coal that can be mined profitably are called coal reserves. Figure 12-2 shows that bituminous coal reserves occur in 37 states in this country. The United States is the largest coal-producing country in the world, followed by the Soviet Union and China. Coal and its products supply about 20 percent of our energy needs.

Figure 12-2 Coal reserves of the United States. Where are most lignite deposits found?

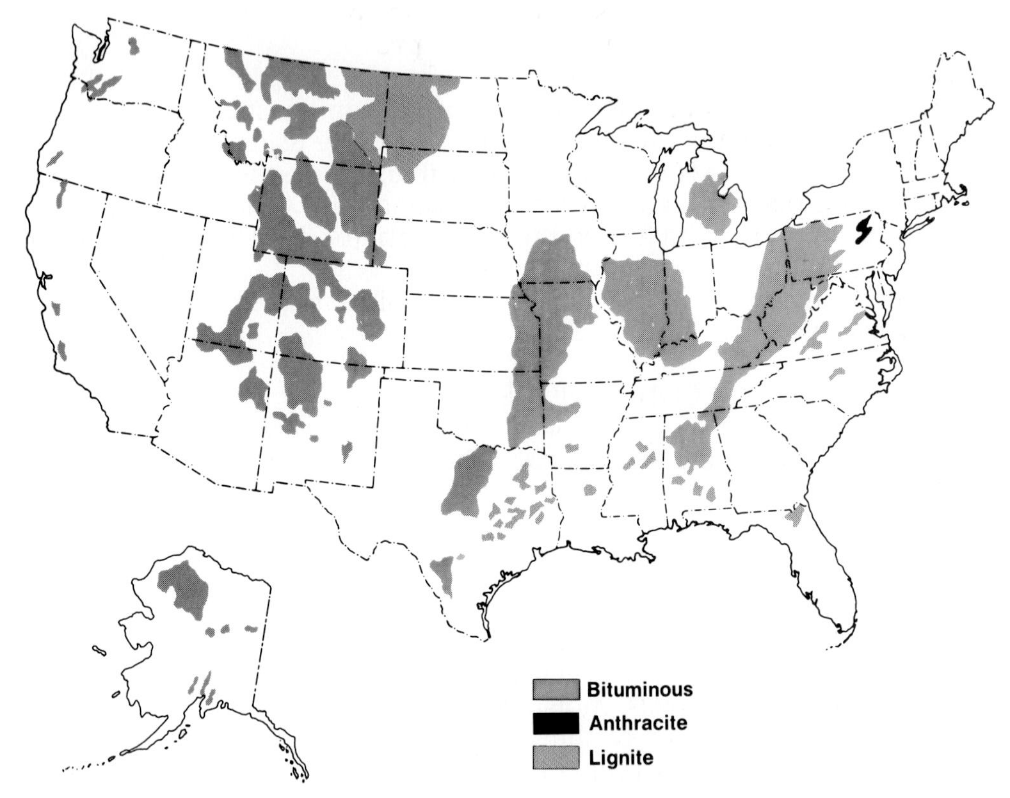

In the United States, coal mines are generally less than 500 m in depth. However, many mines in other countries go to depths of from 1,000 m to 1,300 m. Much of the coal found in this country is less than 100 m deep. This coal is retrieved by strip-mining. In this process, removal of the surface deposits exposes the coal below.

Figure 12-3 Red-hot coke made from bituminous coal is being discharged into a railroad car.

Many by-products come from bituminous coal. Bituminous coal, or soft coal, can produce black smoke and soot if its volatile matter is incompletely burned. If the burning of bituminous coal is carefully controlled, the smoke will be reduced, and more heat will be produced. Today, in many factories and electric power plants, bituminous coal is ground into powder, which burns in a blast of air. In powdered form, coal burns like a gas, with great efficiency.

If bituminous coal is heated to between 700° C and 1,000° C in the absence of air, it breaks down into coal gas, a mixture of liquids, and a solid called coke. Breaking down coal in the absence of air is a process called **destructive distillation**.

Coke is a hard, porous, solid fuel produced by destructive distillation of bituminous coal. It varies in color from gray to black. Coke contains more carbon than coal because most of the volatile matter is driven off during distillation. Thousands of metric tons of this fuel are prepared daily in huge coke ovens like the one shown in Figure 12-3.

Most of the coke produced today is used by the iron and steel industry. Coke is used as a fuel and a reducing agent, as you learned in Chapter 9. Coke ovens also produce many useful by-products. These products include ammonia, which is used in making fertilizers; coal gas, which is used as fuel; and coal tars, used in making dyes, perfumes, flavors, and medicines. Synthetic rubber, antiseptics, nylon, and many thousands of other products are made from materials obtained from bituminous coal.

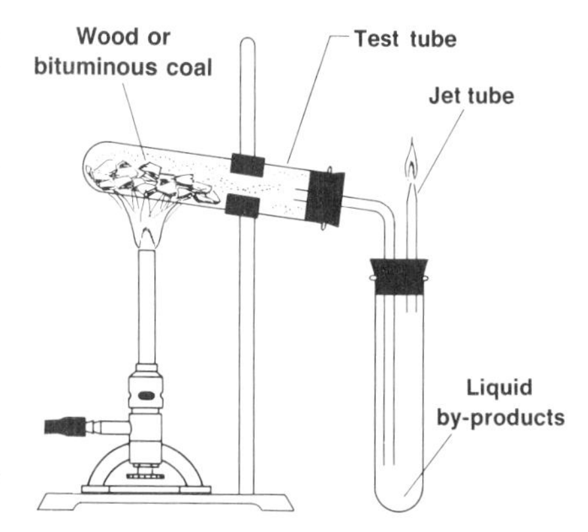

Figure 12-4 The destructive distillation of bituminous coal or wood. In each case, what is the solid product called? **CAUTION: Do not attempt this procedure without the permission and supervision of your teacher.**

Coke and charcoal are made in similar ways. An example of the destructive distillation process described earlier is shown in Figure 12-4. A test tube containing small pieces of bituminous coal is heated. Coal gas is generated through the jet tube and can be ignited. A tar-like, liquid by-product containing ammonia, oils, and coal tar condenses in the vertical test tube. With burning, the bituminous coal becomes coke. If wood is used in place of coal, the same procedure will produce charcoal as a solid by-product.

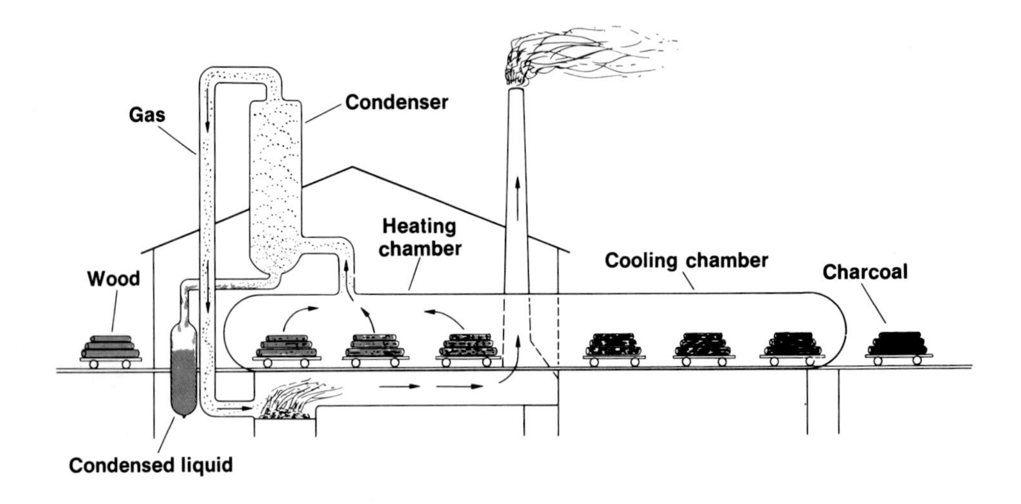

Figure 12-5 Diagram of the commercial preparation of charcoal from wood. Describe this process based on the diagram.

In industry, charcoal is made in furnaces such as the one shown in Figure 12-5. Wood is loaded on steel cars and pushed into a huge oven, or heating chamber. Around the chamber are flues where gas is burned. The heat drives off gases from the wood. The solid that remains is charcoal. Besides charcoal, this process yields wood alcohol, acetic acid, wood tar, pitch, and creosote. These by-products are as useful as the charcoal itself. Ordinary charcoal can be further refined by treating it with steam and air heated to greater than 316° C. This removes all of its impurities, leaving an end-product called activated charcoal.

Like coke, charcoal burns with almost no flame because it contains no volatile matter. Although it gives a hot, smokeless flame, charcoal is too costly to use as a fuel for heating homes, but it is commonly used in outdoor barbecues in the form of briquettes. There are many industrial uses for charcoal, such as the making of special steels. Charcoal is also used in filters that clean dirty water and air.

The heating values of coke, charcoal, and several other fuels are compared in Table 12-1.

TABLE 12-1

Comparison of Fuel Heating Values

Fuel	Heating Value (cal/g)
Wood	5,000
Alcohol	6,400
Bituminous coal	7,300
Coke	7,800
Charcoal	8,000
Gasoline	10,800

12.2 PETROLEUM

Petroleum is formed from marine life. The solid, liquid, or gaseous fuel found beneath the earth's surface and consisting mostly of carbon and hydrogen is called **petroleum**. The word petroleum is derived from the Latin *petra*, meaning rock, and *oleum*, meaning oil. Petroleum includes both oil and natural gas; however, gas will be discussed separately in the next section. In some parts of the world, petroleum occurs as a solid in sands and rocks. Examples include tar sands and oil shales.

Figure 12-6 Workers setting up an advertising road sign for one of the first motor oils made of petroleum. What other products do you know of that come from petroleum?

Geologists believe that petroleum was formed from marine life buried in sediments millions of years ago. Heat and pressure due to burial by layers of sediment, rock, and water compressed this organic matter into layers of sedimentary rock. Bacterial action also acted upon this organic matter, causing chemical changes that helped convert the organic materials into crude oil and gas. Once formed, the petroleum either migrated under pressure to the surface or became trapped as a pool below the surface.

Crude oil is a slippery mixture made up of thousands of compounds. It consists mostly of hydrocarbons (see Chapter 10, page 182), as well as compounds of oxygen, nitrogen, and sulfur. Crude oil is seldom used in its natural state. Instead, it is separated into a number of products by refining. Lubricating oils, fuel oils, greases, gasoline, asphalt, kerosene, and tar are only a few of the thousands of products that come from this resource. These products have abundant uses in our society. Many people heat their homes with fuel oil that comes from petroleum. The invention of the automobile made petroleum products such as gas and motor oils essential. See Figure 12-6.

Obtaining petroleum from the ground is a costly process. Petroleum has been used by people for thousands of years, although it was not commercially produced until the nineteenth century. The first North American oil well, which was only 21 m deep, was dug by Col. Edwin L. Drake in 1859 near Titusville, Pennsylvania (Figure 12-7). The well produced a few barrels of oil a day for many years.

Figure 12-7 The first oil well dug in this country barely resembles its modern equivalent.

Today, oil and natural gas together provide about 70 percent of this country's energy needs. The Soviet Union, the United States, and Saudi Arabia are the leading oil-producing countries. Texas, Louisiana, California, and Alaska produce most of the oil in this country.

Petroleum production is expensive. One reason is that many drilling operations fail to produce oil. About one drilling in eight hits oil or gas. The other seven are called "dry" holes, or "dusters," and are a total financial loss.

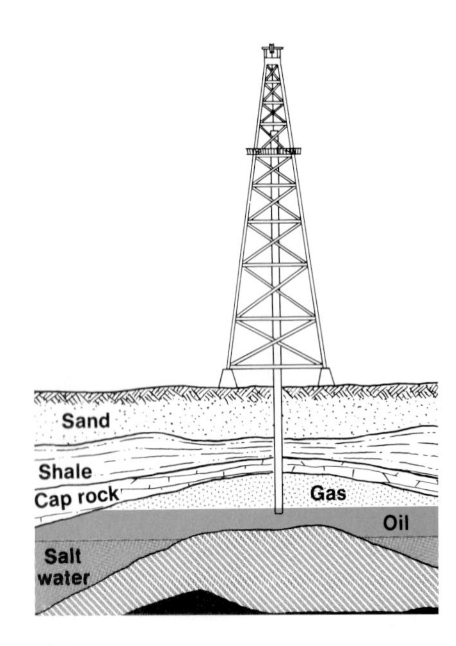

Figure 12-8 Crude oil, natural gas, and salt water saturate the oil-bearing sands. Approximately 3 out of 100 wells that strike oil produce enough oil to be profitable.

Another reason for the great expense is that petroleum is deep in the earth. Oil, natural gas, and salt water fill the oil-bearing sands under domes of cap rock. Cap rock is a layer of impenetrable rock, such as shale, that keeps the petroleum from migrating to the surface (Figure 12-8). A drill must be used to reach the deeply buried, oil-bearing sands. When the sands are penetrated by the drill, the pressure of the natural gas forces the oil upwards to the surface. The oil may be forced out violently as a gusher. An arrangement of valves, known as a "Christmas tree," is used to control the pressure and regulate the flow of crude oil and natural gas (Figure 12-9). As the pressure of the natural gas decreases, pumps are used to obtain the remaining oil.

Oil wells are usually drilled with rotary bits. A heavy bit equipped with a diamond or hard steel tip is fastened to a length of steel pipe called a drill pipe. The drill pipe is then attached to a round drill table. The power to rotate the table, pipe, and bit is supplied by an engine. Additional lengths of pipe are attached to the main shaft as the bit bores its way down. Each time the worn-out bit is replaced, the entire shaft has to be raised and taken from the hole. Rock cores removed from the hollow steel pipe are used to identify the kind of rock encountered below the surface.

During the drilling, a fluid called drilling mud is pumped down through the drill pipes to flush out the cuttings. This drilling mud also serves to cool the drill. The mud can also help control a gusher.

The crude oil must be transported from the well to a refinery or to a port where it can be loaded on ocean-going tankers. A network of pipelines carries the oil thousands of kilometers. Pumping stations on land, located every 130 km to 145 km, force the crude oil through the pipes. Branch lines connect different oil fields with main lines, forming a huge underground network. Similar pipelines transport the natural gas.

Figure 12-9 The "Christmas tree" is a series of valves designed to control the flow of oil from the well.

Petroleum is refined by distillation. Each of the many hydrocarbons in the mixture of petroleum has a different boiling point. As you can see in Table 12-2, the smaller hydrocarbon molecules with fewer carbon atoms boil at lower temperatures. These differences in boiling points can be used to separate the components in petroleum.

The process of separating a mixture of liquids having different boiling points is called **fractional distillation**. The initial refining of petroleum into various products is done by this process. The apparatus used to refine petroleum is divided into three main parts: a pipe still, a fractionating tower, and a water condenser. Locate these three parts in Figure 12-10.

TABLE 12-2

Petroleum Products*			
Number of Carbon Atoms	Boiling Temperature Range	Name of Product	Uses
1 to 5	below 40° C	gas	fuel
6 to 10	40° C to 180° C	gasoline	fuel
11 to 12	180° C to 230° C	kerosene	fuel
13 to 17	230° C to 300° C	light gas, oil	diesel fuel
18 to 25	300° C to 405° C	heavy gas, oil	lubricant stock
26 to 60	405° C to 515° C	residue	wax, residual oil, asphalt

*from *Chemistry and Petroleum,* American Petroleum Institute

The crude oil is heated under pressure in a furnace, called a pipe still, to between 370° C and 430° C. Pressure in the tubes keeps the crude oil from becoming a vapor. The hot oil flows from the pipe still to the bottom of a tall fractionating tower. Here the pressure is reduced and the liquid becomes a vapor.

The vapors rise up the tower, which may be more than 30 m high. Hydrocarbons with a low boiling point, such as those in gasoline, move to the top of the tower. In the condenser, the vapors are cooled. Each vapor becomes a liquid at a different level of the tower. Shelves collect the condensed liquids and hold the liquids until they are drained off into separate storage tanks as shown in Figure 12-10. For better separation, each of the liquids obtained is distilled again. In addition, each liquid is chemically treated to remove unwanted materials.

Figure 12-10 The basic parts of a fractional distillation plant for petroleum refining. What is the function of each part?

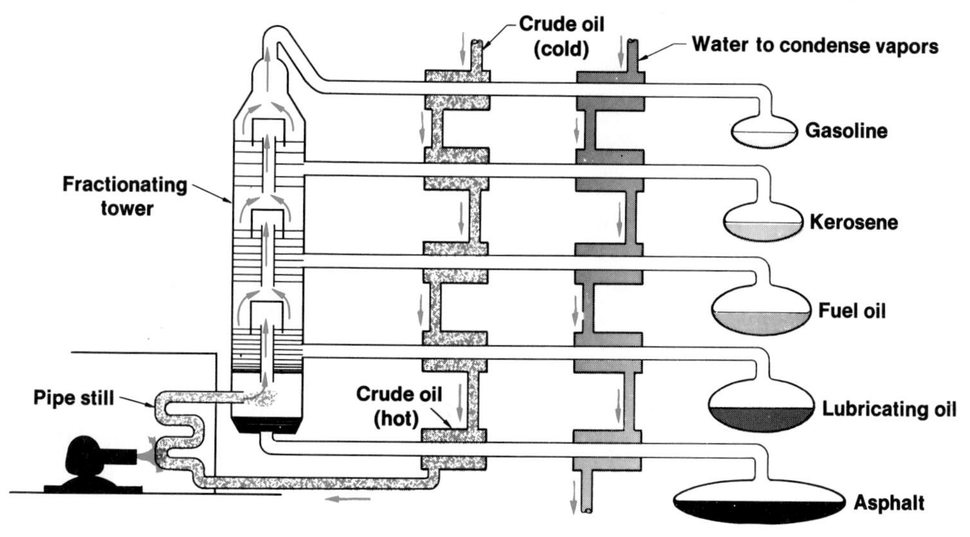

12.3 GASEOUS FUELS

Natural gas is one of our most important resources. Many homes and factories in the United States use natural gas as a fuel for cooking and heating. Natural gas is an excellent fuel, giving more heat per cubic meter than any other fuel gas. Natural gas is convenient to use, leaves no ash, and burns with almost no smoke. As stated previously, natural gas and oil formed millions of years ago and became trapped deep beneath the earth's surface. Natural gas is almost entirely made up of the hydrocarbon called methane (CH_4). Methane also forms when plant life breaks down underwater. Methane is sometimes called marsh gas because it is found bubbling from the water in warm, marshy areas. Natural gas may also include propane (C_3H_8) and butane (C_4H_{10}). Additionally, these two gases may be obtained from the refining of crude oil.

Figure 12-11 Gas pipelines have to be constructed over hills and mountains. Why do you think these pipes are often buried?

Once refined, natural gas is supplied to the consumer by a network of pipelines over 650,000 km in length (Figure 12-11). Propane, butane, or a mixture of the two gases can be compressed into a liquid and stored in steel cylinders. Gas stored in this way is known as bottled gas, or LPG (Liquefied Petroleum Gas). LPG provides a fuel of high heating value to places that are far from regular gas-line service. The liquid fuel changes to a gas when the pressure is reduced as the fuel leaves the tank through the outlet valve. The tank shown in Figure 12-12 may be refilled from a delivery truck or it may be replaced when empty.

The use of natural gas in this country increased steadily during this century. Because of this increase and the limited amount of natural gas, people became concerned about possible gas shortages, particularly during the 1970s when gas supplied to consumers became scarce. The use of natural gas has decreased in recent years, as people and industry have tried to use gas more efficiently.

Figure 12-12 With bottled or tank-stored gas, homes located away from city gas lines can have the same gas service. What fuel gas might be in this tank?

Water gas is an industrial fuel.

Water gas is an industrial fuel. The existence of a fuel called water gas may seem unlikely, since water will not burn. Yet water in the form of steam combined with carbon in the form of coke produces a useful fuel. This fuel is made up of hydrogen and carbon monoxide.

First, the burning coke is heated white-hot by air forced through it. This process is called the air run. Then, the air is cut off and steam is allowed to pass through the hot coke, known as the steam run. After a short time, the steam is shut off and the coke is reheated by another blast of air. Air and steam runs are alternated until all the coke is used up. Water gas is produced only during the steam run. When the coke and steam (water) react, carbon monoxide and hydrogen are given off:

$$C(s) + H_2O(g) \rightarrow CO(g) + H_2(g)$$

Since both carbon monoxide and hydrogen can burn, this mixture can be used as a fuel. Carbon monoxide and hydrogen can also be separated for other consumer uses.

The heating value of water gas is often enriched by the addition of propane (C_3H_8) or butane (C_4H_{10}). Although water gas is easy to store (see Figure 12-13), the availability of natural gas from pipelines has made natural gas a more convenient fuel to use.

Figure 12-13 These tanks are efficient for storing butane, propane, and other fuel gases under pressure. Why do you think these tanks are spherical?

A burning hydrocarbon provides energy.

A burning hydrocarbon provides energy. As you have seen, liquid and gaseous fuels are made up mostly of hydrocarbons (hydrogen and carbon). Remember from Chapter 7 that when a substance burns or oxidizes, its atoms of hydrogen and carbon join with oxygen in the air to form new compounds. When a hydrocarbon burns completely, the products formed are carbon dioxide and water. Heat and light energy are also released. When a fuel joins with enough oxygen to complete the burning process, **complete combustion** takes place. For example, when methane burns completely, carbon dioxide, water, and energy are the products formed:

$$CH_4(g) + 2O_2(g) \rightarrow CO_2(g) + 2H_2O(g) + \text{energy}$$

On the other hand, when there is a shortage of oxygen for the burning of a hydrocarbon, **incomplete combustion** occurs. The resulting products are carbon monoxide, water, and energy:

$$2CH_4(g) + 3O_2(g) \rightarrow 2CO(g) + 4H_2O(g) + \text{energy}$$

Whether there is complete or incomplete combustion, energy is always released during the burning of fuel.

PRODUCTS FORMED IN THE BURNING OF A HYDROCARBON

activity

OBJECTIVE: Perform an experiment to identify the products formed by burning a hydrocarbon.

PROCESS SKILLS

In this activity you will *observe* the formation of products produced by burning a hydrocarbon and *hypothesize* as to how they are produced.

MATERIALS

candle
matches
beaker
tongs
safety goggles

PROCEDURE

1. Put on your safety goggles. Light the candle and, using tongs, hold the open end of a dry, cold beaker about 5 cm above the flame for a few seconds. Record your observations.
2. Hold the bottom of the cold beaker in the yellow part of the flame for several seconds. Record your observations.

OBSERVATIONS AND CONCLUSIONS

1. When you held the open end of the beaker over the candle, what did you observe? Explain your observations. Why must the beaker be cold?
2. What did you observe on the bottom of the beaker the second time you held the beaker in the flame? Explain your observations.

ENERGY

our need for more

These wind turbines harness the wind for producing electricity.

You have learned that oil and natural gas provide about 70 percent of all our energy needs. There is a limited amount of this fuel left in the earth. Coal provides most of the rest of our energy, and some experts say that there is enough to provide for our needs for the next 300 years. Coal, however, is expensive and can be hazardous to mine. In addition, burning coal adds to air pollution. The shortages and hazards of using fossil fuels have taught us that we must use existing sources carefully. However, it is more important to find new sources of energy.

Wind and tidal power will probably not become major power sources because of limitations imposed by weather patterns and geography. Solar energy has become a popular source for small energy needs. Many solar collectors installed on buildings use the sun's energy to heat tap water or the building's interior. Some solar collectors use photovoltaic cells that convert the sun's energy into electricity. Solar energy is a clean energy source, but night and cloudy weather limit the energy solar collectors can provide.

Heat from the earth's interior, or geothermal energy, can be used to satisfy some of our energy needs. Geothermal energy is harnessed by diverting steam from heated underground streams into steam turbines. In areas where streams do not exist, engineers can inject water into cracks in the hot rock. This method of water injection and steam recovery is being studied in a number of research projects around the world.

The most controversial of the new energy sources is nuclear fission. As you learned, the splitting of the nucleus of certain heavy radioactive elements releases tremendous amounts of energy. When this nuclear reaction is controlled by scientists, energy is produced that can be used to generate electricity. More than 200 nuclear power plants around the world now produce 11 percent of the world's electricity. There are many advantages to nuclear energy, but there are also hazards involved, including the accidental release of radioactive material.

FOR FURTHER RESEARCH AND DISCUSSION

Should we continue to rely on fossil fuels for most of our energy needs? Are there enough practical and safe alternatives?

CHAPTER REVIEW 12

SUMMARY

1. Fuels are substances that can be burned to provide heat at a reasonable cost.
2. Coal is formed from the buried remains of plants and is composed mostly of carbon.
3. Coke is made by the destructive distillation of bituminous coal; charcoal is made by the destructive distillation of wood.
4. Fractional distillation is the process by which petroleum is refined and involves the separation of a mixture of liquids having different boiling points.
5. Petroleum is formed from marine life acted upon by bacteria, heat, and pressure.
6. Crude oil and natural gas consist mostly of hydrocarbons.
7. When a hydrocarbon burns completely with enough oxygen, the products formed are water, carbon dioxide, and energy (heat and light).

VOCABULARY

Match the item in the left column with the best answer in the right column. Do not write in this book.

1. anthracite
2. bituminous coal
3. coke
4. complete combustion
5. destructive distillation
6. fossil fuels
7. fractional distillation
8. fuel
9. heating value
10. incomplete combustion
11. lignite
12. peat
13. petroleum

a. first stage in coal formation
b. process of separating a mixture of liquids having different boiling points
c. fuel source below the earth's surface composed mostly of hydrocarbons
d. formed by destructive distillation of bituminous coal
e. hard coal
f. second stage in coal formation
g. soft coal
h. process of breaking down coal or wood in the absence of air
i. group of fuels produced from organic matter
j. substance that produces heat when burned
k. burning of a fuel in sufficient oxygen
l. burning of a fuel in insufficient oxygen
m. heat produced by burning a given amount of fuel

REVIEW QUESTIONS

1. What are some traits of a good fuel?
2. What are the four stages in the formation of coal?
3. Why did the plant matter of the coal-forming period not decay completely?
4. Compare the carbon and moisture content of anthracite with that of bituminous.
5. How is charcoal made? List three by-products in making charcoal.
6. Name two widely used household fuels in your community.
7. List advantages and disadvantages of bituminous coal as a fuel.
8. Where do anthracite deposits occur in the United States?
9. What group of compounds is found in petroleum?
10. Discuss the theory geologists use to explain petroleum formation.
11. Name the three main parts of the distillation apparatus in a refinery and describe their functions.
12. Why is fractional distillation particularly useful in refining petroleum?
13. List five major products obtained directly from the distillation of petroleum in order of increasing boiling point.
14. Which states lead in the production of petroleum? Which countries are the leaders?
15. How are oil wells drilled? In what way is crude oil transported to the refineries?
16. Describe why the oil sometimes "gushes" from newly drilled wells. How is this prevented?
17. What is the composition of natural gas? Discuss some of its advantages as a fuel.
18. How is water gas prepared? How is it enriched?
19. Why is the "air run" needed in the making of water gas?
20. What is the composition of bottled gas? Why is it used in rural areas?
21. What products are formed by the complete burning of a hydrocarbon fuel? by the incomplete burning of a hydrocarbon fuel?

CRITICAL THINKING

22. Explain why burning fuels without ventilation can be dangerous.
23. Biomass is plant or animal material used to provide energy. Examples of the use of biomass are burning of wood and alcohol made from grains. Why is biomass renewable, whereas petroleum and coal are not?
24. Explain why it is possible to increase the world's coal reserves.

FURTHER READING

Cross, Wilbur. *Coal.* Chicago: Children's Press, 1983.

Cross, Wilbur. *Petroleum.* Chicago: Children's Press, 1983.

Fogel, Barbara R. *Energy: Choices for the Future.* New York: Franklin Watts, Inc., 1985.

Kraft, Betsy H. *Oil and Natural Gas,* revised ed. New York: Franklin Watts, Inc., 1982.

RUBBER, PLASTICS, AND FIBERS

Can you imagine how your life would be different without plastics? Today, plastics are used in so many ways: for building materials, furniture, appliances, and even clothing. How many things in your classroom are made from plastics?

CHAPTER 13

SECTIONS

OBJECTIVES

- ☐ Describe the process by which crude rubber is made from latex of the rubber tree.
- ☐ Distinguish between the properties of natural and synthetic rubber.
- ☐ Describe the differences between the making of thermoplastic and thermosetting plastics.
- ☐ Identify some properties of and uses for the various families of plastics on the commercial market.
- ☐ Summarize the chemistry of natural and synthetic fibers.
- ☐ Describe bleaching and dyeing processes of fabrics.

13.1 RUBBER

Natural rubber is a product from the sap of a tree. When European explorers came to the New World, they discovered that some Central-American Indians made rubber shoes by dipping their feet in latex and allowing the latex to dry. They also found Indians in Central and South America playing games with balls made of raw rubber. Today, rubber is much more than a plaything. In the United States alone, between 40,000 and 50,000 different products are made from rubber.

Figure 13-1 To obtain latex from the rubber tree, a worker makes a V-shaped cut on the tree trunk.

Rubber is easy to stretch. It is a nonconductor of electricity, making it a good cover for electric wiring. Rubber is waterproof, it absorbs shocks, and it is airtight, making it a suitable material for car tires. Special types of rubber resist heat, cold, and the corrosive action of many chemicals.

Natural rubber comes from the rubber tree. This tree grows wild in the Amazon Valley of South America but is also planted in tree farms in Southeast Asia. Over 90 percent of the world's supply of natural rubber comes from Southeast Asia. A rubber tree must grow for five to eight years before it is large enough to tap. A warm climate and an average rainfall per year of 170 cm to 250 cm are needed for its proper growth. Rubber trees grow to a height of 16 m to 21 m and will produce rubber for about 40 years.

A sap, called latex, drips from V-shaped cuts made in the trunk of the rubber tree, as shown in Figure 13-1. **Latex** is a milky fluid that consists of 30 percent to 35 percent pure rubber and 60 percent to 65 percent water. Other materials, such as resins, proteins, sugar, and mineral matter are also present. Have you ever broken or cut the stem of a goldenrod or a dandelion? Did you notice the sticky, white juice that came from the cut stem? This liquid is like latex, the raw material that comes from the rubber tree.

The rubber in latex appears as small particles scattered in the liquid, somewhat like butterfat in cow's milk. The latex is processed into crude rubber as soon as possible after tapping. First, it is strained to remove any dirt or twigs collected during the tapping process. Next, the liquid passes into special tanks, where acetic acid is added. The acid causes the strained latex to clump, forming particles of rubber. The solid mass of rubber is then washed and made into sheets, which are smoked, dried, and baled (Figure 13-2).

Figure 13-2 At many rubber plantations in Malaysia, much of the work is done by hand. Here, wringers are being used to remove moisture from sheets of rubber.

Rubber is a hydrocarbon. Rubber is composed of carbon and hydrogen, and the simplest formula for rubber is $(C_5H_8)_x$. The *x* in the formula is a number that is unknown at the present time. Some chemists believe that the value of *x* may be as high as 400,000. Rubber molecules are giants compared with simple molecules like H_2O and CH_4. Chemists call the C_5H_8 unit a **monomer** (MAHN-*uh*-MUR), or a single unit of a whole molecule. When many monomers link together, as in the case of $(C_5H_8)_x$, they form a giant molecule, called a **polymer** (PAHL-*uh*-MUR), meaning many smaller molecules joined together.

Figure 13-3 shows the structure of natural rubber, which is a polymer of C_5H_8. The molecules form long zigzag chains. This property is probably one reason why refined rubber is so elastic.

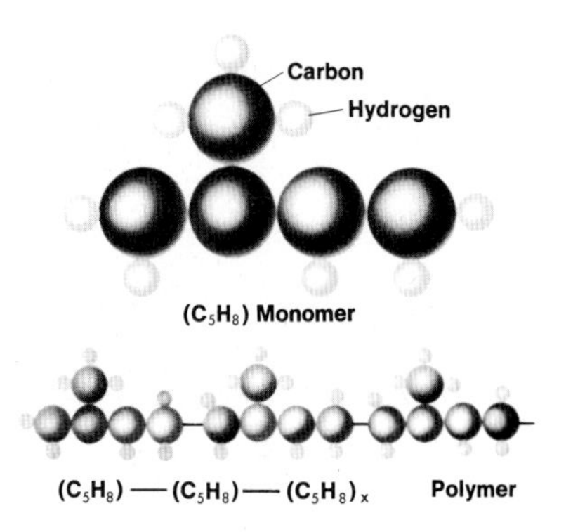

Figure 13-3 Rubber monomers link together to form a polymer.

Crude rubber can be improved through chemistry. Crude rubber is tough and sticky, but it is not strong and elastic. It is also affected by temperature extremes, becoming sticky in hot weather and brittle in the cold. In 1839, Charles Goodyear accidentally discovered a method of making rubber less sticky, more elastic, and resistant to heat and cold.

By heating small amounts of sulfur with crude rubber, Goodyear developed what became known as vulcanization. **Vulcanization** is the process by which sulfur joins with the polymer chains, forming rubber with desirable properties such as strength and hardness. This process launched the modern rubber industry, supplying rubber products for thousands of uses.

The properties of vulcanized rubber depend upon the amount of sulfur that is added. Auto tires and rubber bands usually contain about 4 percent sulfur. A harder, less elastic rubber contains a higher percent of sulfur.

Used, or reclaimed, rubber from worn-out auto tires or other rubber products may be added to a batch of new rubber. Used rubber may be treated to get rid of the fibers or fillers added for specific purposes. Rubber goods made largely from reclaimed rubber are not as flexible as those made from pure natural rubber. However, reclaimed rubber goods are tougher, longer lasting, and less costly than natural rubber goods.

Organic catalysts are added to rubber to speed up the vulcanizing process. Other compounds are added to rubber to keep it from getting stiff and brittle and to protect it from the harmful effects of sunlight, oxygen in the air, and petroleum products.

Figure 13-4 Newly made, vulcanized tires are sorted for shipment. Cords of cotton, rayon, and nylon are embedded in the soft rubber for strength.

Most rubber is used in making auto tires. A great deal of rubber is used in making cars. In fact, there are close to 300 rubber parts in a typical car. About two-thirds of all the rubber used in the United States goes into making car tires (Figure 13-4).

Car tires contain large amounts of carbon black, a form of pure carbon, for added strength and longer life. The carbon black accounts for the color of tires. Carbon black also gives longer life to rubber footwear and other articles as well. In less costly rubber goods, clay is used as a filler. Zinc oxide is sometimes added to make rubber white.

Figure 13-5 White foam rubber, after being whipped with air, is poured into a mattress mold.

Foam rubber has many uses. Thousands of tiny air bubbles in foam rubber make it an excellent material for use in sofas and chairs. Blocks of foam rubber, covered with leather, cloth, or plastic sheet, make excellent cushions because they hold their shapes well, even after long use. Some mattresses, for example, are thick blocks of foam rubber covered with cotton cloth.

One method of making foam rubber consists of whipping air into the latex, as is done when whipping cream. During vulcanization, the trapped air expands, inflating the mass into foam rubber. In Figure 13-5, the whipped latex is being poured into a mattress mold.

REMOVAL OF RUBBER FROM LATEX

activity

OBJECTIVE: To perform an experiment to observe the properties of raw rubber.

PROCESS SKILLS

In this activity, you will *observe* the formation of rubber and *draw conclusions* about its properties.

MATERIALS

safety goggles
2 100-mL beakers
graduated cylinder
latex
dilute acetic acid
stirring rod
hot water
tongs
paper towels

PROCEDURE

1. Put on your safety goggles. Fill a beaker with about 25 mL of latex. Add 5 mL of acetic acid. Stir the mixture with a stirring rod and observe the lumping of raw rubber.
2. Using tongs, wash several lumps thoroughly under running water to remove the acid. Dry the lumps and observe what happens when you squeeze and stretch them.
3. Heat a portion of the sample by dropping it into a beaker of hot water. Using tongs, remove the sample and examine its elastic properties.
4. Cool another sample of the raw rubber in a refrigerator for several hours. Examine its elastic properties.

OBSERVATIONS AND CONCLUSIONS

1. What are the properties of the heated sample of rubber?
2. What are the properties of the refrigerated sample of rubber?
3. What do your observations tell you about the limitations of raw rubber?

Another method of making foam rubber is to add ammonium carbonate [$(NH_4)_2CO_3$] to a batch of raw rubber. When heated, ammonium carbonate changes into gaseous ammonia, carbon dioxide, and water vapor:

$$(NH_4)_2CO_3(l) \xrightarrow{\text{heat}} 2NH_3(g) + CO_2(g) + H_2O(g)$$

The heat produced in the vulcanizing process breaks down the ammonium carbonate into gases, which inflate the dough-like mass of rubber.

Organic compounds are used to make synthetic rubber. A **synthetic** substance is one that does not occur naturally but is produced artificially by assembling chemically simpler substances. For many years, scientists were unable to make a good substitute for rubber. In 1931, neoprene, a suitable synthetic rubber, appeared on the market. The main raw material for making neoprene is the first member of the alkyne series of hydrocarbons, called acetylene. If you compare the structural formulas of natural rubber and neoprene, you can see that they are quite similar:

$$\left[\begin{array}{ccccccc} & H & & & & H & \\ & | & & & & | & \\ - & C & - & C = C & - & C & - \\ & | & & | \quad | & & | & \\ & H & & CH_3 \quad H & & H & \end{array}\right]_x \qquad \left[\begin{array}{ccccccc} & H & & & H & H & \\ & | & & & | & | & \\ - & C & - & C = & C & - C & - \\ & | & & | & & | & \\ & H & & Cl & & H & \end{array}\right]_x$$

natural rubber — neoprene rubber

Oil and greases, which destroy natural rubber, have little effect on neoprene. Therefore, neoprene is useful for making gasoline-pump hoses, gaskets, tank linings, and many other items in common use. There are now over 100 types of synthetic special purpose rubber made to resist oils, fuels, air, and extreme temperatures.

Manufacturers also produce general purpose rubber. One very vital type of general purpose rubber that is made in large amounts is styrene-butadiene (STIE-*reen-byoo-tah*-DY-EEN) rubber, better known as SBR (Figure 13-6). Styrene and butadiene are two hydrocarbon compounds.

During World War II when the Japanese cut off the supply of natural rubber from Southeast Asia, a number of SBR plants were built. Since that time, the synthetic rubber industry has grown such that most of the rubber in use today is synthetic. The main rubber used today for making car tires is SBR.

Figure 13-6 This worker inspects the SBR rubber in a synthetic rubber plant in Houston, Texas.

13.2 PLASTICS

Plastics are synthetic products. The word plastic has two meanings. If used as a property, it refers to a material's ability to be shaped or molded into a desired form. Glass, clay, wax, and concrete, therefore, are considered "plastic" materials under this general meaning. However, the term plastic, as used by chemists, refers to synthetic products made from coal, oil, or related raw materials that have been molded or shaped into final forms by heat and pressure. Synthetic rubber is a good example of a plastic.

In 1868, while trying to find a substitute for ivory used in billiard balls, John W. Hyatt treated cellulose nitrate with camphor. This reaction formed a plastic material called celluloid. The discovery of celluloid marked the beginning of the plastics industry.

Some 40 years later, a second major step forward was taken. Dr. Leo Baekeland mixed phenol with formaldehyde. This reaction produced the first phenolic plastic, Bakelite.

The plastics industry is one of the few billion-dollar industries in the United States. The growing use of plastic objects in almost every part of society is due to their range of colors, durability, weight, and ease of mass production.

Plastics are thermoplastic or thermosetting. Plastics can be classified as thermoplastic or thermosetting, depending on the kind of chemical reaction that occurs while the plastic is being heated.

A **thermoplastic** product results when chains of molecules are formed by head-to-tail addition of monomer units. These materials melt easily when gently heated, but harden as they cool. Thermoplastics can be melted and shaped again and again.

A **thermosetting** product results when new bonds are formed between chains of molecules and monomer units, like a cross-linking. Thermosetting plastics take their permanent shape when heat and pressure are applied during the forming process. Reheating will not soften these materials. Thermosetting plastics are difficult to melt.

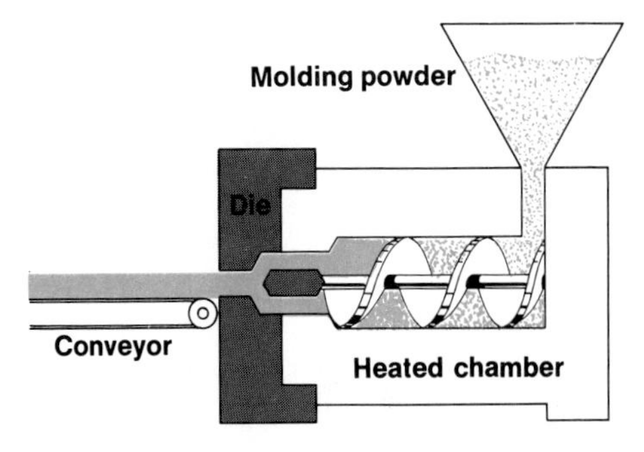

Figure 13-7 Extrusion is similar to squeezing toothpaste out of a tube. A variety of openings are used to shape a number of finished products.

Several methods are used in making plastic articles. Plastics are made from substances called synthetic resins. These resins are made from chemicals that come from coal, petroleum, limestone, salt, and water.

Thermoplastic materials are shaped into rods, tubes, and sheets by a process called extrusion (Figure 13-7). First, a dry powdered resin is loaded into a hopper. Then a turning screw feeds the powder into a heated chamber, where it melts. The melted resin is then squeezed through an opening in a die onto a conveyer belt where it is cooled.

Another method by which thermoplastic articles are made is injection molding. In this process, machines squeeze melted resins into a closed mold. As soon as the plastic cools, the finished product is taken from the mold.

Blow molding is a process that consists of blowing a plastic material against a mold, as illustrated in Figure 13-8. Air is blown into the plastic, as into a balloon, to force the material onto the sides of the mold.

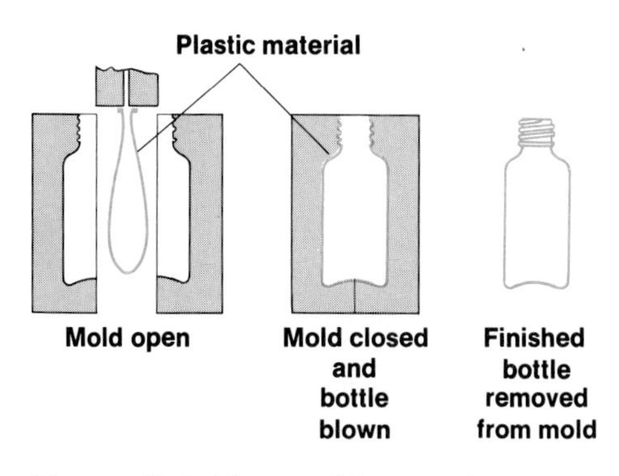

Figure 13-8 Blow molding produces hollow items such as bottles. How does the process of blow molding compare with glass-blowing?

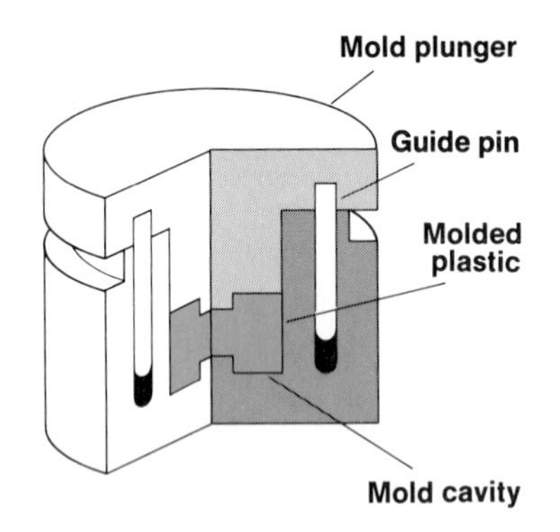

Figure 13-9 Compression molding can be compared to making waffles. Many identical plastic articles can be made from this compression mold.

Compression molding is used for thermosetting plastics. Thermosetting resins are placed into the heated chamber of a giant press, shown in Figure 13-9. When the press closes, heat and pressure from the mold plunger are applied, causing chemical changes that make the resin harden in the mold cavity. Thermosetting plastics cannot be remelted.

Another method of producing thermosetting products is by laminating. In this process the article is built up in layers or plates. Sheets of cotton cloth or paper are dipped in Bakelite varnish and then piled one on top of another to any desired thickness. The layers are then put in a press, and heat and pressure are applied. This process turns out a very hard, tough sheet of material.

Thermoplastic and thermosetting articles are plentiful. Because both types of plastics have a variety of properties, including flexibility, transparency, and durability, they have been adapted to unlimited uses (Figure 13-10). In industry, for example, plastics make tough machine parts and electrical insulation. With the addition of pigments, plastics are

TABLE 13-1

Kinds of Thermosetting Materials		
Plastic	**Properties**	**Uses**
Epoxy	resists water and weather, hardens quickly, bonds tightly	protective coatings, bonding of metals, ceramics, glass and other plastics
Melamines	provide glassy surface, resist scratching and heat	dinnerware, buttons, table tops, electrical parts, handles
Phenolics (Bakelite)	strong and hard, resist heat and cold, do not readily absorb water or chemicals	paints, adhesives, printing plates, electrical devices, cabinets, telephone cases, car distributor heads, insulation
Polyesters	strong, harden quickly; highly resistant to most solvents, acids, bases, and weather	saturation of cloth, paper, and mats of glass fiber; luggage, swimming pools, boats, chairs, automobile bodies
Polyurethanes	tough, resist chemicals	insulation, furniture finishes

Figure 13-10 A single worker can easily handle this new polyethylene pipe. What properties make polyethylene particularly suited for making pipes and hoses?

TABLE 13-2

Kinds of Thermoplastic Materials		
Plastic	**Properties**	**Uses**
Acrylic (*uh*-KRIL-*ik*) (trade names: Lucite, Plexiglas)	resists weather and chemicals, softer than glass, easily scratched	camera lenses, windows, signs, car taillights, furniture, surgical instruments, costume jewelry, domed roofs
Cellulose acetate	tough, transparent, does not burn readily	vacuum cleaner parts, combs, toys, knobs, handles, packaging, eyeglass frames, tapes, film, lampshades
Nylon	tough, flexible, absorbs very little water; resists common chemicals, greases, solvents, and abrasives; has good electrical qualities	fabrics, gears, bearings, ropes, brush bristles, hardware, electrical appliances, fishing line, faucet washers
Polyethylene	flexible, tough, waxy, lightweight, moisture-proof, resists most solvents	ice-cube trays, moisture-proof freezer bags, bottles, packaging, electrical insulation, pipes
Tetrafluoro-ethylene (trade name: Teflon)	stickless material that produces almost no friction; resists heat, cold, and quick temperature changes	frying pans, self-oiling piston rings, car bearings, ball joints, parts for artificial veins and arteries
Vinyl	tough, strong; resists chemicals, water, heat, and cold	raincoats, door frames, water toys, garden hoses, floor coverings, shower curtains

made to look like marble, wood, and tile for use in home building. In medicine, doctors use plastics in the form of screws and plates to join broken bones. Plastics are even used to replace bones, parts of the intestines, and faulty heart valves. Because of their resistance to damage from chemicals and breakage, plastics are widely used in scientific equipment such as beakers, test tubes, and instrument cases. Tables 13-1 and 13-2 summarize the properties and some additional uses of several thermoplastic and thermosetting products.

13.3 NATURAL FIBERS

Fibers can be classified into four major groups. Animal, plant, and mineral fibers are all natural in origin. Synthetic fibers are manufactured. Both natural and synthetic fibers contain organic compounds. All animal fibers are proteins; they are compounds that contain carbon, hydrogen, nitrogen, and oxygen. Plant fibers are mostly cellulose, a carbohydrate compound. Coal and petroleum are the basic materials for making synthetic fibers. Fibers have a wide variety of uses besides textiles for clothing. These include home furnishings, insulation materials, and many products for industrial use such as rope, nets, book bindings, and adhesive tapes.

Wool and silk are animal fibers. A fiber that comes from the fleece of sheep, llama, alpaca, goat, or camel is called **wool.** The curly hairs of these animals mat together, trapping air. This insulating property makes wool an ideal cold weather fiber.

Once cut from the animal, wool fibers are cleaned and combed, then drawn and twisted into thinner strands. These strands are spun into the yarn that is used to weave cloth.

Wool is made up of polymers, which are large molecules made of repeating smaller units. Chemically, wool is very much like hair, feather, and horn. Wool is easily dissolved by strong bases, such as sodium and potassium hydroxides. It burns poorly, with an unpleasant odor, and leaves a black charred mass as residue.

Silk comes from a caterpillar. The fiber **silk** starts as a liquid that comes from tiny glands below the mouth of a silkworm caterpillar. When the fluid comes in contact with air, it changes into a solid and becomes a strand of silk. The caterpillar builds its cocoon by winding this strand around itself again and again. If permitted to live, the silkworm would undergo metamorphosis from caterpillar to pupa to moth, finally breaking out of the cocoon. However, to harvest an unbroken strand of silk, silk farmers use heat to kill the insect before it breaks out of the cocoon.

The cocoons are then treated with heat to soften the gummy material on the outside. The fibers are unwound and joined with fibers from other cocoons to form a single strand of silk. As many as 3,000 to 5,000 cocoons are needed to produce a kilogram of raw silk.

Linen and cotton are plant fibers. Cotton plants are grown from seeds that are planted yearly. The length, color, and cleanliness of cotton fibers vary with soil and weather conditions. Once mature, cotton plants are harvested mechanically (Figure 13-11) and their fibers are separated, cleaned, and manufactured into yarn.

Under the microscope, cotton fibers look much like flattened, twisted tubes, quite unlike wool. Body moisture is readily absorbed by cotton fibers and then lost through evaporation. This makes cotton garments comfortable for summer clothing. However, perspiration in contact with cotton fiber produces a weak acid reaction that is harmful to cotton. Like other natural fibers, cotton is often used in blends with synthetic fibers.

Cotton is nearly pure cellulose, having a formula of $(C_6H_{10}O_5)_x$. The x in this giant molecule is not known, but it is believed to be as high as 500,000.

Figure 13-11 Cotton forms the foundation for the employment of more than nine million Americans. The Cotton Belt includes 18 southern states, stretching from Virginia to California, and is home to an extensive cotton industry.

Linen comes from the flax plant. Linen is the oldest fiber known, dating back many thousands of years. Linen is much smoother, more lustrous, and more soil resistant than cotton. Cloth made from linen readily absorbs moisture, making it popular for warm-weather clothing.

The flax plant provides two main products: linen fiber from the stems and seeds from the flowers. To obtain linen fiber, the crop must be cut before the seeds mature.

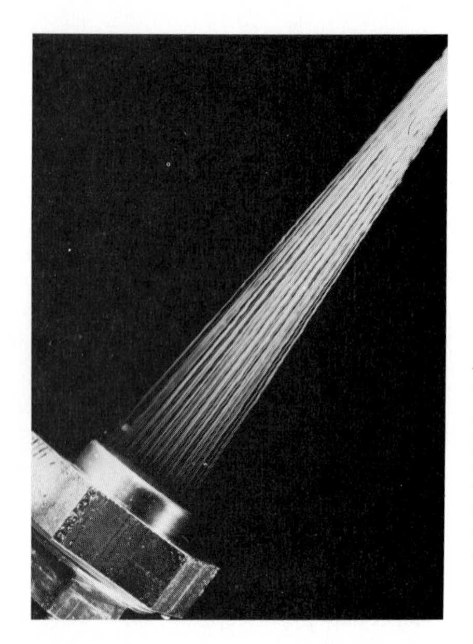

Figure 13-12 The spinneret is used in making long strands of synthetic fiber. Large mills have hundreds of these spinnerets arranged in rows.

The flax plants to be made into linen are pulled up and spread out in the sun to dry. The stems have a thin covering of bark, which must be removed. Linen fibers are nearly a meter in length, very soft, and yellow in color. These fibers, which are much stronger than cotton, tend to lie parallel. Linen is nearly pure cellulose with chemical properties similar to those of cotton.

Asbestos is the best known mineral fiber. Asbestos fiber occurs in certain types of rock that are mined and then processed to remove the fiber. Asbestos is useful because of its resistance to heat and chemicals. It has been used in fireproof clothing, insulation material, curtains, and automobile parts. Its use has recently declined, with the discovery that asbestos inhalation can cause health problems.

13.4 SYNTHETIC FIBERS

Most synthetic, or manufactured, fibers are made from cellulose or petrochemicals. Rayon and acetate are the most common fibers made from the cellulose of wood pulp or cotton. Nylon, acrylic, and polyester are manufactured from petrochemicals, which are chemicals made from crude oil, coal, and natural gas. These fibers are made by joining together monomers to form polymers. This process is called polymerization.

Most synthetic fibers are made by forcing liquids through tiny holes in a metal plate and allowing them to harden. The plates, called spinnerets, as shown in Figure 13-12, are about the size of a sewing thimble and have from 10 to 150 small openings, depending on the thickness of the strand wanted. The great variety of synthetic fibers is produced by the wide range of liquids used.

Each synthetic fiber has two names. One is generic, or the group name, like those mentioned previously. The other is the brand name, used by the maker of the fiber as a copyrighted trademark. Some of these brands that may be familiar to you are Dacron, Kodel, Orlon, Acrilan, and Qiana.

Rayon and acetate are the oldest of the synthetic fibers. Cellulose from purified wood pulp is the basic material for making rayon fiber. The pulp is formed into sheets and treated with sodium hydroxide. It is then shredded into crumbs and dissolved through a series of chemical reactions. The liquid is then forced through the openings of a spinneret. The long filaments then pass through a dilute acid solution for hardening.

Rayon absorbs moisture, making it suitable for clothing. However, rayon has two weaknesses: perspiration weakens rayon fibers, and its fibers lose strength when wet.

Acetate is another well-known fiber made from wood pulp. The reaction between cellulose and acetic acid is the basis for this product. Acetate is made of fibers that do not wrinkle or shrink as much as rayon.

Acetate fibers have two weaknesses: a very hot pressing iron will destroy the fibers, and some dry-cleaning solvents will dissolve the fibers. Acetate melts when burned.

Nylon, acrylic, and polyester are later inventions. The basic materials for making nylon come from chemicals derived from coal and petroleum. These chemicals are heated under pressure to form chains of huge molecules, or polymers. The liquid is then squirted through spinneret holes to form nylon threads. The strands are then stretched to about four times their starting length. This stretching forces the molecules to line up, giving nylon increased strength and making it more elastic. This elastic property makes nylon a popular fiber for socks, stockings, and other clothing. Its strength makes it popular for use in nonwearable products (Figure 13-13).

Figure 13-13 Nylon ropes are strong and water resistant. Industries also use nylon to make bearings, gears, and machine parts.

Acrylic and polyester were developed in the early 1950s. These fibers are easy to launder, are light in weight, dry quickly, and resist wrinkles. They are difficult to dye, do not absorb moisture well, and produce static electricity. Like all noncellulose fibers, they are thermoplastic. Thermoplastic fibers require low to moderate ironing heats. These fibers blend well with natural materials to make fabrics that combine the best features of both types of fibers. Polyesters are used to manufacture shatter-proof soft drink bottles. These bottles can be reworked later into fabrics, insulation materials, and auto parts.

13.5 BLEACHES AND DYES

Chemicals are used to bleach cloth. All natural fibers have a yellow tinge. These yellow fibers must be bleached to produce white yarn or cloth. Sunlight is effective for bleaching moist plant fibers, such as cotton and linen. However, sunlight tends to yellow animal fibers, such as wool and silk. Most synthetic fibers do not need bleaching.

Figure 13-14 A bowl of sodium hypochlorite solution bleaches the color out of a dark cotton shirt.

Sulfur dioxide and hydrogen peroxide are used to bleach wool and silk. A dilute solution of sodium hypochlorite, otherwise known as chlorine bleach, is commonly used for home-bleaching of cotton fabrics (Figure 13-14). Chlorine bleach cannot be used on wool or silk because it destroys these fibers.

Chlorine is a poisonous, greenish-yellow gas. It reacts with many metals and with hydrogen. In the bleaching process, chlorine reacts slowly with water, uniting with the hydrogen and setting oxygen free. This freed oxygen then reacts with some dyes in the cloth, causing a loss of color. As indicated in the following equations, the oxygen does the actual bleaching of the cloth:

1. $2Cl_2(g) + 2H_2O(l) \rightarrow 4HCl(aq) + O_2(g)$
2. oxygen (O_2) + dye in fabric $\rightarrow$ colorless compound

Sulfur dioxide is a dense, colorless, choking gas. This gas unites with water to form sulfurous acid (H_2SO_3), a very weak acid. Sulfurous acid readily unites with atoms of oxygen to form sulfuric acid (H_2SO_4). In the bleaching process, the sulfurous acid removes the color from some materials by removing oxygen. In other words, while sulfur dioxide bleaches by removing oxygen, chlorine bleaches by adding free oxygen.

Hydrogen peroxide is prepared in large volumes by passing an electric current through cold, dilute sulfuric acid. You have learned that hydrogen peroxide is an unstable liquid that breaks down readily, forming water and free oxygen:

$$2H_2O_2(l) \rightarrow 2H_2O(l) + O_2(g)$$

The oxygen that is set free can remove the color from some materials. Besides wool and silk, hydrogen peroxide is commonly used to bleach paper, feathers, and hair.

Dyes are mostly complex organic compounds. Until the 1850s, dyes were made from natural sources such as roots, barks, leaves, flowers, and berries. The dyes in use today are made by chemists. There are over 5,000 dyes now in use, but probably over 100,000 have been made at one time or another.

Dyes consist of molecules of bright-colored substances. Inorganic compounds are combined with organic compounds to produce many shades of color.

Coal tar is the raw material used for most dyes. Coal tar is a black, sticky liquid obtained as a by-product from coke ovens. While the tar does not contain dyes, it does provide benzene. Benzene is used to make a colorless liquid called aniline. Aniline is the starting point in the making of many synthetic dyes.

Dyes vary in their ability to hold color. A **fast dye** is one that clings to the fabric, does not lose its color, and does not wash away easily.

If a fiber takes and holds the dye with little help, the dye is called a **direct dye.** Direct dyes can be used for all natural fibers. Common salt added to the dye bath forces the dye to leave the solution and attach itself to the cloth.

Acid dyes are made with acids and are sometimes used to help silk and wool hold bright colors. These dyes do not cling to linen and cotton. Fabrics that have been colored with acid dye must not be washed with alkalis. Acid is neutralized by alkali, and the cloth cannot hold the dye.

Certain dyes will not adhere to cotton unless a chemical, called a mordant, is used first. The mordant combines with the dye molecules and fixes them firmly within the cotton fibers. Dyers often use different mordants with the same dye to make a variety of shades. Aluminum hydroxide is a widely used mordant.

Certain solvents can remove stains from cloth. Before you try to remove a stain from cloth, it helps to know what caused the stain. Also, it is helpful to know the kind of cloth that is stained. Soap and water are effective in taking out many kinds of stains. However, some stains need solvents other than soap and water. A little testing may be needed to find the right solvent. You must be careful to use a solvent that will not harm the fabric. Some dry-cleaning solvents, for instance, destroy acetate fibers. Dilute acids must be washed out of cotton and linen because they weaken the fiber. Bleaches may not only take out the stain; they may bleach the dye out of the cloth as well. Table 13-3 contains data on the removal of some common stains from fabrics.

TABLE 13-3

Removal of Stains from Clothing

Stain	Type of Cleaner
Acids	ammonia; rinse with water
Alkalis	vinegar; rinse with water
Blood	cold water and salt, or soap and cold water
Coffee	boiling water
Fruit juices	water
Grass	alcohol; wash with soap and water
Grease	dry-cleaning solvent, benzene, or kerosene
Ink, iron rust	detergent and water, oxalic acid, or lemon juice
Lipstick, blush	soap and water, or glycerin
Paint (oil)	turpentine or mineral spirits

CAREERS

THE TEXTILE INDUSTRY

from design to sales

The textile industry, and closely related industries, are large and cover a wide range of job opportunities. The textile industry provides fibers, cloth, dyes, and cloth products, employing over 3.4 million people and producing sales of over 50 billion dollars per year. In the United States, the making of textiles and clothing accounts for one manufacturing job in eight.

The textile industry can be subdivided into three broad areas: textile manufacturing, garment manufacturing, and marketing. In textile manufacturing, fabric designers suggest the makeup of all types of fabrics for clothing, draperies, carpets, upholstery materials, and household linens. This job requires both artistic talent and a knowledge of fibers, materials, and textile machinery.

Knitting machine operators inspect the material when it is made into cloth. The operators must know how to start and stop knitting machines and replace the various types of yarn on needles, as well as understand how high-quality cloth is made for various consumer uses.

In garment manufacturing, fashion designers create new styles and clothing ideas, also choosing the fabrics and colors that will be used in their manufacture. Pattern makers translate these designs into paper patterns. In most manufacturing of ready-to-wear garments, a succession of workers assemble each garment. Cutters cut the cloth from the pattern and sewers use machines to stitch the pieces of cut material together. Finishers add all finishing touches and pressers iron the completed garment. In some companies, tailors assemble the entire garment by hand and machine.

Once textiles and garments are manufactured, wholesalers and retailers market their products using artists, photographers, and promotional writers. In retail stores, salespeople and floor and window display artists bring the products to the attention of the consumer.

A fabric designer or fashion designer may require a college degree or courses in textile design, manufacturing, and merchandising. Other positions require technical knowledge of the tools of the trade.

For more information about careers in the textile and clothing industries, contact

American Apparel Manufacturers Association
2500 Wilson Boulevard, Suite 301
Arlington, VA 22201

CHAPTER REVIEW

SUMMARY

1. Natural rubber is a hydrocarbon made from the sap of rubber trees.
2. A rubber molecule is a polymer of C_5H_8 monomers linked together in a twisted zig-zag chain.
3. Neoprene, the first successful synthetic rubber, resists the harmful action of oils and greases better than natural rubber.
4. Thermoplastic articles become soft when heated and harden when cooled.
5. Thermosetting plastics take a permanent shape when heat and pressure are applied.
6. Cellulose, vinyl, acrylic, polyethylene, nylon, and fluorocarbons are families of thermoplastics.
7. Thermosetting families include phenolics, melamines, polyesters, epoxies, and polyurethane plastics.
8. Natural fibers are animal, plant, or mineral in origin, and include wool, silk, linen, cotton, and asbestos.
9. Synthetic fibers are made by squirting certain liquids through tiny holes in a metal plate called a spinneret.
10. Nylon, acrylic, and polyester are made by linking together small molecules (monomers) into large molecules (polymers).
11. Chlorine bleaches cotton and linen fibers by adding oxygen to a dye to make the cloth colorless.
12. Sulfur dioxide bleaches wool and silk by removing oxygen from some dyes to make the cloth colorless.

VOCABULARY

Match the item in the left column with the best answer in the right column. Do not write in this book.

1. acid dye
2. direct dye
3. fast dye
4. latex
5. monomer
6. polymer
7. silk
8. synthetic
9. thermoplastic
10. thermosetting
11. vulcanization
12. wool

a. sap from a rubber tree
b. single, repeated unit of a large molecule
c. not occurring in nature
d. fiber from the fleece of sheep, llama, alpaca, goat, or camel
e. requires little or no help from another chemical to hold color
f. many smaller molecules joined together
g. from the liquid emitted by a caterpillar
h. can be melted and shaped again and again
i. holds color well by clinging to fabric
j. acidic and used to color silk and wool
k. takes permanent shape when heat and pressure are applied
l. heating sulfur with crude rubber

REVIEW QUESTIONS

1. What are five properties of rubber?
2. What is the simplest formula for rubber?
3. How is latex processed to obtain crude rubber?
4. Describe the difference between a monomer and a polymer.
5. Why is the vulcanization process so important in the making of rubber?
6. What is the effect of adding reclaimed rubber to a batch of new rubber?
7. How is foam rubber made?
8. What is neoprene? What is the main difference in the structural formula of a natural rubber molecule and that of a neoprene molecule?
9. What does the word plastic mean as used by chemists?
10. List four properties of plastics.
11. What raw materials are used in making celluloid?
12. Describe the differences between thermoplastic and thermosetting plastics.
13. Tell how a plastic article is made by compression molding.
14. Describe how thermoplastic articles are made.
15. List several uses of acrylic plastics and vinyl plastics.
16. What are some of the uses of polyurethane plastics?
17. Discuss the properties of Teflon. What are some of its uses?
18. List two general groups of synthetic fibers. Give two examples of each group.
19. What are the advantages and disadvantages of acetate fibers?
20. What basic raw materials are needed for making nylon?
21. Name four properties of chlorine gas.
22. Is it the chlorine in a bleaching solution that bleaches cotton? Explain.
23. Describe the bleaching action of hydrogen peroxide and sulfur dioxide.
24. Can any dye of suitable color be used for cotton? Explain.
25. What is the difference between acid dyes and direct dyes?
26. How would you remove (a) a coffee stain, (b) grease, and (c) paint?

CRITICAL THINKING

27. What type of plastic would you recommend for (a) housing for a computer, (b) a garden hose, (c) adhesive bonding, (d) frying and baking pans, (e) squeeze bottles, (f) dinnerware, (g) radio and TV consoles, (h) recording tape, and (i) fishing lines? State a property that makes each plastic appropriate for the recommended use.
28. You are given two textile samples that appear to be identical and are told that one is wool and one is acrylic. Using the standard materials in your school laboratory, how would you tell the two fabrics apart?

FURTHER READING

Addy, J. *The Textile Revolution*. New York: Longman, 1976.

Goodman, Sidney H., ed. *Handbook of Thermoset Plastics*. Park Ridge, N. J. : Noyes Publications, 1986.

Jambro, D. *Manufacturing Processes: Plastics*. Englewood Cliffs, N. J.: Prentice-Hall, 1976.

Morton, Maurice, ed. *Rubber Technology*, 3rd ed. New York: Van Nostrand Reinhold, 1987.

Twigg, John. *Looking at Plastics and Other Big Molecules with Carbon*. North Pomfret, Vt.: David and Charles, Inc., 1987.

ENVIRONMENTAL POLLUTION

Impurities in the air that burn your eyes and cloud your vision represent just one form of environmental pollution. What other types of pollution are you aware of? What steps does your community take to control pollution?

CHAPTER 14

SECTIONS

OBJECTIVES

- ☐ Describe the causes and effects on the environment of water, air, and land pollution.
- ☐ Identify environmental pollutants and how to control their effects.
- ☐ Explain the role of weather in scattering air pollutants.
- ☐ Describe the various methods used in solid waste disposal.
- ☐ Identify radiation as a potential hazard and describe the disposal of radioactive waste.

14.1 WATER POLLUTION

Pollutants are harmful to living organisms. A **pollutant** is any substance added to a natural system in larger amounts than can be disposed of by that system. A substance becomes a pollutant when there is too much of it to be absorbed by our water, air, or land, such that it becomes harmful to living organisms. Even "natural substances," such as the radiation we receive from the sun or the bacteria we breathe in the air, can become pollutants when their levels build up to a point where they cause harm or illness.

However, in this chapter, the discussion will be limited to pollutants and their buildup resulting from the activities of people.

Figure 14-1 This unsightly mess was caused by solid waste materials. What are other causes of water pollution?

Water that is free of pollutants is essential to our survival. Clean water is needed for human consumption, and thus for continued good health. It is vital to aquatic life forms, which are an important link in the food chain. Large amounts of clean water are also needed on farms to grow our food and by industries to produce our clothing and shelter.

Up to a certain point, the water of flowing rivers and streams cleans itself by exposure to air and sunlight. Bacteria in the water help to break down wastes into nutrients. Also, the movement of water dilutes and scatters pollutants. Unfortunately, this natural cleaning system is no longer able to effectively handle the excessive volume of water pollutants occurring today.

What materials pollute earth's waters? Water pollutants fall into three categories: sewage, industrial wastes, and agricultural wastes.

In Chapter 8, you learned that sewage includes human waste, detergents, and other organic and inorganic materials. You learned that when untreated sewage enters lakes or rivers it destroys the ecological balance (Figure 14-1). How does this take place? Aerobic bacteria use dissolved oxygen to break down the sewage into nutrients for aquatic plants. As waste continues to enter the lake or river, bacteria further deplete the oxygen supply. The cycles of nutrient production and waste breakdown are thrown off balance. These cycles can also be destroyed if too many nutrients enter the body of water as phosphates from sewage or as nitrates from fertilizers. Now examine some of the pollutants generated by industrial and agricultural wastes.

Oil pollutes water. Have you ever visited the beach and found a dark, oily film coating the sand? Most of the oil that washes up on beaches comes from the normal operation of ships. Sea water is pumped into empty fuel tanks to make ships more stable. However, the tanks usually retain small amounts of unused fuel oil. When the ships are ready for refueling, the oily sea water is pumped out, covering much of the surface water with an oily film. Also, large amounts of crude oil are spilled, flushed, or leaked from ships. Laws against this practice exist, but they are hard to enforce. Although the sinking of an oil tanker or an underwater oil leak may cause great damage, most of the oil in the ocean is the result of the day-to-day operation of thousands of ships.

The oily film that covers the water reduces the water's ability to absorb oxygen from the air. This, in turn, affects the health of fish and other sea life. Even if they are not killed by the lack of oxygen, fish caught in these oily waters are not fit to be eaten. Water birds whose feathers become matted with oil, such as the one in Figure 14-2, are unable to swim or fly and soon die.

Figure 14-2 Fish and water birds can be killed as a result of water pollution. Here, volunteers try to clean the feathers of a bird trapped by an oil spill. Why do fish and birds die if they are covered with oil?

Mercury compounds can pollute water. Mercury and mercury compounds are used in some manufacturing processes. They are used to make chlorine, lye, plastics, and paper. Mercury compounds are also used by farmers as a coating for seeds and as a spray to kill fungus. Despite its usefulness, mercury is dangerous, causing a wide range of ailments including paralysis and mental disorders.

Mercury and mercury compounds find their way into rivers and streams through runoff and waste discharge. Because mercury is a dense metal, it sinks to the bottom of a stream. However, bacteria in the water act on the mercury, changing it internally to a compound that can be absorbed. The bacteria are then eaten by tiny forms of water life, which in turn are eaten by small fish. In this way, mercury compounds accumulate in higher concentrations up through the food chain. At the end of the food chain, where the concentration of poisonous substances is highest, irreversible damage to animals and human beings may result.

Fertilizers and pesticides can enter the water supply. Preparing soil for crops involves the addition of fertilizers and pesticides. These chemicals help provide healthy crops, but some seep into groundwater or wash into rivers and streams as agricultural wastes, causing water pollution. The pesticide DDT has been used on crops for over 30 years to eliminate insect pests. Unfortunately, DDT decays slowly and may affect organisms far removed from the farms. As DDT is washed from fields into streams, it ultimately moves up the food chain to larger fish and fowl, and finally to human beings. This pesticide has been found in the fat tissue of many animals, including man. There is evidence that its use affects the reproductive cycle of many animals and may cause similar problems in human beings. It is also a nonspecific insecticide, killing beneficial as well as harmful insects. In 1972, the federal government banned almost all uses of DDT, but it is still used in other countries. Scientists are seeking new ways to protect crops that include chemical pesticides that break down within a few days and biological methods of insect control.

Waste heat may cause thermal pollution. Power plants use large amounts of heat to generate electricity. To get rid of the excess heat, water from a stream or lake is used to cool the main steam condensers of the power plants. This process, diagrammed in Figure 14-3, heats the water, which is then returned to the stream or lake. The discharge of hot water from fossil fuel and nuclear power plants may cause an increase in heat, known as **thermal pollution**, in some lakes and rivers.

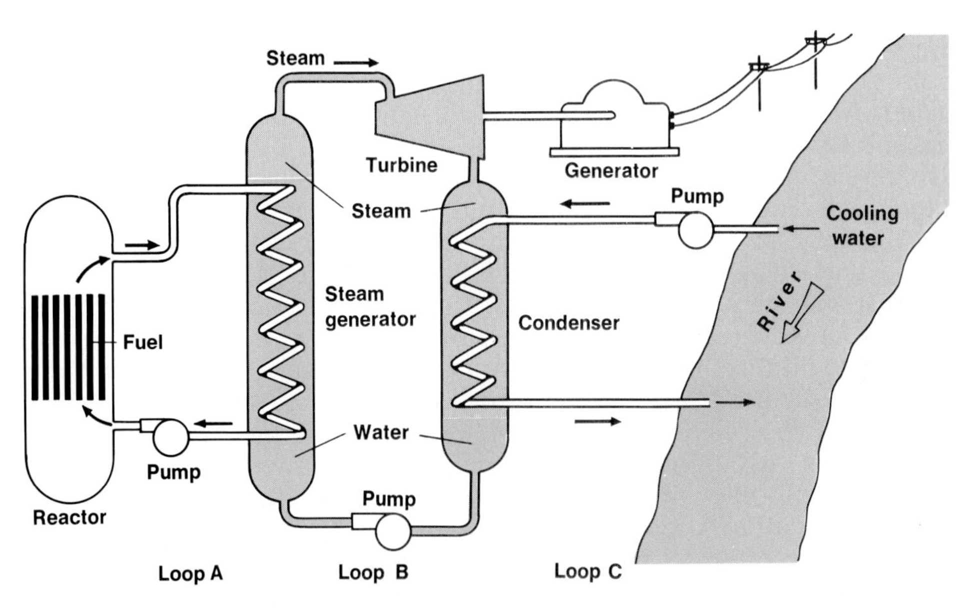

Figure 14-3 Cool water from the river condenses the exhaust steam in loop B. The heated water is returned to the river in loop C. Describe what happens in each loop.

In the discharge areas, the added heat drives some of the oxygen out of the water. This alters the life functions of certain animals and changes the ecology of the immediate area. Warming the water does not always create harmful effects. The heated water from fossil fuel and nuclear power plants could be used as a source of energy. Warmer pond or lake water can also cause fish to grow faster. Where excessive heat might damage the environment, cooling towers, air cooling, and spray ponds can be used to cool the heated water before it is returned to the main stream.

Controlling water pollution is costly. Because of public demand and government regulations, many industries are trying to stop or reduce water pollution. This process is expensive and it may affect jobs and raise the cost of consumer goods. Many cities and towns reduce water pollution by building better sewage treatment plants.

Some environments provide a natural biological means of removing pollutants. In a cattail marsh, for example, water filters through the plants and solid pollutants sink to the bottom. Bacteria in the marsh's ooze transform the pollutants into simple nutrients that nourish the plants. As the plants grow, their leaves release oxygen into the air and water. The end result is water as clean as, or cleaner than, water passing through a high-cost treatment plant. In some small towns and rural areas, people have used this model to build low-cost artificial wetlands to treat wastes (Figure 14-4). Animal wastes and pesticides in the water runoff from farmland have been successfully treated using a wetland system. The same results have been obtained with water used in refining petroleum. Although this method may become the most efficient treatment for agricultural pollution in rural areas and small towns, it would not be practical in urban environments where land on which to build wetlands is limited.

Figure 14-4 Cattail marshes represent a model for the low-cost treatment of polluted water. What are the limitations of such a method?

SEARCHING FOR WATER POLLUTANTS

activity

OBJECTIVE: Examine water samples to find water pollutants.

PROCESS SKILLS

In this activity, you will *observe* materials found in the water of streams and ponds, which will help you to *draw conclusions* about the pollutants in that water.

MATERIALS

safety goggles
jars or bottles containing water samples
from nearby ponds and streams
microscope
medicine dropper
microscope slides
coverslips
hand magnifier
gas burner
crucible
ring
ringstand
wire gauze

PROCEDURE

1. Obtain several samples of water from nearby streams or ponds. Using an eyedropper, slides, and cover slips, create slides of the water samples and examine them under a microscope. On a separate sheet of paper, sketch what you see under the microscope.
2. Put on your safety goggles. Set up the gas burner, ring, ringstand, and wire gauze. Pour several milliliters of one sample in the crucible. Evaporate the sample until almost all of the water is gone.
3. Allow the crucible to cool. Examine the residue in the crucible under a hand magnifier and record your observations. Repeat Steps 2 and 3 for each sample.
4. Check with your local health department to find out what steps are taken to prevent pollution of the local water supply.

OBSERVATIONS AND CONCLUSIONS

1. What did you observe in the water under the microscope?
2. What did you find in the residue that remained after evaporation?
3. Describe the quality of your water samples.

14.2 AIR POLLUTION

Maintaining clean air is a difficult but essential job. Clean water and air are perhaps our most important natural resources. The struggle for clean air is not new. For thousands of years, people have complained about smoke, soot, and odors. The Romans objected to the smell of burning coal and the deposit of soot it left on their clothes. During the Middle Ages, and even during the colonial period of this country, people were faced with these same problems.

In the past two hundred years, there has been a great increase in the amount of fuel burned in homes, factories, and cars. This increased burning of fuel has added pollutants such as excessive levels of CO_2 gas to the air and set the stage for the greenhouse effect. As you learned in Chapter 7, this effect results in heat being trapped in our atmosphere and could lead to increased global temperatures.

Air pollutants are highly concentrated where they originate, near large cities and industrial areas. Clouds of solids, liquids, and gases are constantly being added to the air above these areas. Even the air in sparsely populated areas is not clean, since pollutants spread out from the air above cities to the countryside.

In order to stop polluting the air, people must stop or greatly reduce those activities that cause pollution. People may not be able to live without sources of pollution such as cars or power plants, but they can change their lifestyles to reduce some of their needs.

Auto emissions pollute the air. The exhaust from cars, trucks, and other machines with internal combustion engines is one of the major sources of air pollution, as indicated by the graph in Figure 14-5. Motor vehicles account for an estimated 60 percent of the air pollutants in a large

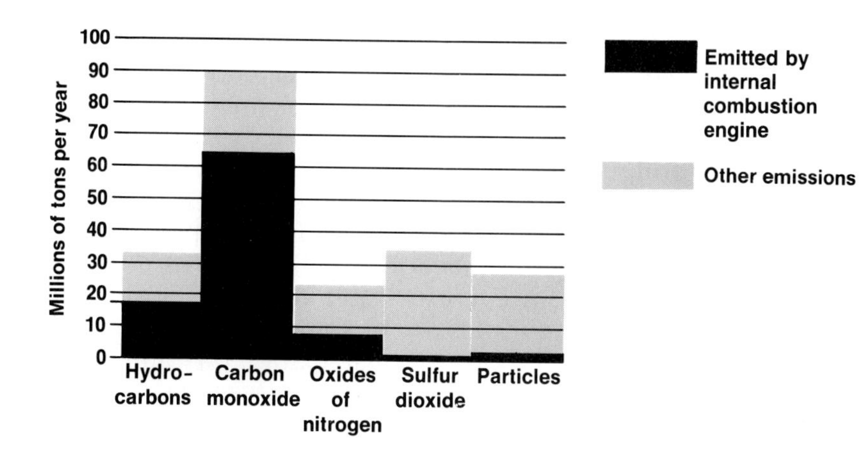

Figure 14-5 Major pollutants of the air in the United States, resulting from emissions by internal combustion engines and from other sources. What is the major source of carbon monoxide?

city. The waste products of motor vehicles come from the incomplete burning of fuel. Exhaust gases include carbon monoxide, oxides of nitrogen, and unburned hydrocarbons.

The word smog was once used to describe a mixture of smoke and fog. The word was later applied to pollutants in the Los Angeles area, even though neither smoke nor fog was the real problem there. Scientists now refer to this type of pollution as photochemical smog.

Figure 14-6 Heavy traffic in large cities promotes the formation of photochemical smog. In what way can this type of pollution be controlled?

Photochemical smog, pictured in Figure 14-6, is formed when unburned hydrocarbons react with oxides of nitrogen in the presence of ozone (a form of oxygen, O_3) and sunlight. Since the chief source of these pollutants is the exhaust from motor vehicles, modern cars are now equipped with devices that greatly decrease these harmful gases. One such device, called a **catalytic** (KAT-*uh*-LIT-*ik*) **converter**, changes the hydrocarbons and carbon monoxide in exhaust to harmless water vapor and carbon dioxide before the gases leave the vehicle. Catalytic converters also change the harmful oxides of nitrogen back into oxygen and nitrogen gases.

Another pollutant from auto exhaust comes from the use of a compound called tetraethyl lead, which was added to gasoline. This lead in the gasoline reduced knocking in engines and increased gas mileage (see Chapter 10). Unfortunately, tetraethyl lead, even in small amounts, is also an air pollutant. New car engines are designed to use lead-free gasoline, but many older cars still use leaded gasoline. Car makers are now trying to build a pollution-free engine that will also provide better mileage.

Pollution comes from factories, power plants, and homes. Another major source of air pollution is the burning of fuels that produces smoke and harmful gases. The smoke comes from the partial burning of fossil fuels, especially bituminous coal. Smoke contains free carbon particles that can darken the air and cover objects with soot. The harmful gases include sulfur dioxide, carbon monoxide, and others. Sulfur dioxide comes mostly from burning bituminous coal and to a lesser extent from burning some fuel oils. Sulfur dioxide harms the nose and throat, and can cause diseases of the lungs. Sulfur dioxide reacts with oxygen in the air to form sulfur trioxide, which in turn reacts with moisture in the air to form weak levels of sulfuric acid.

Some industries pollute the air with bad smells. Hydrogen sulfide (H_2S) gas, which has an odor like rotten eggs, is given off by some industrial processes. Chemicals such as ammonia and chlorine that are irritating or poisonous are other by-products of industry. Liquids from fermenting and decaying processes often have foul odors.

Dust and other particles also pollute the air. They are caused by grinding, cutting, sanding, mixing, crushing, and blasting. Whenever a solid is broken down into finer particles, dust is released. If the particles are small enough, they may drift in the air for a long time and be inhaled. Major industries that release dust include grain and feed, lumber, cement, mineral and rock, metalworking and mining. A problem at construction sites and in old buildings is the release of asbestos. As you learned in Chapter 13, asbestos is a mineral fiber that has many practical uses but has been identified as a cause of lung disease and cancer.

Factory wastes pollute the environment. Fluorides of sodium and calcium come mostly from the making of steel and ceramic products, and to a lesser extent from the making of aluminum and superphosphate fertilizer. When fluoride dust settles on food eaten by livestock, it can make the livestock ill.

Other industrial pollutants include waste products of living things, or organic refuse (from dairies, canneries, and meat-packing plants), arsenic compounds (produced by tanneries), and sulfur compounds and wood fibers (from pulp mills).

Acid rain results from air pollution. Rain is not pure water. As rain falls to earth, it picks up the pollutants in the air, including dust particles and gaseous combustion wastes. These gaseous wastes include oxides of carbon, sulfur, and nitrogen. When these gases react with water vapor, carbonic, sulfuric, and nitric acids are formed.

Pure water has a pH of 7, which is neutral. However, impurities can change the value to below 7, which is on the acidic side of the pH scale. As you learned in Chapter 8, acid rain is precipitation with a pH of less than 5.6. In extreme cases, rain has reached a pH as low as 3.0.

It is widely believed that acid rain harms some forms of plant and animal life. Some lakes, for instance, in the Adirondack Mountains of New York State, are so acidic that fish life is endangered. Evergreen forests in many parts of the world are being destroyed by acid rain. Acid rain has been blamed for damage to sculptures, buildings, and the paint on cars.

Pollutants may be affecting the ozone layer. Ozone in the atmosphere shields the earth from harmful ultraviolet radiation. Recently, a hole has been detected in the ozone layer above Antarctica. Development of the hole has been linked to the presence of chlorofluorocarbons in the air. **Chlorofluorocarbons** (CFCs) are synthetic organic compounds, used and emitted by many industries, that scientists believe to be harmful to the ozone layer. In 1978, the U.S. government banned certain chlorofluorocarbons because of this potential danger.

Noise can also be harmful. People who live near airports, factories, and busy traffic areas are exposed to loud noises. City noise, and even loud music, can damage hearing after long exposure. Noises can disturb other bodily functions as well. Loud noises may cause adrenalin to be released into the bloodstream, thereby increasing blood pressure and putting a strain on the heart. Because noise is potentially so harmful, it is considered a pollutant.

It is the degree of loudness of a sound that qualifies it as noise pollution. A **decibel** is a unit of measure of sound loudness (energy). A noise level of 100 db can cause temporary deafness. The sound level of a rock band has been measured at 120 decibels (db). Compare these values with the level of very soft music (about 30 db) and normal talking (about 65 db). (See Table 21-2 in Chapter 21 for a list of some common sound levels in decibels and Table 21-3 for some limits to noise exposure.)

Air pollution can be affected by air movement. You have already learned that precipitation can affect air pollution, as in the case of rain picking up dust and gaseous wastes as it falls to earth. Wind speed, wind direction, and air turbulence are also factors that affect air pollution, by either spreading or concentrating pollutants in the air. If

FINDING SOLID PARTICLES IN THE AIR

activity

OBJECTIVE: Perform an experiment to find solid particles suspended in the air.

PROCESS SKILLS
In this activity, you will *collect data* that will help you to *draw conclusions* about the kinds of particles that are in the air.

MATERIALS
tank-style vacuum cleaner
several pieces of filter paper
bright light
hand magnifier

PROCEDURE
1. Hold a piece of filter paper over the end of the hose of a tank-style vacuum cleaner.
2. Turn on the switch and allow the motor to run for several minutes while you hold the paper-covered hose in midair.
3. Turn off the switch and examine the filter with a magnifier under bright light. Compare the filter paper with an unused piece of filter paper.
4. Repeat Steps 1, 2, and 3 indoors and outdoors several times at different hours and on different days. Record your findings.

OBSERVATIONS AND CONCLUSIONS
1. What did you find on the filter paper after the first use?
2. How did your findings change at different times and dates? when you tested outdoors?

there is no wind to move the air above the ground, pollutants can concentrate in one area. However, a strong breeze will blow pollutants away. The amount of air pollution around you depends on whether the wind is blowing the pollutants away from or toward you.

The wind also tends to mix and scatter pollutants. Winds that blow over level ground usually move parallel to the ground in a relatively constant direction. However, winds that blow over tall buildings or uneven terrain tend to be deflected by the irregular topography and take on irregular motion, as illustrated in Figure 14-7, thus scattering pollutants with turbulence.

Figure 14-7 Turbulence and mixing of the air and its pollutants caused by uneven terrain or surface features. How does this wind motion compare with that over level land areas?

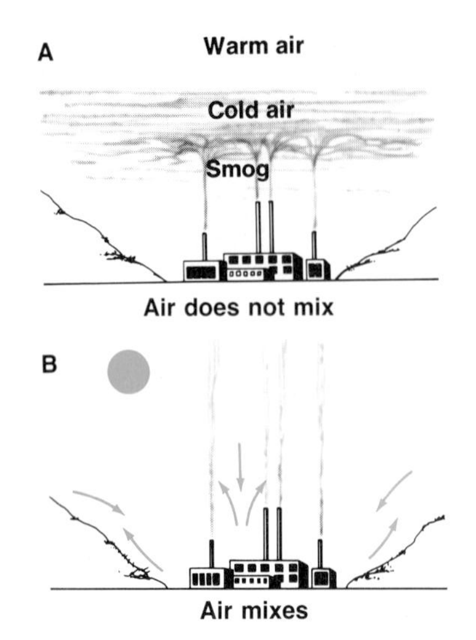

Figure 14-8 Temperature inversions tend to trap air pollutants near the surface (A). Under normal conditions, mixing of the air (shown as arrows) due to warming of the ground by sunlight and due to wind motion prevents inversions from forming (B).

Another kind of mixing of the air is caused by differences in temperature between layers of the air. On a clear day, the sun heats the ground and warms the layer of polluted air near the surface. The warm air is less dense than the cold air and rises, while cleaner, cooler air sinks from above. This mixing action creates an unstable condition and helps to scatter a polluted air mass.

However, on a clear night, the earth's surface cools quickly by radiating its heat into space. This radiating process cools the air near the ground while the air above this layer remains warm. This atmospheric condition where a layer of cool air stays near the ground and warmer air occurs above it is called a **temperature inversion.** Because cold air is denser than warm air, pollutants become trapped beneath the inversion (Figure 14-8A).

How long will a temperature inversion last? That depends on how quickly the earth cools during the night and warms up during the morning. Clouds, landforms such as mountains or valleys, and seasons of the year all affect the duration of the inversion. Temperature inversions tend to occur more often and last longer during the fall and winter months than during the spring and summer months. An inversion over a large city may remain for three or four hours after sunrise. In this case, the pollutants from the morning rush-hour traffic become trapped, resulting in smog and a serious health hazard. Inversions can be particularly hazardous in cities located at the base of mountains that block winds and prevent mixing of the air.

Figure 14-9 Research shows that when plants are exposed to polluted air, they do not grow as well as plants grown in clean air. The discoloration of this grape leaf is the result of exposure to polluted air.

Air pollution affects health. When you inhale, your lungs take in some of the oxygen in the air. Your body needs oxygen to work properly. If you breathe too much polluted air for too many years, impurities can remain in your lungs so that your lungs become less able to take in oxygen. Your chances of getting certain diseases, including cancer and pneumonia, will increase.

Tobacco smoke adds to air pollution in enclosed areas. Carbon monoxide levels in smoke-filled rooms can reach levels much higher than those found along busy highways and in traffic tunnels. Carbon monoxide presents a special danger to people with lung and heart problems. Heavily polluted air may also harm or kill animals and plants as shown in Figure 14-9.

Doctors have found that polluted air causes watery eyes, stuffy noses, coughs, sneezes, headaches, and sore throats. The list of ailments that may be caused by polluted air includes asthma, emphysema, tuberculosis, bronchitis, and certain types of lung cancer. Allergenic substances such as pollens and organic dusts can initiate or aggravate hay fever and other respiratory ailments.

Nitrogen compounds are plentiful in the air, and they play a key role in the life processes of all living things. Most nitrogen compounds come from the decay of organic matter caused by bacteria. However, nitric oxide and nitrogen dioxide are pollutants formed by manufacturing processes. Nitrogen dioxide, the chief irritant in smog, can cause damage to the lungs.

High levels of lead from leaded gasoline have been found in people living near areas of auto traffic. Lead poisoning affects the central nervous system.

Air pollution can be controlled. The best way to control air pollution is to keep impurities from getting into the air in the first place. Most cities and towns forbid open burning of trash and leaves. The catalytic converter is one way that the auto industry meets air-quality standards set by the federal government. Other industries are aware of their contributions to pollution, and are spending billions of dollars yearly on research and on pollution-control equipment. The cost of research and equipment is passed on to the consumer.

Figure 14-10 The photographs document the difference in emission of dust particles before (top) and after (bottom) the installation of a Cottrell device. How does this device work?

In many industries, the Cottrell device is used to remove smoke and dust particles from waste gases that pass through chimneys into the air. Wires in this device are charged with a high voltage. Dust particles become charged with electricity when they pass near the wires. The particles then collect in another part of the device that has an opposite charge. In Figure 14-10 you can see how effectively this works. (See Chapter 24, Figure 24-12 for a cross section of a Cottrell device.)

Smaller filters of the same type have been developed for homes and offices. One such device gives the dust particles passing through it a positive charge. These positively charged dust particles are then attracted to negatively charged collector plates. The dust is held firmly to the plates by a sticky coating that removes about 90 percent of the dust from the air as it passes through.

Another method of cleaning waste gases from a smokestack is by the use of scrubbers. A fine spray of water is directed into a chamber to catch and wash away the soluble gases from the exhaust. The gases can later be reused. However, the resultant liquid residue, along with lime used in the scrubbing process, forms an enormous, toothpaste-like mass. This substance itself poses a disposal problem. Other methods of removing gases include adsorption by charcoal and filtering. Special machines are also used to separate materials of different densities.

What can be done to eliminate air pollution entirely? In an industrial country such as the United States, this goal is probably unrealistic. All industrial production cannot be stopped in order to stop pollution. Without industry, there would be insufficient food, clothing, and shelter. However, there must be a balance between the adverse effects of pollution and the benefits obtained from industry. Pollution of the atmosphere can be controlled at the very least. Since 1970, levels of air pollutants such as sulfur dioxide, carbon monoxide, and dust particles have been greatly reduced in some areas. However, these and other pollutants continue to pose a serious problem.

14.3 SOLID WASTES

Vast amounts of solid waste are generated. Consider that with only 6 percent of the world's population, the United States consumes about 40 percent of all the world's resources. In one year alone in this country, people discard about 8 million cars, 20 million metric tons of glass, 180 million tires, 40 million metric tons of paper, and 80 billion cans. Paper makes up about 60 percent of the litter scattered along roads and highways in the United States.

Junk cars present a major disposal problem for many cities and towns. It can cost more to reuse materials from junk cars than the materials are worth. Many auto parts, such as rubber tires and plastic dashboards, do not decompose naturally. Burning tires can produce air pollutants. Buried tires cannot be compacted and tend to rise slowly to the surface.

Many plastics cannot be broken down by natural causes. They cannot be burned either because they melt and tend to clog the burners. Some plastics give off poisonous gases when burned. The plastic polystyrene, which is used to make containers for storing food, is a major environmental contaminant. Many disposable items are now made of biodegradable materials. **Biodegradable** materials can be broken down into simpler compounds by bacteria, sunlight, or other natural processes.

Cleaning up the debris of a consumer society and of industry has only recently been addressed. Until the 1970s little attention was devoted to disposal of hazardous by-products. The problem came to a head in 1978 when Love Canal in Niagara Falls, New York, made national headlines. Illegally buried barrels of dangerous chemicals began to leak and the chemicals began to surface. In 1983, Times Beach, Missouri, was evacuated when it was discovered that oil contaminated with dioxin, a highly poisonous waste product, had been used to coat dusty dirt roads. Some chemicals and other wastes, when inadequately disposed of, pose a long-term threat to life.

Most solid wastes are burned or buried. Two common methods of getting rid of solid waste are by burning or burying. Some waste, such as paper, is combustible. However, burning these materials can cause air pollution.

One advantage of the burning method is that the heat released during the burning process can be used to produce steam to make electricity. This heat can also be used to heat homes and other buildings. Another advantage to burning all solid wastes is that after combustion is completed, nonburnable materials, such as pieces of glass, iron, and other metals, are easily sorted out. These materials can be used again to make new products.

Currently, the most popular way of disposing of waste is by burying it in a sanitary landfill. A **sanitary landfill** is a community-operated landsite at which the disposal of wastes is regulated. Garbage and trash are spread on the ground and a thin layer of clay is added as a cover. Then the soil containing the waste is compacted by a bulldozer and covered with fresh soil. This prevents rodents and insects from breeding in the refuse. Other layers of waste are added until the area is completely filled. Unsanitary landfills, or open dumps, like the one pictured in Figure 14-11, often have a messy appearance and produce offensive odors. Unless properly located, water seeping through either kind of landfill can pollute underground water supplies. Also, landfills must be vented to allow the escape of methane, an explosive gas that builds up as the waste decomposes.

Figure 14-11 What are the advantages and disadvantages of disposing of waste in open garbage dumps?

Some waste materials can be recycled. The process by which wastes are reused to make new products is called **recycling**. By recycling, some substances in solid waste can be made into usable items again, thus saving raw resources and reducing the amount of litter. Recycling is essential since our resources are being used up rapidly and many, such as mineral resources, cannot be replaced.

Some materials can be recycled whereas others cannot. Organic matter, for example, is broken down naturally by bacteria into compounds that enrich the soil. Used aluminum cans are collected, melted down, and used to make more cans or other aluminum products. Old newspapers are made into usable paper products and used in manufacturing building materials such as plasterboard, roofing, and insulation. Copper, glass, steel, and iron are also recycled. Rubber and plastics can be recycled, but not easily. About 75 percent of our solid waste is recyclable; however, recycling does not reach this level due to limitations of technology, public cooperation, and available markets. Many American communities now require mandatory recycling of aluminum cans, glass bottles, and newspapers.

It is clear that our consumer society needs a solid waste management system that minimizes environmental risk at a price we can afford. This system should include resource conservation, recycling, and new landfill technology to dispose of waste safely.

14.4 RADIATION EXPOSURE

Radiation comes from several sources. One source of radiation is the earth itself. Radioactive substances are scattered through the entire crust of the earth. They are found as solids, liquids, and gases. Any hole that is drilled or carved into the earth is likely to hit some radioactive matter. Water, oil, and natural gas from wells are likely to contain some radioactivity. Likewise, mines for coal and minerals contain some radioactive matter.

An example of radioactive matter from the earth is radon. **Radon** is a naturally occurring radioactive gas considered to be a pollutant when concentrated. It is produced by the decay of radium in rocks and soil. In the 1980s, radon was recognized as a health hazard at certain locations when it was discovered in many buildings at high concentration levels. The gas seeps into homes from the rocks or soil below, becoming trapped and concentrated. Scientists believe that health problems associated with radon can be avoided simply with adequate building ventilation.

Another major source of radiation is outer space. Cosmic rays constantly bombard the earth. Taken together, radiation from the earth and space produce the natural **background radiation** that all people receive. There is not much that one can do to reduce such natural radiation, which is not part of the problem of radioactive exposure that is discussed in this chapter.

Radioactive wastes come from a number of human activities. By far the greatest source of radiation caused by humans is from the activities connected to the medical field. In fact, radiation exposure from all sources other than natural background and medical uses amounts to less than 2 percent of the total. Table 14-1 lists the various sources of radiation.

Radiation doses are measured in REMs. Radiation doses can be measured in several different ways. The most accurate way of measuring the effect of radiation on the human body is by a unit called a **REM**. A small fraction, only one-thousandth of a REM, is called a millirem. The average

TABLE 14-1

Sources of Radiation	Percent of Total
Natural background	67.6
Medical	30.7
Fallout from bomb testing*	0.6
Miscellaneous uses	0.5
Occupational exposure	0.45
Nuclear industry	0.15

*testing conducted in the atmosphere prior to 1963

radiation dose received by individuals in the United States is 250 millirem (0.25 REM) per year. The annual dose varies greatly from place to place and depends on the activities of each person. Average doses and sources as well as doses based on special activities are listed in Table 14-2.

As can be seen from Tables 14-1 and 14-2, medical uses produce by far the greatest amounts of radiation exposure from nonbackground sources. Also note that the medical uses listed in Table 14-2 deal with diagnosis, that is, the use of X-rays to find out what may be wrong with a person. Radiation doses in treatment of diseases, such as cancer, are many times higher than the doses used in diagnosis.

TABLE 14-2

Source or Activity	Radiation Dose in Millirem (annual levels, unless stated otherwise)
Cosmic rays at sea level*	35
Cosmic rays at 1,000-m altitude*	50
Air*	5
Building materials*	34
Food*	25
Ground*	11
Coast-to-coast jet flight (6 hr)*	5
One chest X-ray (bone marrow dose)**	44
One chest X-ray (skin dose)**	1,500
One fluoroscopic X-ray**	20,000
From nuclear power plant (within 80 km)*	0.01
From coal-fired power plant (within 80 km)***	180 to 380

Sources: *Atomic Industrial Forum, Inc.
**United Nations: *Ionizing Radiation: Levels and Effects,* 1972.
***Environmental Protection Agency Report: *EPA 520-1-77009.*

Nuclear power plants must ensure safety. In Chapter 6, you learned about the use of radiation in nuclear reactors to generate energy. An atomic power plant has many layers of protection built into it so that unforeseen events, human error, equipment failure, or accidents will not easily lead to emission of high levels of radiation. However, the Three Mile Island accident in 1978 and the destruction at the Chernobyl nuclear reactor in the Soviet Union in 1986 where many people were killed and the environment was contaminated show that all precautions are not foolproof.

Since the fuel rods are located below ground and are surrounded by layers of concrete, less radiation is released to the public from the normal operation of a nuclear power plant than from most building materials. In fact, people living near a nuclear power plant receive less radiation than those living near a coal-fired power plant. Coal, like most substances, contains small amounts of radioactive matter. This matter is released into the air along with smoke and gases when the coal is burned in a power plant.

Radioactive wastes must be disposed of safely. In a typical nuclear power plant, about one-third of the fuel is removed as waste every year and replaced by fresh fuel. The used fuel rods still contain much fuel but are no longer effective enough to be used in the reactor. The total volume of waste from a large nuclear power plant amounts to about 2 m^3 per year. By comparison, the waste from a modern coal-fired power plant fills an area of 2 km^2 to a depth of more than 1 m per year.

The U.S. government continues to work on guidelines for the permanent disposal of nuclear waste. Originally, it was assumed that the used fuel would be reprocessed to be used again and that the waste remaining after reprocessing would be a small percentage of the spent fuel. However, in 1977, reprocessing was prohibited in the U.S. and spent fuel rods continued to accumulate.

In 1982, the Nuclear Waste Policy Act was established by the U.S. Congress as a framework for setting up a national waste storage and disposal program. As outlined by this policy, the federal government is in the process of selecting and creating permanent waste repository sites. Scientists recommend that these sites be located deep below the ground in a geologic environment where the waste is not likely to be disturbed and damaged by earthquakes or groundwater. At present, spent fuel is encased in special containers and temporarily placed in on-site storage pools where they await transport to a national repository.

NUCLEAR HAZARDS

the fine line between safe and unsafe

Steel containers built to house radioactive materials generated by nuclear facilities reduce the risk of radiation exposure.

On April 26, 1986, an explosion destroyed the Chernobyl nuclear power plant near Kiev, in the Soviet Ukraine. Radioactive material released by the explosion spread across Europe, polluting the skies with radiation up to 100,000 times the normal levels. Radioactive fallout contaminated air, water, and food. At least 31 people died that year as a direct result of the explosion. Scientists estimate that between 6,500 and 40,000 more Soviets may die over the next 30 years from cancer brought on by exposure to Chernobyl's fallout and consumption of contaminated resources.

Many countries rely on nuclear power plants like Chernobyl to provide electricity, and some have nuclear weapons facilities to provide materials for bombs and nuclear warheads. There have been nuclear accidents at some of these facilities, though none as serious as the Chernobyl incident. Even if another accident involving high-level radiation never occurs, there are still reasons to be concerned about risks from nuclear facilities.

One such concern is low-level radiation. Several nuclear weapons plants have released excessive amounts of low-level radioactive gases into the air or have allowed radioactive water or refuse to seep into water supplies. Some people living near such plants have argued that, as a result of this contamination, there are higher-than-normal levels of cancer and birth defects in their communities. Unfortunately, scientists have only limited information on the long-term effects of exposure to low levels of radiation. Some scientists believe low levels are not hazardous to most people, whereas others believe exposure to any amount of radiation can be dangerous.

The ill effects of high-level radiation are undeniable. The effects of low-level radiation are still controversial. Whether or not the benefits of nuclear power and weapons outweigh the risks to our health is an issue for an informed public to decide.

FOR FURTHER RESEARCH AND DISCUSSION

Should we continue to rely on nuclear power for our energy needs? Are there alternative sources of energy that are safer and as practical?

CHAPTER REVIEW 14

SUMMARY

1. A pollutant is any substance added to a natural system in larger amounts than can be disposed of by that system.
2. Water pollutants include improperly treated sewage, industrial wastes, and agricultural wastes such as pesticides and fertilizers.
3. Air pollutants include carbon monoxide, sulfur dioxide, oxides of nitrogen, unburned hydrocarbons, and dust particles, and come from motor vehicles, factories, power plants, industrial plants, and homes.
4. Catalytic converters, Cottrell devices, scrubbers, and special filters are used to prevent pollutants from getting into the air.
5. Wind, temperature inversion, and precipitation affect air pollution.
6. Prolonged exposure to air pollutants may cause emphysema, tuberculosis, asthma, bronchitis, and certain types of lung cancer.
7. Solid waste disposal can be controlled by proper burning or by burying in a sanitary landfill.
8. Some radiation exposure is caused by background radiation from the earth or from space, but most comes from medical uses.

VOCABULARY

Match the item in the left column with the best answer in the right column. Do not write in this book.

1. background radiation
2. biodegradable
3. catalytic converter
4. chlorofluorocarbons
5. photochemical smog
6. pollutant
7. radon
8. recycling
9. REM
10. sanitary landfill
11. temperature inversion
12. thermal pollution

a. any substance added to a natural system in larger amounts than can be disposed of by that system
b. radioactive gas that is dangerous when concentrated
c. formed when unburned hydrocarbons react with oxides of nitrogen in the presence of ozone and sunlight
d. chemicals affecting the ozone layer
e. unit measure of radiation doses
f. cool air layer forms near the ground while air above remains warm
g. device to decrease harmful exhaust
h. materials that can be broken down naturally into simpler compounds
i. landsite for solid waste disposal
j. caused when hot water from power plants is discharged into some lakes and rivers
k. radiation from the earth and space
l. process in which wastes are reused to make new products

REVIEW QUESTIONS

1. What is a pollutant?
2. How can water in streams and rivers become polluted?
3. In what way do oil spills contribute to water pollution?
4. How do mercury compounds get into the human body?
5. Why is it necessary to find pest control methods other than DDT?
6. What is the cause of thermal pollution? What effects can it have?
7. What are the major sources of air pollution? List five air pollutants.
8. What pollutants are given off by the incomplete burning of coal?
9. How is photochemical smog formed?
10. How does sulfur dioxide get into the air?
11. How is the formation of acid rain connected to air pollution?
12. Is noise a pollutant? Why?
13. What is a temperature inversion? How does it affect air pollution?
14. What effect does polluted air have on people? What diseases can be caused by prolonged exposure to polluted air?
15. What are three devices used to control air pollution?
16. What are scrubbers? What problem is created by their use?
17. Why is it hard to dispose of most plastic materials?
18. When is a material biodegradable? Are all plastics biodegradable?
19. Name three methods of solid waste disposal.
20. What is a sanitary landfill? Describe how it works.
21. What is meant by recycling? Why is it important?
22. Why is radon considered a pollutant in certain areas?
23. What is background radiation? How does it compare to radiation from human sources?
24. What U.S. congressional act governs the safe disposal of radioactive waste from nuclear reactors?
25. Compare the volume of wastes from a nuclear power plant with the volume of wastes from a coal-fired power plant.
26. What do scientists believe to be the best environment for disposal of radioactive wastes?

CRITICAL THINKING

27. Why is the natural cleaning system no longer able to handle the problem of water pollution?
28. Would it be practical to suggest that artificial wetlands be built to alleviate New York City's water pollution problems? Explain your answer.
29. Recall that when sulfur dioxide is produced from the burning of fossil fuels, it reacts with oxygen in the air to form sulfur trioxide, which in turn reacts with water in the air to form sulfuric acid. Write the chemical equations for these two reactions.

FURTHER READING

Bright, Michael. *Pollution and Wildlife.* New York: Gloucester Press, 1987.

Magnuson, E. "The Poisoning of America." *Time* 116 (December, 1980): 58–69.

How the Sciences Work Together

The Process in Physical Science: From Sewage to Drinking Water

As cities grow, they demand more water. In many arid regions—parts of the Western United States for example—city leaders are asking how they can use limited water supplies to serve booming populations. One answer may be recycled sewage.

A supply of clean drinking water is something we depend on everyday.

An experimental facility for converting sewage into drinking water has been operating in Denver, Colorado, since 1985. So far, tests have indicated that sewage treated at this facility may be as clean as any city water supply in the country.

Nonetheless, much testing needs to be done before this recycled water is allowed to come out of kitchen taps. For instance, more than 60,000 organic chemicals are used in industry. Many of these are toxic, and some may find their way into sewage. Scientists on the Denver project have decided to test recycled wastewater on living organisms before recommending it for human consumption.

The Denver plant treats wastewater that has been treated in conventional facilities. Conventional treatment turns sewage into wastewater that meets cleanliness standards for release into lakes and rivers. This wastewater, however, contains living organisms, and organic and inorganic

Facilities like this one in Oakland, California, treat sewage so that it can be released into local waterways.

chemicals that make it unfit to drink.

The Connection to Biology

Bacteria, viruses and other disease-causing organisms, collectively called *pathogens,* remain in wastewater after conventional treatment. To combat pathogens, the wastewater is first mixed with lime. This raises alkalinity to levels not tolerated by most organisms, killing 99 percent of the pathogens. Injections of ozone, a strong disinfectant, destroy more microorganisms. Chlorine dioxide is added in the final step to prevent growth of new pathogens.

The Connection to Physics

Lime is added to wastewater and causes waste particles to combine into larger particles—a process called *flocculation.* The mixture is then pumped into a tank where these large particles settle out under the force of gravity. Any remaining suspended particles are removed by filtering the water through tanks containing sand and crushed coal.

In a step called *reverse osmosis,* the tendency of water to diffuse through a thin barrier, or membrane, at equal rates in either direction is disrupted. Pressure is applied on one side of the membrane causing more water to diffuse in one direction than the other. Since only water passes through the membrane, contaminants are diluted on one side of the membrane. By applying this process a number of times, the concentration of contaminants can be greatly reduced. Ammonia for instance, which can cause blood disorders and dissolve copper pipe used in most homes, is removed this way.

The Connection to Chemistry

Another step is *carbon adsorption* which removes organic compounds by passing the water through granular carbon which attracts organic compounds to its surface. One gram of granular carbon has a surface area of about 1,000 square meters and will *adsorb* organic compounds until it is totally covered.

In an *air-stripping* process, air is forced through a water spray. The water then picks up dissolved gases, mainly carbon dioxide. Carbon dioxide forms carbonic acid in water which lowers its alkalinity. Air stripping along with bubbling carbon dioxide directly through the water lowers the high alkalinity caused by the addition of lime in an earlier stage.

So far the Denver plant is the only one of its kind. But if the treated water it produces continues to pass purity tests, many thirsty cities may have one more source of water.

These columns store activated carbon for use in the carbon adsorption stage of wastewater recycling.

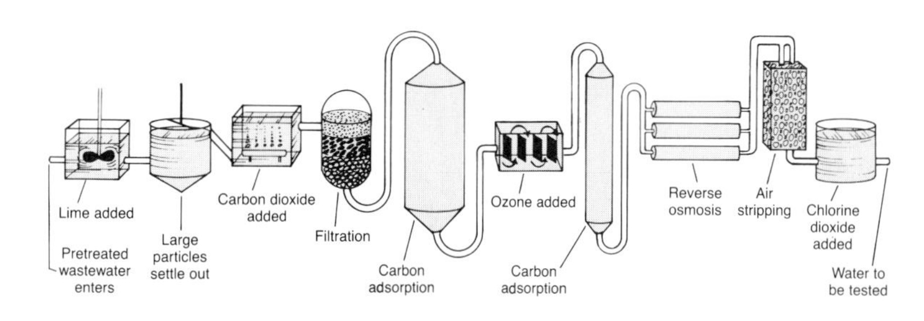

Pretreated wastewater goes through numerous purification stages at an experimental wastewater recycling facility in Denver, Colorado.

MOTION, FORCES, AND ENERGY

UNIT 4

Power—it means something different to each of us. What things would you describe as being powerful? Several thousand tons of water roaring down a steep cliff? A charging elephant? A lighted stick of dynamite? Or is power a 247-pound football player preparing to tackle? Power is often thought of as great strength, force, speed, and energy, or the ability to accomplish work. But what makes a waterfall more powerful than a meandering brook, or a football player more powerful than an infant? What is power and how can it be measured?

Understanding power begins with an understanding of the energy needed to do work, the motion that represents the work done, and the forces that produce the motion. In this unit, you will learn about energy, motion, and forces, and how these three concepts are related. You will gain an understanding of machines and engines that use force and energy to do work. You will also discover how heat and pressure affect different phases of matter.

CHAPTERS

MOTION AND ITS CAUSES

How can you tell from this still photograph that some kind of motion has just taken place? What causes such motion?

CHAPTER 15

SECTIONS

OBJECTIVES

- ☐ Define motion, speed, velocity, and acceleration.
- ☐ Differentiate between balanced and unbalanced forces.
- ☐ Describe forces including those of friction and gravity.
- ☐ Interpret force diagrams.
- ☐ Analyze and solve problems dealing with torque and parallel forces.
- ☐ Explain Newton's three laws of motion.

15.1 MOTION IN OUR WORLD

Objects are in motion all around you. You blink your eyes, turn a page, cross your legs. The hands of a clock move as time passes. A car drives past the window of your classroom. What is different about these motions? What is the same? What definition fits all of these motions? **Motion** is movement of an object from one place to another. When you blink, your eyelids move from one position to another and back again. Your fingers and the page of this book move from one place to another when you turn the page.

Now suppose you wanted to describe motion more precisely to someone. If you said, "The car is in motion," the other person would not have a complete idea of what you meant. For example, the person would not know how fast the car was moving or in what direction. The term motion, by itself, does not tell everything about how an object moves. There are more complete ways to describe movement. For example, you might say, "The car is moving at 30 km/hr." When you say how fast something is moving, you are giving the speed of that object. **Speed** is the distance an object moves in a certain amount of time. In this example, the speed of the car was 30 km/hr.

The speed of an object can only be detected and measured if it is compared to an object that is not moving. The object that is not moving, such as a tree, is called a reference point. Recall the examples of motion given above. What might be some reference points for an eyelid blinking, a page turning, or a car moving at 30 km/hr?

Velocity and acceleration describe motion. What else is needed to describe the motion of the car in the above example? You know that the car is moving. You know how fast it is going. It would also help to know the car's direction.

The statement, "The car is going north at 30 km/hr," gives the car's velocity. **Velocity** is a measure of the speed and direction of an object.

Figure 15-1 How can you teli that the flume car is accelerating?

What if the speed and direction are not constant? A driver starts her car and gradually brings it to a speed of 30 km/hr. For a while she drives down a straight road. She then comes to a sharp curve, slows down, and turns around the curve. This change is called acceleration.

In common use, acceleration means that an object speeds up. In science, the term has a specific meaning. **Acceleration** is any change in speed or direction. An object is accelerating when it slows down as well as when it speeds up or turns a corner (Figure 15-1). Acceleration is expressed as a change in velocity per unit of time.

SAMPLE PROBLEM

A car moving on a straight, level road changes speed from 35 km/hr to 45 km/hr in 5 sec. Find its acceleration.

SOLUTION

Step 1: Analyze

You are given that the initial velocity of the car is 35 km/hr, the final velocity is 45 km/hr, and the time elapsed during the change in speed is 5 sec. The car's acceleration is the unknown.

Step 2: Plan

From the definition of acceleration, you can derive the formula

$$a = \frac{v_f - v_i}{t}$$

where a is the acceleration, v_f is the final velocity, v_i is the initial velocity, and t is the elapsed time.Since you know v_f, v_i, and t, you can use the equation to find the acceleration. Since the three values do not have the same time units, you must convert seconds to hours

$$5 \text{ sec} \times \frac{1 \text{ hr}}{3{,}600 \text{ sec}} = 0.0014 \text{ hr}$$

Step 3. Compute and Check

Substitute all values in the equation.

$$a = \frac{45 \text{ km/hr} - 35 \text{ km/hr}}{0.0014 \text{ hr}}$$

$$= \frac{10 \text{ km/hr}}{0.0014 \text{ hr}}$$

$$= 7{,}142.9 \text{ km/hr/hr or } 7{,}142.9 \text{ km/hr}^2$$

Since the acceleration is expressed in velocity (km/hr) per unit of time (hr), your solution is complete.

15.2 FORCES IN OUR WORLD

Forces cause motion. So far you have learned about motion and differences in motion. What causes these motions in the first place? What causes the car to start, or your eyelids to blink? What makes the pages of the book turn? In every case, a push or pull of some kind caused the motion. One set of muscles in your eyelids pulls the lids down. Another set of muscles pulls them up. Your finger pulls the page of the book upward and pushes the page to turn it over. The car engine pushes the car forward. Even though all the pushes and pulls come from different sources, each one is called a force. Recall from Chapter 1 that a force is not something you can see. A force is a name given to any push or pull.

There are many kinds of forces. The easiest to understand are those in which you can actually see something pushing or pulling something else. Think of a baseball hitting a catcher's mitt, or a bowling ball hitting the pins. You can see the baseball pushing the mitt and the bowling ball pushing the pins. You can see the mitt and the pins move. Objects can also exert pulling forces. When a cowhand ropes a running steer, the steer pulls against the rope. Meanwhile, the cowhand and the horse are pulling on the other end of the rope in the opposite direction.

It is easy to imagine the forces discussed above because you can see that they cause things to move. But what about an object that does not move when you push against it? Is there a force acting upon the object? There may not appear to be any, but the forces are there. Suppose two people pick up a rope and begin a tug-of-war. They tug on the rope but neither can "out-pull" the other. The forces are exactly equal and opposite, and the rope does not move. In other words, when they are combined, the forces acting on the rope exactly balance each other.

Friction is a common force. Suppose you put a cardboard box on the gym floor and give it a push. You apply a force to the box and the box slides across the floor. What happens to the sliding box? Will it keep on sliding forever? Of course not. You know from experience that the box will come to a stop. Again, your experience tells you that friction caused the box to slow down and stop. **Friction** is resistance to motion, caused by one surface rubbing against another. To start the box moving again, you must overcome a kind of friction called static friction between the box at rest and the floor. Once the box starts moving, you need to overcome another force, called sliding friction, an example of which is shown in Figure 15-2. Whereas static friction occurs between stationary objects, sliding friction occurs between objects that are sliding with respect to each other.

Friction is affected by the smoothness of the sliding surfaces. If the surfaces are polished, the object will slide with less friction. For instance, a playground sliding board has a very smooth surface so children can slide down easily.

Another way to reduce the force of friction is to change sliding friction to rolling friction. If you attach wheels to the bottom of the box and again give it a push across the gym floor, how far would it move? You would discover that the box moves much farther with wheels than without them. In other words, it is easier to roll an object than to slide it. Friction can also be decreased by using some form of lubrication. Oil is an excellent lubricant. It makes the surfaces very slippery. Wax is another example of a good lubricant used on skis to reduce friction between the skis and the snowy slopes.

Figure 15-2 What kind of friction must the bobsled overcome in order to stay in motion?

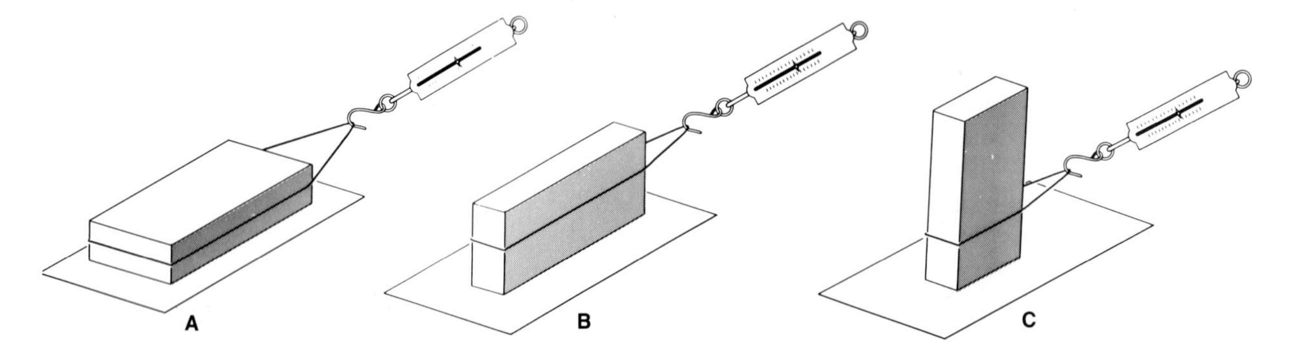

Figure 15-3 How is the force of friction different on each of the three sides of the brick?

You have learned that a number of factors affect friction. Now take a look at Figure 15-3. It reveals that friction is independent of the size of the surface areas that are in contact. A large surface area does not produce more friction than a small surface area. What in the illustration supports this idea? What is the variable in the experiment shown?

Another kind of friction is fluid friction. (A fluid is either a liquid or a gas.) For example, the wind that pushes against a moving car or airplane produces fluid friction. In water, fluid friction tends to reduce the speed of moving objects such as boats or fish.

Although you might feel that friction is generally unwanted, there are circumstances when it is very helpful if not essential. For example, a car turning a corner needs a great deal of friction between the tires and the road. If this friction is not present, the car will skid off the road. Likewise, much friction is needed for a car to make a quick stop (Figure 15-4). What purposes do snow tires and tire chains serve? Think about a second example. By spreading sand over an icy sidewalk, you alter the nature of the icy surface your feet come in contact with. By making the icy sidewalk rougher, you increase the friction between your feet and the sidewalk. This is because friction depends on the nature and smoothness of the materials in contact. Why is it impossible to walk without friction? Why do nails need friction in order to hold?

Figure 15-4 This graph shows stopping distances of a car at 48 km/hr. The bottom three bars represent icy road conditions. How does the amount of friction between the road and the car tires differ in each case?

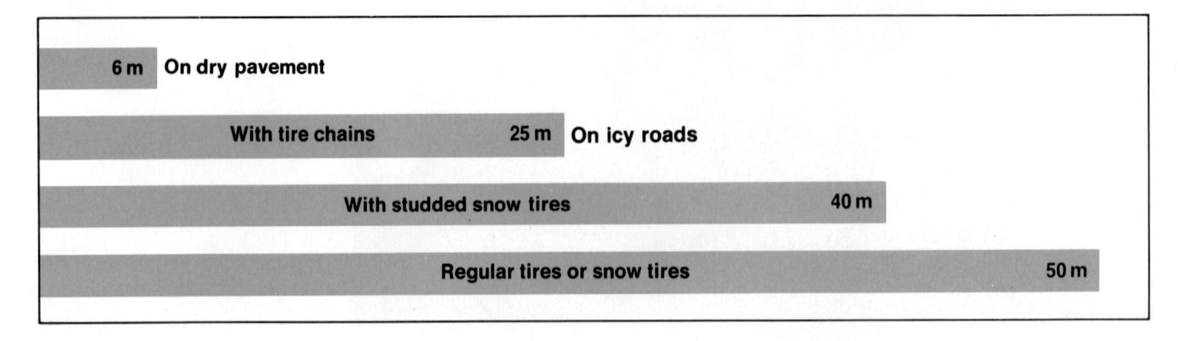

MEASURING THE FORCES OF FRICTION

activity

OBJECTIVE: Determine how sliding friction can be reduced.

PROCESS SKILLS
In this activity you will *measure* force and *collect* and *analyze data* to solve a problem about the force of friction.

MATERIALS

spring scale
3 bricks
board
lubricant (oil, soapy water, or waxed paper)

PROCEDURE

1. Attach a spring scale to a brick and place it on a board, as shown in Figure 15-3A. Pull the brick across the board at an even rate. Repeat the activity with the brick on its side and then on its end as shown in Figures 15-3B and 15-3C. Then put a second brick on top of the first, and finally a third brick on top of the other two. Be sure to keep the scale level.

   In your notebook, record your readings in a table similar to the one shown. Do not write in this book. Find and record the area of the brick's sliding surfaces in each trial.

Arrangement of Bricks	Area of Sliding Surface	Reading of Spring Scale			
		No Lubrication		With Lubrication	
		To Start	To Pull	To Start	To Pull
1 Brick (flat)					
1 Brick (side)					
1 Brick (end)					
2 Bricks (flat)					
3 Bricks (flat)					

2. After completing the trials on a dry board, repeat them using a lubricant. If you find oils too messy, you can use soapy water or waxed paper as a lubricant. Record your readings in the table.

OBSERVATIONS AND CONCLUSIONS

1. What effect does an object's weight have on friction?
2. What effect does a lubricant have on friction?
3. What relationship did you find between sliding area and friction?

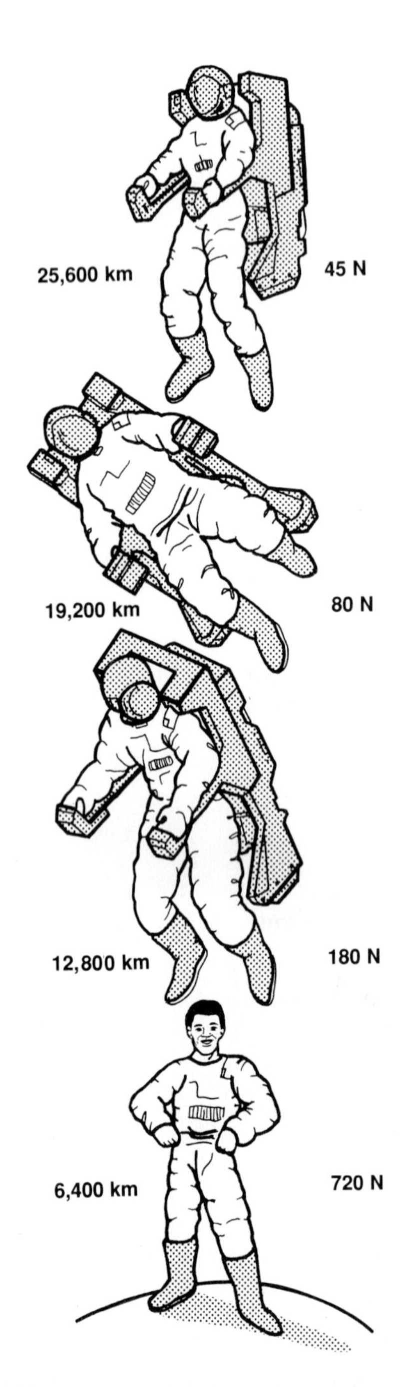

Figure 15-5 The force of gravity on an astronaut at various distances from the center of the earth. How does the inverse-square rule apply in this diagram? What would be the force of gravity on the astronaut at 32,000 km above the earth's center?

Gravity is another common force. You know that when you throw a ball straight up in the air, it does not keep going up forever. It stops and falls toward the ground again. Think about the up and down motion of the ball. You applied a pushing force in an upward direction to start the ball moving. What force slowed and stopped the ball's upward movement? Then, what caused the ball to come back down? The force that caused these effects is gravity. In Chapter 1, you learned that gravity is the force of attraction possessed by all masses. In the example given above, the masses of the earth and the ball pull them towards each other. Since the mass of the earth is much greater than that of the ball, the earth does most of the pulling. Gravity gives objects the property we call weight. Put another way, the weight of an object depends on the force of gravity acting on it.

Sir Isaac Newton, an English physicist of the seventeenth century, worked out the law of universal gravitation. This law states that every mass in the universe attracts every other mass. Newton found that the greater the masses of two objects, the greater the force between them. A 1,000-kg mass on the surface of the earth is attracted toward the center of the earth by twice as much force as is a 500-kg mass.

Newton's law also states that the farther apart the centers of two masses are, the lesser is the force of attraction between them. If the distance between two objects is doubled, the force between them becomes four times as weak. If the distance is tripled, the force becomes nine times as weak, and so on. Newton expressed the relationship this way: The force of gravity varies inversely as the square of the distance between the centers of the masses. This concept became known as the inverse-square rule.

To see how this rule works, examine Figure 15-5. When on the surface of the earth, a person is 6,400 km above the center of the earth. If that person moves to an altitude of 6,400 km the person would now be 12,800 km above the center of the earth. Thus, at 12,800 km the person would be twice as far away from the center of the earth as originally, and so would weigh only one fourth as much as before.

Gravity causes uniform acceleration of any free-falling bodies. Galileo, the Italian astronomer and physicist, was one of the first to study acceleration. He timed a metal ball as it rolled down a groove in a sloping plank. Galileo found that the speed of the ball increased at a regular rate. In other words, as the ball rolled, it accelerated at a uniform rate.

Instead of using a plank, scientists now study the motion of objects as they fall freely. These studies have shown that at the end of one second, objects falling in a vacuum attain a speed of 9.8 m/sec. At the end of two seconds, they fall at a speed of 19.6 m/sec. This means that the acceleration of a freely falling object is 9.8 m/sec for each second that it falls, or 9.8 m/sec/sec (9.8 m/sec^2).

This acceleration, caused by the earth's gravity, is called 1 *g*. For objects near earth's surface, *g* equals 9.8 m/sec^2. An acceleration of 5 *g*'s, therefore, means a change in speed, or an acceleration, of 49 m/sec^2.

Since the downward speed of a falling object begins at zero and is up to 9.8 m/sec after one second, the average speed during the first second of the fall must be (0 m/sec + 9.8 m/sec)/2, or 4.9 m/sec. Since the object moved at an average speed of 4.9 m/sec for one second, it must have fallen 4.9 m. If the object falls for two seconds, its speed at the end of two seconds will be 19.6 m/sec. (Remember, the object gains 9.8 m/sec each second.) The average of the object's initial speed (0 m/sec) and its final speed (19.6 m/sec) is (0 m/sec + 19.6 m/sec)/2 = 9.8 m/sec. If an object moves at an average speed of 9.8 m/sec for two seconds, it will go a distance of (2 sec × 9.8 m/sec), or 19.6 m.

Notice that an object falls four times as far in two seconds as it does in one second (19.6 m/4.9 m = 4). So when the time is doubled, the distance becomes four times as great. This can be stated in another way: The change in distance is proportional to the square of the change in time.

Galileo noticed that if the time of fall were tripled (times three), the distance covered would be nine times as great. He expressed this as a law, stating that the distance covered by an object that is uniformly accelerated depends on the square of the elapsed time. Thus, to find how far an object will fall in a given amount of time, square the number of seconds that it falls and multiply by one half the acceleration caused by the force of gravity. The formula that expresses this relationship is

$$s = \frac{at^2}{2}$$

where *s* is the displacement (or distance), *a* is the acceleration, and *t* is the time.

Air gives resistance to freely falling bodies. In the discussion about free-falling bodies, we assumed that there was no air resistance to slow down the objects. In other words, we assumed that the objects were falling in a vacuum where they all accelerate at the same rate, regardless of their size, shape, or density. The earth's atmosphere, however, is not a vacuum. Figure 15-6 illustrates how air resistance in normal atmospheric air affects the rate at which a feather and a steel ball fall.

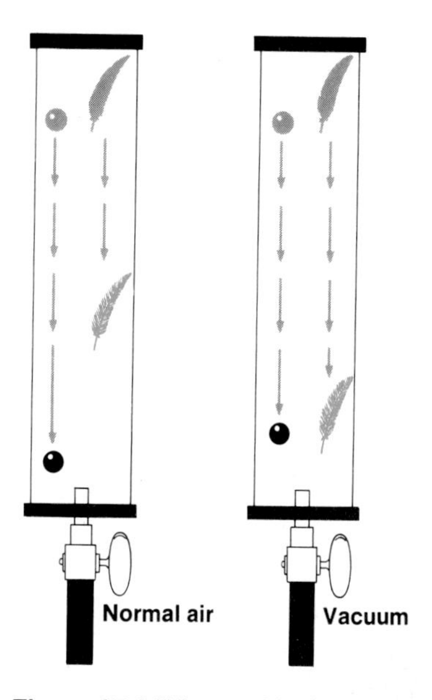

Figure 15-6 Why would a feather and a steel ball fall at the same speed in a vacuum?

SAMPLE PROBLEM

An object starts from rest and falls freely under the force of gravity. (a) What is its velocity at the end of 5 sec? (b) What is the average velocity during the 5-sec fall? (c) What distance has the object fallen during this time?

SOLUTION TO (a)

Step 1: Analyze

You are given that the initial velocity is zero and the elapsed time is 5 sec.The final velocity is the unknown.

Step 2: Plan

You know that the acceleration owing to the force of gravity is 9.8 m/sec^2. Since you know three of the four values in the equation

$$a = \frac{v_f - v_i}{t}$$

you can find the final velocity by solving for v_f.

$$v_f = (at) + v_i$$

Step 3: Compute and Check

Substitute all values in the equation

$$v_f = (9.8 \text{ m/sec}^2 \times 5 \text{ sec}) + 0$$
$$= 49 \text{ m/sec}$$

Since the unit m/sec is correct for velocity and the value appears to be within an acceptable range, your solution is complete.

SOLUTION TO (b)

Step 1: Analyze

You are given that the initial velocity is zero. You calculated the final velocity as 49 m/sec. The average velocity is the unknown.

Step 2: Plan

From the information in this chapter, you can derive the equation for the average velocity (v_a) as

$$v_a = \frac{v_f + v_i}{2}$$

Another example shows that free-falling objects do not accelerate indefinitely. A sky diver jumps from a high-flying plane and delays opening the parachute. The diver's speed will increase quickly, but, as the speed increases, the friction of air resistance also increases. At some point, the force of

Step 3: Compute and Check

Substitute all values in the equation.

$$v_a = \frac{49 \text{ m/sec} + 0 \text{ m/sec}}{2}$$
$$= 24.5 \text{ m/sec}$$

Since 24.5 m/sec is within the acceptable range between 0 m/sec and 49 m/sec, your solution is complete.

SOLUTION TO (c)

Step 1: Analyze

You are given that the elapsed time is 5 sec. You know the acceleration is 9.8 m/sec^2. The distance fallen is the unknown.

Step 2: Plan

Since you know the time and acceleration, you can use the following equation to find the distance:

$$s = \frac{at^2}{2}$$

Step 3: Compute and Check

Substitute all values in the equation.

$$s = \frac{(9.8 \text{ m/sec}^2) \times (5 \text{ sec})^2}{2}$$
$$= 122.5 \text{ m}$$

Since distance equals the velocity multiplied by the time, you can check your solution by noting that in 5 sec an object with an average speed of 24.5 m/sec will move 122.5 m.

friction of the air becomes equal to the downward force of gravity; that is, it is equivalent to the diver's weight. At this point the forces are balanced and no further acceleration occurs. The diver will continue to fall at a constant speed. The diver's actual speed depends on a number of factors but

Figure 15-7 What force is slowing down the sky diver for a safe landing?

may be as high as 250 km/hr. When the parachute opens, it increases air resistance, reducing the rate of fall to a point that will permit safe landing (Figure 15-7).

Torque tends to cause an object to rotate. As a child, you and a friend learned how to play on a seesaw, which is a kind of lever. A **lever** is a rigid bar that is free to turn on its point of support, or its fulcrum. The direction in which a seesaw moves depends on two factors: (1) how much you and your friend weigh and (2) how far each of you sit from the fulcrum. [The weight of an object on the earth's surface is found by multiplying its mass in kilograms by 9.8 m/sec^2. The product of this calculation is expressed in newtons (N).] A child that weighs 200 N can balance a child weighing 400 N by sitting twice as far from the fulcrum.

In Figure 15-8, the arrows show that there is a twisting movement around the fulcrum. As one side goes up, the other goes down. A twisting movement caused by one or more forces is called a **torque** (*TORK*). To find the torque, you multiply the force times the distance from the fulcrum. If you multiply the force on the left in Figure 15-8 by the distance from the fulcrum, the product is 800 newton-meters (N-m). The torque to the left of the fulcrum tends to turn the seesaw in a counterclockwise motion. Torque is always measured in force units (newtons) multiplied by units of distance (meters) from the fulcrum (or N-m).

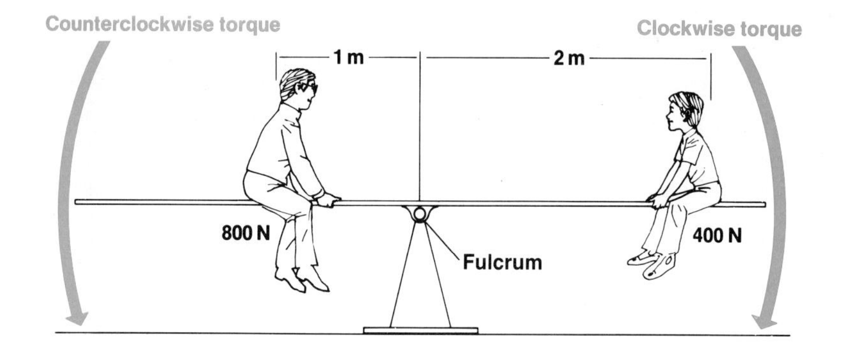

Figure 15-8 What would happen if the man moved away from the fulcrum?

If a seesaw is to balance, the clockwise torque must equal the counterclockwise torque. A force of 400 N placed 2 m to the right of the fulcrum gives a torque of 800 N-m. The seesaw, or lever, is then balanced.

To prevent rotation of a body or change in rotation, the sum of the clockwise torques about any axis must equal the sum of the counterclockwise torques about the same axis.

Parallel forces act together. Two or more forces acting in either the same or in the opposite direction are called parallel forces. Think of a bridge that is held up by two piers. The piers exert parallel upward forces. Objects on the bridge exert parallel downward forces.

Suppose a truck weighing 10 kilonewtons (kN) is resting on the middle of a 25-m long bridge. If the truck is exactly in the middle, its weight will be supported equally by the two piers. However, if the truck moves to a spot 15 m from one end, as shown in Figure 15-9, how much weight will each pier then support? (In this example, do not count the weight of the bridge itself. Also assume that all the weight of the truck is not located at the wheels but at its center of gravity.

Figure 15-9 How much of the truck's weight is supported by each pier?

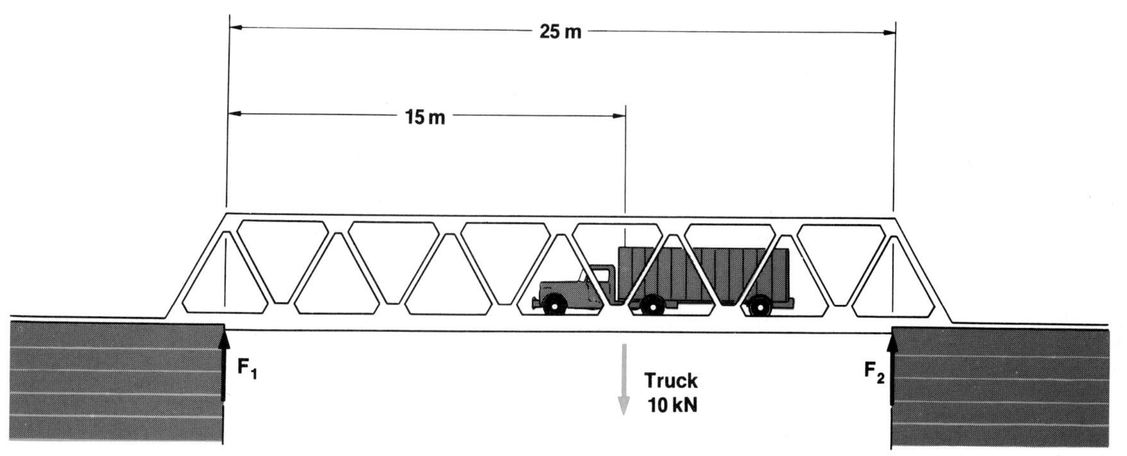

SAMPLE PROBLEM

A weight of 5 N hangs from a 100-cm uniform lever, 20 cm to the right of its midpoint fulcrum. Twenty centimeters to the left of the fulcrum is a weight of 2.5 N and 50 cm to the left of the fulcrum is a weight of 1 N. Is the lever balanced? Explain.

SOLUTION

Step 1: Analyze

You are given that the three forces are 5 N, 2.5 N, and 1 N. You are also given that their distances from the fulcrum are 20 cm right, 20 cm left, and 50 cm left, respectively. (It may help you to make a drawing of the lever, showing the forces and torques acting on it.) You need to determine if the clockwise torque equals the counterclockwise torque.

Step 2: Plan

Since you know the torque equals the force times the distance, you know the total torque equals the sum of all torques on one side of a fulcrum, and you know the force and distance for all three weights, you can calculate both the clockwise and counterclockwise torques.

Step 3: Compute and Check

$$\text{Clockwise torque} = 5 \text{ N} \times 20 \text{ cm} = 100 \text{ N-cm}$$

$$\text{Counterclockwise torques} = (2.5 \text{ N} \times 20 \text{ cm}) + (1 \text{ N} \times 50 \text{ cm}) = 100 \text{ N-cm}$$

Since the clockwise torque equals the sum of the counterclockwise torques, the lever is balanced. You can check your results in the laboratory using a meter stick as a lever. Place the fulcrum at its midpoint and hang known weights on it at appropriate points.

The **center of gravity** is the point where all the weight of an object may be considered to be concentrated.)

The problem can be worked out if you think of the bridge as a lever, with one pier (F_1) acting as the fulcrum and the other pier (F_2) pushing upward. Thus, 10 kN × 15 m = 150 kN-m of clockwise torque. Pier F_2 at the end of a 25-m lever must therefore give an equal counterclockwise torque of 150 kN-m. Since F_2 is 25 m from F_1, then 150/25 = 6 kN. Therefore, the second pier (F_2) supports 6 kN of the truck's weight. The remaining 4 kN must be supported by the first pier (F_1).

Forces can be diagrammed. One way you can describe a force is by its size or amount. The size of a force is measured in newtons. For example, you can say you pushed on a box with a force of 10 N. However, a force cannot be fully described by its size alone. The direction in which the force acts is also important. Any quantity that has both size and direction is called a **vector**. Since forces have both size and direction, forces are vectors. A vector quantity may be shown as an arrow. The head of the arrow shows the direction. The length of the arrow shows the size. Suppose you want to use a vector to show a force of 10 N pushing toward the east. As a scale, let 1 cm equal 2 N. Thus, a line 5 cm long should be drawn with the arrowhead pointing toward the right. This arrow would show a force of 10 N in an eastward direction.

Forces can be added or subtracted. You know that when you push on something, it tends to move in the direction you push it. But what happens when you apply two forces in different directions on the same object? You can use force vectors to answer this question.

When two or more forces act together, their effect on an object is a combination of the forces. This combined effect is called the sum or **resultant** of the forces.

Suppose, for example, that one person is pulling on a rope with a force of 5 N. Another person is pulling on a second rope (attached to the same object, and pulling in the same direction) with a force of 10 N. The resultant force is the same as if one person were pulling on a single rope with a force of 15 N. This force is shown in Figure 15-10A.

Now suppose that one of the people goes to the opposite side of the object. Then both people pull again with the same forces as before but in opposite directions. What is the resultant? In what direction will the object move? For the answer, see Figure 15-10B. Now suppose that the people are still pulling in opposite directions but with equal forces of 10 N. The forces cancel each other out and the resultant is zero, as shown in Figure 15-10C. When two or more vectors are acting at the same time on a point, they can be combined into a single resultant vector. The effect on the point is the same as if the resultant force alone were acting on the point.

Figure 15-10 Which points in the diagram are acted on by balanced forces? by unbalanced forces?

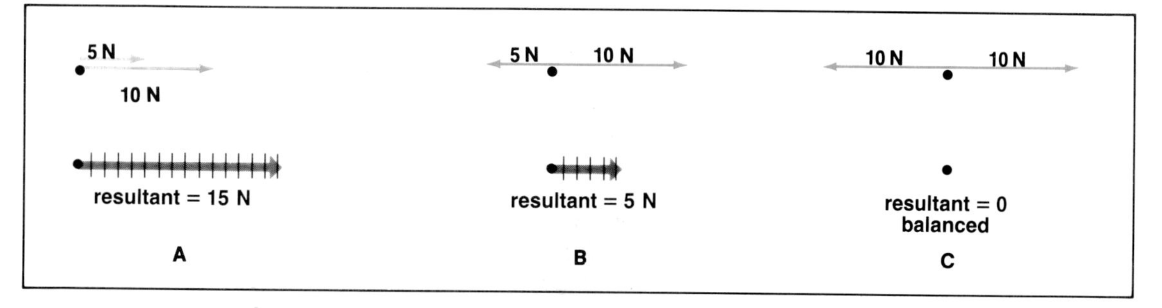

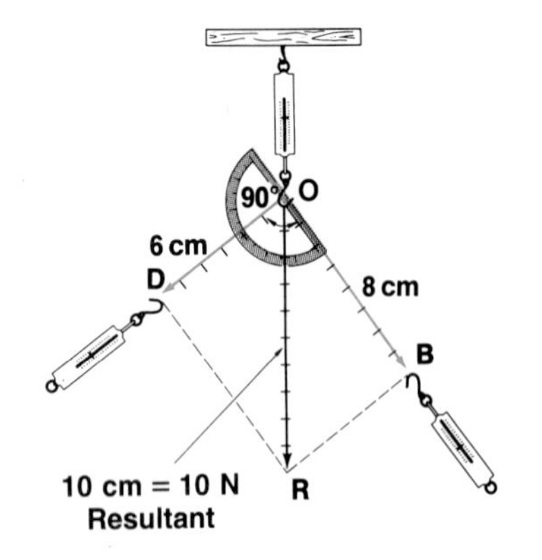

Figure 15-11 When the vectors are added, what is the value of the resultant? Angle BOR is read using the protractor.

What happens when two forces are not acting in the same line? In Figure 15-11, two strings are shown pulling on a spring scale fastened to a board. The strings are pulled at an angle of 90° to each other. Other spring scales show that the force along OD is 6 N and the force along OB is 8 N.

You can show the 6-N pull by making OD 6 cm long, with a scale of 1 cm for each newton. Make OB 8 cm long to show the 8-N pull. Complete the drawing as shown in Figure 15-11. You will have a four-sided figure in which each pair of opposite sides is parallel. Use a protractor to find angle DOR and BOR. Measure line OR of the diagram. What is its length? Its value gives the size and direction of the resultant of the two forces.

In this example, the resultant of the forces OD and OB is 10 N. The spring balance at O should read the same as the force obtained by measuring line OR. The single resultant force has the same effect and can be used in place of the 6-N and 8-N forces at 90° to each other.

Picture the forces acting on a corner post of a wire fence. What might happen to the corner post if it were pulled too hard by wires at right angles to each other?

SAMPLE PROBLEM

If each of two wires at 90° to each other exerts a 100-N pull on a fence post, find the size and direction of the resultant force.

SOLUTION

Step 1: Analyze

You are given the size (100 N) and direction (90° relative to each other) of the two forces acting together. The size and direction of the resultant force are the unknowns.

Step 2: Plan

Since you know the size and direction of the forces, you can draw a scaled vector diagram by selecting a convenient scale such as 1 cm equals 10 N.

Step 3: Compute and Check

On a sheet of paper, lay out two vectors 10 cm in length, at a 90° angle to each other. Complete the parallelogram and measure the diagonal, which represents the resultant force vector. It should be slightly over 14 cm long, or equal to 140 N. A protractor will show that the angle made by the resultant with each force vector is about 45°.

Since the direction of the resultant is between that of the initial forces and since the size of the resultant is greater than either force as expected, your solution is complete.

The size and direction of the resultant of any two forces can always be found by drawing a force diagram. Sometimes three or more forces act at different angles upon the same point. How would you find the resultant of three or more forces acting at the same time upon the same point?

15.3 THE LAWS OF MOTION

Newton's first law of motion describes inertia. You have learned to describe how objects move in terms of speed, velocity, and acceleration. You have also learned that all motion and changes in motion are caused by forces. But just how are forces and motion related? How does an object behave if no forces act on it? These are some of the questions about motion Sir Isaac Newton tried to answer.

How does an object behave when it appears to have no forces acting on it? Remember in the discussion of friction, you saw that if you put a box on the gym floor and gave it a push, it would start moving. However, the box would not move forever. It would soon come to a stop. The force of friction between the box and the ground would cause this change in motion. Now suppose there were no friction forces acting on the box. Would the box slow down? Would it ever stop? Newton did many tests to find the answer to this question. He concluded that if there is no force acting on it, an object will continue to do whatever it has been doing. If an object is moving in a certain direction at a certain speed, it will continue to do so. If the object is not moving, it will not start moving.

This conclusion is known as Newton's first law of motion. It is stated as follows: If an object is at rest, it tends to stay at rest; if it is moving, it tends to keep on moving at the same speed and in the same direction. Newton used the word inertia to describe this tendency to resist changes in motion. (See also the discussion of inertia in Chapter 1.) An example of inertia is shown in Figure 15-12. Here are some other examples.

If you are standing facing forward in a bus and it starts suddenly, what happens? The bus begins to move, but your body is still at rest. So you tend to fall backwards. There is an opposite effect if the bus stops too fast. Your body tends to keep moving even though the bus has stopped. Both cases are examples of Newton's first law.

Inertia tends to throw a person off a fast-turning merry-go-round. In baseball a base runner "rounds" the bases instead of making a sharp 90-degree turn at each base. Inertia makes it hard for the runner to change direction.

Figure 15-12 What happens to the coin when you flick away the card? What would happen to the coin if you pulled the card away slowly?

Even the turning earth shows the effect of inertia acting on it. The effect is so great that the earth has a slight bulge at the equator. In fact, if the earth turned 18 times as fast as it does now (or fast enough to give us a day of only 90 min), loose objects at the equator would be thrown into space.

Inertia is also a factor that must be considered in order to safely drive a car. The faster a car moves, the greater is its tendency to stay in motion. Compare the stopping distances of cars traveling at different rates of speed, as shown in Figure 15-13. Note that at twice the speed, four times the stopping distance is needed. At a speed of 96 km/hr, or three times as fast as 32 km/hr, it takes nine times the distance to stop. Remember this when you are tempted to drive too fast or too close to the car ahead.

Tests have shown that the mass and the inertia of a body are directly proportional. That is, if one object has twice the mass of another, it will also have twice the inertia.

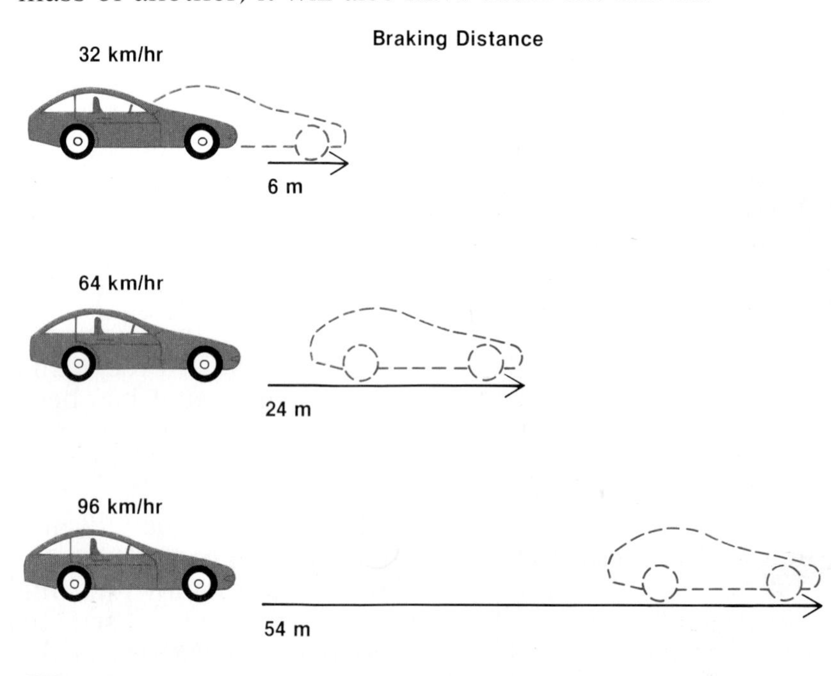

Figure 15-13 What kind of relationship does speed have to braking distance in the examples given?

Balanced forces cause no net change. In the discussion of force vectors, you saw that unbalanced forces cause non-moving objects to move in the direction of the resultant. If the forces are balanced, that is, if their resultant is zero, there is no motion. Although forces are being applied, they are canceling each other out. In such cases, the net force is zero, and the object is balanced.

An object in uniform motion is also balanced. Imagine an airplane traveling a perfectly straight and level course eastward at 200 km/hr. The plane's velocity is constant and there is no change in speed or direction. The only way the plane can keep this uniform motion is if all the forces on it are balanced in such a way that they cancel each other out. If any of the forces acting on the plane change in size or direction, the forces will be unbalanced and the velocity of the plane will change.

Newton's second law of motion shows that force changes motion. The second law of motion shows how force, mass, and acceleration are related to one another. Newton found that the size of the acceleration for a given mass depends upon the size of the unbalanced force applied to it. In order to throw the ball, a baseball pitcher applies a force to the ball. To throw the ball faster, a greater force must be applied.

It is also easy to see that acceleration is related to the mass of an object. The greater the mass of an object, the greater its inertia. Thus, the greater the mass, the greater the force needed to change the acceleration. In other words, the same force gives a greater acceleration to a baseball than it does to a bowling ball.

For example, a racing car has both a powerful engine and a light-weight body (Figure 15-14). The same engine would accelerate the car more than it would a truck.

To summarize the statements above, the acceleration of an object is directly proportional to the net force acting on it and inversely proportional to its mass. This is a statement of Newton's second law of motion. Stated as a formula,

$$a = \frac{F}{m} \text{ or } F = ma$$

where F is the force in newtons, m is the mass in kilograms, and a is the acceleration in meters per second squared.

Figure 15-14 Why is it important for this drag racer to have such a light front end and such a large engine?

SAMPLE PROBLEM

How much force would your arm have to exert to accelerate a bowling ball with a mass of 7 kg down a bowling alley at a rate of 5 m/sec^2?

SOLUTION

Step 1: Analyze

You are given that the mass of the ball is 7 kg and its acceleration is 5 m/sec^2. The force is the unknown.

Step 2: Plan

Since you know the mass and acceleration, you can use the following equation to find the force:
$F = ma$

Step 3: Compute and Check

Substitute all values in the equation.

$F = 7 \text{ kg} \times 5 \text{ m/sec}^2$
$= 35 \text{ kg-m/sec}^2$, or 35 N

Since the unit kg-m/sec^2 can be expressed as newtons, the metric unit for force, your solution is complete.

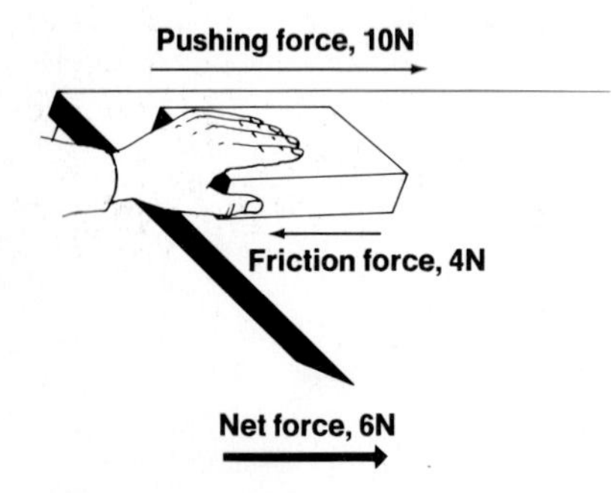

Figure 15-15 The net force is the force that exceeds the balanced force (difference between pushing force and friction force). How much force is needed to balance friction in this picture?

Only the net force is used in finding the acceleration of a body. No matter how hard you push against a standing railroad train, you will not be able to move it. The pushing force of your body would all be lost to friction. Therefore, no net or unbalanced force would be left to accelerate the train. Study the example shown in Figure 15-15. Before a body will accelerate, all opposing forces, including the force of friction, must first be overcome. It is important to note that net force always causes a change in motion, that is, an acceleration.

Newton's third law states that an action causes a reaction. If you place a heavy book on a table, the weight of the book exerts a downward force on the table. From the opposite direction, the tabletop exerts an upward and equal force on the book. Therefore, no net force is acting on the book and the book does not move.

Does it seem strange to think of the table exerting an upward force on the book? Perhaps it becomes clearer if you can imagine trying to support the book on your outstretched hand. To keep the book from moving up or down you must apply a force to your hand that exactly balances the downward force of the book. Your muscles provide this force. But if your muscles are too weak, the forces will be unbalanced and the book will fall to the floor.

In walking across the classroom, your feet exert a backward push against the floor, while the floor exerts an equal and opposing force (friction) against your feet. See Figure 15-16. What would happen if you tried to walk on a floor that was smooth and slippery as ice? Why?

In each example, one force is the action and the other force is the reaction. When one object exerts a force on a second object (action) the second object exerts an equal and opposite force upon the first (reaction). This is a statement of Newton's third law of motion.

Newton's third law explains a gun's kick. The sharp recoil when a gun is fired is another example of Newton's third law of motion. The action and reaction are between the bullet and the gun. The burning powder pushes just as hard on the gun as it does on the bullet. This push gives the bullet momentum in one direction and the gun equal momentum in the other direction. **Momentum** is the product of the mass (m) and velocity (v) of a body. The momentum of the bullet ($m_b \times V_b$) equals the momentum of the gun ($m_g \times v_g$). If any three of these four values are known, you can find the fourth value.

Look at the gun and the bullet in another way. Since the gun and the bullet were at rest with respect to each other before the shot was fired, the total momentum of the gun and the bullet was zero. In other words, the velocity of each object was zero. Mass times zero velocity gives zero momentum. What were the values after the shot was fired? If you think of velocities in one direction as positive and those in the opposite direction as negative, then the total momentum after firing again adds up to zero. The total momentum is the same both before and after the firing.

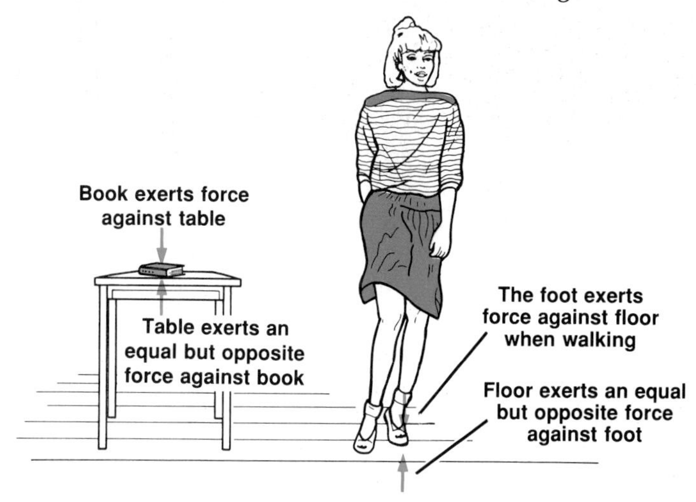

Figure 15-16 Action and reaction are pairs of forces that apply to two bodies. How can motion occur if action and reaction are equal and opposite?

SAMPLE PROBLEM

A gun having a mass of 5,000 g fires a 20-g bullet at a velocity of 500 m/sec. Find the recoil velocity of the gun.

SOLUTION

Step 1: Analyze

You are given that the mass of the gun is 5,000 g, the mass of the bullet is 20 g, and the velocity of the bullet is 500 m/sec. The velocity of the gun is the unknown.

Step 2: Plan

Since you know that the momentum of the gun equals the negative momentum of the bullet (negative for movement in opposite directions), or

$$m_g v_g = -m_b v_b$$

and since you know three of the four values in this equation, you can use the equation to find the velocity of the gun by solving for v_g:

$$v_g = \frac{-m_b v_b}{m_g}$$

Step 3: Compute and Check

Substitute all values in the equation.

$$v_g = \frac{-(20\text{ g} \times 500\text{ m/sec})}{5{,}000\text{ g}}$$
$$= -2\text{ m/sec}$$

Since the unit m/sec is correct for velocity and the negative value indicates that the gun and the bullet move in opposite directions, your solution is complete.

Total momentum is conserved in any reaction. Let us explore Newton's third law in greater detail. Suppose a man and a boy are facing each other, each standing on a pair of roller skates as shown in Figure 15-17. They push each other, and each one moves away in the opposite direction. Each person has the same force acting on him, but one moves away faster than the other. The boy, having a smaller mass, moves away with a greater velocity.

In this example, suppose the boy has a mass of 40 kg and the man 100 kg, and the man moves backward at a velocity of 0.4 m/sec. The boy will move off in the opposite direction at a velocity of 1 m/sec. Notice that when you multiply the mass and the velocity of the boy (40 kg × 1 m/sec), you get

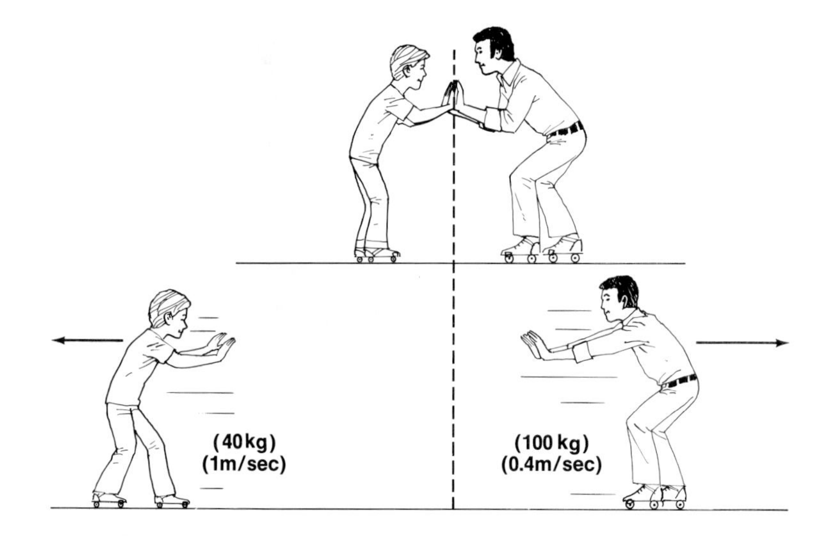

Figure 15-17 Why does the boy move away from the starting point faster than the man?

the same product as when you multiply the mass and velocity of the man (100 kg × 0.4 m/sec). This example shows the law of conservation of momentum, which states that in any reaction between masses, there is no net change in total momentum.

In any reaction between two objects, the change in momentum produced in one body is equal and opposite to the change in momentum in the other. In the example given, mv (man) = $-mv$ (boy).

There are many examples of the reaction principle. Jets and rockets are other examples of Newton's third law. Remember that action and reaction must take place between two different masses. The jet plane or rocket is one mass, and the exhaust gases are the other mass. Think of the rocket as a gun firing bullets of high-speed molecules. The molecules go one way and the rocket goes the other way. All movement in water, air, or space rests upon the reaction principle.

You have seen other examples of the reaction principle. In a rowboat, the oars push against a mass of water. This gives the boat a push in one direction and the water a push in the opposite direction. If you leap from the front of a rowboat, the boat will be kicked back with as much momentum as your body develops jumping forward. You will probably land in the water! What would happen if you were to leap for shore from a large ship? Why?

The rotary water sprinkler is another good example of the reaction principle. The jet of water squirted out by the nozzle gives a certain momentum. As a result, the nozzle receives the same "kick" in the opposite direction. The reaction causes the sprinkler to move in a circular pattern as shown in Figure 15-18.

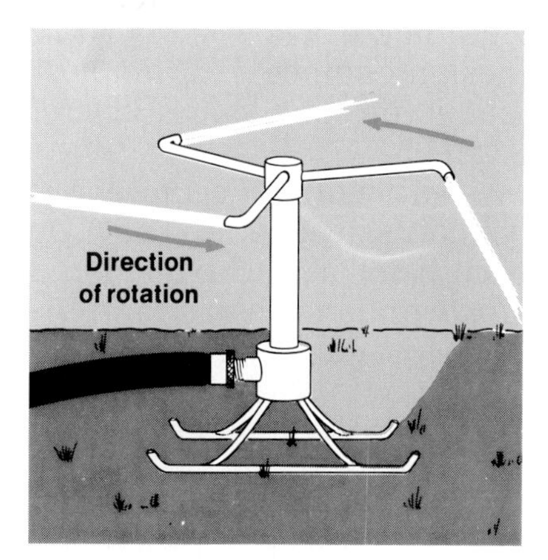

Figure 15-18 If the rate of flow of water through the nozzle increases, what happens to the speed of the sprinkler head? Why?

AUTOMOTIVE DESIGN ENGINEERING

building cars of the future

This engineer is using lasers to check the air velocity in an automobile combustion chamber.

Futuristic cars—with their streamlined looks, their internal computers, and their state-of-the-art safety features—are brought to you "before their time," courtesy of automotive design engineers. These engineers design and test the bodies of cars, trucks, and off-road vehicles, as well as the engines and fuels such vehicles use.

Innovations in a car's chassis, brakes, carburetor, and all the interior equipment a driver can reach are also the responsibility of automotive design engineers. Examples of contemporary innovations include aerodynamic designs, computerized dashboard displays, and computerized fuel systems.

Automotive design engineers also develop safety features to protect a vehicle's occupants. They concentrate not only on common features such as seat belts and air bags, but also on break-away steering columns, energy-absorption cushions, and break-away poles by the side of the road.

Most automotive design engineers are trained as mechanical engineers, but they may also be electrical engineers, aerospace engineers, chemical engineers, computer engineers, or civil engineers. To become an automotive design engineer, you need a college degree with a solid background in math, physics, and chemistry. Other courses that are useful to engineers include English, a foreign language, shop, and drafting.

For more information on a career in automotive design engineering, contact

Engineer's Council for Professional Development
United Engineering Center
345 East 47th Street, New York, NY 10017

CHAPTER REVIEW 15

SUMMARY

1. Speed is the distance an object moves in a certain amount of time.
2. Velocity is speed in a given direction.
3. Acceleration is any change in speed or direction of an object.
4. Force is a push or pull.
5. Forces include friction and gravity.
6. Newton's first law states that objects in motion tend to stay in motion and objects at rest tend to stay at rest.
7. Newton's second law states that the acceleration of an object is directly proportional to the net force acting on it and inversely proportional to its mass.
8. Newton's third law states that for every action there is an equal and opposite reaction.
9. Mass times velocity equals momentum.

VOCABULARY

Match the item in the left column with the best answer in the right column. Do not write in this book.

1. acceleration
2. center of gravity
3. friction
4. lever
5. momentum
6. motion
7. resultant
8. speed
9. torque
10. vector
11. velocity

a. resistance to motion, caused by one surface rubbing against another
b. distance traveled in a given unit of time
c. speed in a definite direction
d. any quantity that has both size and direction
e. movement of an object from one place to another
f. point where all the weight of an object appears to be concentrated
g. combined effect of two or more forces acting together
h. product of the mass and the velocity of an object
i. change in velocity per unit of time
j. twisting movement caused by one or more forces
k. rigid bar that is free to turn on its point of support

REVIEW QUESTIONS

1. What is the difference between speed and velocity?
2. What forces are usually acting on a body at rest?

3. Forces on an object in uniform motion and an object at rest are both in balance. Why?
4. Define friction.
5. Which has less resistance, rolling or sliding friction? Give examples.
6. What is gravity?
7. What is torque?
8. Define inertia and give three examples.
9. What is acceleration? What causes it?
10. State Newton's second law of motion. Give an example of this law.

CRITICAL THINKING

11. A car starts from a rest position, is accelerated to 60 km/hr, and then is held at 60 km/hr. At what steps in this sequence of events are the forces on the car in balance?
12. Is friction ever desirable? Explain.
13. Why would a person weigh less on the moon than on the earth?
14. If you doubled the net force acting on a moving object, how would its acceleration be affected?
15. If twice the net force results in twice the acceleration, why doesn't a 10-kg stone fall twice as fast as a 5-kg stone?
16. What happens to the distance needed for stopping a car if its speed is increased from 10 km/hr to 40 km/hr?
17. What happens to acceleration if the mass is tripled and the force remains the same?
18. Why does a slowly moving freight train have more momentum than a high-speed bullet?
19. Would the acceleration of a falling object located 1,000 km above the earth's surface be more or less than 9.8 m/sec^2? Explain.
20. Newton's third law of motion involves two objects and two forces. Explain what this means in terms of action and reaction.

PROBLEMS

1. A car is slowed down from 50 km/hr to 20 km/hr in 6 sec. What is the deceleration in kilometers per hour squared?
2. An object is dropped from a high building. (a) How far (in meters) will the object fall in 5 sec? (b) Neglecting air resistance, what would its velocity be at the end of this time? (c) What would its average velocity be during the fall? (d) If it took 7 sec for the object to hit the ground, find the height of the building.
3. A man weighing 800 N sits on one side of a seesaw, a distance of 2 m from the fulcrum. How far on the other side of the fulcrum must a 500-N boy sit to balance the seesaw?
4. A truck weighing 12 kN is located 20 m from the east end of a 60-m bridge. Find the weight supported by each of the two piers. (Ignore the weight of the bridge itself.)

5. Find the resultant of two forces, one a force of 20 N acting east, and the other of 5 N acting on the same point in the same direction. What would be the resultant if they acted in opposite directions?
6. There is a 10-N pull on an object in as easterly direction. On the same object, another pull of 15 N acts to the south. What is the size and direction of the resultant force?
7. What force is needed to accelerate a 1,000-kg car at 3 m/sec^2?
8. If a 5-kg gun fires a 25-g bullet at a speed of 500 m/sec, what is the recoil speed of the gun?

FURTHER READING

Ander, Mark. "Renoodling Newton." *U.S. News and World Report* 105 (August 15, 1988): 14.

Ford, Barbara. *The Elevator.* New York: Walker, 1983.

Gardener, Robert, and David Webster. *Moving Right Along: Book of Science Experiments and Puzzles About Motion.* New York: Doubleday and Co., Inc., 1978.

Goswami, Amit, with Maggie Goswami. *The Cosmic Dancers: Exploring the Physics of Science Fiction.* New York: McGraw-Hill, 1985.

Ipsen, D. C. *Isaac Newton: Reluctant Genius.* Hillside, N.J.: Enslow Publishers, 1985.

Watson, Philip. *Super Motion.* New York: Lothrop, Lee, and Shepard Books, 1983.

USING FORCE AND MOTION

Machines, like these pulleys on a ski lift, make our work easier for us. They do so by helping us to change force and motion. How is force changed by this system of pulleys in order to lift the skiers up the mountain?

CHAPTER 16

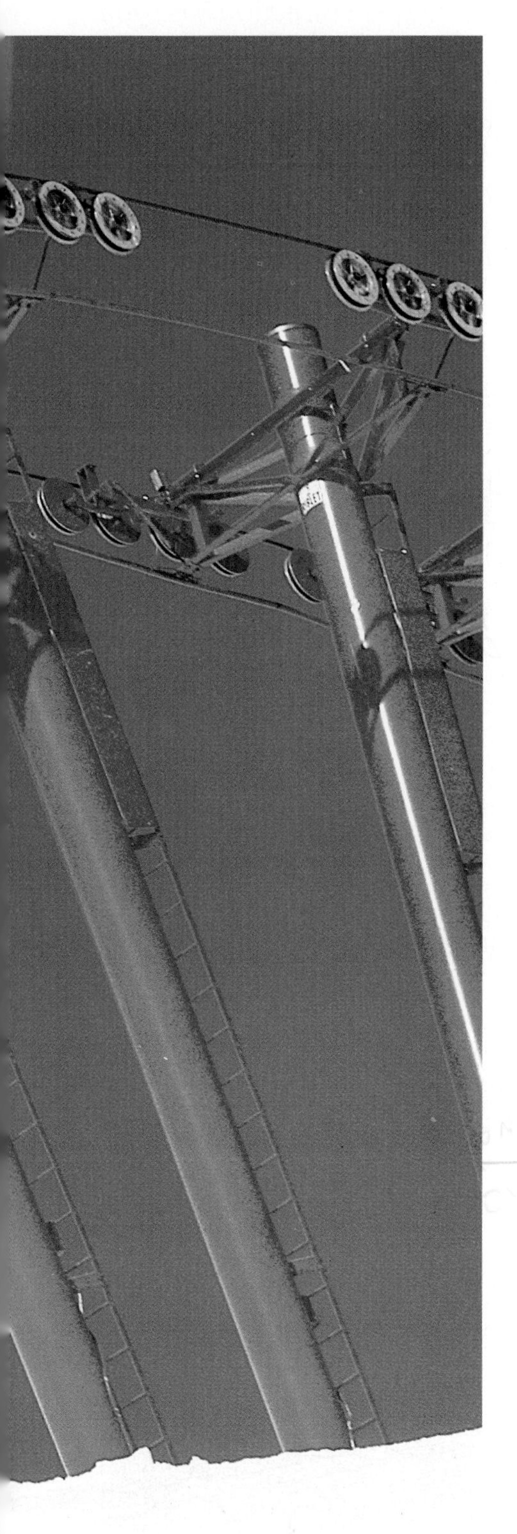

SECTIONS

OBJECTIVES

- ☐ Describe how forces operate in simple machines.
- ☐ Discuss how force and motion are related in flight through the air.
- ☐ Explain the forces and motions that make space flight possible.
- ☐ List the forces and motions that keep satellites in orbit.

16.1 MACHINES

Machines are used to change force. All machines make things easier to do. They do this by changing a force in any of three ways. (1) A machine can change the size of the force. If you press down on the handle of an automobile jack, you can lift a car that weighs more than a ton with little effort on your part. (2) It can change the direction of the force. If you push down on one end of a seesaw, you can lift someone at the other end. (3) It can change the speed with which the force is applied. If you slowly turn the crank of a hand-driven eggbeater, the blades turn much faster than the crank.

No doubt you could think of other ways in which machines make it easier to do things. For example, is there more force, greater speed, or a change in direction if (a) you push down on a bicycle pedal; (b) you hit a ball with a baseball bat; (c) you push down on a crowbar to pry the lid off a box?

There are two main kinds of machines. A simple device, such as a knife or a screwdriver, is a machine. Maybe you think of a machine as something more complex, such as a typewriter, an airplane, or a car. In science, however, the word machine includes all these things. A knife and a screwdriver are as much machines as are cars. As a matter of fact, cars are made up of many different machines.

All the machines ever invented are classed as either simple machines or complex machines. **Simple machines** are devices that change the force used to do work. They include the lever, pulley, inclined plane, wedge, screw, and wheel and axle. **Complex machines** are devices made of two or more simple machines. Regardless of how complex a machine is, it is still made up only of various simple machines. See Figure 16-1.

Figure 16-1 Is this crane a simple or complex machine?

A machine can give a mechanical advantage. The **mechanical advantage** (*MA*) of a machine tells how much your effort is multiplied by the machine. Mechanical advantage is sometimes expressed as the ratio of two distances. For example, suppose that you use a rope and pulleys to lift a weight. Any weight to be moved is referred to as the resistance (*R*). You pull the rope 2 m and the resistance is lifted 1 m. The 2 m is the effort distance (*ED*), or the distance over which your effort force acts. The 1 m is the resistance distance (*RD*), or how much the weight moves. If you divide the effort distance by the resistance distance, using the equation $MA = ED/RD$, you have a ratio of 2 to 1 (2:1), or a mechanical advantage of 2. This number actually represents the **ideal mechanical advantage** (*IMA*), or the mechanical advantage a machine has if there is no friction. The **actual mechanical advantage** (*AMA*) is the mechanical advantage a machine has if there is friction. Since friction is present in all machines, the *AMA* is always less than the *IMA*.

The *AMA* can be found by dividing the resistance of the lifted object by the effort put into the machine ($AMA = R/E$). The number you get tells you just how much a machine multiplies the force you put into it. For example, a long bar is used to lift a resistance of 1,000 N with an effort of 40 N. The bar has a mechanical advantage of 1,000 N/40 N, or $MA =$ 25. This number means that for every newton of force that is applied, this machine produces a force of 25 N.

All machines waste effort. Some people are surprised to learn that a machine does not increase the amount of work that is done. You have seen that a machine can change the size, speed, or direction of a force. However, this does not mean that it increases the amount of work done. To scientists, work is equal to the force times the distance through which the force moves. In fact, the work that comes out of a machine is always less than the work that goes into it. Why is this true? All machines have parts that rub against each other. Therefore, some of the work that goes into a machine is used to overcome friction. Some machines cause less friction than others. If a machine is very efficient, less work is needed to operate it.

The **efficiency** of a machine is a comparison of the work that goes into a machine (called input work) with the work that comes out (called output work) as follows:

$$\text{efficiency} = \frac{\text{output work}}{\text{input work}}$$

For example, if you put 4,000 N-m into a machine and get 4,000 N-m out, as in Figure 16-2, the machine has an efficiency of 100 percent.

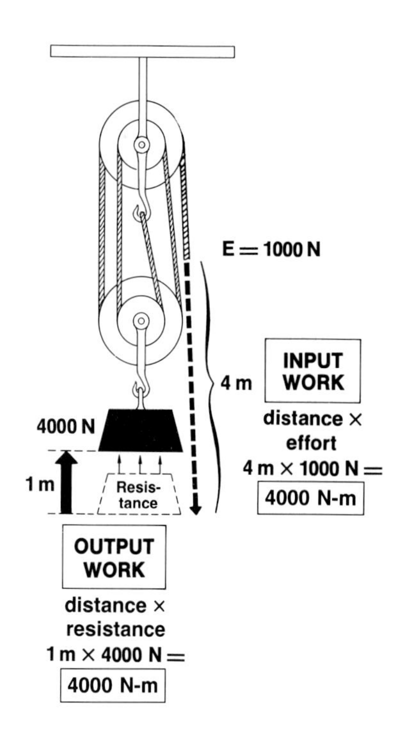

Figure 16-2 This system of pulleys, called a block and tackle, would ideally provide an efficiency of 100 percent (as shown). However, because of friction, the output work is always less than the input work, and efficiency is less than 100 percent.

SAMPLE PROBLEM

A machine similar to the one in Figure 16-2 has an efficiency of 75 percent. If 4,000 N-m is put into the machine, how much work will the machine produce?

SOLUTION

Step 1: Analyze

You are given that the efficiency of the machine is 75 percent and the input work is 4,000 N-m. The output work is the unknown.

Step 2: Plan

Since you know the efficiency and the input work, you can solve for the output work using the following equation:

$$\text{efficiency} = \frac{\text{output work}}{\text{input work}}$$

$$\text{output work} = \text{efficiency} \times \text{input work}$$

Step 3: Compute and Check

Substitute all values in the equation.

$$\begin{aligned}\text{output work} &= 0.75 \times 4{,}000 \text{ N-m} \\ &= 3{,}000 \text{ N-m}\end{aligned}$$

Since the efficiency is less than 100 percent and the calculated output work is less than the input work as expected, your answer is complete.

You can also find the efficiency of a machine by comparing the actual and ideal mechanical advantage as follows:

$$\text{efficiency} = \frac{AMA}{IMA}$$

For example, if the actual mechanical advantage of a machine is 2 and the ideal mechanical advantage is 2.5, the efficiency is 80 percent (2/2.5 = 0.8).

If a machine were 100-percent efficient, the input work would be equal to the output work. In addition, the *AMA* would be equal to the *IMA*.

There are three classes of levers. Now that you know what is meant by mechanical advantage, input work, and output work, you can see how these ideas apply to many simple machines. The lever, which you have already learned about in Chapter 15, is a rigid bar that is free to turn about its point of support, the fulcrum. The torques and parallel forces discussed in Chapter 15 can be applied to any lever.

The loaded wheelbarrow shown in Figure 16-3 is a lever. A load of 1,000 N is placed so that the center of gravity for the wheelbarrow and its load is 0.8 m from the fulcrum (F). This distance from the center of gravity of the load to the fulcrum is called the resistance arm. How much effort force (E) must be applied upward at the handles in order to lift the wheelbarrow?

Look at the torques about F acting on the wheelbarrow. The clockwise torque produced by R equals 1,000 N $\times$ 0.8 m = 800 N-m. The effort force E needed can also be found. Since the effort arm, or the distance that the effort is exerted from the fulcrum, is 2 m, the counter-clockwise torque equals $E \times 2$ m. Equating the torques:

$$E \times 2 \text{ m} = 1{,}000 \text{ N} \times 0.8 \text{ m}$$
$$E = 400 \text{ N}$$

Note that even though the bottom of the wheelbarrow is sloping, all distances are measured at right angles to the forces of lift.

The wheelbarrow is an example of a second-class lever, as shown in Figure 16-4B. With a second-class lever, the resistance R is between the fulcrum and the effort, and the mechanical advantage (MA) is always greater than 1. The MA can be increased by moving the resistance toward the fulcrum or by lengthening the effort arm.

Figure 16-4A is a diagram of a first-class lever. The fulcrum is located between the resistance and the effort. The MA may be greater or less than 1 in a first-class lever. To increase the MA, the fulcrum must be moved nearer to the resistance. Scissors and seesaws are examples of first-class levers.

A third-class lever, shown in Figure 16-4C, always has an MA of less than 1. A third-class lever, such as a fishing rod, is used to increase speed.

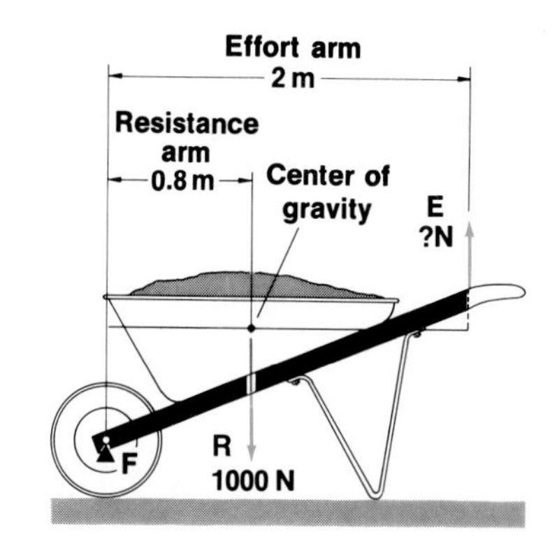

Figure 16-3 Does the wheelbarrow change the amount, the direction, or the speed of the effort force? How much effort would you need to lift this wheelbarrow?

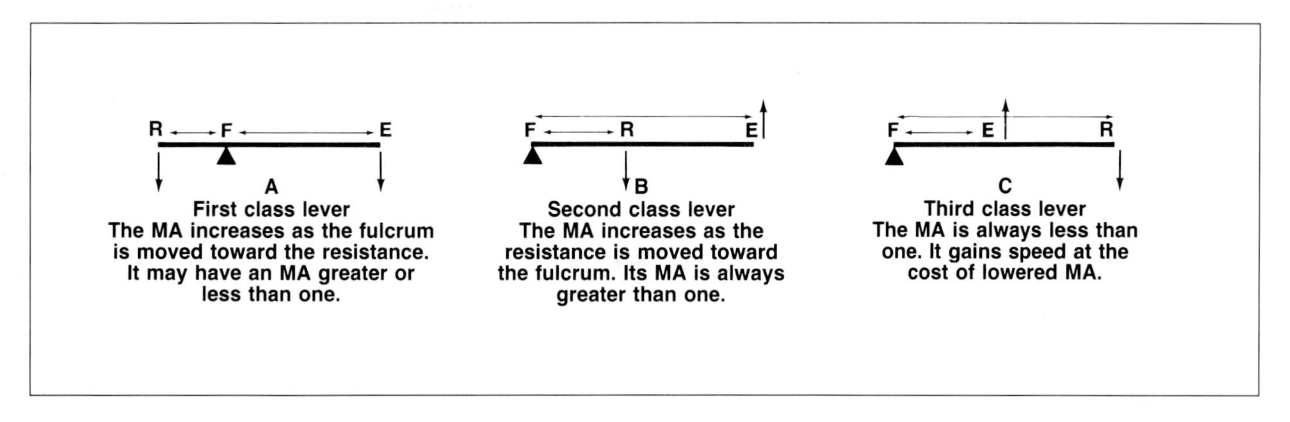

Figure 16-4 How does the effort arm compare with the resistance arm in each lever?

With your elbow on the table, raise your book in your hand. You have used your forearm as a lever. Levers are used very often in daily life. Shovels, brooms, piano keys, and nutcrackers are all levers. You will work with levers in the following activity.

INVESTIGATING THE MECHANICAL ADVANTAGE OF LEVERS

activity

OBJECTIVE: Determine which class of lever should be used to most easily remove a bottle cap.

PROCESS SKILLS

In this activity, you will *measure* and *collect data* on two classes of levers, which will help you to *infer* which one provides a higher mechanical advantage.

MATERIALS

2 bottles with non-twist caps
bottle opener

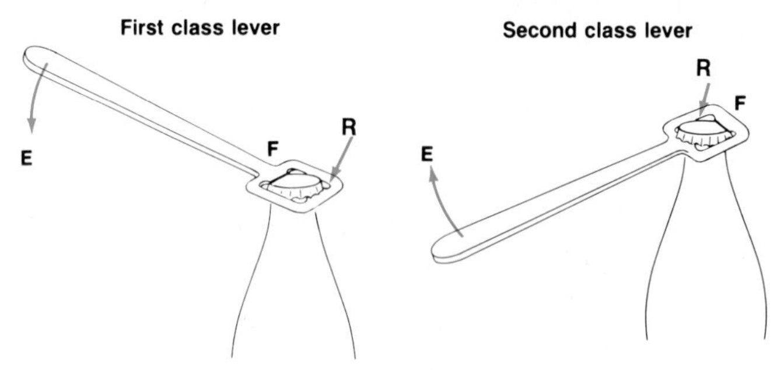

PROCEDURE

1. Place the opener over the bottle cap as shown in the diagram on the left. Measure the length of the effort arm and the resistance arm.
2. Find the mechanical advantage.
3. Now place the opener over the bottle cap as shown in the diagram on the right. Measure the length of the effort arm and the resistance arm.
4. Find the mechanical advantage.

OBSERVATIONS AND CONCLUSIONS

1. Why would you classify the bottle opener in the diagram on the right as a second-class lever?
2. Which class of lever should more easily remove the cap? Explain.
3. Have you found the *AMA* or *IMA* in this activity? Explain.
4. Use each class of lever to open one of the two bottles. Which method is easier? Does this match your prediction in Question 2?

Pulleys are one type of simple machine. It is much easier to use a pulley at the top of a flagpole to raise a flag than it is to climb the pole to raise it. With a single fixed pulley there is no increase in the effort force or in the distance through which the effort moves. However, by using a fixed pulley, you can change the direction of the effort force. Changing the direction of the effort force means you can stand on the ground, pull down on a rope, and raise a flag to the top of a pole.

Compare a fixed pulley, used to change the direction of a force, with a single movable pulley as shown in Figure 16-5. With the movable pulley, the force is applied in the same direction as the effort. There is no change in direction, but there is a decrease in the effort needed to lift the weight.

Piano movers may use a block and tackle to raise a piano to a second-story window. A **block and tackle** is a system of pulleys used to raise a heavy object with a small amount of effort. Though a piano may have a weight of approximately 3,000 N, one person can raise the piano by simply pulling down on the rope.

In the pulley system shown in Figure 16-2, the 3,000-N piano would be the load to be raised. (Of course, the lower, movable block also has some weight, and it too is part of the load. The weight of the movable block is not included in the following example.)

Note that the weight and pulley in Figure 16-2 hang from four strands of rope. The total upward force must equal the total downward force of 3,000 N exerted by the piano. Therefore, each of the four strands must lift 750 N and the force on the effort rope must be 750 N. If friction is added, a pull of more than 750 N will be needed. Why is a pulley a kind of lever?

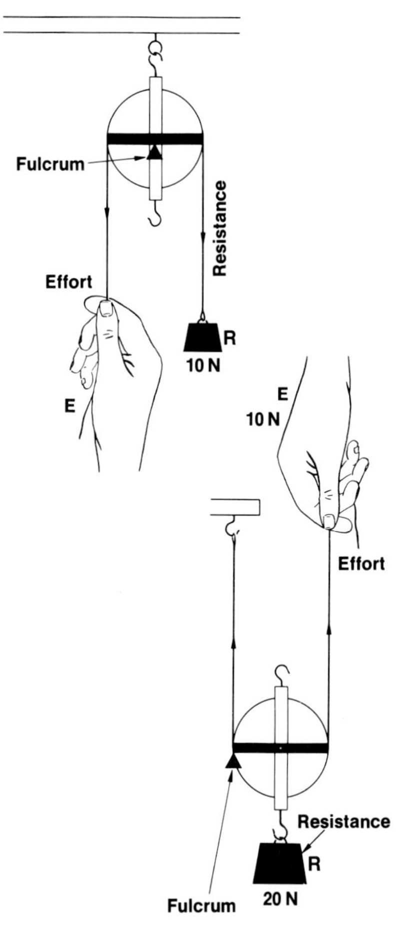

Figure 16-5 How does the mechanical advantage of the single fixed pulley compare with the mechanical advantage of the movable pulley?

A wheel and axle is a lever. Have you ever tried to open a door by turning the rod with your fingers when the door-knob was missing? If you have, you can see how a doorknob makes the job easier. The doorknob acts like a wheel and axle. The knob is the wheel, and the rod to which the knob is attached is the axle. The screwdriver and the hand drill are other common examples of the wheel and axle. The steering wheel of a car is another good example. Just a little effort on the rim of the steering wheel is enough to change the direction of the front wheels of the car. You can see that although the "wheel" and the "axle" have different sized radii, they are rigidly attached to each other and move as one piece. When you turn the wheel, the axle turns too. When you rotate the axle, the wheel also rotates.

In Figure 16-6 you can see a wheel and axle in which line AF is the resistance arm and BF is the effort arm. From your knowledge of torques you know that $R \times \text{AF} = E \times \text{BF}$. The ideal mechanical advantage, or *IMA*, in this wheel and axle equals BF/AF. Actually, the *IMA* of any wheel and axle equals the radius of the wheel divided by the radius of the axle. You can increase the *IMA* by increasing the radius of the wheel or by decreasing the radius of the axle.

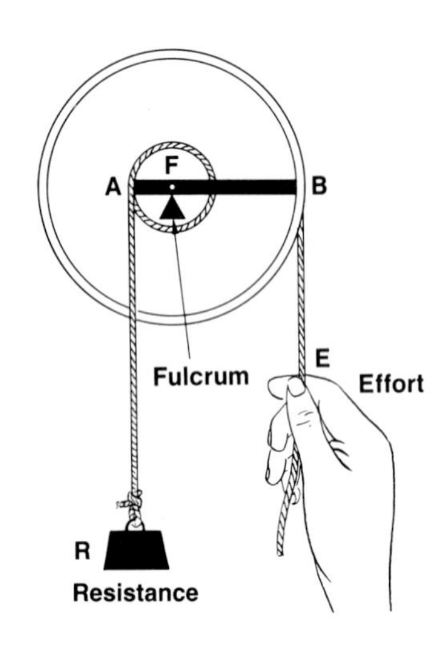

Figure 16-6 How is the wheel and axle like a lever? How is it different from a pulley?

An inclined plane increases force. If you have ever watched a moving van being loaded or unloaded, you have seen how an inclined plane works. The movers do not lift a heavy box straight upward, which would require a force equal to the weight of the box. Instead, they may push or wheel the box up a long ramp, or inclined plane. This requires much less force. Notice how this works in Figure 16-7. If you do not count friction, a force of 250 N is sufficient to roll a weight of 1,000 N up the plank. Therefore, the mechanical advantage is 4 (R/E = 1,000 N/250 N = 4). You can also see that you have to use a plank 80 cm long in order to raise the load a height of 20 cm. Thus, you see that the four-times gain in force is obtained by a four-times increase in distance.

The *IMA* of an inclined plane can be found by dividing the length of the plank by its height above the ground, 80 cm/20 cm = 4. Of course, when friction is counted, the effort E must be greater than 250 N.

If a road rises 4 m along a stretch of 100 m, then it is an inclined plane with an *IMA* of 100 m/4 m = 25. This means that a car engine must exert a force equal to only 1/25 of the car's weight to climb the hill at constant speed. In addition, the car must also have enough force to overcome friction.

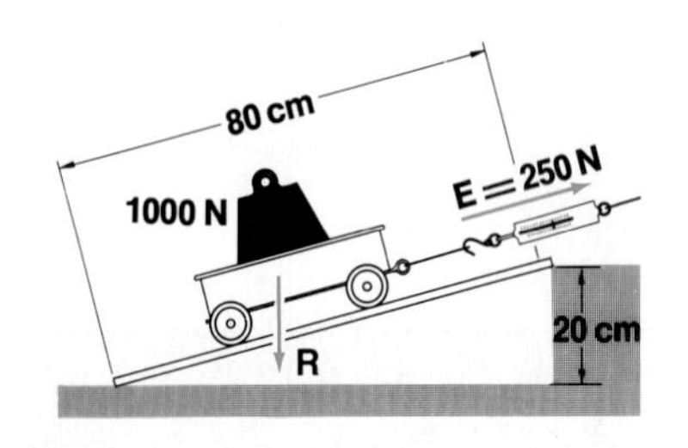

Figure 16-7 In what way can you increase the mechanical advantage of the inclined plane?

Other inclined planes are the wedge and the screw. It would be easy to split a log with a thin wedge if you struck the wedge with a heavy hammer. A wedge looks like a small, ramp-style inclined plane. The thinner the wedge, the easier it is to split a log. Also, the thinner the wedge, the higher the mechanical advantage.

Wedges are not very efficient because much of the effort is used up in friction. A knife is a form of wedge, and the sharper the knife, the less force is needed to cut.

A screw is an inclined plane that is very long compared to its height. You can see in Figure 16-8 that a screw is an inclined plane wrapped in a spiral around a cylinder. The wood screw, threaded bolt, and the bench-vise are examples of screws that have a high mechanical advantage. This

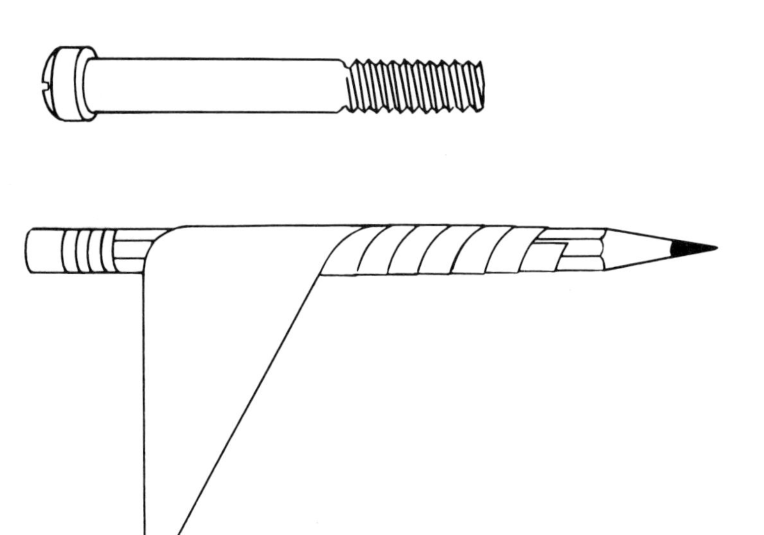

Figure 16-8 How does this diagram show that a screw is like an inclined plane?

makes screws very valuable machines. For example, the difficult task of raising one side of a small building is accomplished with the help of a jackscrew, which is a combination screw and lever. See Figure 16-9.

The *IMA* of the screw is found by dividing the distance the effort moves in one full turn by the **pitch**, or distance, from one thread to the next. How does Figure 16-8 show that this is so? Keep in mind that a great deal of friction is produced as a screw bores through wood and other materials. This means the screw has a low efficiency.

16.2 FLIGHT THROUGH THE AIR

Thrust gets a plane moving. Have you ever stood behind a propeller-driven plane when the engine was started? If you have, you know that air is driven backward. Recalling Newton's third law, you know that this backward force on the air will produce a forward force on the airplane. The backward force on the air is the action, and the forward force on the airplane is the reaction. In a jet airplane, the movement of gases through the engine produces the same effect.

The force pushing the airplane forward is called **thrust**. Airplane bodies are designed with a slim shape in order to

Figure 16-9 A jackscrew, which is a combination screw and lever, makes it possible for one person to raise the side of a small building.

increase thrust. When thrust is great enough to overcome inertia, the plane begins to move down the runway. As the plane gathers speed, another force called drag begins to act on the plane. **Drag** is a force caused by the friction of the air that is pushed aside by the moving plane. The faster the plane flies, the greater the drag acting on the plane becomes. At first, when the plane is accelerating, the drag is smaller than the thrust. Later, however, when the plane is flying at a constant speed, the drag equals the thrust. What would happen to a plane in flight if the drag were greater than the thrust? The resulting net force would cause the plane's speed to decrease. As the speed decreases, the drag decreases. Eventually, the thrust and the drag become balanced, and the plane returns to a constant speed.

Lift gets the plane off the ground. As the plane goes faster and faster down the runway, another force called **lift** begins to take effect. Air pressure builds up under the wings until they lift the plane off the runway. How do the wings act to cause lift? To understand what happens, think of holding out your hand, palm down, from the window of a moving car. What happens when you turn your hand slightly so that the air strikes the underside of the hand? The air pushes your hand upward. This is one way that moving air acts to push the airplane up and this push is called the **kite effect**.

Lift is also produced by air going over the wing. How can this happen? Look closely at a cross section of a wing in Figure 16-10. The wing is shaped so that it is curved on top. When the airstream hits the front edge of the wing, it divides. Part of the air goes above the wing and part of it goes below, as you would expect. The air above the wing speeds up as it goes over the raised curve. Surprising as it may seem, when the speed of air increases, its pressure on a surface goes down. This rather startling fact was discovered by Daniel Bernoulli in the early eighteenth century. The effect is summarized in the **Bernoulli** (*ber*-NOO-*lee*) **principle**, which states that when the speed of a fluid (either a liquid or gas) is increased, the fluid's internal pressure is decreased.

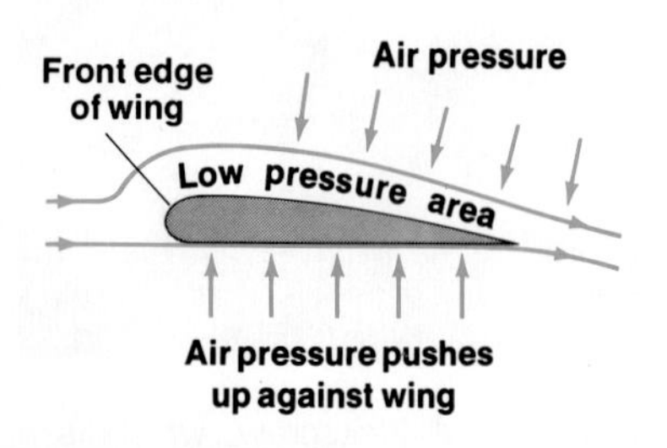

Figure 16-10 Where is the air moving fastest? How does air speed affect lift?

To review what happens, just think of three steps. First, the air speeds up as it goes over the raised curve of the wing. Second, as the air speeds up, its pressure goes down. Third, the air passing under the wing does not speed up. As a result, the pressure under the wing stays the same. Therefore, the pressure is greater under the wing. The combination of these effects produces the upward lift that gets the plane off the ground and helps to keep it aloft. What force opposes lift?

DEMONSTRATING HOW BERNOULLI'S PRINCIPLE WORKS

activity

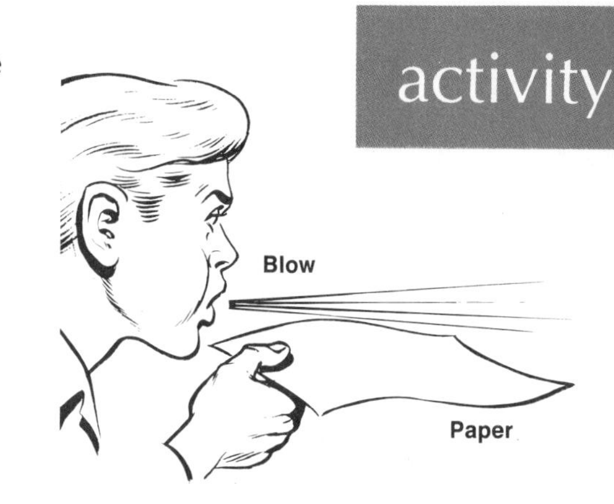

OBJECTIVE: Perform an experiment to demonstrate Bernoulli's principle.

PROCESS SKILLS
In this activity, you will *model* the Bernoulli principle and *predict* the outcome of your experiment.

MATERIALS
sheet of paper

PROCEDURE
1. Hold a sheet of paper tilted slightly downward in front of you and blow against the lower half of it.
2. Predict what would happen if you blew on and slightly over the rounded top curve of the paper as shown in the illustration. Now do this and compare the result with your prediction.

OBSERVATIONS AND CONCLUSIONS
1. What happened when you blew against the lower half of the paper?
2. What happened when you blew over the rounded top curve of the paper?
3. What conclusion can you draw from this experiment?

Forces keep the plane steady. An airplane is designed so that it is steady while in flight. That is, it does not tilt to the right or left, to the front or back, or swing its nose right or left. It will even correct itself if something is done to upset its balance. It is more like the flight of an arrow than that of a stick. A stick will just tumble in its path if you throw it. However, when an arrow is shot from a bow, its tail feathers act to keep it steady in flight. An airplane needs something like tail feathers too. It has horizontal and vertical **stabilizers** to keep the plane steady in flight.

The wings of many airplanes are angled upward. That is, the leading edge of the wing is higher than is the trailing edge. This is called an upsweep. This upsweep tends to keep the airplane from rolling and swinging from side to side.

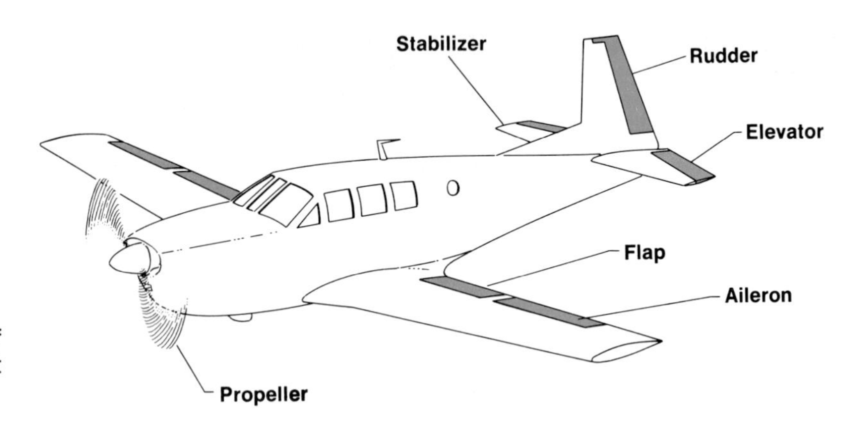

Figure 16-11 What is the function of each control surface on this light plane?

Forces change a plane's motion. An airplane has moving parts as seen in Figure 16-11 that cause it to change speed, turn left or right, move up or down, and roll clockwise or counterclockwise. These moving parts, or controls, cause the forces of flight acting on the plane to become unbalanced. Thus, the speed or direction or both can be changed by using these controls.

The thrust or forward speed of the airplane comes from the propeller (or from the exhaust gases if it is a jet plane). The front surface of a propeller blade is curved like the surface of a plane's wing. As the propeller spins, a low pressure area develops in front of the blades. Thrust is produced as the high pressure area behind the blades moves toward the low pressure area in front.

Up and down movements of the airplane's nose are controlled by adjusting the **elevators**. To turn the nose downward, the elevators are turned downward. Air striking the bottom surface of the elevators pushes the tail up, thus tilting the nose down.

When the pilot pulls back on the control wheel, or control stick, the elevators tilt up. The air blows against their upward tilted surfaces, forcing the tail down and the nose of the plane up. With extra power, the plane is able to climb.

The **rudder** is used to move the nose of the plane to the left or right. When the rudder swings to the right, the airstream pushes the tail to the left. The plane swings about, with its nose pointing to the right. However, just turning the rudder is not enough to make a smooth turn. Although the plane is facing a new direction, it is still moving on its original path. The plane is said to be skidding. It is somewhat like a car skidding on an icy curve. Even though the car's nose is pointed in a new direction, the car tends to continue moving straight ahead. Therefore, turning the rudder is not a good way to make an airplane change its direction of flight. Instead, the pilot must also bank the plane, that is, roll it slightly. In order to turn the plane, the pilot turns the control wheel or the control stick to adjust the **ailerons** (AY-*luh-rahns*), movable parts of an airplane's wings that impart a rolling motion to the airplane and so provide lateral control. One wing goes up and the other goes down. The plane now turns just as smoothly as a car on a banked highway curve.

Most planes also have surfaces called **flaps**, which aid in controlling lift, speed, and turns. Two flaps are hinged to the rear edge of the wing. Since the flaps on each wing turn downward at the same time, the plane does not bank, as it would if the flaps operated like ailerons. In order to slow down for a landing, the flaps are used as airbrakes to increase drag. The flaps also increase the lift, so that less speed is needed to keep the plane from falling. Landings are therefore possible at safer speeds and on shorter runways than would be needed without the flaps. Flaps are also used in takeoffs to increase lift.

Keep in mind that all of these controls are simply ways of changing the forces acting upon a plane. For a plane to start to climb, the lifting force must be greater than the plane's weight. To increase the plane's speed, the thrust must be greater than the drag.

Some jet planes fly faster than the speed of sound. You have seen that the way air moves over the surfaces of an aircraft is very important. If a plane is flying at less than the speed of sound, the air tends to flow more or less smoothly over the surfaces of the plane. When the plane goes faster than the speed of sound, however, air flow changes. At these speeds, the air piles up along the front edges of the aircraft to form a shock wave. The shock wave, in turn, spreads out behind the aircraft much like water waves spread out from a moving boat. If the edges of the shock wave were visible, the shock wave would look like a cone behind the aircraft. See Figure 16-12.

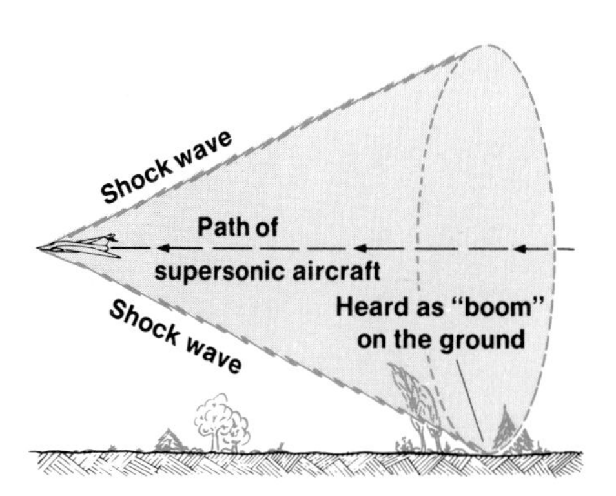

Figure 16-12 Where is a supersonic airplane when the sonic boom is heard by someone on the ground?

Shock waves cause sonic booms. What happens when the shock wave of a supersonic aircraft reaches the ground? If the shock wave is strong enough, it can be heard as a **sonic boom**. This loud noise is caused by a jump in air pressure as the shock wave hits the ground. The greater the pressure of the wave, the louder the boom. A pressure jump of 15 N/m^2 is heard as a distant boom that would hardly be noticed. A pressure jump of up to 50 N/m^2 is generally not

TESTING AN AIRPLANE'S CONTROLS

activity

OBJECTIVE: Make a model airplane to find out how the controls affect its flight.

PROCESS SKILLS:
In this activity, you will *experiment* with a model airplane and *observe* how the controls can be used to change the motion of the plane.

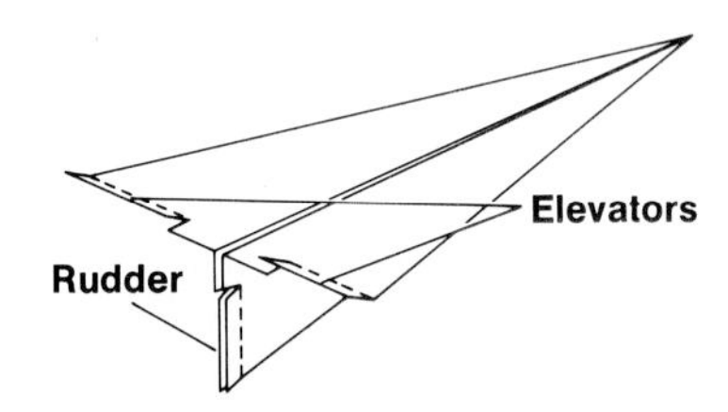

MATERIALS
sheet of paper
safety goggles

PROCEDURE
1. Build a paper airplane as shown in the diagram and test its flying ability a number of times. Make adjustments so that its flight is as stable as possible. CAUTION: *All students in the vicinity of the flying airplane models should wear goggles.*
2. Cut slits into the back edges of the wings to make elevators. Also make a rudder, as shown in the figure.
3. Make small adjustments in the angles of the elevators and rudder and see how this affects the flight of your plane. Record your findings in your notebook.

OBSERVATIONS AND CONCLUSIONS
1. How did the plane fly when (a) the elevators were raised slightly, (b) the elevators were lowered slightly, (c) the rudder was turned slightly?
2. What combination of adjustments produced the best flight?

harmful. However, a pressure increase of over 50 N/m^2 sounds like a loud thunderclap and can cause damage to glass windows.

The loudness of the sonic boom is affected by the speed, size, and height of the plane above the ground. A large airplane flying at twice the speed of sound at a height of 25 km would cause a pressure jump on the ground of about 20 N/m^2. If it were flying at 16 km, the pressure would jump by 50 N/m^2. At a height of 10 km, the pressure jump would be 120 N/m^2, and at a height of 3 km the figure would be 420 N/m^2. What would the total force be on a large store window measuring 3 m by 6 m in each of these cases? Hint: The area of the window will provide the m^2 factor for each calculation. Why is there some objection to low-level supersonic flights across land areas?

16.3 FLIGHT THROUGH SPACE

Getting off the launching pad takes a great deal of force. Which is a better engine to boost a 5,000-kN spaceship into orbit: one that gives 1,000 kN of thrust for 1 hr, or one that gives 10,000 kN of thrust for 3 min?

Many people would choose the engine that gives the lower thrust because it provides thrust for an hour. However, they would be wrong. Why? Just think how high you can lift a 5,000-kN weight (the spacecraft) with a 1,000-kN thrust. Even though the thrust is applied for a full hour, the spaceship would not even get off the launching pad. To lift it off the pad, the rocket engine must develop a thrust that exceeds the weight of the entire space vehicle.

Stages are needed for many space flights. Suppose you are all alone in the desert 300 km from town with a truck, a jeep, and a motorcycle. Suppose also that each vehicle has a sealed gas tank with enough gas to go 100 km. You cannot transfer gas from one vehicle to another. How would you travel the 300 km out of the desert to safety?

The answer is to put the motorcycle into the jeep and the jeep into the truck and drive the truck 100 km until it runs out of gas. Then continue on with the jeep and motorcycle until the jeep runs out of gas. Finally, go the last 100 km on the motorcycle. This is a costly way to take a trip, but the "payload" (the driver) gets out of the desert.

The desert problem can be compared to the most widely used method of launching objects into orbit: using "stages." So much energy is needed to put a payload into orbit that launchers are built in steps, or stages, each riding "piggyback" on the other as shown in Figure 16-13. The first stage gets the spacecraft off the pad. Then when the first stage burns out its fuel, it drops off and the second stage ignites. The second stage, in turn, pushes the spacecraft farther into space, and when it burns out it too drops off. Finally, the small third stage finishes the job of getting the payload to its target destination.

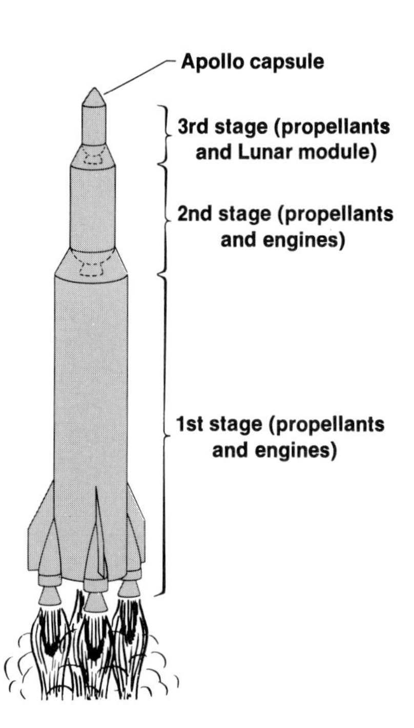

Figure 16-13 Why is the first stage of a three-stage rocket so large in comparison to the other stages?

The Saturn V rocket, used to send astronauts to the moon in the late 1960s and early 1970s, was a three-stage rocket. If you were to lay a Saturn V rocket flat on a football field, the tip would reach beyond the goal post at one end of the field and the engines would reach beyond the goal post at the other end of the field.

Gravity affects satellites in orbit. When a satellite is in orbit around the earth, it is affected by inertia and gravity. Remember Newton's first law: an object in motion tends to stay in motion. That motion, or inertia, would take the satellite in a straight line into space. But the satellite does not do that. What pulls it into a curved path around the earth?

The satellite is not only affected by inertia. It is also affected by the force of gravity. Gravity is always pulling the satellite toward the earth. In an orbiting body, the force of gravity pulling the body toward the earth is balanced by inertia.

At any given height a satellite must travel at a certain speed in order to stay in orbit around the earth. The lower the orbit, the greater the speed must be. Why? Gravity is stronger closer to the earth, so the inertia must be greater to balance the gravity. The greater inertia comes from greater speed. In an orbit 200 km above the earth, the satellite must move at a speed of 29,000 km/hr. In an orbit 36,000 km above the earth, a satellite needs to go only 5,500 km/hr.

At still greater distances from the earth, satellite speeds are even slower. For instance, the moon, 384,000 km from the earth, travels at only 3,500 km/hr. See Figure 16-14.

Flight to the moon and planets is difficult. There are many problems in traveling to the moon. You have already seen that just getting into earth orbit takes a great deal of force using three-stage rockets. After the spaceship gets into earth orbit, it coasts for a while until it is lined up for a flight to the moon. When it is in just the right place, the engines are fired again and the extra force pushes the spaceship away from the earth's gravity. The spaceship follows a long curved path through space. It finally reaches a place where the gravity of the moon is stronger than the gravity of the earth.

If the spaceship is properly aimed, it will swing into an orbit around the moon. Its engines are fired again to slow the ship enough so that it can be "captured" by the gravity of the moon. Further careful steps are needed to send a "moon lander" to the moon's surface, get it back to the spaceship, and get the spaceship back to earth.

Flights to the planets are even more complex. The distances are much greater, and the trips take much longer.

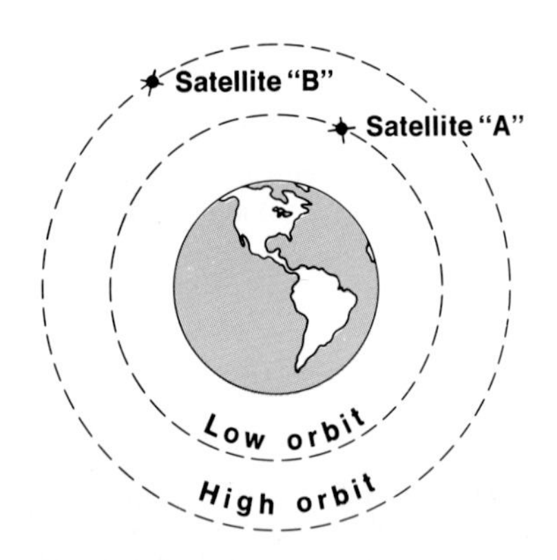

Figure 16-14 Which satellite travels faster, the one in low orbit or the one in high orbit? Explain.

Furthermore, when the spaceship leaves the earth-moon area, the sun's gravity has a greater influence on the spaceship than the gravity of the earth or moon.

To get to another planet, a spaceship must (1) leave the gravity field of the earth, (2) become a satellite of the sun, (3) enter the gravity field of the target planet.

Surprising as it may seem, it does not take very much more speed and fuel to go to the planets than to go to the moon. Most of the fuel is needed just to get off the ground. Once in orbit, somewhat more fuel is needed to get to the moon, but very little extra is needed to go beyond.

Re-entering the earth's atmosphere produces heat. When any object enters the earth's atmosphere, whether it is a tiny lump of metal or rock—a meteor—or the space shuttle orbiter, it rubs against gas molecules in the air. This rubbing, or friction, produces great amounts of heat and extremely high temperatures. The amount of heat and the temperatures produced depend to a large extent on the speed at which the object enters the atmosphere. For example, the space shuttle orbiter streaks back into the earth's atmosphere at about 25,800 km/hr. At this speed, the orbiter's wings can heat up to about 1,510° C, hot enough to melt many metals. In order to protect the orbiter from disintegrating, it is covered with more than 25,000 ceramic tiles. These tiles are extremely heat-resistant and will not melt or burn away as the orbiter dives through the earth's atmosphere (Figure 16-15).

Figure 16-15 Although seared by enough heat to melt many metals, the space shuttle orbiter makes its way safely through the earth's atmosphere thanks to 25,000 heat-resistant ceramic tiles.

JOHANNES KEPLER

solving the puzzle of the planets

In a sense, Johannes Kepler (1571–1630), a German mathematician and astronomer, found the key to the motion of the planets by not looking at them. As it turns out, he studied objects in space with nothing more than pencil and paper. Due to the effects of a childhood illness, Kepler's eyesight was so poor that he was unable to make useful observations of the night sky. However, he could read and was fortunate to have access to the notes and data of the great Danish astronomer, Tycho Brahe, for whom Kepler worked during the last year of Brahe's life. Over many years Brahe had made accurate measurements of the positions of planets and stars. These data provided all Kepler needed to discover how the planets (those known at the time) orbited the sun.

Astronomers of Kepler's time could not account for the slight differences that were noted in the orbits and speeds of the planets. Kepler tried all the accepted theories, but none seemed to fit Brahe's observations. Kepler, like many great scientists before and after him, had great persistence. He was particularly interested in the irregular motion of Mars and worked on that planet's orbit for four years. He worked many more years on the other planets.

Finally, Kepler came to the conclusion that the paths of the planets are not circular, as had previously been thought. Being a mathematician helped Kepler to see that an oval-shaped, or elliptical, orbit for each of the planets would fit Brahe's data. This became known as Kepler's first law and led to two conclusions, which are as true today as they were in Kepler's time: sometimes a planet is closer to the sun than at other times; as a planet draws closer to the sun, it speeds up in its orbit; likewise, when a planet draws farther from the sun, it slows down. Kepler had solved the mystery of the orbits and motions of the planets! And he had done so without directly observing the planets themselves.

CHAPTER REVIEW 16

SUMMARY

1. Machines change the size, direction, or speed with which a force is applied.
2. The ideal mechanical advantage of a machine is a ratio of the effort distance to the resistance distance.
3. The actual mechanical advantage of a machine is a ratio of resistance to effort.
4. All machines waste effort through friction.
5. The ratio of output to input work gives the efficiency rating of a machine.
6. Levers (including pulleys and wheels and axles) and inclined planes (including screws and wedges) are simple machines.
7. The forces acting on a flying airplane are thrust, drag, lift, and gravity.
8. Flight can be controlled by using a rudder, elevators, ailerons, and flaps.
9. In order to send a spaceship into orbit, the thrust of the rocket engine must exceed the weight of the entire space vehicle.
10. The inertia, or tendency of an orbiting satellite to continue moving outward into space, is balanced by gravity, or the force pulling the satellite back to earth.

REVIEW QUESTIONS

1. What are three functions of a simple machine?
2. Explain the difference between high mechanical advantage and high efficiency.
3. How do you find the ideal mechanical advantage of a pulley system? inclined plane? wheel and axle?
4. Name four parts of a plane used in controlling its flight.
5. Explain the action of a propeller.
6. What is meant by "staging" a spacecraft?
7. What keeps a satellite from flying off into space?
8. What keeps a satellite from crashing into the earth?
9. Into what two groups can the six simple machines be placed?
10. List four main forces acting on a plane in flight.

CRITICAL THINKING

11. How are the single pulley and wheel and axle related to the lever?
12. Explain how you would determine the efficiency of a pulley system.
13. Diagram a pulley system that you could use to lift a 3,000-N load by pulling down with a 500-N force (assume no friction).

14. Of the machines discussed in this chapter, which do you think is the most efficient? least efficient? Why?
15. Why has no one succeeded in inventing a perpetual motion machine?
16. How can the lift of an airplane be increased?
17. Why is a satellite traveling faster at the low point of its orbit than at the high point?
18. What causes space capsules to become hot when they re-enter the earth's atmosphere?

VOCABULARY

Match the item in the left column with the best answer in the right column. Do not write in this book.

1. actual mechanical advantage
2. ailerons
3. Bernoulli principle
4. block and tackle
5. complex machine
6. drag
7. efficiency
8. elevators
9. flaps
10. ideal mechanical advantage
11. kite effect
12. lift
13. mechanical advantage
14. pitch
15. rudder
16. simple machine
17. sonic boom
18. stabilizers
19. thrust

a. ratio of the output work to the input work
b. control up and down movements of an airplane's nose
c. device that changes the amount, speed, or direction of a motion
d. maintain an airplane steady in flight
e. help a plane roll
f. force pushing an airplane forward
g. ratio of resistance distance to effort distance with friction
h. friction caused by air on a moving airplane
i. caused by a jump in air pressure as a shock wave hits the ground
j. made of two or more simple machines
k. lift caused by air striking the underside of the wing
l. upward force on an airplane
m. used to move the nose of an airplane to the left or right
n. performance of a machine as if it were frictionless
o. used as air brakes during landing
p. how much a machine multiplies your effort
q. system of pulleys for raising a heavy object with little effort
r. distance from one screw thread to the next
s. as fluid speed increases, internal pressure decreases

PROBLEMS

1. What is the *AMA* of a machine that lifts a weight of 100 N with an effort of 20 N?
2. What is the efficiency of a machine that does 160 N-m of output work with 200 N-m of input work?
3. The movable pulley block of an oil rig is supported by eight strands of wire cable. (a) What is the ideal mechanical advantage? (b) Neglecting friction, how much weight could be supported by an effort of 500 N? Refer to Figure 16-2, if necessary.
4. What is the efficiency of a block and tackle that lifts a 500-N resistance a height of 1 m when an effort of 80 N moves through a distance of 10 m? What is the input work? output work?
5. What is the *IMA* of an inclined plane 20 m long and 2 m high?
6. A 500-N cart is rolled up a 20-m plank to a platform 5 m off the ground by an effort of 150 N parallel to the plank. Find (a) the ideal mechanical advantage, (b) the actual mechanical advantage, (c) the input work, (d) the output work, and (e) the efficiency.
7. What is the weight that could be lifted by a second-class lever 60 cm long if the fulcrum is at one end and a 15-N upward force is applied at the other? Assume that the weight is placed 20 cm from the fulcrum. What downward effort would be needed if the fulcrum and the weight were to exchange places?
8. What weight is lifted by a lever with a 3-m resistance arm when 24 N is applied to the 5-m arm?
9. A plane has a wing area of 64 m^2. The average lift on the wings is 800 N/m^2. Find the total lift.
10. A satellite is orbiting quite close to the earth at a speed of 28,000 km/hr. About how long will it take to make one revolution? (Assume the orbit diameter is 12,800 km.)

FURTHER READING

Adkins, J. *Moving Heavy Things*. Boston: Houghton Mifflin, 1980.

Dalton, S. *Caught in Motion*. New York: Van Nostrand Reinhold, 1982.

Kerrod, R. *The Way It Works: Man and His Machines*. New York: W. H. Smith Publishers, 1980.

Lemonick, M. D. "Back to Earth Unscathed." *Time* 131 (February 1, 1988): 48.

FORCES IN SOLIDS, LIQUIDS, AND GASES

When divers swim along the bottom of the ocean, the water exerts pressure on them. How does this pressure change as they swim toward the surface of the ocean? Why do our bodies tend to float rather than sink?

CHAPTER 17

SECTIONS

OBJECTIVES

- ☐ Describe some of the useful properties of matter that are exploited by people in the making of important products.
- ☐ Compare cohesive and adhesive forces in solids and liquids.
- ☐ Use forces and areas to calculate the pressures on solids and in liquids.
- ☐ Identify the relationship between buoyancy and specific gravity.
- ☐ Discuss the relationship between pressure and volume of confined gases.

17.1 FORCES AFFECT SHAPE

Some forms of matter are elastic. Why is steel used in car springs? Why is steel used in a watch spring? Why does a rubber band snap? Which do you think is more elastic, steel or rubber?

When the ends of a coiled steel spring are pulled apart, the applied forces cause the spring to stretch. If the forces are increased, the spring will stretch farther. When the forces are released, the spring snaps back to its original shape. A rubber band acts in the same way. It can be stretched, and when released, it returns to its former shape.

The property that allows matter to return to its original shape after being distorted by a force is called **elasticity**. However, if a spring is stretched too far, it will not return to its former shape; thus the spring has been stretched beyond its **elastic limit**.

Because steel is very elastic, it tends to return to its original shape when the stretching force is removed. Tests show that steel is more elastic than rubber; that is, a greater force is required to stretch steel and it returns to its beginning shape with more force. This makes steel ideal for use as car springs and watch springs. On the other hand, rubber can be stretched farther than steel without reaching its elastic limit. This property is called resilience. Some substances, such as putty or modeling clay, are not elastic at all. They lose their shape easily and do not regain it.

Spring scales make use of the elastic properties of matter. For example, when a given weight is hung on a spiral spring, the spring stretches by a certain amount. When a second, equal weight is added to the first weight, the spring stretches twice as far. If three equal weights are used, the spring stretches three times as far, and so on. In other words, the distance the spring stretches is proportional to the weight pulling on it. However, this step-by-step increase does not continue indefinitely, as seen in Table 17-1. At what point is the spring scale pulled beyond its elastic limit?

TABLE 17-1

Elastic Limit of a Spring	
Weight on Spring (newtons)	**Stretch of Spring (centimeters)**
0	0.0
100	1.5
200	3.0
300	4.5
400	6.0
500	8.1
600	10.7

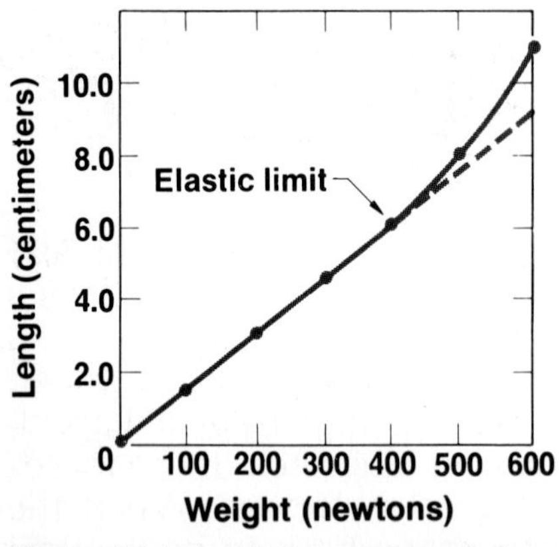

Figure 17-1 How is the solid graph line different above and below the elastic limit?

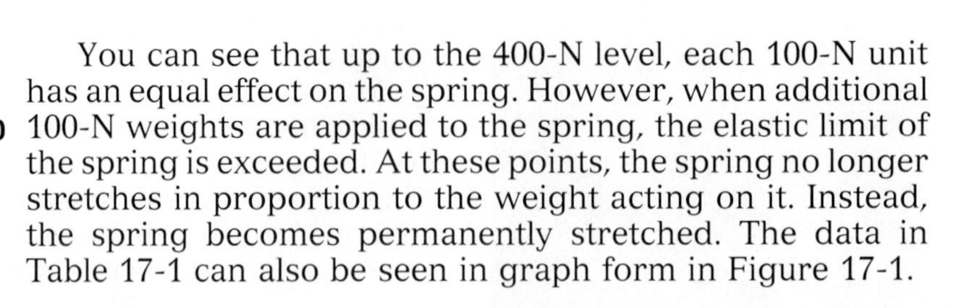

You can see that up to the 400-N level, each 100-N unit has an equal effect on the spring. However, when additional 100-N weights are applied to the spring, the elastic limit of the spring is exceeded. At these points, the spring no longer stretches in proportion to the weight acting on it. Instead, the spring becomes permanently stretched. The data in Table 17-1 can also be seen in graph form in Figure 17-1.

You have learned how force is related to elasticity in terms of a coil spring and rubber band. Forces may also stretch, twist, and bend steel that is forged in different shapes, such as those of beams in buildings and drive shafts in cars. However, if the elastic limit of the steel is not exceeded, no permanent damage results from the forces.

Cohesion holds like molecules together. Why doesn't a steel beam fall apart? Why doesn't a concrete wall crumble? What holds solids together? The property that holds matter together is called cohesion (*coe*-HEE-*zhun*). **Cohesion** is a tendency for the same kind of particles of matter to stick together. Of course, cohesion is quite strong in a steel beam.

What about liquids? Do they also have a cohesive force? If you observe water carefully, you see that water does have a cohesive force. Other liquids also have cohesion, but the strength varies with the substance. Mercury, a liquid metal, has a strong cohesive force as compared to other liquids, but its cohesive force is still weak when compared to that of solids.

Adhesion holds unlike molecules together. The attraction between molecules of two or more different substances is called **adhesion**. A good example of adhesion is the attraction between adhesive tape and skin or that between glue and paper.

Not all substances are equally cohesive. For example, after you have finished washing your hands, some of the water still sticks to your skin by adhesion. Yet if you were to dip your finger in mercury, no mercury would stick to your finger. This is because the force of attraction between the molecules of mercury (cohesion) is greater than the force of attraction between mercury and your skin (adhesion). CAUTION: *Do not touch mercury, which is a poison.* What does this reveal about the relative cohesiveness of water and mercury?

If you look carefully at a beaker or test tube filled with water, you see that the water is not level where it touches the sides of the glass. As you can see in Figure 17-2, the attraction between water and glass is strong enough to pull some of the water up the sides of the glass. Mercury, on the other hand, is pulled down from the walls of the glass. The cohesive force of mercury is stronger than is the attraction of mercury for glass.

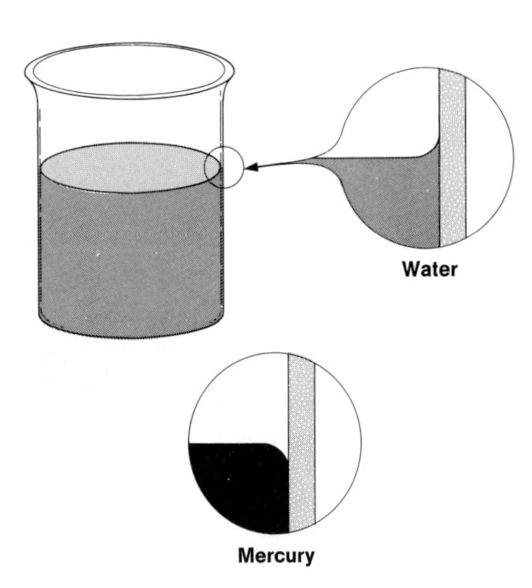

Figure 17-2 Why does water climb up the side of the glass whereas mercury draws away from it?

There are special forms of cohesion and adhesion. Cohesion between molecules on the surface of a liquid causes them to draw together. This effect does not exist below the

INVESTIGATING THE FORCES OF COHESION AND ADHESION

activity

OBJECTIVE: Perform experiments to investigate the forces of cohesion and adhesion.

PROCESS SKILLS
In this activity, you will *observe* and *collect data* that will help you to *hypothesize* about the forces of cohesion and adhesion.

MATERIALS

waxed paper	water
glass plate	soap solution
cork	paper
beaker	pencil
toothpick	medicine dropper

PROCEDURE
1. Place a few water drops on the waxed paper. Start with a small drop and make each succeeding drop larger until the largest is about 10 mm across.
2. Study the drops carefully. Make a diagram showing a side view of each drop. Notice which water drop is the highest and which is the roundest. Then dip a toothpick into a soap solution and touch the tip to each drop. Again, draw side views of the water drops.
3. Place several drops of water on a clean piece of glass. Compare these drops with those on waxed paper.
4. Place a cork into a beaker half full of water. Try to get the cork to stay in the middle. Notice what happens.

OBSERVATIONS AND CONCLUSIONS
1. What effect did the soap solution have on the water drops?
2. How did the water drops on the glass compare with the drops on the waxed paper? How do you account for any differences that you noticed?
3. What happens when you try to make the cork stay in the middle? Why do you think this happens?

surface where all the cohesive forces on a molecule are balanced. At the surface, however, molecules are attracted only by the molecules below and at the sides. This attraction is shown by the force vectors in Figure 17-3. The attraction causes the surface to act as if it were covered by a thin elastic film, and this effect is called **surface tension**. You can see the effect of surface tension in a number of ways. For instance, you can pour water carefully into a glass until the liquid rises just above the rim. The surface tension of the water is strong enough to keep the water from overflowing. However, if more water is added, the surface tension breaks and the water spills over.

Mercury shows this property even more clearly. It may pile up as much as 0.6 cm above the rim of a vessel before spilling over the edge.

Have you ever wondered why paper towels soak up spilt milk or a bath towel soaks up water? Perhaps you have seen water rise in a glass tube with a very small opening. It does this by capillary action. The word capillary comes from the Latin word for hair. Capillary action takes place in fine hair-like tubes and is a form of adhesion.

If you take a glass tube with a small inside opening and dip it into water, how high will the water rise inside the tube? The height of the water level in capillary tubes varies inversely with the tube's inside diameter. That is, the smaller the inside of the tube, the higher the water level will rise. This relationship is illustrated in Figure 17-4.

In some liquids, such as mercury, with stronger cohesive than adhesive forces, the reverse happens. Instead of rising inside the tube, the liquid is pulled down below the level of the liquid outside the tube, as shown in Figure 17-5. The surface tension is so great that the liquid is kept from rising up the tube.

Capillary action occurs in places other than in fine glass tubes. For instance, blotting paper and paper towels soak up water. This is due partly to the tiny, hair-like passages that exist between the fibers of the paper.

If soil is tightly packed, capillary action brings water to the surface more easily than if the soil is loosely packed. At the surface, the water evaporates and the soil dries up. To conserve water, farmers till the soil so that it is crumbly and loose.

Capillary action also plays a large part in the growth of plants by lifting water and nutrients from the roots to the stems and leaves. Water from the soil enters a plant through the roots by the process of osmosis, which you learned about in Chapter 8. Recall that osmosis is the movement of water through a thin membrane from an area of higher concentration to an area of lower concentration. Once inside the plant, the water enters narrow tubular cells that act like capillary tubes to draw the water upward through the plant. Although capillary action, or adhesion, tends to pull the water up the cell walls, the main force behind water movement in plants is actually cohesion. Each water molecule that evaporates from a leaf exerts an upward force on the water molecules below. Thus, through cohesion and adhesion, a steady supply of water flows through the plant along an interconnected passage from the roots to the leaves.

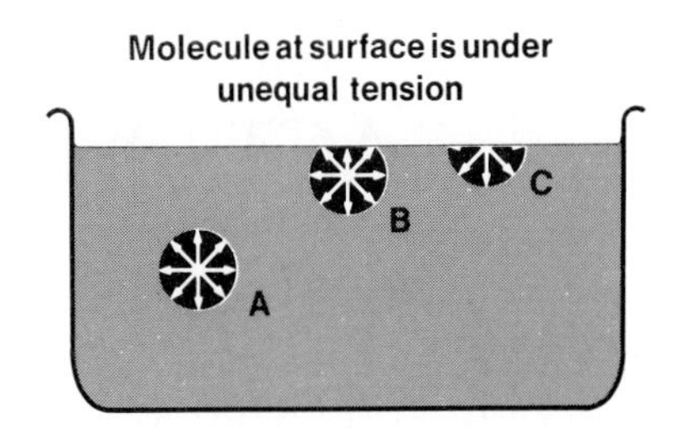

Figure 17-3 The vectors on the three liquid molecules represent external forces exerted on each molecule by neighboring molecules. Molecules B and C have a net downward force that draws the surface molecules together.

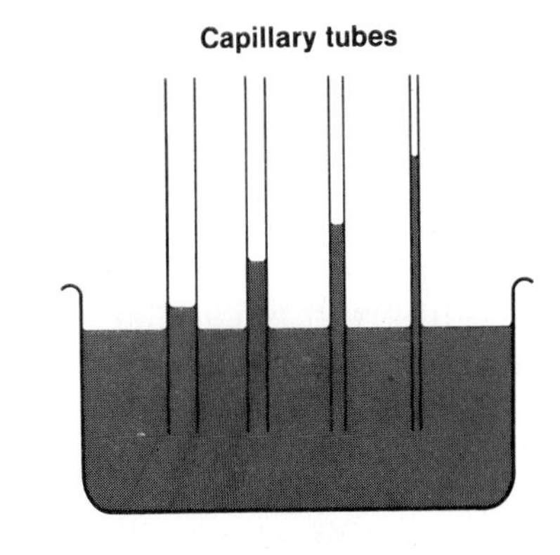

Figure 17-4 The water level rises as the diameter of the capillary tubes gets smaller.

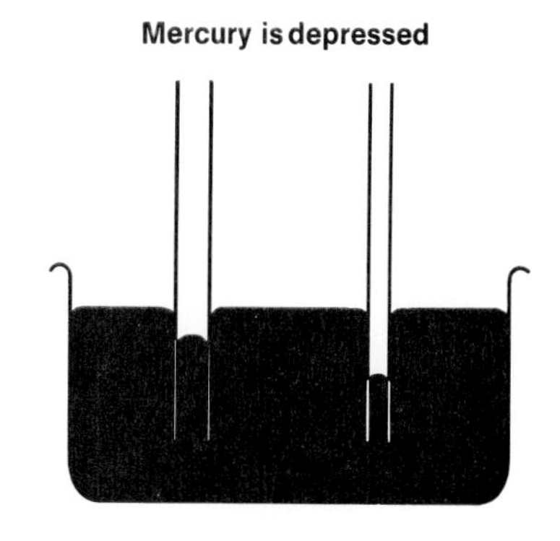

Figure 17-5 What force holds down the mercury inside the tubes? Why is the surface of the liquid curved down where the mercury touches the glass?

MODELING CAPILLARY ACTION IN SOIL

activity

OBJECTIVE: Use a model to show how capillary action works in tightly and loosely packed soils.

PROCESS SKILLS

In this activity, you will *model* and *observe* capillary action that will help you to *draw conclusions* about how this process works in soil.

MATERIALS

sugar cube
container with a small amount of ink
powdered sugar

PROCEDURE

1. Place a sugar cube into a container holding a small amount of ink. Pour some powdered sugar on top of the cube. Observe what happens to the sugar cube and to the powdered sugar.
2. Compare this capillary action to what happens to water in tightly and loosely packed soils.

OBSERVATIONS AND CONCLUSIONS

1. What happened to the sugar cube? to the powdered sugar?
2. How is this activity like capillary action in tightly packed soils? in loosely packed soils?

17.2 PRESSURE: FORCE ON A UNIT AREA

Solids exert pressure. It is much easier to cut meat with a sharp knife than with the side of a fork. All of the force on the knife is focused on a small area of meat beneath the thin edge of the blade. The same force applied to the fork is spread out over a larger area of meat because the fork has a larger surface area touching the meat than has the knife edge. Force acting on a unit area of surface is **pressure**. When the same amount of force is applied in both cases, the pressure exerted by the knife is much greater than the pressure exerted by the fork. To measure the pressure against any surface, the force is divided by the area upon which the force acts.

SAMPLE PROBLEM

A rectangular block measuring 20 cm by 10 cm by 4 cm with a weight of 40 N is placed on a table. Find the pressure exerted on the tabletop when the block is placed in positions A, B, and C shown in the diagram below.

SOLUTION

Step 1: Analyze

You are given that the block measures 20 cm by 10 cm by 4 cm with a weight of 40 N. The block's pressure on the tabletop is the unknown.

Step 2: Plan

Since you know the dimensions of the block and can compute its area for each of the three positions, and since you know the weight of the block, you can use the following equation to find the pressure:

$$\text{pressure} = \frac{\text{force}}{\text{area}}$$

Step 3: Compute and Check

Calculate the block's area in positions A, B, and C and then substitute all values in the equation.

Position A: $10 \text{ cm} \times 20 \text{ cm} = 200 \text{ cm}^2$

$$\text{pressure} = \frac{40 \text{ N}}{200 \text{ cm}^2} = 0.2 \text{ N/cm}^2$$

Position B: $20 \text{ cm} \times 4 \text{ cm} = 80 \text{ cm}^2$

$$\text{pressure} = \frac{40 \text{ N}}{80 \text{ cm}^2} = 0.5 \text{ N/cm}^2$$

Position C: $10 \text{ cm} \times 4 \text{ cm} = 40 \text{ cm}^2$

$$\text{pressure} = \frac{40 \text{ N}}{40 \text{ cm}^2} = 1.0 \text{ N/cm}^2$$

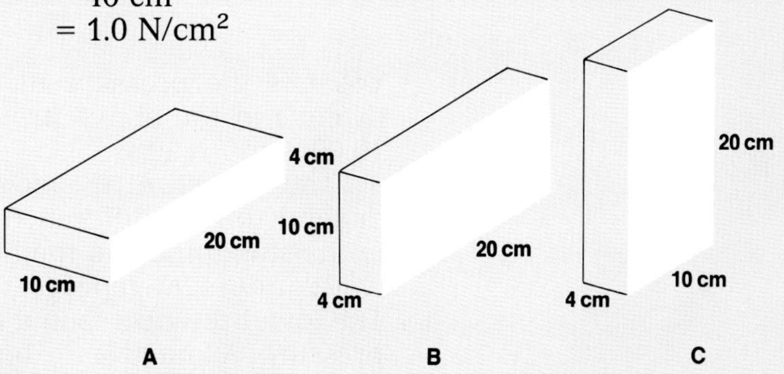

Study the sample problem. It shows that, although the force against the tabletop remains the same in each position, the pressure is five times as great in position C as in position A. You can see that pressure can be increased by decreasing the area of contact. It is also increased by increasing the force.

Liquids exert pressure. Because it has weight, water pushes down on the bottom of a river bed or a water tank. The deeper the water, the greater the pressure. Many cities have water tanks to provide the needed pressure for their water supply systems.

How do you determine the pressure of water in such a tank? Suppose that a city has a tank 30 m high with a bottom area of 50 m^2. The volume of water that a full tank can hold is 50 $m^2 \times 30$ m, or 1,500 m^3. The density of the fresh water is 10,000 N/m^3 (or 10 kN/m^3). Therefore, the water in the tank weighs 1,500 $m^3 \times 10$ kN/m^3, or 15,000 kN. Of course, this weight or force is spread evenly over the entire bottom of the tank. Since the bottom has an area of 50 m^2, the pressure on the bottom is the force divided by the area, or 15,000 kN/50 m^2 = 300 kN/m^2. In other words, each square meter on the bottom of the 30-m-high tank has a force of 300 kN pressing upon it.

The pressure exerted by a liquid can be found in another way. The pressure is the product of its depth and its weight/volume, or

$$\text{pressure} = \text{depth} \times \frac{\text{weight}}{\text{volume}}$$

Since the tank in the example is a column of water 30 m deep,

$$\begin{aligned}\text{pressure} &= 30 \text{ m} \times \frac{15{,}000 \text{ kN}}{1{,}500 \text{ m}^3} \\ &= 300 \text{ kN/m}^2\end{aligned}$$

What is the pressure in newtons per square centimeter (N/cm^2) in the above problem? Remember that there are 10,000 cm^2 in 1 m^2.

The pressure at any given tap or faucet in the city depends on the difference between the height of the water in the standpipe and the height of the faucet. As a result, a water tank is often put on top of a hill to give it more height. The lower a faucet is in a building, the greater is its water pressure. A faucet in the basement delivers water with more force than one on the top floor. Also, water tanks are kept nearly full at all times to insure maximum pressure.

Liquid pressure acts in all directions. Liquids push not only against the bottom of a tank but against the sides as well. Figure 17-6 illustrates the equality of pressure from all directions at a given level. Note that the pressure exerted by the liquid in the tank on the liquid in tubes A, B, and C, although applied in three different directions, is equal because the tubes are at the same level. If you pulled the plug out of a small hole in the side of a barrel filled with water, the water would spurt out because of pressure. The pressure at any given point on the side of the barrel is equal to the downward pressure at that point.

If you push an empty glass down into water, you feel an upward push of the water against the bottom of the glass. The deeper you push the glass, the greater is the upward force. This demonstrates that the upward pressure at any depth is proportional to depth. The sides of deep tanks are made stronger near the bottom because the pressure is greater there than near the top.

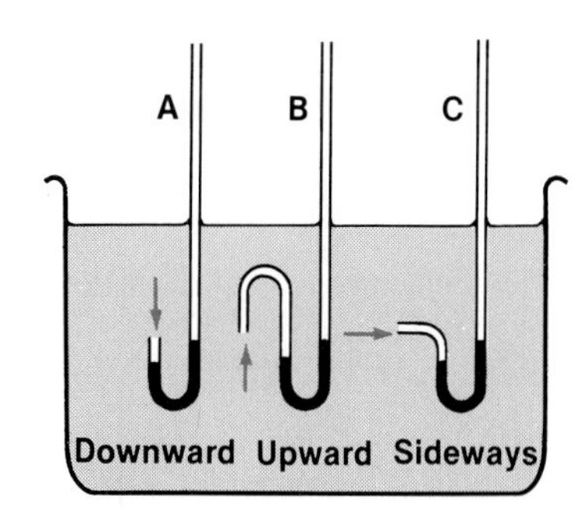

Figure 17-6 How does the upward, downward, and sideways pressure exerted by the liquid in the tank compare at any given point?

Liquid pressure does not depend on the shape of the container. Figure 17-7 shows four various-shaped vessels filled with water and connected by a level pipe. The diagram shows the truth of the saying, "Water seeks its own level." You will now learn why this is so.

Liquids flow in a pipe because of a difference in pressure. The flow is always from a point of high pressure to one of low pressure. If the liquid is not flowing along the tube, the pressure throughout its length must be the same.

The height of the water in each vessel is the same. You already know that the pressure depends on the depth of the liquid. If the water level in one vessel were lower than in the others, the pressure in that vessel would also be less. Water would then flow from the other vessels through the pipe toward the low-pressure point until all levels were the same.

In what way is this concept applied? Before building a dam, engineers find the total force of water that will press against its wall. They begin by finding the pressure half-way down from the top. The pressure at this point is the average water pressure. The average water pressure multiplied by the area of the whole wall of the dam gives the total force exerted by the water. Dams and other structures are always built to resist a force several times greater than what is expected.

Surprising as it may seem, the length of the lake produced by a dam has no effect on the force on the dam. The force is the same whether the lake is 100 km long or 100 m long. The push against the dam depends only on the size of the dam and depth of the water that touches the dam.

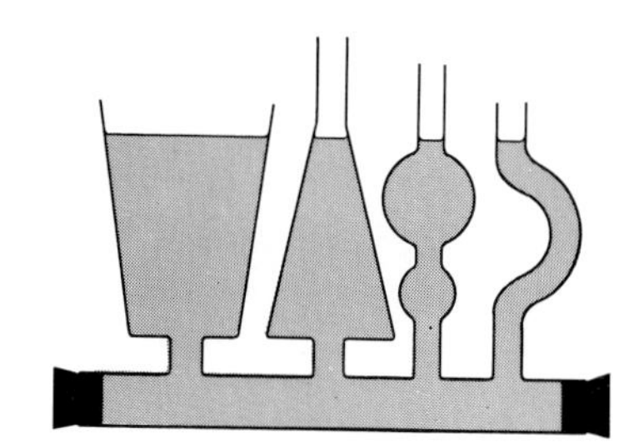

Figure 17-7 What is the relationship between the shape of the container and the pressure at the bottom of the container?

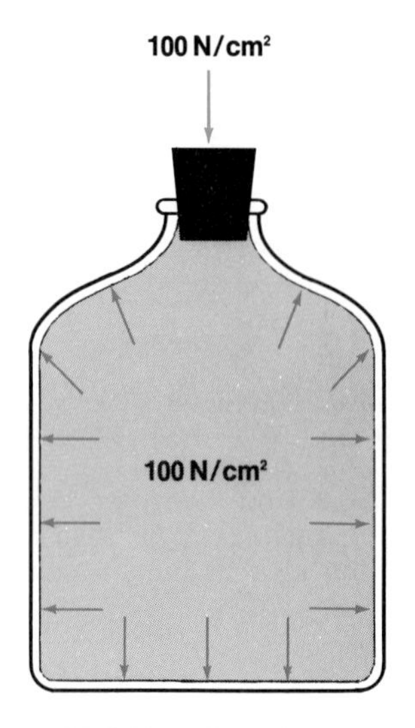

Figure 17-8 How does the pressure on the stopper affect the force against the sides of the bottle?

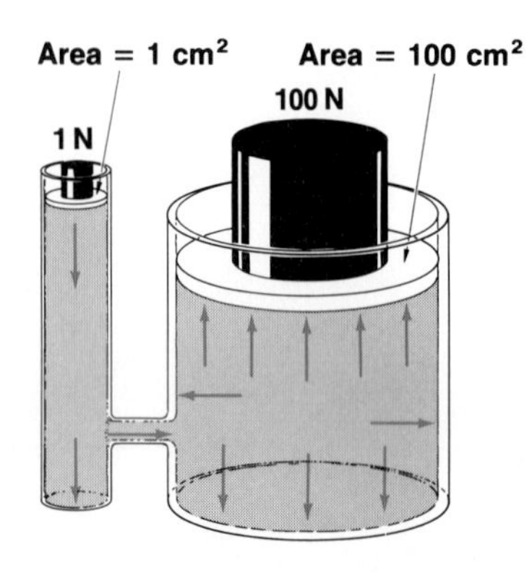

Figure 17-9 What is the relationship between the force applied to the small piston and the weight lifted by the large piston?

Figure 17-10 The hydraulic brakes in this airplane use fluids to apply pressure to the wheels. Upon landing, this pressure creates friction that causes the wheels to stop or slow down on the runway.

Pressure is multiplied in a confined liquid. An increase in depth increases the pressure of a liquid in an open vessel. Now consider the effect of extra pressure applied to a liquid in a closed space.

A liquid-filled glass jug can be smashed by striking a sharp blow on its stopper. The principle that explains why the jug cracks so easily was explained over 300 years ago by the French scientist, Blaise Pascal. **Pascal's law** states that pressure applied to a confined fluid is transmitted equally in all directions.

If a pressure of 100 N/cm^2 is applied to the stopper of a jug containing a confined liquid, as in Figure 17-8, each square centimeter of the jug is pushed out with an added force of 100 N. If the area of the surface of the jug is 1,600 cm^2, a force of 100 $N/cm^2 \times 1{,}600\ cm^2$, or 160 kN, pushes against its inner surface—enough to shatter the glass jug. (Remember that force = pressure × area.)

A number of devices make use of Pascal's law, including hydraulic presses and pneumatic drills. The word **hydraulic** refers to any device that works by putting pressure on a liquid. When pressure is put on a gas, the gas is said to be compressed. A device operated by a compressed gas is a **pneumatic** (*n(y)ooh*-MAT-*ik*) device. Both kinds of machines multiply forces greatly.

A hydraulic press has two pistons. The pressure, or force, acting on each square centimeter, is the same for both pistons. However, one piston is much larger than the other. Since total force is equal to the area times the pressure, the total force acting on the larger piston will be much greater than that acting on the smaller piston.

The smaller piston in Figure 17-9 has an area of 1 cm^2, and the larger piston has an area of 100 cm^2. A downward force of 1 N on the smaller piston produces an upward force of 100 N on the larger piston ($1\ N/cm^2 \times 100\ cm^2 = 100\ N$). Now suppose a 100-N push is applied to the smaller piston such that a force of 10,000 N acts against the larger piston ($100\ N/cm^2 \times 100\ cm^2 = 10{,}000\ N$). Without friction, this device could support a resistance of 10,000 N by exerting a force of only 100 N. What is the formula for the ideal mechanical advantage (*IMA*) of this machine (see Chapter 16)? Would the kind of liquid used (water, oil, or alcohol) make any difference in the *IMA* or in the actual mechanical advantage (*AMA*)?

Hydraulic machines are used in many industries where large forces are needed. Hydraulic presses are used for baling cotton, squeezing oils from seeds, shaping car bodies, punching holes in steel plates, and molding products from plastics. Other uses of Pascal's law include hydraulic brakes on cars and airplanes (see Figure 17-10), automobile lifts used in service stations, and barber chairs.

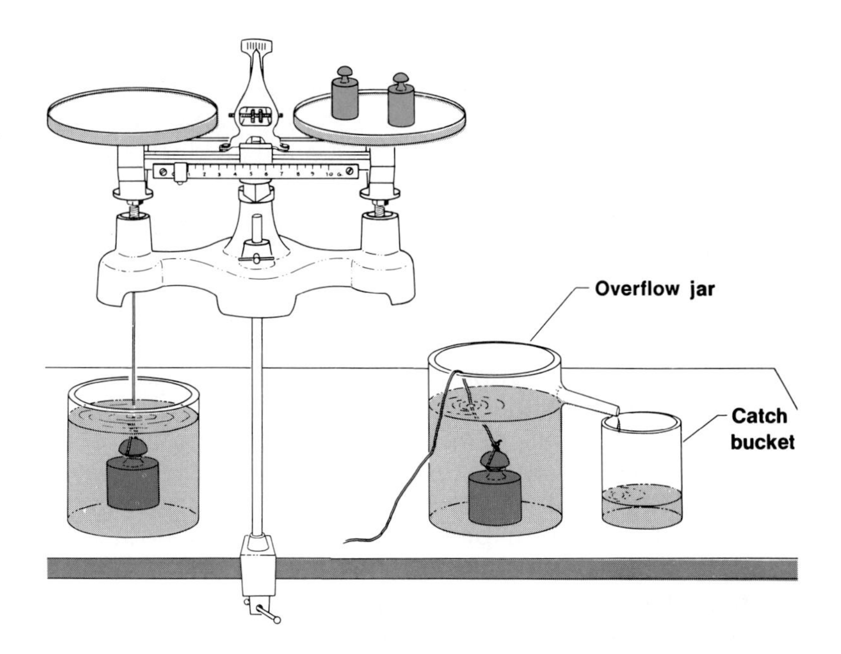

Figure 17-11 What is the relationship between the object in the water and the water in the catch bucket?

Fluids exert a buoyant force. If a piece of cork is dropped into water, some upward force pushes it to the surface. This upward force that fluids exert on objects immersed in them is called **buoyancy**. Remember that by fluids, we mean both liquids and gases.

Maybe you have noticed that it is easier to lift a heavy object when it is underwater than when it is in the air. The ease in lifting the object is due to the upward buoyant force of the water.

You can measure this upward force by performing a simple experiment. First weigh an object on a balance. Then attach a string from the scale to the object. Lower the object into a beaker of water as shown in Figure 17-11. Now weigh the object again while it is suspended underwater. Suppose the object weighs 10 N in the air but only 7.5 N underwater. The object's apparent loss of weight equals the difference between the two values, or 2.5 N. This value of 2.5 N is equal to the buoyant force.

Next, carefully lower the object into an overflow jar that is full of water. Weigh the overflow of water in the catch bucket. If this is done carefully, an amount of water weighing 2.5 N will overflow. This experiment shows that the apparent loss of weight of the object is equal to the weight of the liquid displaced.

Similar experiments were performed by the Greek scientist Archimedes (AHR-*kih*-MEE-*deez*) over 2,000 years ago. He found that an object immersed in a fluid is pushed up with a force that equals the weight of the fluid it displaces. Buoyancy is the force that pushes an object up and makes it seem to lose weight in a fluid.

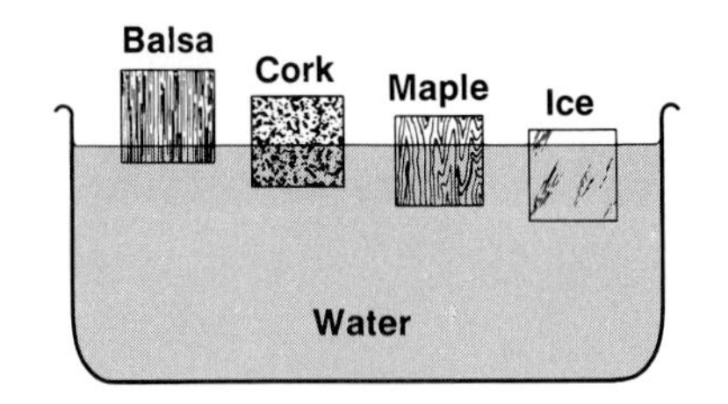

Figure 17-12 What is the relationship between the weight of each block and the water each displaces?

Buoyancy causes some objects to float. Suppose the object in Figure 17-11 were made of wood instead of metal. Wood is less dense than metal. And almost all kinds of wood are less dense than water. Suppose also that the wooden object weighed only 2 N. Because the water can supply an upward force, in this case of up to 2.5 N, the object will float instead of sink. In other words, if the density of the object is less than that of the fluid, the object floats; if the density of the object is greater than that of the fluid, it sinks.

Equal-sized cubes of balsa, cork, maple, and ice, shown in Figure 17-12, all float in water. Notice, however, that some cubes float more deeply in the water than do others. The greater the density of the floating object, the more liquid it displaces, and the deeper it sinks. A floating object sinks deep enough to displace a weight of liquid equal to its own weight, as shown by the following experiment.

Weigh a wooden block on a platform balance and then lower it into an overflow jar. Weigh the water that overflows into the catch bucket. Note that the weight of the block is equal to the weight of the water that overflows. What happens to a block if its density is the same as that of the liquid into which it is placed?

If you replaced the water with a less dense liquid, the block would sink deeper and cause more liquid to overflow. However, the weight of this displaced liquid still equals the weight of the block. Since the liquid is less dense, it takes more of it to equal the weight of the block.

Gases as well as liquids exert buoyant forces. Since fluids include both gases and liquids, buoyant forces are exerted by gases as well as liquids. For example, an inflated balloon rises in air if the weight of the balloon is less than the weight of the air it displaces. The displaced air exerts a net upward force, causing the balloon to rise. For this reason, the balloon shown in Figure 17-13 is filled with low density gas such as helium or hot air. Weights such as sandbags are used to hold the balloon down until it is ready for flight. When the weights are released, the balloon rises.

To return the balloon to earth, the gas inside the balloon is released or compressed into a tank. This compression reduces the volume of the gas. The balloon decreases in size and displaces less air, so the buoyancy decreases.

A submarine works on the same principle. In order to dive, a submarine takes on sea water and becomes heavier. It rises by forcing out the water, becoming lighter.

Density can be expressed as specific gravity. It is often useful to compare the density of a solid or liquid with that of

Figure 17-13 Hot air balloons rise because the heated air in them is less dense than the cooler air around them.

water. This comparison, which is called specific gravity, is found by dividing the weight of a substance by the weight of an equal volume of water. In other words, the **specific gravity** of a substance is a number that compares the density of that substance with the density of water.

Table 17-2 gives the specific gravities of some common substances. Using the data given in the table, decide which substances float in water. Give reasons for your answers.

TABLE 17-2

Specific Gravities of Some Common Substances

Substance	Specific gravity	Substance	Specific gravity
Butter	0.87	Ice	0.92
Copper	8.9	Iron, steel	7.79
Cork	0.25	Lead	11.3
Diamond	3.5	Mercury	13.6
Gasoline	0.7	Water	1.00
Gold	19.3	Wood (oak)	0.85
Human body (lungs inflated)	0.98–1.01	Wood (pine)	0.5

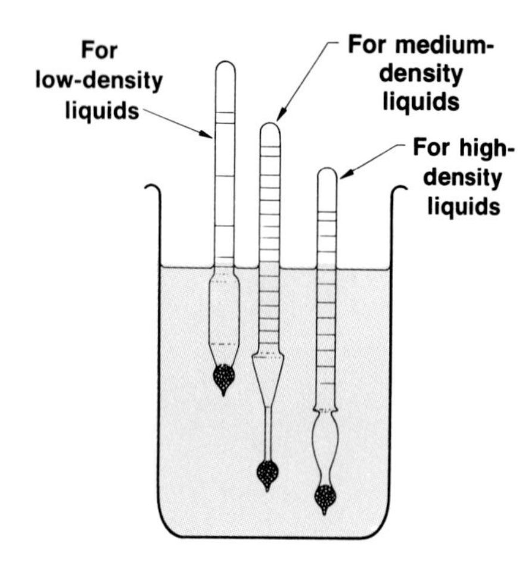

Figure 17-14 Which hydrometer is heaviest? Which has the greatest density?

The easiest way to find the specific gravity of a liquid is to use a floating device called a hydrometer (*hie*-DRAHM-*uh-tur*). A **hydrometer** is a hollow glass tube weighted at the lower end so that it floats upright. A hydrometer sinks into a liquid until it displaces its own weight. Of course, a hydrometer sinks deeper in liquids of low density than in liquids of high density, as illustrated in Figure 17-14. The specific gravity is read directly from the scale on the glass tube.

Many industries, such as those making gasoline, salt, sugar, and soap, use specific gravity readings to help determine the quality of their products. If the specific gravity differs from the true value, the substance is assumed to be impure. An auto mechanic uses a hydrometer to help determine the concentration of antifreeze in the cooling system and acid in the battery. This is possible because the specific gravity of these substances varies with concentration.

Gases exert pressure. We live at the bottom of an ocean of air made up of a mixture of gases. The air in an average classroom weighs several thousand newtons and all of the air around the earth weighs many billions of kilonewtons. This weight of air exerts a pressure of 10 N/cm^2 at sea level. The following experiment demonstrates the effects of air pressure.

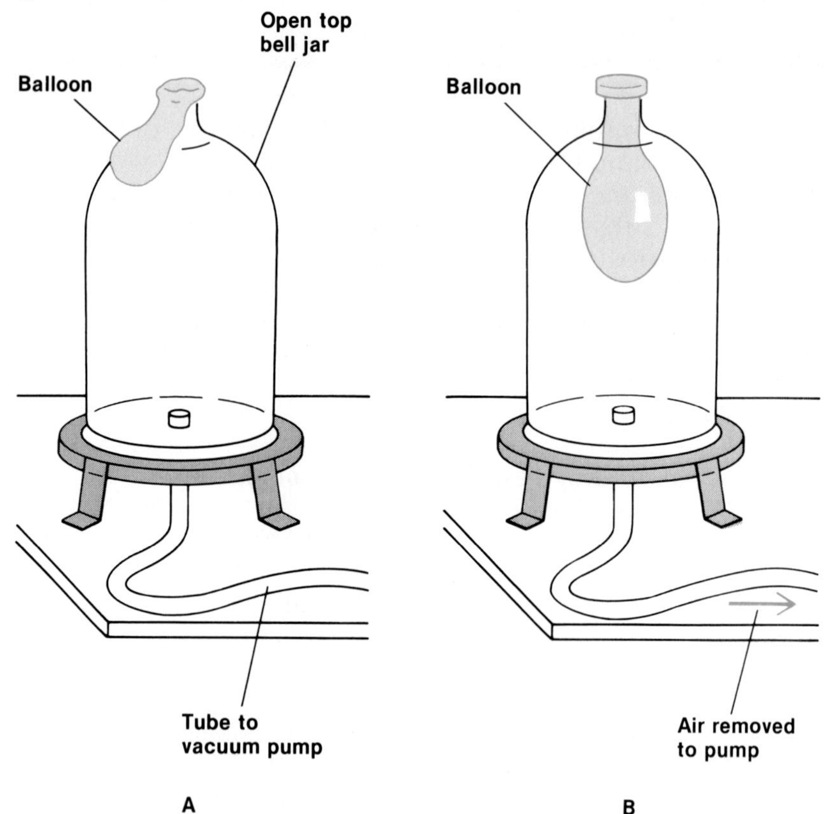

Figure 17-15 In diagram A, the pressures inside and outside the bell jar are equal. As air is removed from the jar (B), the pressure inside the jar becomes lower than that outside, which forces the balloon into the jar.

A thin rubber balloon is tied over the top of an open-top glass bell jar as shown in Figure 17-15. The bell jar sits on an evacuation plate that is attached to a rubber tube leading to a vacuum pump. As air is pumped out of the jar, the outside air presses with more and more force on the balloon, pushing the balloon farther and farther into the jar.

Before pumping begins, the pressure of the air inside the jar equals the pressure of the air on the outside. Then, as air inside the jar is removed, the inside pressure decreases. The outside pressure, however, remains unchanged. This outside air pressure pushes the balloon in. A practical application of this principle is shown in Figure 17-16 and illustrates how you can drink through a soda straw. In this case, you lower the air pressure inside the straw by drawing out the air. The higher outside air pressure pushes on the surface of the liquid and forces the liquid through the straw.

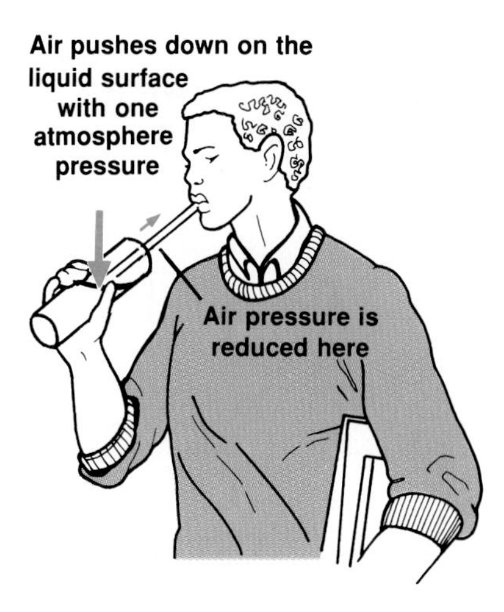

Figure 17-16 How does air pressure make it possible for you to drink through a straw?

Air pressure is measured with a barometer. More than 300 years ago, the Italian scientist Evangelista Torricelli performed an experiment that led to the invention of the **barometer**, an instrument used to measure air pressure.

Torricelli took a glass tube about 1 m long, closed one end, and filled it with mercury. Placing his thumb over the open end, he turned the tube over, dipping its open end below the surface of mercury in a dish. When Torricelli took his thumb away, some mercury ran out, but a column about 760 mm high stayed in the tube. He reasoned that it was air pressure pushing on the mercury in the dish that supported the mercury in the tube.

It was left to French mathematician and physicist Blaise Pascal to carry Torricelli's work further. In 1648, Pascal had Torricelli's experiment repeated at the top of a mountain. He found the height of the column of mercury was lower on the mountain top than on the ground. Why? It was obvious that the air pressure was lower at that altitude. As a result, it could not support as high a column of mercury. Figure 17-17 illustrates different barometer readings at various altitudes.

Torricelli also noticed that the height of the mercury varied from day to day even at the same place. This meant that the air pressure varied from day to day, or even hour to hour. Today, we know that such pressure changes indicate changes in weather. The barometer, an instrument invented more than 300 years ago, is used today to help predict weather.

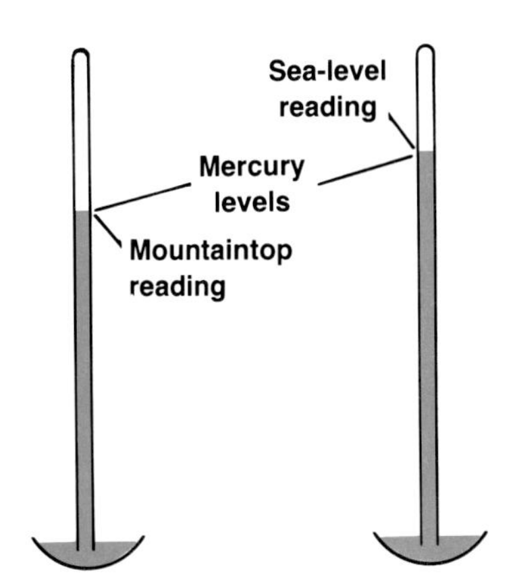

Figure 17-17 Pascal compared a mercury barometer reading at sea level with a reading taken on a mountain top. Why was there a difference in the readings?

Gases obey Boyle's law. How does the volume of a confined gas change when you change its pressure? Suppose a gas is placed in a cylinder with a movable piston, such as

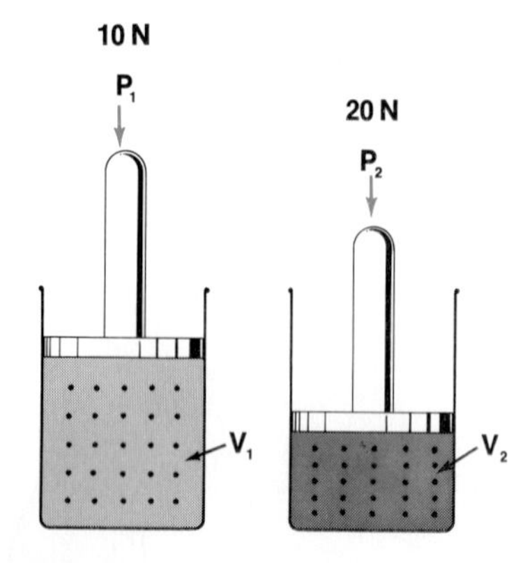

Figure 17-18 The volume of a gas decreases as the pressure increases, providing the temperature remains the same. How does the increase in pressure affect the density of the gas?

that shown in Figure 17-18. The initial volume of the gas is V_1 and its pressure is P_1. When the pressure on the piston is increased to a new pressure, P_2, its volume is decreased to a new value, V_2. If you keep increasing the pressure, you get a set of data such as those shown in Table 17-3. These data are plotted on a graph in Figure 17-19.

TABLE 17-3

Pressure-Volume Relationships

Pressure P (N/cm^2)	Volume V (cm^3)	Pressure × Volume PV
15	12	180
30	6	180
45	4	180
60	3	180
90	2	180
180	1	180

In 1662, the English scientist Robert Boyle discovered that, if the temperature remains unchanged, the pressure of a gas varies inversely with its volume. That is, as one goes up, the other goes down and vice versa. This relationship came to be known as **Boyle's law.** When any two quantities are inversely proportional, their product is a constant.

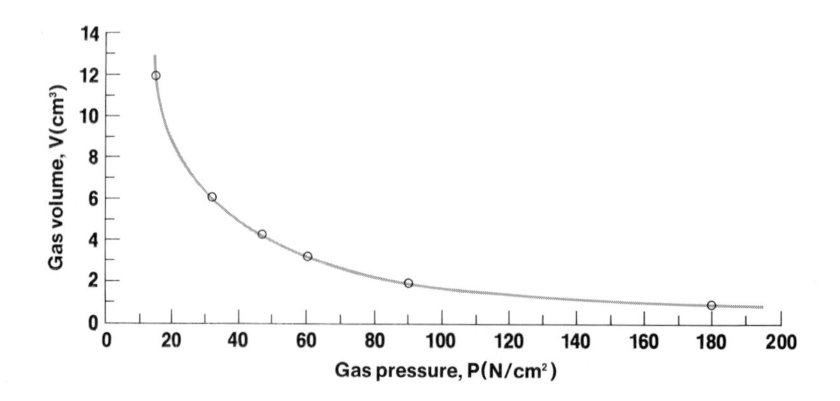

Figure 17-19 What is the relationship between volume and pressure?

Therefore, we can conclude that the pressure of a gas multiplied by its volume (PV) is a constant, as long as the temperature is unchanged. Note that in Table 17-3, $PV = 180$.

The following formula expresses Boyle's law:

$$P_1V_1 = P_2V_2$$

If any three of the quantities in the formula are known, the fourth can be found.

You know that when the pressure of a confined gas is increased, the gas molecules are crowded closer together, and the volume is decreased. If the pressure is doubled, for example, the volume of the gas is reduced to one-half of its former volume. What effect does the increase in pressure have on the density of the gas?

SAMPLE PROBLEM

When 300 cm^3 of a gas under a pressure of 15 N/cm^2 is compressed to a volume of 20 cm^3, what will be the new pressure of the gas? Assume that there is no change in temperature.

SOLUTION

Step 1: Analyze

You are given that the volume of the gas is 300 cm^3, its pressure is 15 N/cm^2, and its new volume after it is compressed is 20 cm^3. Its new pressure is the unknown.

Step 2: Plan

Since you know the gas's original volume and pressure and its new volume, you can use the following equation to solve for the new pressure:

$$P_1 \times V_1 = P_2 \times V_2$$

$$P_2 = \frac{P_1 \times V_1}{V_2}$$

Step 3: Compute and Check

Substitute all known values in the equation.

$$P_2 = \frac{15\ N/cm^2 \times 300\ cm^3}{20\ cm^3}$$

$$= 225\ N/cm^2$$

Since the pressure increased as the volume decreased as expected, your solution is complete.

(Note: Gases do not follow Boyle's law exactly, especially when they are near their boiling points, that is, close to liquefying.)

ARCHIMEDES

solving the mystery of the king's crown

Would you believe that the theory of buoyancy was discovered in a bathtub? There is a story that in ancient Greece during the third century B.C., King Hiero of Syracuse had ordered a crown from a local goldsmith, who swore that every bit of metal in the crown was pure gold. King Hiero, a skeptical man, suspected that the gold had been mixed with silver, and wanted proof of its purity. If there was silver in the crown, however, it would be invisible. To make matters worse, the king decreed that any analysis of the crown would have to be done without damaging it.

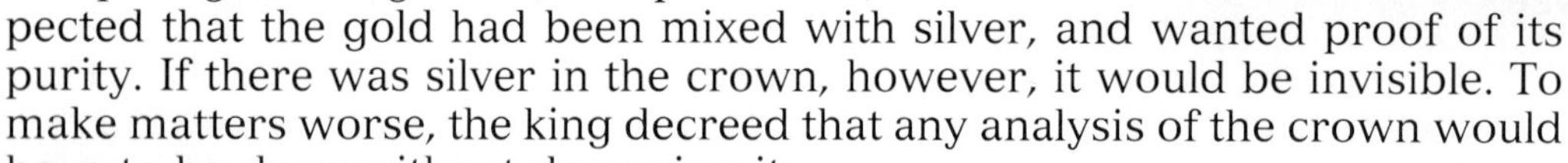

The king knew that this was a problem worthy of only the greatest scientific mind of the time, the mathematician and engineer Archimedes. Archimedes, who was destined to become the father of scientific experimentation, seemed delighted with the challenge. But he could not come up with a way to test the crown for purity, that is, not until he was in the process of taking his regular bath.

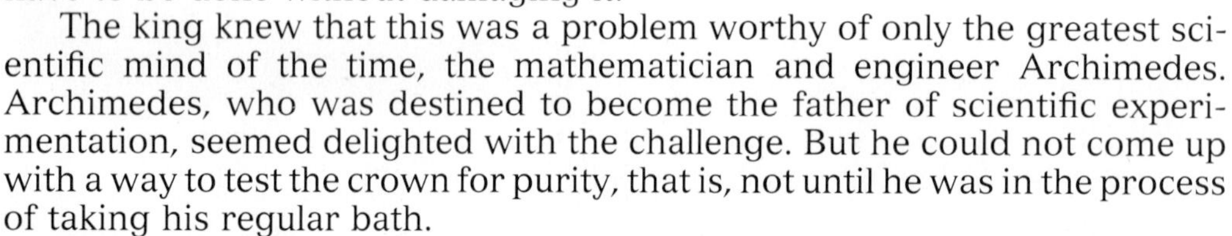

Archimedes had already determined that the volumes of equal weights of pure gold should also be equal. That is, if he got a piece of pure gold that weighed the same as the crown, both the crown and the piece of pure gold should have the same volume. It was not difficult to measure the volume of a regularly shaped cube of pure gold. But how could the volume of the irregularly shaped crown be determined?

This may have been the thought that occupied the mind of Archimedes as he stepped into a tub, filled to the brim with water. As he settled into the tub, the water flowed out onto the floor. As the story goes, Archimedes realized that the amount of water spilling out of the tub must be equal to the amount of his volume that rested in the water.

Archimedes decided to lower the crown into a vessel filled with water and measure the volume of water that overflowed. He would then lower an equal weight of pure gold into a similar vessel filled with water and measure the volume of water that overflowed. If the volumes were equal, he could conclude that the crown was pure gold. As it turned out, the volumes were not equal. The goldsmith, who undoubtedly thought he had committed the perfect crime, was punished for his overconfidence.

CHAPTER REVIEW 17

SUMMARY

1. Elastic materials tend to return to their original shapes after being stretched or deformed.
2. Cohesion holds like molecules together.
3. Adhesion holds unlike molecules together.
4. Pressure is the force acting on a unit area of surface.
5. Pascal's law states that pressure applied to a confined fluid acts equally in all directions.
6. Fluids exert a buoyant force on objects immersed in them.
7. Archimedes' principle states that an object immersed in a fluid is pushed up with a force that equals the weight of the displaced fluid.
8. Specific gravity of a substance is its density compared to the density of water.
9. Gases generally obey Boyle's law: $P_1V_1 = P_2V_2$

VOCABULARY

Match the item in the left column with the best answer in the right column. Do not write in this book.

1. adhesion
2. barometer
3. buoyancy
4. Boyle's law
5. cohesion
6. elasticity
7. hydraulic
8. hydrometer
9. Pascal's law
10. pneumatic
11. pressure
12. specific gravity

a. instrument used to measure specific gravity
b. compares the density of a substance with the density of water
c. instrument used to measure air pressure
d. device operated by compressed gas
e. force acting on a unit area
f. states that pressure on a confined fluid is transmitted equally in all directions
g. upward force on objects in a fluid
h. force of attraction of like molecules
i. device that works by putting pressure on a confined liquid
j. tendency of matter to return to its original shape
k. force of attraction between unlike particles
l. states that at constant temperature, pressure varies inversely with volume

REVIEW QUESTIONS

1. Which is less resilient, steel or rubber?
2. What is elasticity? Is all matter elastic? Explain.

3. Both cohesion and adhesion are forces between molecules. What is the difference between them?
4. Why does mercury not "wet" your finger?
5. How does the height to which water rises in a capillary tube vary with the diameter of the tube?
6. What happens to the level of mercury in a capillary tube?
7. Why does the tilling of topsoil tend to keep it from drying out?
8. Explain what is meant by the following statement: Liquid pressure does not depend on the shape of the container.
9. How could you show that there is a pressure against the side of a barrel filled with a liquid?
10. Why does a piece of iron float on mercury but sink in water?
11. What instrument is used to find the specific gravity of a liquid?
12. If a solid substance has a specific gravity of 0.9, will it sink or float in water? Why?
13. If the pressure of a confined gas at a constant temperature is tripled, what happens to the volume?

CRITICAL THINKING

14. Does Archimedes' principle apply to both gases and liquids? Give examples.
15. Determine from Table 17-2 whether lead will float in mercury.
16. Why doesn't the mercury run out of the open lower end of a barometer tube?
17. Does Boyle's law apply to liquids as well as gases? Explain.
18. Explain why the bristles of a camel's-hair brush spread apart when slipped into water but come together when the brush is removed from the water.
19. Explain how kerosene is drawn up a wick in a kerosene lamp.
20. Why is it possible for a steel needle or paper clip to float on the surface of water, even though the density of steel is much greater than that of water?
21. Which exerts more pressure at the bottom of a tank, gasoline or water? Explain.
22. How can a small force on one piston of a hydraulic press produce a large force on the other piston?
23. Without tasting them, how could you find out which of two sugar solutions was more concentrated?
24. Describe how you would find the specific gravity of lead.
25. Explain why balloons can float at specific levels in air, whereas anything that sinks below the surface in water goes all the way down.
26. How can you "weigh" a ship too massive to be placed on a scale?

PROBLEMS

1. A coil spring hanging from a hook stretches 3 cm when a 2-N weight is attached. (a) How much would the string be stretched by a 1-N weight? (b) a 3-N weight? (c) What weight would be required to lengthen the spring a total of 7.5 cm? (d) What assumption did you make in finding your answers?
2. A rectangular block, measuring 10 cm (base) by 5 cm (side) by 2 cm (end) and weighing 100 N, is placed on a table. Find the pressure exerted on the table when the block is placed on its base, its side, and its end.
3. One piston of a hydraulic press has an area of 1 cm^2. The other piston has an area of 25 cm^2. If a force of 150 N is applied on the smaller piston, what will be the total force on the larger piston?
4. When the deep-sea vessel *Trieste* reached a depth of 11,000 m, what was the water pressure in newtons per square centimeter? (Note: The weight of sea water is 10,000 N/m^3.)
5. A ship displaces 25,000 m^3 of sea water. What is its weight?
6. A 30-kN tank weighs 48 kN when filled with kerosene and 60 kN when filled with the same volume of water. Compute the specific gravity of the kerosene.
7. A pine raft is 4 m long, 2 m wide, and 20 cm thick. How much of a load can this raft carry without sinking in water? (Hint: use Table 17-2.)
8. If the pressure on 1 L of a gas is decreased from 105 N/cm^2 to 15 N/cm^2, what will be its new volume? Assume that the temperature remains the same during the process.
9. What is the total amount of force exerted by air pressure on the floor of your classroom? Why doesn't the floor cave in?

FURTHER READING

Ardley, Neil. *Force and Strength.* New York: Franklin Watts, Inc., 1985.

Arnold, Ned, and Lois Arnold. *The Great Science Magic Show.* New York: Franklin Watts, Inc., 1979.

Challand, Helen J. *Activities in the Physical Sciences.* Chicago: Children's Press, 1983.

Daintith, J., ed. *A Dictionary of Physical Sciences.* Totowa, N.J.: Littlefield, Adams & Co., 1983.

Goswami, Amit, with Maggie Goswami. *The Cosmic Dancers: Exploring the Physics of Science Fiction.* New York: McGraw-Hill, 1985.

WORK, ENERGY, AND POWER

During an important race, every member of the crew team must apply her energy to do the most work possible to power the boat across the finish line. What kind of energy is transferred from each team member to the boat? How are work, energy, and power related?

CHAPTER 18

SECTIONS

OBJECTIVES

- ☐ Differentiate between kinetic and potential energy and give examples of each.
- ☐ Describe how potential energy can be changed to kinetic energy.
- ☐ Explain how energy can be changed from one form, such as chemical, electrical, or heat energy, to another.
- ☐ Define work, energy, and power, and show how they are related.

18.1 ENERGY AND WORK

Work can be done only if energy is applied. In the discussion on machines in Chapter 16, you learned that the energy put into and generated out of a machine can be referred to as input work and output work. **Work** is defined as what is done when a force causes an object to be displaced. In other words, work results from a force acting through a distance. In this chapter, you will learn about various forms of energy, all of which have the ability to do work. The metric unit that identifies the work done by energy is the joule (JOOL). A **joule** (J) is the work done when one newton (N) of force acts through a distance of one meter (m). This can be expressed as 1 J = 1 N-m. In later chapters, you will see that the joule is also used to measure energy, particularly heat energy and electrical energy.

Kinetic energy is energy of motion. You undoubtedly know that a huge boulder rolling down a steep hill, a flying airplane, a speeding train, or a rolling ball all have a certain amount of energy. But how much does each have?

To begin a search for the answer, you must first know the kind of energy each object possesses. Recall from Chapter 4 that moving objects all possess kinetic energy, which is the energy of motion. The amount of energy they have depends on two factors: (1) the mass of the moving object and (2) the speed of the object.

Suppose two cars are moving in a straight line down the road. If their speeds are the same, the car that has the greater mass will have the greater amount of kinetic energy. If these cars were to strike an object, the more massive car would cause more damage to the object than would the less massive car.

Now suppose that two cars having the same mass are moving at different speeds. The car moving at the greater speed will have the greater amount of kinetic energy. The kinetic energy of an object increases as the speed of the object increases.

Potential energy is stored energy. You know that when a huge boulder is rolling down a hill, it has kinetic energy. But did the boulder have any energy before it began rolling?

Figure 18-1 How much work was done to lift the skier to the top of the slope?

To answer this question, consider another example. A box lies on the floor. You want to lift the box to the top of a table. It takes energy to lift the box. What happens to this energy? It is stored in the box. Stored energy is called **potential energy** because, if conditions change, it can be used to do work. A boulder on top of a hill has potential energy. When it starts to roll, under the force of gravity, it has kinetic energy that can do work. Another example of potential energy includes the water held back by a dam, which can be used to produce electricity as the water falls through the pipes. Also, the wound-up mainspring of a clock has potential energy. As it uncoils, the stored energy in the spring is released to move the gears of the clock.

The potential energy of a body is equal to the work that it can do. In the case of a lifted object, the potential energy is also equal to the work done on the body to lift it. For example, you can find the potential energy of a box on a table by determining the work done to lift the box from the ground onto the table. This is done by multiplying its weight (force) by the height (distance) it was lifted (potential energy = work = weight × height).

The skier shown in Figure 18-1 also has potential energy due to her position. If the skier weighs 755 N, and the top of

SAMPLE PROBLEM

What is the potential energy acquired by a 200-N box when it is placed on a table 1 m above the ground? See the figure at the right.

SOLUTION

Step 1: Analyze

You are given that the box weighs 200 N and that it is on a table 1 m high. The box's potential energy is the unknown.

Step 2: Plan

Since you know the weight of the box and the height of the table, you can calculate the box's energy by using the equation

potential energy = weight × height

Step 3: Compute and Check

Substitute all known values in the equation.

potential energy = 200 N × 1 m
= 200 N-m or 200 J

Since your calculated answer is in newton-meters, which can be expressed as joules (the unit that identifies work done by energy), your solution is complete.

the slope is 100 m high, how much potential energy does the skier have? (755 N × 100 m = 75,500 J) In the examples of the box and the skier, potential energy is energy that an object has because of its position. Potential energy can also be the result of an object's condition. A stick of dynamite has potential energy because the stick can do work when it explodes. A lump of coal has potential energy because it can give off heat when it burns. Thus, chemical energy is also a form of potential energy.

Energy can be changed from one form to another. The water behind a dam has potential energy. The potential energy changes to kinetic energy as the water gathers speed in its fall to the river below.

Suppose a student throws a ball straight up into the air. The ball goes up, stops briefly, and comes back down. In this up-and-down path, there is a constant change of energy from kinetic to potential and back again to kinetic. The ball starts up with the kinetic energy given to it by the student's arm. Gravity reduces the ball's speed (and the kinetic energy) to zero as the ball reaches the highest point of the throw. At this point, the ball has its greatest amount of potential energy. This energy, in turn, is changed back to kinetic energy as the ball falls back toward the earth. The ball returns to the student's hand with almost the same speed and kinetic energy as it had when it was thrown. (Some of the ball's energy is changed into heat energy because of air resistance.)

Another example of the transfer of energy occurs in a swinging pendulum. The trapeze artist in Figure 18-2 acts like a pendulum. When on the platform—because of its height—he has the maximum amount of potential energy (position A). When he swings away from the platform, his potential energy decreases. It reaches its lowest point at position B. However, he does not stop there, but moves

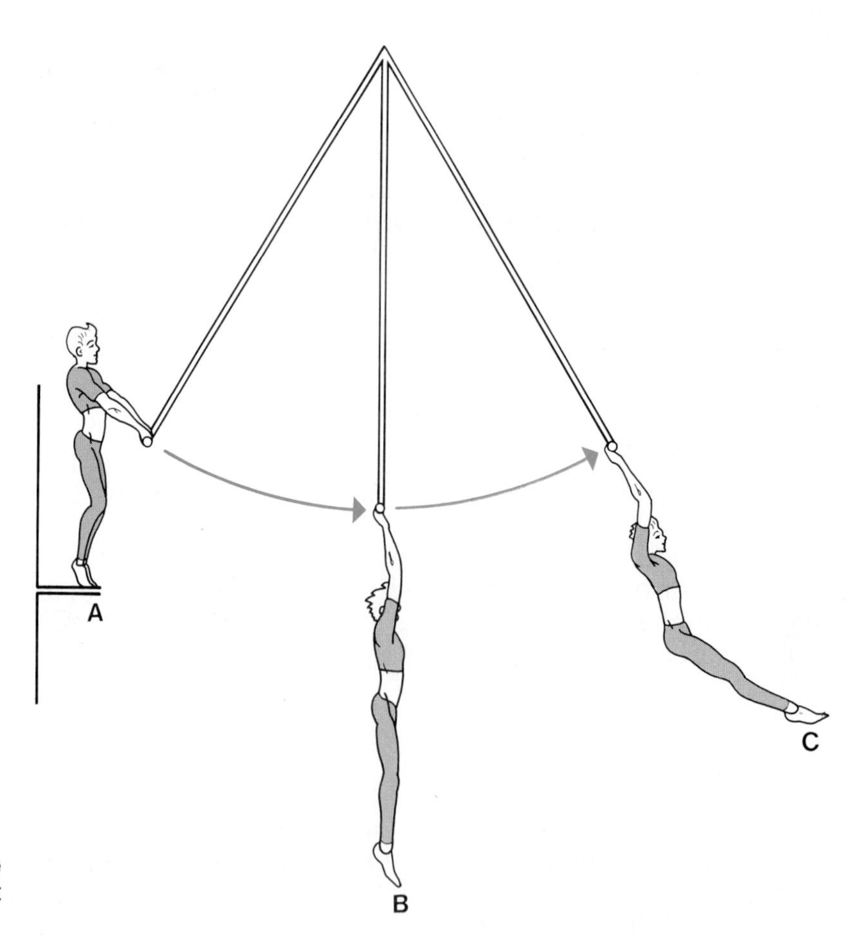

Figure 18-2 At what point does the trapeze artist have the greatest amount of potential energy? kinetic energy?

right past B. Why? Although the trapeze artist has lost all of his potential energy at B, he now has gained his greatest amount of kinetic energy. The kinetic energy is enough to lift him to position C. By the time he reaches C, all of the trapeze artist's kinetic energy has been changed back to potential energy (assuming that there are no losses due to friction).

These kinds of changes keep repeating as the trapeze, or any pendulum, swings back and forth. At the high points of the swing, the trapeze has its greatest amount of potential energy but no kinetic energy. At the lowest point of the swing, the trapeze has its greatest amount of kinetic energy but no potential energy. Despite this constant transfer from one kind of energy to another, the total amount of energy does not change. The sums of the potential and kinetic energies are the same at any point on the curved path.

How do the forms of energy change on a roller coaster? Where is the potential energy greatest on the roller coaster? Where is the kinetic energy greatest? Where is the kinetic energy least?

Figure 18-3 How does the height of the dam affect the potential energy of the water behind the dam?

Falling water provides energy. In the early years of human history, water was used to turn millstones, which ground grain into flour. Dams were built to hold back the water and make mill ponds. Water was led from the pond through a channel to a water wheel. The turning waterwheel did the work needed to grind the grain.

Have you ever seen water pouring over the top of a dam or a spillway? This water does no useful work. Note that, in Figure 18-3, only the water that goes through the power station at the base of the dam does useful work. In the power station, falling water makes electricity by turning a rotary motor called a **turbine.**

Large volumes of falling water have large amounts of energy. At Niagara Falls, for instance, most of the water goes over the falls, but some of it is channeled off to run the turbines to produce electrical energy.

From points above the falls, some of the water is sent through a large channel to a basin located near the top and to the side of the falls. Next, the water drops approximately 60 m through large pipes to the water turbines, as shown in Figure 18-4. These turbines are modern versions of old-time waterwheels.

The falling water in the pipe causes the turbine to spin. The shaft of the turbine is coupled to the rotor of an electric generator. Energy that is obtained from falling water is called **hydroelectric energy**.

Energy in the form of electricity is generated on both the American and Canadian sides of Niagara Falls. The cost of electrical energy in this region is lower than in many other

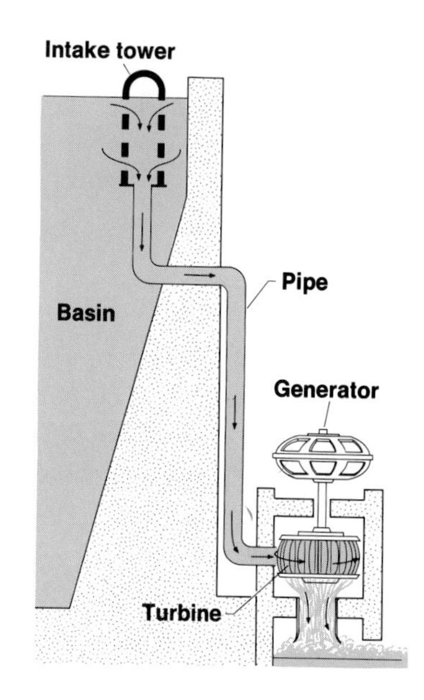

Figure 18-4 Water is diverted from the falls to an intake tower. Here it is channeled to turbines to produce energy. What kind of energy does the water have at the intake tower? at the turbine?

parts of the country because the waterfall is a natural one. Where there is no natural waterfall, people have had to build dams to create a source of water for hydroelectric energy. The cost of the dam, therefore, increases the cost of the energy that is produced.

Energy cannot be made or destroyed. A moving car has kinetic energy. This energy comes from fuel that is burned in the car's engine. When fuel burns in the engine, chemical energy is changed into heat energy. The burning gases expand with great force against the pistons. In this way, heat energy is changed into mechanical energy, which moves the car.

The **law of conservation of energy** states that energy cannot be made or destroyed but can only be changed from one form to another. In the case of a car, falling water, or a ball thrown into the air, energy is not made or destroyed. It is only changed from one form to another. Remember that when a ball is thrown straight up, there is a change of energy from kinetic to potential and back to kinetic.

This law does not apply in the case of nuclear energy. A large amount of energy can be produced from a small amount of matter. As you learned in Chapter 6, this energy is released when atoms of matter are split or fused.

Because matter can be converted into energy, the law of conservation of energy and the law of conservation of matter are often combined. The total amount of matter and energy in the universe does not change. This far-reaching principle is one of the most useful in all of science. Many tests have confirmed this statement, and there are no known exceptions to this law.

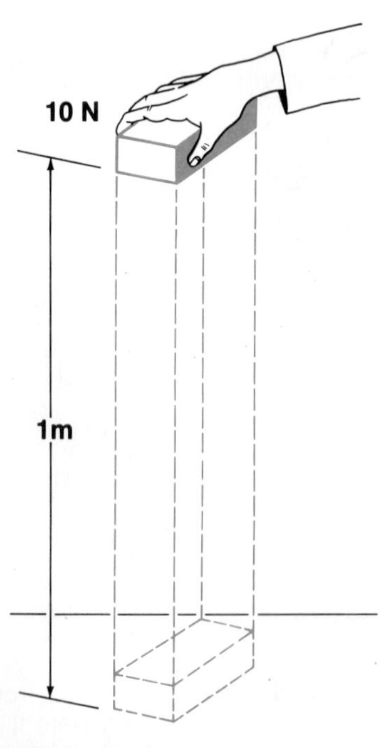

Figure 18-5 How much work is done if the brick is lifted to a height of 1 m? to a height of 2 m?

18.2 WORK AND POWER

There is no work without motion. Applying energy is not enough to produce work. The energy must cause something to move. For example, if you push hard against a box but it does not move, you have done no work! However, if the box moves, you have done work. The amount of work depends on the force you use and how far you move the box. Work is equal to force times the distance through which the force moves as shown in the following equation:

$$W = Fd$$

In this formula, W is the work done, F is the force applied, and d is the distance the object moves under the influence of this force. Look at the example in Figure 18-5.

If you use a force of 10 N to lift the brick 1 m, you do 10 J of work. Notice that the force was applied in an upward direction, and the brick moved upward. The force must be in the direction of motion. Work is not done by a force that acts at right angles to the direction of motion.

SAMPLE PROBLEM

A trunk weighing 500 N rests on the floor. How much work is done in pushing it a distance of 5 m using a force of 150 N? See the figure below.

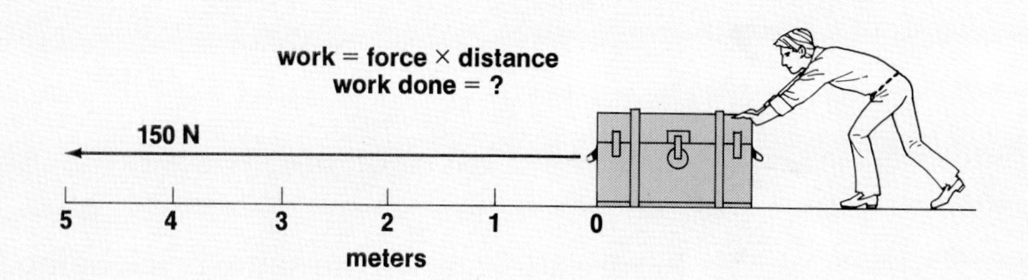

SOLUTION:

Step 1: Analyze

You are given that the trunk weighs 500 N and that it is pushed 5 m using a force of 150 N. The amount of work done is the unknown.

Step 2: Plan

Since you know the force used to move the trunk and the distance the trunk is moved, you can use the following equation to calculate the work done:

$$W = Fd$$

Step 3: Compute and Check

Substitute all known values in the equation.

$$W = 150 \text{ N} \times 5 \text{ m}$$
$$= 750 \text{ N-m or } 750 \text{ J}$$

Since your answer can be expressed in both newton-meters and joules, your solution is complete.

Think again of a hydroelectric power plant. When the water falls down through pipes to turn the turbine blades, it does a certain amount of work. The kinetic energy of falling water is changed to mechanical energy in the turbine. This mechanical energy, in turn, is changed to electrical energy. Work always involves a transfer of energy.

Power is the rate of doing work. Look again at the figure in the previous sample problem. What is the difference in the amount of work done if this job is done in 20 sec or in 45 sec? Since work is independent of time, there is no difference in the amount of work done. However, there is a difference in the amount of power used.

The rate at which work is done is called **power.** The greater the amount of work to be done in a given time, the greater the power needed. Power depends on the following three factors: (1) the force applied, (2) the distance through which the force moves, and (3) the time during which the force is applied. The formula for power is

$$\text{power} = \frac{\text{work}}{\text{time}}$$

$$\text{or } p = \frac{W}{t} = \frac{Fd}{t}$$

Figure 18-6 shows the same brick being lifted at two different rates. When 50 N-m of work is done in 5 sec, the power is 10 N-m/sec (50 N-m/5 sec). If, on the other hand, the same amount of work is done in 25 sec, the power needed is 2 N-m/sec (50 N-m/25 sec). When more time is taken to do the work, less power is used. Review the sample problem involving the work done on a trunk moved across the floor. A force of 150 N moved the trunk 5 m. Now consider the time factor. Suppose it took 5 sec to move the trunk. How much power was applied?

$$p = \frac{W}{t}$$

$$= \frac{750 \text{ N-m}}{5 \text{ sec}}$$

$$= 150 \text{ N-m/sec}$$

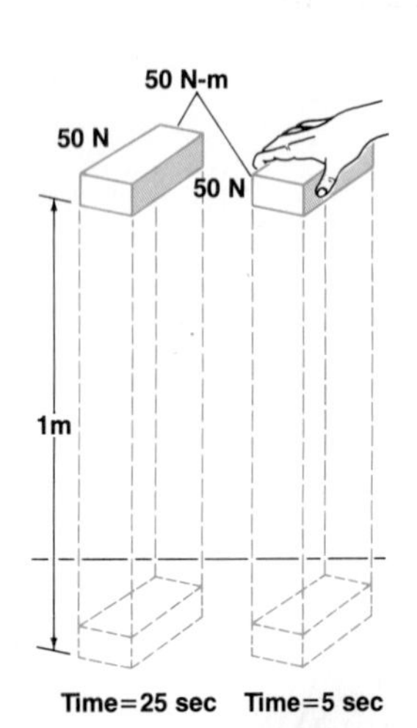

Figure 18-6 Which of the two diagrams represents more work being done? Which shows more power?

Horsepower and watts are units of power. You have probably heard the unit **horsepower** (HP) used to rate the strength of engines. One horsepower is 746 J of work being done per second. An engine of 2 HP can do 1,492 J of work in 1 sec (2 × 746). How many joules per minute are equal to 1 HP?

The units of power in the metric system are the **watt** (W), which is 1 J/sec, and the kilowatt (kW), which is equal to 1,000 W. One horsepower equals 746 W, or 0.746 kW. It may be useful to remember that one horsepower equals about 3/4 kW.

DETERMINING YOUR HORSEPOWER

activity

OBJECTIVE: Perform an experiment to find your strength in horsepower.

PROCESS SKILLS
In this activity, you will *measure* data that will help you to calculate your own horsepower and *communicate* your results to other members of the class.

MATERIALS
stopwatch
paper
pencil

PROCEDURE
1. Select a set of stairs and measure its height in meters.
2. Using a stopwatch, have another student record the time in seconds that it takes you to climb from the bottom to the top of the stairs. You must begin from a standing start and climb vertically, not diagonally. CAUTION: *Never run down stairs.*
3. Make several trials and record the average time.
4. Find and record your weight in newtons.
5. Use the recorded data to calculate your own horsepower. Compare your results with the values obtained by other members of your class.

OBSERVATIONS AND CONCLUSIONS
1. What three measurements were needed to find your horsepower?
2. Why were you asked to make several trials?
3. What equation did you use to find horsepower?
4. What is your strength in watts? in horsepower?

DAMS VS. THE ENVIRONMENT

who wins?

On September 25, 1979, President Carter signed legislation allowing the Tennessee Valley Authority's Tellico Dam on the Little Tennessee River to be completed. The construction of that dam destroyed the natural habitat of the tiny endangered fish known as the snail darter.

Heated discussions surrounded this decision. Environmentalists felt the protection of the snail darter was more important than the building of the dam. Those who felt the dam was needed to control flooding argued that, since snail darters that had been moved to the Hiwassee River in southeastern Tennessee were surviving there, the dam was not a threat to the species' survival. Before the dam was finished, snail darters still living in the Little Tennessee River were removed and placed in the nearby Holston River.

Dams serve as barriers to stop a river's flow. They are used to create artificial lakes or reservoirs. Dams can also be used to generate hydroelectric energy or prevent flooding. They can help reroute the natural flow of water to irrigate farmlands. Although the construction of dams can endanger or destroy certain species of wildlife, dams can sometimes be used to keep natural streams below the dam flowing, thereby helping to preserve the habitats of other species. Dams also serve humans by providing recreational areas. The building and maintenance of dams create jobs for many people.

FOR FURTHER RESEARCH AND DISCUSSION

What sorts of issues should people be aware of before they begin building a dam? Is building a dam always the best solution for the problems in a given area? Are there other ways to solve the same problems? What compromises should be considered if a dam threatens the survival of one or more species of animals and plants?

CHAPTER REVIEW 18

SUMMARY

1. Work is the result of a force acting through a distance. The unit that identifies work done by energy is the joule.
2. Kinetic energy depends on the mass and speed of an object.
3. Potential energy is energy that is stored until called upon for later use.
4. Energy can be changed from one form to another.
5. The law of conservation of energy states that energy cannot be made or destroyed.
6. Power is the rate of doing work and it depends on the force applied, the distance through which the force moves, and the time during which the force is applied.
7. Horsepower and watts are units of power.

VOCABULARY

Match the item in the left column with the best answer in the right column. Do not write in this book.

1. horsepower
2. hydroelectric energy
3. joule
4. law of conservation of energy
5. potential energy
6. power
7. turbine
8. watt
9. work

a. states that energy cannot be made or destroyed
b. rate at which work is done
c. rotary engine turned by falling water to make electricity
d. done when a force causes an object to be displaced
e. kind of electrical energy generated by falling water
f. 1 J of work being done per second
g. unit of work done when 1 N of force acts through a distance of 1 m
h. 746 J of work being done per second
i. stored energy that an object has due to its position or condition

REVIEW QUESTIONS

1. Upon what factors does the kinetic energy of an object depend?
2. Which of the following two stones of equal weight has more potential energy: one on top of a mountain or one in a neighboring valley? Explain.

3. How is gasoline used to produce kinetic energy?
4. Describe how kinetic energy can be changed into potential energy.
5. How can falling water do work?
6. Why is it less expensive to generate electricity at a waterfall than at a dam?
7. Arrange the following words in the order in which they are connected in a power plant: turbine, generator, basin, pipes.
8. How do water turbines develop electric power?
9. How much work do you do if you push against a car but cannot move it? Explain.
10. Explain the difference between work and power.
11. What is measured by horsepower?
12. How are the many forms of energy related?
13. If a force acts on a body at a right angle to its motion, is any work done by this force? Explain.

CRITICAL THINKING

14. Explain the kinetic and potential energy conditions of a drop of water thrown straight up into the air by a water fountain. Describe the conditions when the water drop (a) is released from the fountain, (b) is at the peak of its flight, and (c) returns to the ground.
15. Discuss the transfer of energy in the swing of a pendulum.
16. State the type of energy (kinetic or potential) that each of the following items has: (a) a tank of propane gas, (b) a diver standing on a diving board, (c) a diver in midair, (d) a soaring bird, and (e) a match.

PROBLEMS

1. How much work is done if you push a 200-N box across a floor with a force of 50 N for a distance of 20 m?
2. How much work is done if you lift a 20-N box up to a shelf 2 m high?
3. How much power is needed to carry a 100-N box up three flights of stairs with a vertical height of 12 m in 20 sec?
4. How much power is needed to push a box with a force of 200 N a distance of 10 m in 25 sec?
5. Find (a) the work done and (b) the power needed when a 400-N student climbs a 3-m ladder in 4 sec.
6. A force of 200 N is needed to drag a 1,000-N box a distance of 25 m along a level roadway. (a) How much work is done? (b) If the box is lifted by a crane to a height of 25 m, how much work is done? (c) What is the kinetic energy at a height of 25 m? (d) What is the potential energy at 25 m?
7. Suppose you exerted a force of 60 N in order to push a 300-N box a distance of 12 m in 20 sec. You then lifted the box 1 m into a truck in 3 sec. (a) How much work did you do when you pushed the box? (b) when you lifted the box? (c) How much power did you exert when you pushed the box? (d) when you lifted the box?
8. If a 600-N student climbed a 4-m flight of stairs in 8 sec, what horsepower did the student generate?

FURTHER READING

Asimov, Isaac. *How Did We Find Out About Energy?* New York: Avon Books, 1981.

Berger, Melvin. *Energy*. New York: Franklin Watts, Inc., 1983.

Purcell, John. *From Hand Ax to Laser: Man's Growing Mastery of Energy*. New York: Vanguard, 1983.

Smith, Norman. *Energy Isn't Easy*. New York: Coward-McCann, 1984.

Weitzman, David. *Windmills, Bridges and Old Machines: Discovering Our Industrial Past*. New York: Charles Scribner's and Sons, 1982.

HEAT

The lighter colors in this thermogram of a boy and a dog indicate higher temperatures. Can you tell whether the boy or the dog generates more heat energy? How do heat and temperature differ?

CHAPTER 19

SECTIONS

19.1 Heat and Temperature

19.2 Some Effects of Heat

19.3 Heat Transfer

OBJECTIVES

- ☐ Distinguish between heat and temperature.
- ☐ Describe how heat and temperature are measured.
- ☐ Demonstrate how matter expands when heated and contracts when cooled.
- ☐ Explain how heat is absorbed or given off during a change of phase.
- ☐ Compare the transfer of heat by conduction, convection, and radiation.

19.1 HEAT AND TEMPERATURE

Heat and temperature mean different things. You have learned that matter can change from one form to another. Solids can change into liquids, liquids can change into gases, and one compound can change into another. When such changes occur, energy is either absorbed or given off. Usually, this energy is in the form of heat. The terms heat and temperature are often interchanged in everyday speech. However, the two words have different meanings.

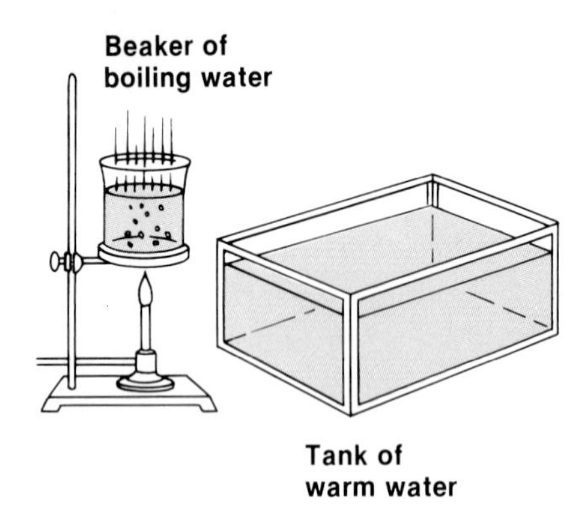

Figure 19-1 Why does having a higher temperature not necessarily mean that the water in the beaker has more heat than the water in the tank?

One way of showing the difference between heat and temperature is to compare a beaker of boiling water with a large tank of warm water, as shown in Figure 19-1. You can dip your finger into one container of water, but not into the other. Why? The answer is obvious. The boiling water is too hot. It has a very high temperature.

Now consider another question. Which of the two examples has more heat energy, the beaker of boiling water, or the tank of water? The tank of water has more heat energy.

In order to understand this answer, you must know what happens to molecules as a substance gets hotter or colder. When a substance is cold, its molecules vibrate slowly. When heat energy is added, however, the kinetic energy of the molecules increases and the molecules speed up. The hotter a substance becomes, the faster its molecules vibrate. **Temperature** can be defined as the measure of the average motion (average kinetic energy) of the molecules.

The faster the molecules vibrate, the higher the temperature of the substance. The molecules in the boiling water vibrate very fast, so the water has a high temperature.

Heat, however, depends not only on the average motion of the molecules, but also on the number of molecules involved. **Heat** is the total amount of energy that can be transferred from one object to another because of a difference in their temperatures. Heat depends on the following two factors: (1) the average kinetic energy of the molecules (the temperature) and (2) the number of molecules present (the mass of the substance).

As a result, the beaker of boiling water has a much higher temperature than the tank of water. However, there is so much warm water in the tank that it has more heat than the much smaller amount of boiling water. Why does 100 g of water have 10 times as much heat energy as 10 g of water at the same temperature?

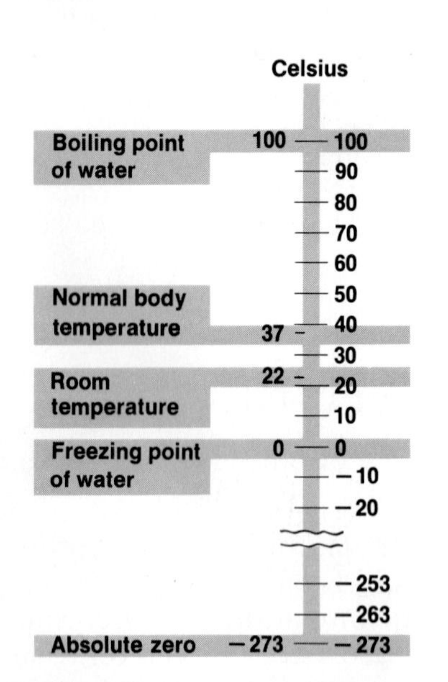

Figure 19-2 How many degrees are there between the boiling and freezing points of water on the Celsius scale?

Temperature is measured in degrees. A day in summer may be said to be hot and a day in winter cold, but these terms are not very exact. A thermometer gives a more exact reading of how hot or cold it is. It measures the temperature in degrees.

The Celsius (SEL-*see-us*) (C) scale is used throughout the world to measure temperature. On the **Celsius scale,** water freezes at 0° and boils at 100°. Normal body temperature is 37° C, and a comfortable room is about 22° C. Refer to the temperature scale in Figure 19-2.

Notice that "zero" on the scale does not represent the lowest point that temperature can reach. Instead, it represents the point at which pure water freezes. In winter, the

CONTRASTING HEAT WITH TEMPERATURE

activity

OBJECTIVE: Demonstrate how heat differs from temperature.

PROCESS SKILLS

In this activity, you will *measure* and *collect data* to help you *draw conclusions* about the difference between heat and temperature.

MATERIALS

safety goggles
2 polystyrene coffee cups
2 thermometers
beaker
water
tongs
large bundle of nails tied with string
single nail
gas burner
ring
ringstand
wire gauze

PROCEDURE

1. Set up the equipment for heating the beaker. Put on your safety goggles. Place a bundle of nails and a single nail into a beaker of water and heat the water until it boils.
2. Fill two polystyrene cups about two-thirds full of cool water. Measure the temperature of the water in the cups and record both temperatures.
3. Using tongs, put the bundle of nails into one cup and the single nail into the other cup. Put a thermometer into each cup and record the readings every 30 sec for 5 min.

OBSERVATIONS AND CONCLUSIONS

1. What was the temperature of the water in each of the two cups before the nails were added?
2. What was the change in temperature in the cup after adding the bundle of nails? the single nail?
3. Since the nails all had the same temperature when put into the cups, how do you account for the temperature differences?

temperature goes well below zero in many parts of the country. Matter can become much colder than zero. As matter gets colder, the motion of its molecules slows down. The point at which the motion of molecules is at a minimum is called **absolute zero.** Absolute zero is 273° below zero on the Celsius scale (−273° C).

Figure 19-3 This rubber ball has been cooled to near −200° C. What causes it to shatter when it is dropped?

Scientists have been able to study matter cooled down very close to absolute zero. They have found that at these temperatures, matter reacts in some surprising ways. For instance, liquid helium becomes a superfluid. That is, it becomes so adhesive that it creeps up the sides of its container. Some poor conductors of electric current become superconductors near absolute zero. Recall in Chapter 9 you learned that a superconductor is a substance with little or no resistance to electric current. Scientists have also recently produced superconductors at higher temperatures. In experiments, thallium superconducts at −148° C and bismuth at −167° C.

At temperatures well above absolute zero but still far below freezing, some common substances react strangely. Lead normally gives off a dull "thud" when struck, but when very cold, it rings like a bell. Mercury, normally a liquid, is a solid when it is very cold. Likewise, many gases change to liquids and then to solids when they are very cold. Rubber, which is normally soft and pliable, becomes hard and brittle when it is cooled to a very low temperature, as shown in Figure 19-3.

A second metric scale is also used to measure temperature. This is the Kelvin (K) scale. The **Kelvin scale** is the same as the Celsius scale except that it has a different zero point. Its zero point is at absolute zero (−273° C). The freezing point of water is 273 K (0° C), and water's boiling point is 373 K (100° C).

Heat is measured in calories and joules. Heat cannot be measured with a meter stick or a balance. It can only be measured in an indirect way. Heat is measured by its effect on matter. Water is most often used as "standard matter" because it is so common.

The amount of heat required to raise the temperature of one gram of water one Celsius degree is defined as one **calorie** (cal). Another unit, the joule (J), which is also used to measure work, can be used to measure heat. One calorie is equal to 4.185 joules.

Sometimes a larger unit called the kilocalorie (kcal), which is equal to 1,000 calories, is used. A kilocalorie is a more useful unit to use to measure large numbers of calories such as the energy content of foods. The Calories you count when you are on a diet are actually kilocalories.

Notice that this Calorie is written with a capital "C," to distinguish it from the calorie. A small baked potato of 100 g has an energy value of 100 Calories, or 100 kcal. Table 19-1 lists the energy content of some other common foods.

TABLE 19-1

Energy Values of Some Common Foods

Food	Quantity of 100 g, roughly equal to:	Energy (kcal)
Hamburger, cooked	1/4 pound	360
Tuna fish, canned	4/5 cup	200
French fries	20 pieces	390
Peanuts	3/4 cup	560
Milk, whole	3/8 cup	70
Milk, skim	3/8 cup	34
Bread, white	4 thin slices	280
Eggs, whole	2 medium	160
Apple, fresh	1 small	60
Tomato, fresh	1 small	20

Users of coal need to know not only the price of coal, but also the number of calories of heat provided by the coal. Coal that supplies 8,000 cal/g will give 6.6 percent more heat than coal that supplies only 7,500 cal/g.

Each substance has its own specific heat. Have you ever taken a heated TV dinner from the oven and noticed that the foil cover is not hot? Why does it remain cool? The answer involves how much heat it took to warm up the foil and the food. Different substances need different amounts of heat to warm them up the same number of degrees. The number of calories that are needed to warm up 1 g of any substance by 1 C° (Celsius degree) is called its **specific heat.** The units for specific heat are calories per gram per Celsius degree, or cal/g C°. Table 19-2 gives the specific heats of some common substances.

It takes 1 cal of heat to warm up 1 g of water by 1 C°. Thus, the specific heat of water is an even 1.00 cal/g C°. Aluminum, with a specific heat of 0.21 cal/g C°, needs only 0.21 cal to warm it 1 C°. Thus, aluminum foil heats easily. Just as the foil warms up quickly, it cools down quickly compared to the moist food in a TV dinner.

Note that the specific heat of water is the highest of those substances listed in Table 19-2. In fact, water has one of the highest specific heats of all matter. Liquid ammonia (NH_3), with a specific heat of 1.2, is one of the few substances with a value higher than that of water.

TABLE 19-2

Specific Heats of Some Common Substances

Substance	Specific Heat (cal/g C°)
Aluminum	0.21
Brass	0.09
Copper	0.09
Ice	0.50
Iron	0.11
Lead	0.04
Steam	0.48
Water	1.00
Wood	0.42
Zinc	0.09

Suppose that you heat 50 g of water from 20° C to 90° C. If 1 cal raises 1 g of water 1 C°, then 50 cal are needed to raise 50 g of water 1 C°. To warm up the 50 g of water 70 C° would take 50 × 70 = 3,500 cal, or 3.5 kcal.

SAMPLE PROBLEM

How much heat is needed to warm up 750 g of iron from 10° C to 130° C?

SOLUTION

Step 1: Analyze

You are given that the mass of the iron is 750 g and that its temperature needs to increase from 10° C to 130° C. The amount of heat required for this temperature change is the unknown.

Step 2: Plan

Since you know the mass of the iron and that the desired temperature change is 120 C° (130° C − 10° C), and since you know the definition of specific heat, you can derive and use the following equation:

$$\text{specific heat} = \frac{\text{heat}}{\text{mass} \times \text{temperature change}}$$

Solving for heat:

$$\text{heat} = \text{specific heat} \times \text{mass} \times \text{temperature change}$$

You can find the specific heat of iron in Table 19-2 (0.11 cal/g C°).

Step 3: Compute and Check

Substitute all values in the equation.

$$\begin{aligned} \text{heat} &= 0.11 \text{ cal/g C°} \times 750 \text{ g} \times 120 \text{ C°} \\ &= 9{,}900 \text{ cal} \end{aligned}$$

Since the correct unit (calories) for heat is derived, your solution is complete.

Now study Figure 19-4 to see what happens when equal masses of some common metals are heated to the same temperature and placed on a block of ice. They melt down to different levels in the ice because they have different specific heat values. (Remember that equal masses do not mean equal sizes or volumes.)

The specific heat of a substance is an important physical property because it describes the suitability of a given substance for a specific purpose. For instance, aluminum pots

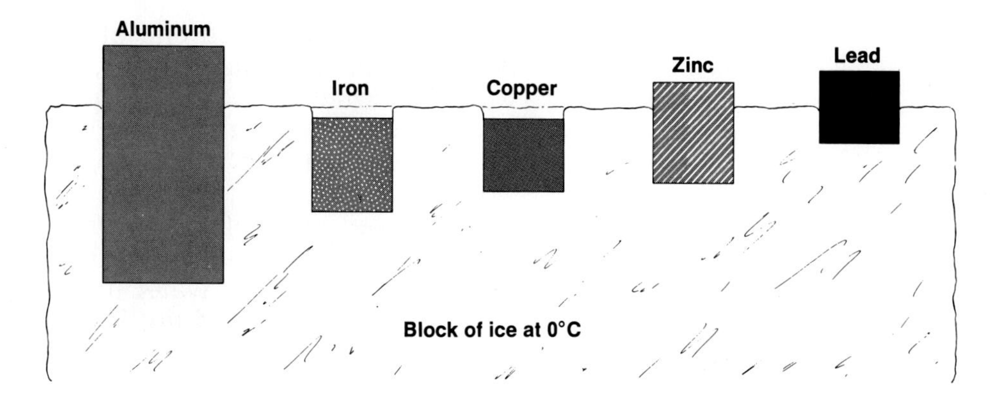

Figure 19-4 These blocks have equal masses. When heated to the same temperature, why do they melt the ice to different depths?

and pans hold approximately twice as much heat as those made of equal weights of iron. You can see this by studying Table 19-2.

The high specific heat of water helps to explain why land near a large body of water is likely to have a milder climate than land that is not near water. The seasonal buildup of heat causes a slow gradual increase in temperature in a body of water. Land near the water also increases in temperature slowly. Dry land, on the other hand, has a much lower specific heat, and reaches a much higher temperature given the same amount of heat. Furthermore, soil is a poor conductor, keeping the heat from going deeply into the ground. Therefore, the heat causes a quick rise in temperature on land far from water. For the same reasons, land areas far from water cool off much faster than do land areas near bodies of water.

19.2 SOME EFFECTS OF HEAT

Heat causes matter to expand. Almost all gases, liquids, and solids expand when heated and shrink when cooled. The expansion is caused by the action of molecules. When heated, the molecules vibrate more rapidly and move farther apart. On the other hand, when matter is cooled, the molecules slow down and move closer together. This slowdown in the motion of the molecules reduces the overall volume of the substance, and so the material shrinks.

The simple experiment illustrated in Figure 19-5 can be used to demonstrate that gas expands when warmed. If you place your hand on the flask for one or two minutes, the air inside the flask is slightly warmed and it expands. As the air

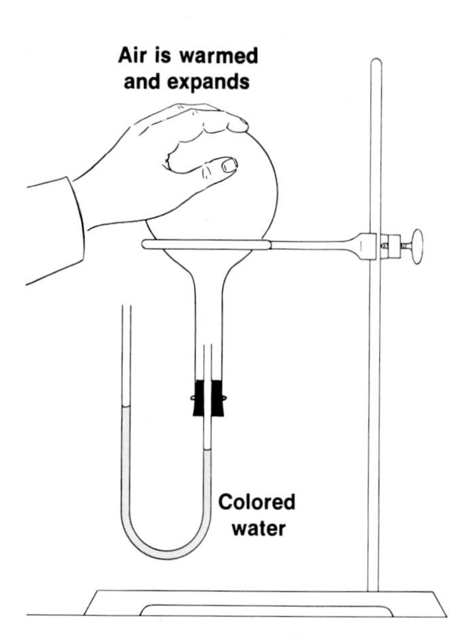

Figure 19-5 What causes the colored water to move when you place your hand on top of the flask?

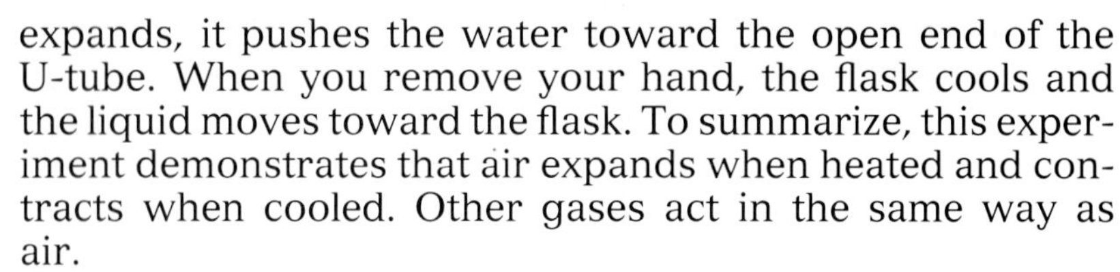

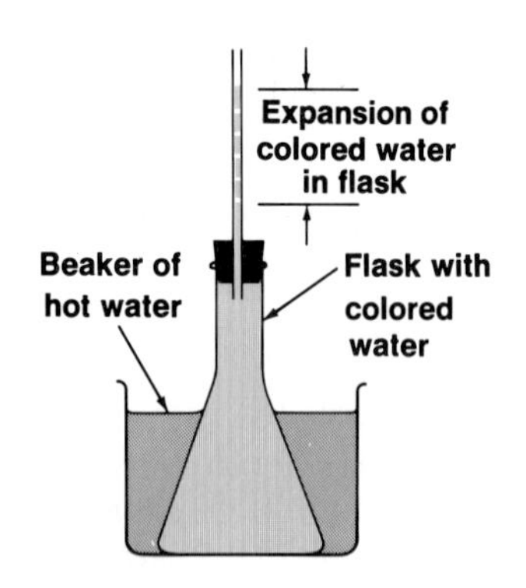

Figure 19-6 In what direction will the water move in the glass tube when the water is heated?

expands, it pushes the water toward the open end of the U-tube. When you remove your hand, the flask cools and the liquid moves toward the flask. To summarize, this experiment demonstrates that air expands when heated and contracts when cooled. Other gases act in the same way as air.

Liquids also expand when heated. If the flask shown in Figure 19-6 were filled with colored water and heated by putting it into a pan of hot water, the colored water in the flask would rise up the glass tube as it warmed up. Water expands when it is heated. How is the size of the glass flask affected by the increase in warmth? Do you think that the flask expands more or less than the water? The difference in the rate of expansion of glass and mercury accounts for the way glass thermometers work. Mercury expands and contracts much more than glass. Therefore, mercury moves up and down the glass tube of a thermometer as it expands and contracts in response to changes in temperature.

The effect of heat on the size of solids can be shown by the ball-and-ring expansion device shown in Figure 19-7. At room temperature, the metal ball will just fit through the metal ring. When the ball alone is heated, it expands and becomes too large to fit through the ring. If the ring is also heated, it too expands and the ball can again fit through the ring. You can see that heating the metal causes it to expand.

All substances do not expand by the same amount. Studies show that gases expand more than liquids and that liquids expand more than solids. In fact, for a given increase in temperature, gases expand more than 10 times as much as water and they expand approximately 100 times as much as steel.

Different liquids expand by varying amounts for a given temperature change. Each solid also has its own rate of expansion that is different from other solids. However, all gases expand and contract by the same amount for a given temperature change.

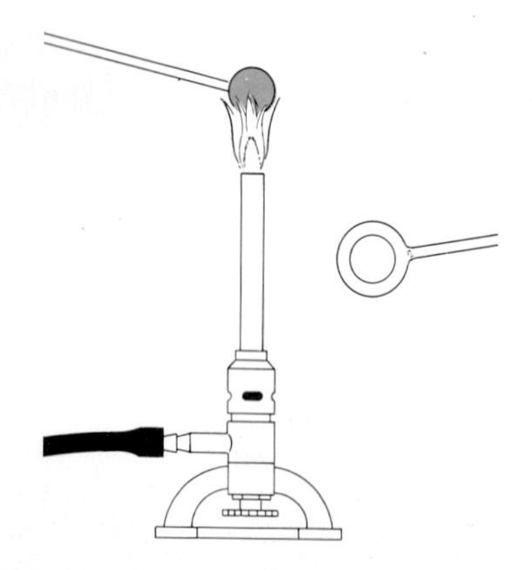

Figure 19-7 When the ball is heated, it no longer fits through the ring. Why? What happens if both the ball and ring are heated?

Substances expand at different rates. When the temperature of a liquid is raised 1 C°, the actual increase in volume of that liquid is a small fraction of the original volume. This fraction is called the **coefficient** (KOE-*uh*-FISH-*unt*) **of volume expansion.**

A 10 C° rise in temperature will produce ten times as much expansion as a rise of 1 C°. One hundred milliliters of a liquid will expand 100 times as much as 1 mL of a liquid.

Thus, if you know the coefficient of volume expansion of a substance, the original volume, and the change in temperature, you can find the change in volume.

The increase is noticeable for large amounts of liquids. For example, a delivery truck holding 6,000 L of gasoline at 32° C empties its tank at a service station. Many hours later, when the gasoline has cooled to 10° C, the volume shrinks by more than 80 L.

Although substances expand in all directions when heated, engineers are often concerned only with changes that occur in the length of solids. An increase in length is called linear expansion. The increase per unit length when a solid is warmed by 1 C° is called the **coefficient of linear expansion.** The rates of expansion for some common solids are shown in Table 19-3.

You can see from Table 19-3 that steel expands when heated. A long steel bridge can expand and contract by as much as 1 m between the hottest and coldest days of the year. How do engineers solve the problem caused by expansion and contraction? One way of allowing for expansion is to use finger-like joints at many points along the bridge roadway. The joints fit loosely into each other and slide back and forth as the bridge warms and cools (Figure 19-8). The next time you go over a long bridge, try to locate these joints.

TABLE 19-3

Substance	Coefficient of Linear Expansion (per Celsius degree)
Pyrex	0.000003
Glass	0.000009
Platinum	0.000009
Iron	0.000011
Steel	0.000013
Copper	0.000017
Brass	0.000019
Aluminum	0.000023
Lead	0.000029

Figure 19-8 Bridges undergo a great deal of expansion and shrinkage during the summer and winter months. How do special joints (shown cutting across the bridge in the foreground) help prevent changes in length from destroying the bridge?

SAMPLE PROBLEM

The steel center span of the Verrazano-Narrows Bridge, the world's longest suspension bridge, is 1,420 m long. If the temperature changes 50 C° from winter to summer, how much will this bridge expand?

SOLUTION

Step 1: Analyze

You are given that the bridge is 1,420 m long, and that the temperature change is 50 C°. Its linear expansion is the unknown.

Step 2: Plan

Since you know the length of the bridge and the temperature change, and since you know from Table 19-3 that the coefficient of linear expansion for steel is 0.000013 per C°, you can use the following equation to calculate the expansion:

$$\text{expansion} = \text{coefficient} \times \text{length} \times \text{temperature change}$$

Step 3: Compute and Check

Substitute all values in the equation.

$$\text{expansion} = 0.000013/\text{C}° \times 1{,}420\ \text{m} \times 50\ \text{C}°$$
$$= 0.9\ \text{m}$$

Since the expansion is in meters and represents a relatively small amount for a long bridge as expected, your solution is complete.

Have you ever cracked a cold drinking glass by pouring some hot water into it? The glass cracks because of uneven expansion of its inside and outside surfaces. Glass is a poor conductor of heat. As the inside surface is heated and expands, the outside surface is still cool and unchanged. The difference in expansion between the inner and outer surfaces causes the glass to crack. If the glass is thin, there is less chance that it will crack. Why?

Pyrex-type glass has become popular for laboratory use and for baking dishes in the home. This type of glass expands only one third as much as common glass, as indicated in Table 19-3. With its smaller rate of expansion, Pyrex resists the strains caused by sudden heating and cooling.

Devices called thermostats control the burning of fuel in many furnaces. These devices usually contain a small, two-layered bar made of two different metals. The two metals expand and contract at different rates when the temperature changes.

The bimetal bar shown in Figure 19-9 is made of a brass strip bonded to a strip of iron. When the bar is heated, the brass expands more than the iron and the bar bends.

In a thermostat, the strips touch a contact point if the temperature gets too cool. This contact switches on an electric current, starting the electric motor or opening the valve that controls the fuel supply to the furnace. The increased heat from the furnace causes the bar to bend away from the contact point, stopping the current. Thus, the temperature of your home stays within a few degrees of any thermostat setting.

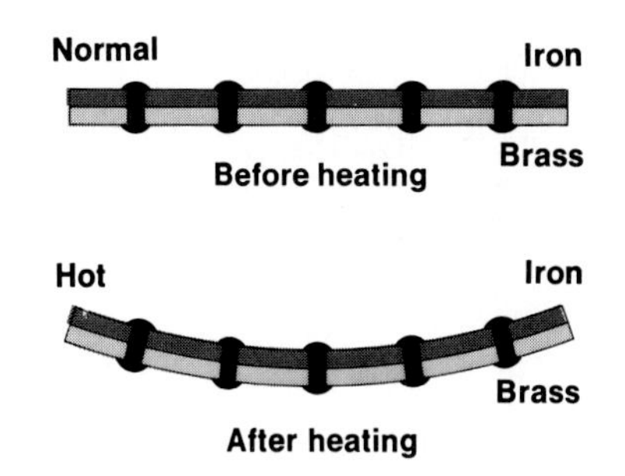

Figure 19-9 Does iron or brass expand more when heated? How can you tell?

Water expands when it freezes. Water, like almost any other substance, will shrink as it is cooled. However, unlike other substances, when water reaches a certain temperature it stops shrinking. Further cooling causes it to expand rather than contract. Water shrinks as it cools to 4° C, and then expands as it cools from 4° C to 0° C.

Because of this odd property, lakes freeze on top instead of on the bottom. As water in a lake cools to 4° C, it shrinks, becoming more dense. But when cooled below 4° C, the lake water becomes less dense and rises to the surface. When it freezes into ice, the water expands still more. The volume of ice is about 1.1 times that of the water from which it was formed. Thus, the ice is less dense than the water, and so ice floats. The ice floating on the surface serves as an insulator that reduces the rate at which the rest of the water freezes.

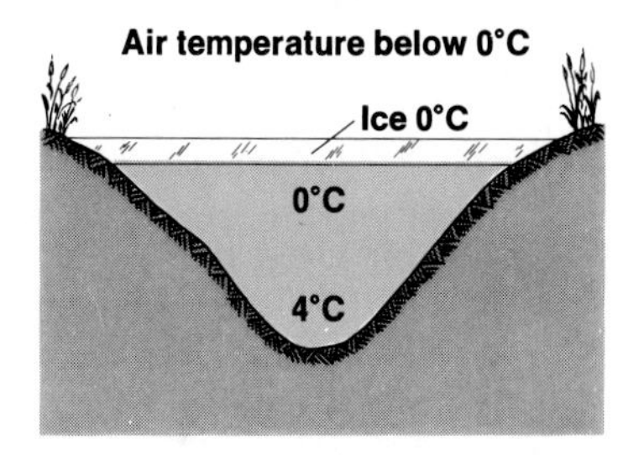

Figure 19-10 Why is it warmer at the bottom of the lake than at the top?

Notice in Figure 19-10 that water close to the surface of an ice-covered lake is at 0° C. Water at the bottom of the lake is at 4° C. Water at its greatest density at 4° C stays at the bottom. If you have ever fished through ice in winter, you know that fish are found mostly at the bottom of the lake where the water is slightly warmer.

If the coldest lake waters sank to the bottom, lakes would freeze from the bottom up. In the summer, warm water would stay near the top of the lake and not reach the ice. As a result, much of the ice would not melt. In time, all lakes and oceans would be completely frozen.

19.3 HEAT TRANSFER

Heat is exchanged during phase changes. Heat causes a solid to change to a liquid and a liquid to change to a gas. For example, water changes to steam when it is heated. Removal of heat from water causes a change of phase from a liquid to a solid.

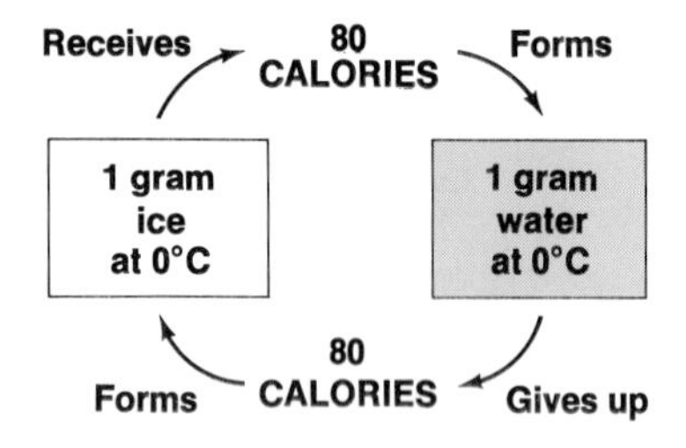

Figure 19-11 In this diagram showing heat of fusion, how much heat is needed to change 100 g of ice at 0° C to water at 0° C?

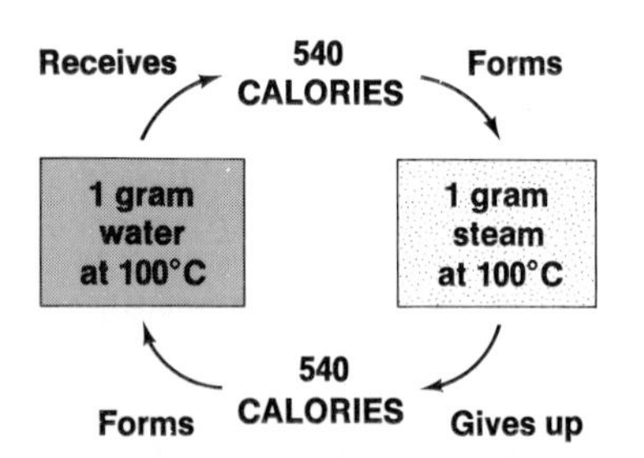

Figure 19-12 In this diagram showing heat of vaporization, what happens when water is changed to steam?

Heat is always gained or lost during a change of phase. Figure 19-11 shows such heat transfers. When ice melts, it gains a certain amount of heat as it changes from ice to water. When water freezes into ice, the same amount of heat that was gained is lost. Likewise, water gains a certain amount of heat when it changes to steam. It loses that same amount of heat when the steam condenses back to water.

It takes 80 cal of heat to change 1 g of ice at 0° C to water at 0° C. Note that the added heat does not change the temperature. The heat is used only for a phase change, to change the ice to water. The heat that is absorbed in the melting process is called the **heat of fusion.**

Ice has a high heat of fusion compared with other substances. Food in an ice chest stays cool because each gram of ice that melts absorbs 80 cal from its surroundings.

Water at 100° C absorbs 540 cal of heat for every gram that changes to steam. The heat needed to change a liquid into a gas is called the **heat of vaporization.** It takes 100 cal to raise 1 g of water from 0° C to 100° C. Figure 19-12 shows that to change this hot water into steam, more than five times as much heat is needed (540 cal/g). Water has a very high heat of vaporization compared to other liquids. For example, the value for ammonia is 341 cal/g, whereas for mercury the heat of vaporization is only 68 cal/g.

When steam condenses into hot water, the heat of vaporization is given off. How does this apply to steam radiators that are used to heat buildings?

Figure 19-13 shows how water changes in terms of temperature and heat as it changes from a solid to a liquid to a

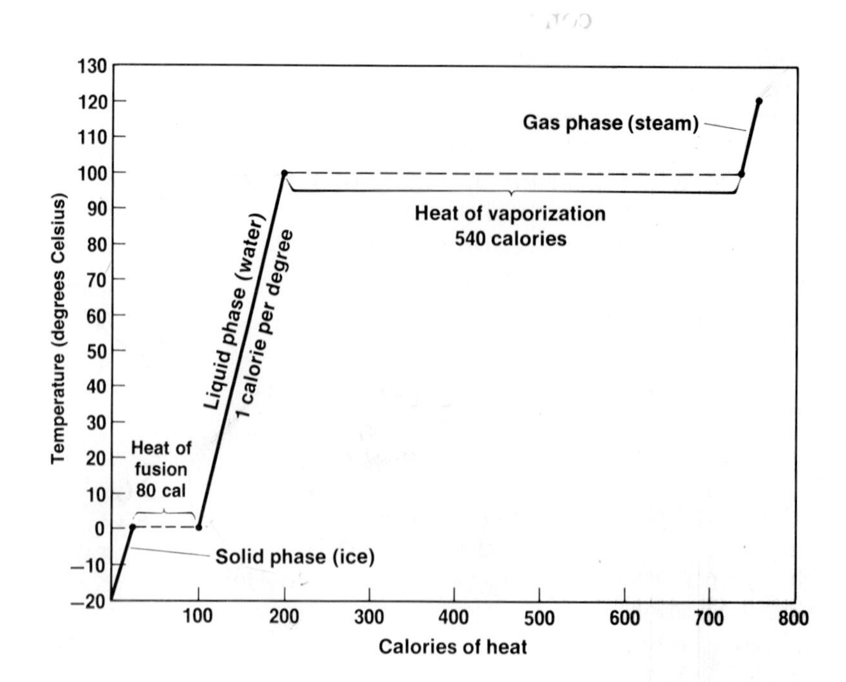

Figure 19-13 What causes the sharp changes in calories of heat at 0° C and again at 100° C?

gas. The dotted lines indicate that the temperature stays constant until enough heat has been added or lost to bring about the change in phase.

A *change of phase can have a cooling effect.* The heat of vaporization of a substance can also be used to produce a cooling effect. The cooling effect is similar to what happens to your hand if you wet it and then let it dry in a breeze. Your hand feels cool as the water dries. Heat is needed to evaporate the water. This heat is removed from your hand, leaving it feeling cool.

Now suppose a scientist places some liquid ammonia in a tube under high pressure. If the liquid ammonia passes through a small opening into a low-pressure space, it changes to a gas. As ammonia changes to a gas, it absorbs heat from its surroundings, causing a cooling effect. This is the process used by cold storage plants to make ice and to cool foods.

Study Figure 19-14 to see how ammonia is used in ice-making. Liquefied ammonia takes heat from the brine (salt water) tank as it becomes a gas. The ammonia gas then goes to a pump where a piston compresses it back into a liquid. This change from a gas to a liquid gives off heat. Water running over the coils absorbs this heat and carries it away. The liquid ammonia is then ready to pass through the needle valve and repeat the process.

Ammonia has been largely replaced in modern electric refrigerators, air-conditioning units, and freezers by another coolant called Freon. Freon is an easily liquefied gas. Air, instead of water, is often used to cool the coils. An electric motor drives the compressor pump, while a fan blows air over the coils.

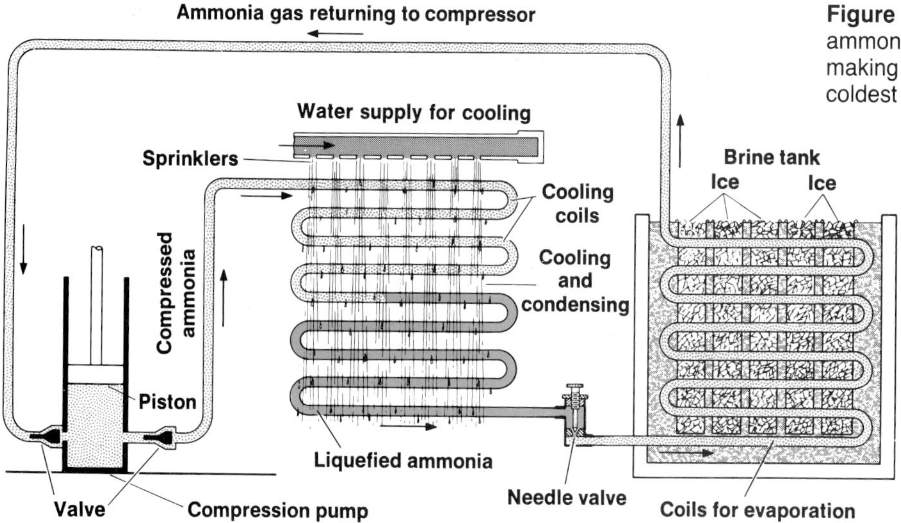

Figure 19-14 Trace the path of the ammonia through the pipes of an ice-making unit. Where are the hottest and coldest points in the system?

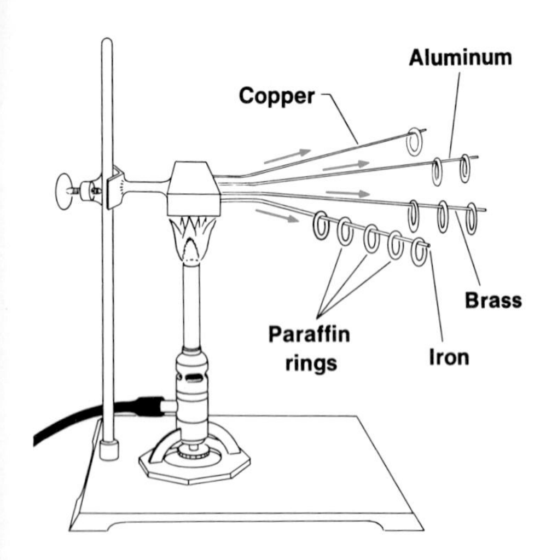

Figure 19-15 Rods made of different metals differ in their ability to conduct heat that will melt the paraffin rings. Which metal is the best conductor of heat? Which is the poorest?

Heat travels in three ways. When two places have the same temperature, there is no movement of heat between them. When there is a temperature difference, however, heat moves to equalize the difference. Heat can move from one place to another in three ways: by conduction, convection, or radiation.

Conduction is the transfer of heat through a substance from molecule to molecule as they collide. Silver conducts heat very well. If you stir a hot liquid with a silver spoon, the handle of the spoon becomes hot almost at once. Heat from the liquid is transferred to the metal. Inside the metal the heat moves from molecule to molecule until the entire spoon is hot. On the other hand, if you stir the hot liquid with a plastic spoon, the plastic handle stays cool because plastic is a poor conductor. Most metals conduct heat much better than nonmetals. The ability of several common metals to conduct heat is shown in Figure 19-15. In this demonstration, an equal number of paraffin rings are placed along several metal rods. As the rods are heated, the paraffin melts, and the rings drop off. As you can see, only one ring is left on the copper rod. How does the ability of iron to conduct heat compare with that of copper?

Solids are better conductors of heat than liquids, and liquids are better conductors than gases. This is true because the molecules of most solids are closer together than the molecules of liquids. The molecules of liquids, in turn, are much closer to each other than those of gases. The demonstration in Figure 19-16 shows how poorly liquid water conducts heat.

Substances that are poor conductors of heat are called **insulators.** Air that does not circulate is a good insulator. Cork, sawdust, and snow have air enclosed between their particles and thus are good insulators. Clothes keep you warm in winter mostly because of the air in the spaces between the fibers.

During the last few years, much progress has been made in home insulation. A number of substances incorporating many air spaces, as well as boards of soft texture, are now widely used as heat insulators. When installed in the outside walls of a building and above ceilings, insulating materials keep heat inside the house in winter and outside in summer. Good insulation saves fuel and increases comfort.

Convection is the transfer of heat by the movement of liquids and gases (fluids). Earlier in this chapter you learned that when a substance is heated, it expands. This expansion makes the substance less dense. Expansion causes gases and liquids to rise when heated. Conversely, cooler liquids and gases sink. Currents of liquids and gases that are set up by this unequal heating are called convection currents. Why is convection not likely to take place in solids?

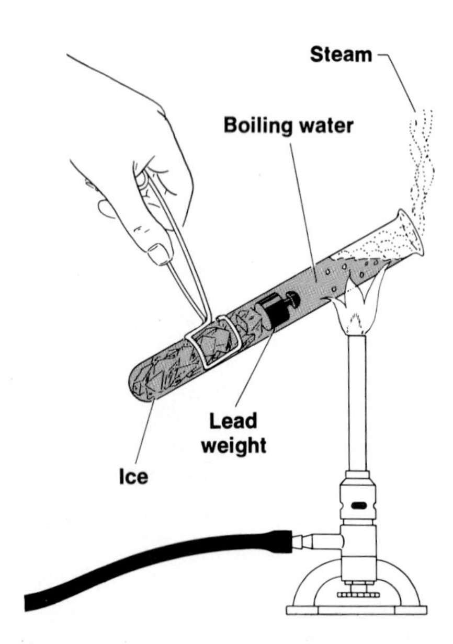

Figure 19-16 Water transmits heat very well by convection but poorly by conduction. Why hasn't the ice melted in this test tube? **CAUTION: Do not attempt this experiment without your teacher's permission and supervision. The test tube should not be pointed at anyone. Steam could explode from the tube if it is heated too rapidly, so be sure to wear goggles.**

Convection in a liquid can be shown by heating water in a beaker. Set the beaker so that just one side of it is over a burner. Drop a few crystals of dye or a few drops of food coloring on the surface of the water close to the side that is being heated. As the dye dissolves, you can see the movement of the colored water in the beaker. This movement, diagrammed in Figure 19-17, shows how convection currents are formed in a heated liquid.

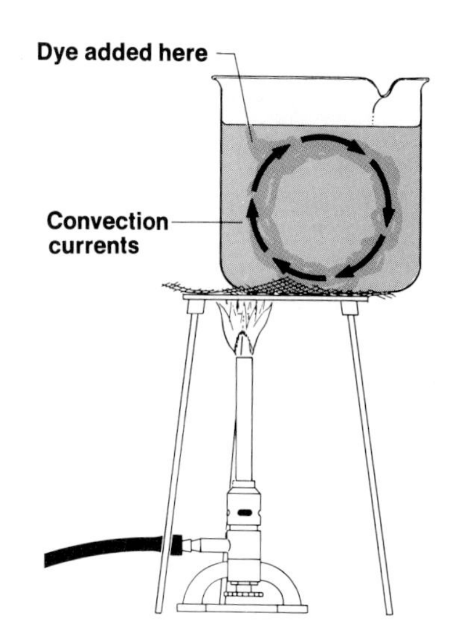

Figure 19-17 What would happen to the direction of the convection current if the source of heat were moved to the right side of the beaker?

A common example of a convection current is the rise of heated gases up a chimney. This rise creates a "draft" of fresh air through the firebox. Some furnaces and heaters also transfer heat by convection. Most modern furnaces use blowers, however, rather than depend on convection currents to heat a house.

In a hot-water heating system, water moves through the pipes by convection. Heat is given up to the cold room, and the cooler, denser water flows back to the furnace, where it is reheated. In some systems, a pump is used to force the water through the radiators.

In some homes, the heating and cooling systems are combined into a single unit. A heat pump is used to bring in heat from outside the house during the winter months and to remove heat from inside during the summer months. It may come as a surprise to learn that there is heat even in "cold" air. You saw an example of how a refrigerator is cooled in Figure 19-14. This same process also gives off heat. With a heat pump, this heat can be used instead of discarded.

Radiation is the transfer of energy in waves through space. If you hold your hand below a lighted electric bulb, you can feel the heat. Since your hand is below the lamp, it is obvious that the heat does not reach your hand by convection. Why not? Conduction will not explain the warmth either, since air is a poor conductor of heat. Your hand is warmed by radiation, a third method of energy transfer.

Radiant energy is best transferred in a vacuum. The earth receives heat from the sun by radiation. The heat travels through the vacuum of space. The other forms of heat transfer, conduction and convection, cannot take place in a vacuum.

Some homes are heated by radiation. In a radiant-heating system, steam or hot water is pumped through pipes in the floor, walls, or ceilings. These warmed surfaces radiate heat into the room.

Dark-colored, rough surfaces are good absorbers of radiant heat. Light-colored, shiny surfaces reflect more radiant heat than they absorb. You can see why light-colored clothes are cooler in hot weather than dark clothes. Thermos bottles have a mirror-like coating that reflects the radiant heat from hot liquids back into the bottle. This helps to keep the liquid hot.

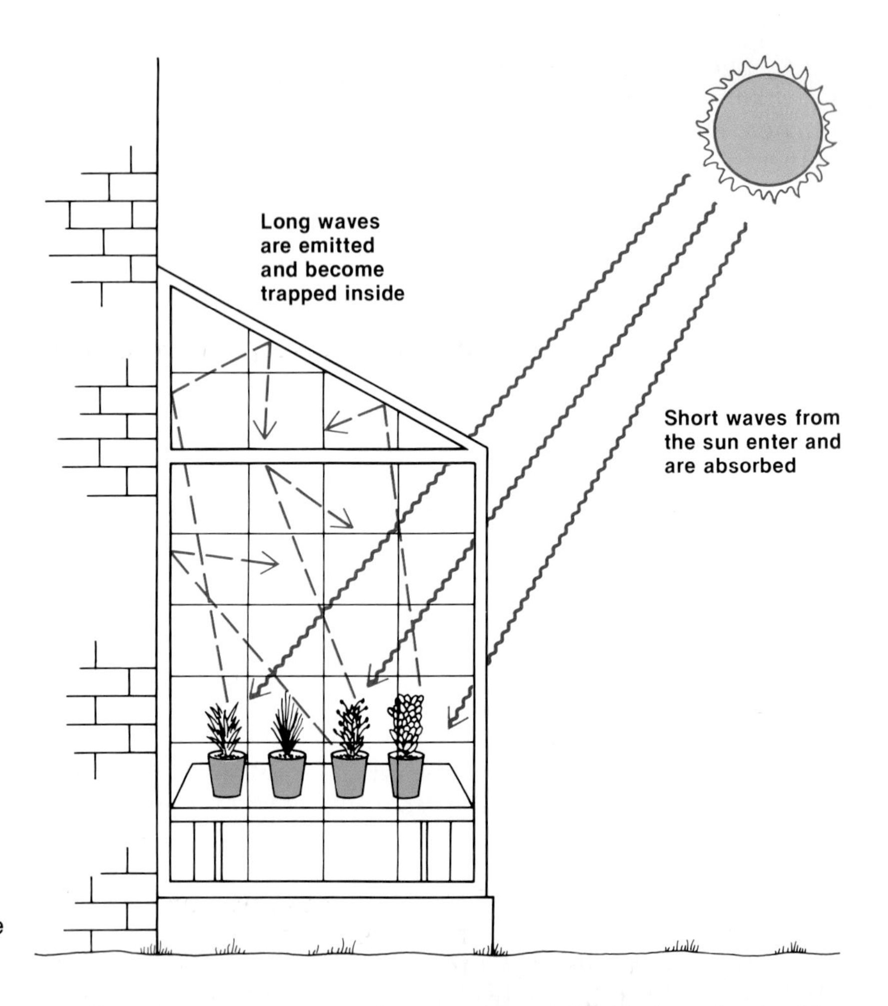

Figure 19-18 How does a greenhouse trap heat?

Have you ever opened a car door to find that the inside of the car was very hot? Where did the heat come from? Short-wave energy from the sun goes through the car windows, warming the surfaces inside. The long-wave heat reflected from the surfaces does not go through the glass and is trapped inside.

Greenhouses are built with large amounts of glass to admit and trap the sun's heat. The greenhouse is a model that resembles how the atmosphere helps keep the earth warm by trapping heat energy from the sun (Figure 19-18). Some of the short-wave radiant energy from the sun goes through the earth's atmosphere, warming the surface of the planet. The long-wave radiation from the surface slowly escapes through the atmosphere, preventing the earth from becoming as hot as its neighbor planet, Venus. Shrouded by a thick cloud cover that traps long-wave radiation, Venus has a surface temperature of 420° C (Figure 19-19). As you learned in Chapter 7, an increase in the amount of radiation trapped by the earth's atmosphere, known as the greenhouse effect, has occurred due to an increase in carbon dioxide.

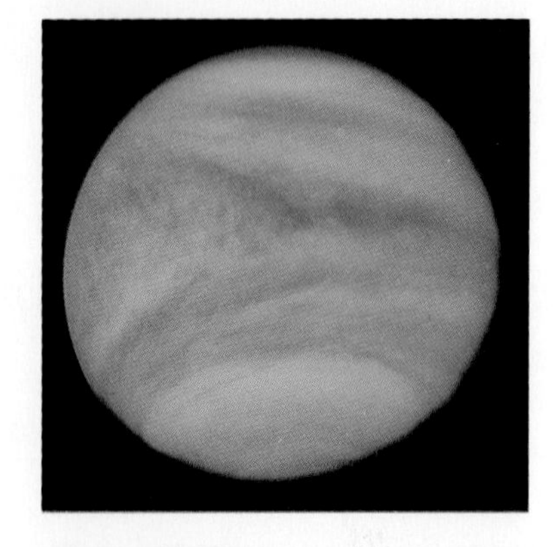

Figure 19-19 Cloud cover of Venus.

HEAT TRANSFER DURING THE MELTING OF ICE

activity

OBJECTIVE: Determine the best way to keep ice from melting.

PROCESS SKILLS

In this activity, you will *communicate* with the members of your team and class, and will *experiment* in order to determine the best method of preventing ice from melting.

MATERIALS

insulating materials
ice cubes
tin cans
graduated cylinder
various containers of glass or plastic

PROCEDURE

1. Your teacher will divide the class into three or more teams and will give each team a tin can with an ice cube in it and one of several kinds of insulating materials. With your team, find a way to keep your ice cube from melting, using only the insulating material you are given and any container you choose. Do not let the insulating material touch the melted ice.
2. At the end of a time period designated by your teacher, carefully pour the melted ice into a graduated cylinder for comparison with those of the other teams.
3. Compare the volume of water obtained by each team and discuss the methods used to preserve the ice cubes. Determine which method works best.

OBSERVATIONS AND CONCLUSIONS

1. Which method of keeping the ice cubes from melting proved most effective? Explain why.
2. How does this activity relate to keeping things cold in a refrigerator or hot in an oven?
3. How does keeping the ice cubes from melting relate to keeping your house cool in summer and warm in winter?

BIOGRAPHY

BENJAMIN THOMPSON (Count Rumford)

physicist and nobleman

Benjamin Thompson (1753–1814) was born in Massachusetts when it was one of the thirteen British Colonies. During the Revolutionary War he served as a spy for the British Government and later commanded a British regiment. He was forced to flee to England in 1776 where he was knighted by King George III. Thompson was also known as Count Rumford, a title awarded him for his work to improve the living conditions of the poor in Munich, Germany.

Thompson devoted himself to the study of physics and helped overturn the then current theory about heat. Scientists thought that heat was an invisible fluid called caloric, which flowed in and out of objects. Scientists believed that an object became warm when it took in caloric. When caloric left the object, the object became cold. Although caloric was supposed to have a fluid form, it was not considered to be matter, since objects weighed the same whether they were hot or cold.

Thompson's studies of heat began through his interest in gunpowder and weapons. While living in Germany, Thompson observed the boring of cannon barrels. In this process, several horses were hitched to a long boom. The boom, in turn, was attached by wheels and gears to metal cutters in a boring machine. As the horses plodded a circular path hour after hour, the cutters slowly bored a hole through the cannon barrel. The cutting action produced a large amount of heat.

According to the accepted theory, the heat "flowed" from the iron barrel as it was being cut. In other words, the metal itself contained the heat, which was not released until the metal was cut. However, Thompson noticed that heat was produced even when the cutters were too dull to cut the metal. He concluded that the heat was produced not by something called caloric, but by the friction between the cutters and the metal. His work helped to establish the theory that heat is energy produced by the motion of molecules.

CHAPTER REVIEW 19

SUMMARY

1. Temperature is the measure of the average kinetic energy of molecules.
2. Heat is the total kinetic energy that can be transferred from one object to another because of a difference in their temperatures.
3. Each substance has its own specific heat, or the amount of heat needed to warm 1 g of the substance by 1 C°.
4. Heat causes matter to expand in both length and volume.
5. When water is cooled from 4° C to 0° C, it expands.
6. Heat is absorbed during melting and given off during freezing.
7. Heat is absorbed during boiling and given off during condensation.
8. Heat can be transferred by conduction, convection, and radiation.

VOCABULARY

Match the item in the left column with the best answer in the right column. Do not write in this book.

1. absolute zero
2. calorie
3. Celsius scale
4. coefficient of linear expansion
5. coefficient of volume expansion
6. conduction
7. convection
8. heat
9. heat of fusion
10. heat of vaporization
11. insulator
12. Kelvin scale
13. radiation
14. specific heat
15. temperature

a. total kinetic energy that can be transferred from one object to another because of a difference in temperatures
b. transfer of heat energy through space
c. heat absorbed by a solid when it melts
d. substance that is a poor conductor of heat
e. heat absorbed by a liquid in changing to a gas
f. thermometer scale with freezing point of water at 0°
g. thermometer scale with freezing point of water at 273°
h. heat needed to warm 1 g of water 1 C°
i. increase per unit length when a solid is warmed 1 C°
j. transfer of heat by movement of fluids
k. measure of the average kinetic energy of molecules
l. transfer of heat from molecule to molecule as they collide
m. increase of volume when the temperature of a liquid is raised 1 C°
n. number of calories needed to warm up 1 g of a substance by 1 C°
o. point at which the molecules have a minimum amount of energy

REVIEW QUESTIONS

1. How is heat energy related to the action of molecules?
2. What is the difference between temperature and heat?
3. The specific heat of aluminum is 0.21 cal/g C°. What does this statement mean?
4. Does a thermometer in a kettle of hot water measure the amount of heat contained in the water? Explain.
5. What is meant by absolute zero?
6. Does water become hotter if it is boiled more vigorously? Explain.
7. Why is the specific heat of a substance an important property?
8. Name two ways in which the expansion of gases differs from the expansion of liquids and solids.
9. When a railroad is built during cold weather, why is a space left at the joints between the steel rail sections?
10. What provision is made in the building of bridge roadways to allow for expansion and contraction?
11. Why do solids expand when they become warmer?
12. As water is cooled from 8° C to 0° C, what happens to its density?
13. Why does a piece of ice float in water?
14. Why does a cold, thick drinking glass often break when boiling water is poured into it?
15. The air above the ice in a pond is −10° C. What is the likely temperature of (a) the upper surface of the ice, (b) the water just beneath the ice, and (c) the water at the bottom of the pond?
16. Why do oceanic islands have a fairly even climate?
17. Does the coefficient of linear expansion depend on the unit of length used? Explain.
18. Why does a car standing in the sun get hotter inside than outside?

CRITICAL THINKING

19. Describe how you could check the accuracy of a Celsius thermometer.
20. Which way will a bimetal bar of copper and steel bend if it is placed in a freezing mixture of water and ice? Explain.
21. A platinum wire can be tightly sealed into the end of a heat-softened glass rod. A copper wire, however, forms a loose seal with the glass. Using Table 19-3, suggest a reason why this is so.
22. Why does steam at 100° C cause more severe burns than hot water at 100° C?
23. Explain why a block of ice in a warm room takes a long time to melt.
24. Why would a hot drink in a metal cup be more likely to burn your mouth than one in a china cup?
25. Why is shiny aluminum foil sometimes used in the walls and ceilings of a building? Explain.

PROBLEMS

1. How much heat could be obtained by burning 6 kg of coal if a test sample shows that the coal releases 8,000 cal/g?
2. How much energy is needed to heat 1,050 g of lead from 25° C to 285° C if the specific heat of lead is 0.03 cal/g C°?
3. Which takes more energy to heat: 40 g of aluminum from 40° C to 120° C, or 60 g of zinc from 40° C to 140° C? How much more energy does this take in calories and in joules?
4. Use Table 19-1 to derive the energy values in kilocalories of the following quantities of food: (a) 200 g whole milk, (b) 1 medium egg, (c) 2 slices of white bread, and (d) 3/8 cup of peanuts.
5. A steel rod is exactly 39 cm long at 0° C. How much longer will it be at 25° C if the coefficient of linear expansion for steel is 0.000013/C°?
6. An oil tank holds 500 L at 30° C. What will be the volume of the oil when it cools to 10° C if the coefficient of volume expansion for oil is 0.001/C°?
7. The tallest building in the world is about 400 m high. How much taller is that building in the summer than in the winter? Assume a temperature difference of 50 C° and base the answer on the expansion of steel.
8. Find the number of calories of heat needed to change 40 g of water at 20° C to steam at 100° C.
9. Find the number of calories of heat needed to change 10 g of ice at 0° C to water at 100° C.
10. A chip of ice weighs 90 g. How many calories will it absorb when it melts? If the water freezes again, how many calories of heat will be released?

FURTHER READING

Adler, Irving. *Hot and Cold: The Story of Temperature from Absolute Zero to the Heat of the Sun*. New York: Thomas Y. Crowell, 1975.

Kavaler, Lucy. *A Matter of Degree: Heat, Life, and Death*. New York: Harper and Row Publishers, Inc. 1981.

Kentzer, Michael. *Cold*. (From the *Young Scientists Series*.) Morristown, N. J.: Silver Burdett, 1976.

Kentzer, Michael. *Heat*. (From the *Young Scientists Series*.) Morristown, N. J.: Silver Burdett, 1976.

Rahn, Joan C. *Keeping Warm, Keeping Cool*. New York: Atheneum, 1983.

Wade, H. *Heat*. Milwaukee, Wis.: Raintree Publishers, Ltd., 1979.

ENGINES

Engines provide us with a way to harness energy to do work. They are found on machines as simple as model airplanes and as complex as space shuttles. What form of energy does the shuttle's engine use for launching?

CHAPTER 20

SECTIONS

OBJECTIVES

- ☐ Describe how engines affect people's lives.
- ☐ Distinguish between rotary and reciprocal engines.
- ☐ Distinguish between internal and external combustion engines.
- ☐ Compare the four-stroke engine with the diesel engine.
- ☐ Describe how jet and rocket engines work.

20.1 POWER FOR A MODERN WORLD

Engines provide power for useful work. Think about how engines affect your life. You probably know that an engine provides the power for a car or a bus. But did you know that engines also help provide you with food and clothing? An **engine** is any machine that uses heat energy to do work. This heat energy is produced by burning a fuel.

How do engines help provide food? If you have ever planted a garden, you know that the soil has to be turned with a shovel and the lumps broken up with a hoe. Then the

Figure 20-1 Farmers use engines to help them grow and harvest crops. How is an engine being used in this photo?

soil has to be worked with a rake. As plants begin to grow, the garden must be kept free of weeds. If the garden is small, you can do this work with hand tools. However, many people find that the use of a power tiller makes these jobs much easier. With the help of engine power, the garden can be made much larger than if all the work were done by muscle power alone.

Farmers, even more than gardeners, need engines. Without large tractors and power equipment it would be difficult to plow and till the soil. Farmers use engines to keep crops weed-free and to add fertilizer. They also use engines to cut and harvest the crops, as shown in Figure 20-1. After it is harvested, much of the food requires some form of processing. For instance, raw wheat and rye have to be made into bread. Engines supply the power used to grind the grain into flour and to mix the dough. Engines also move the loaves of bread to stores for consumers to buy. Without engines, there would be less food for everyone.

Now think of the clothes you are wearing. It is unlikely that someone gathered the raw material and wove the cloth by hand. Engines were used to gather raw materials, to process them into usable cloth, and to cut and sew the cloth into garments. Finally, engines were used again to transport the clothes to stores.

How are engines used in your home? The water you drink is pumped using engine power. Engine power is also needed to make the electric current that is sent to most homes. What about the walls and framework of your home? The walls are most likely large sheets of dry wall nailed to a framework of lumber. Both the dry wall sheets and the lumber are made by using engine power. In what other ways do engines affect your life?

The Machine Age began with the steam engine. Hundreds of years ago, human and animal muscle power were used to do work. Later, wind and water power also had some limited uses. But it was not until the steam engine was invented that chemical energy from burning fuel was changed to mechanical energy to move the parts of large machines.

James Watt, an eighteenth century Scottish engineer, is usually given credit for the invention of the steam engine. Although other steam engines had already been built, Watt, in the late 1700s, improved these early models and made them more practical. It was his steam engine that marked the beginning of the Machine Age. And it was for this same James Watt that the metric unit of power, the watt, was named (see Chapter 18).

Have you seen steam push up the cover on a tea kettle? The steam engine works on the same principle. Instead of pushing up against a cover, the steam pushes against a sliding device, called a **piston**, inside a cylinder. Notice that in Figure 20-2 a sliding valve causes the steam to push the piston, first one way and then the other. The piston is attached by a pivoting rod to a rotating crankshaft. In this way, back-and-forth movement is changed to rotary motion. The back-and-forth movement of the piston is characteristic of a **reciprocating** (*rih*-SIP-*ruh*-KAYT-*ing*) **engine.** The movement is jerky and causes vibrations. Locomotives use reciprocating engines.

In contrast, **steam turbines** are engines containing a set of rotating blades that produce a steady, constant motion. Steam turbines work smoothly, with almost no vibration. Turbines waste less energy than reciprocating engines, so they are more efficient. Turbines are also more compact. A turbine takes up less space than a reciprocating engine that produces the same amount of power. Large electric power plants and large ships use turbines rather than reciprocating engines.

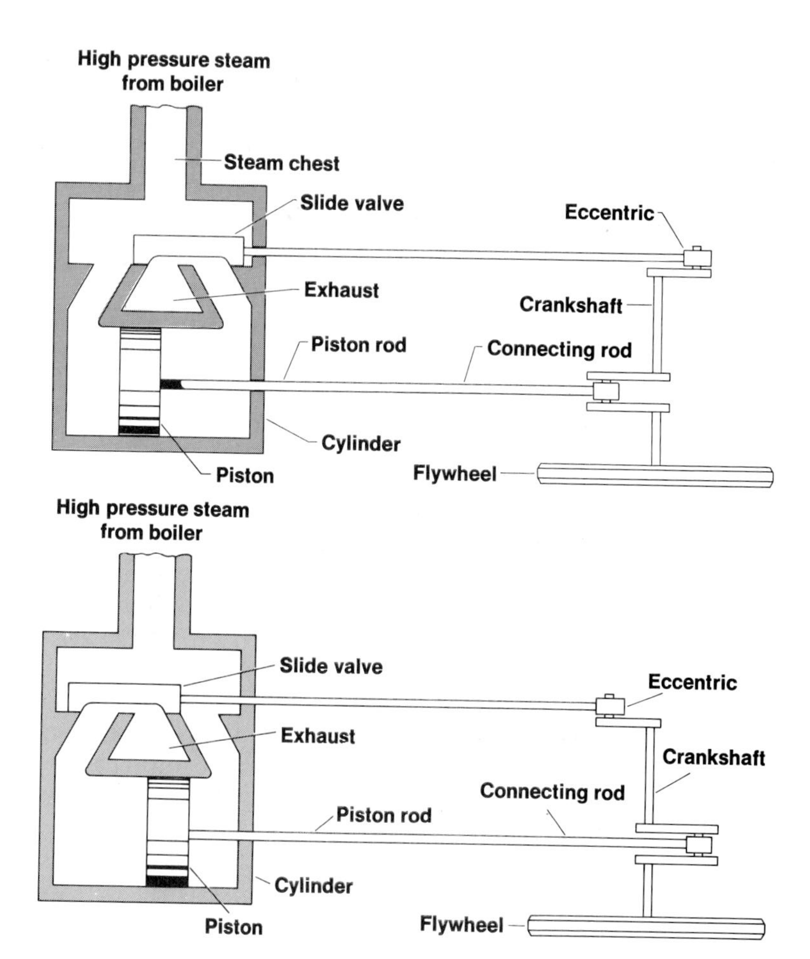

Figure 20-2 Views of the inside of Watt's simple slide-valve steam engine (A) after the piston has moved to the left and (B) after the piston has moved to the right. How does the slide valve determine which way the piston moves?

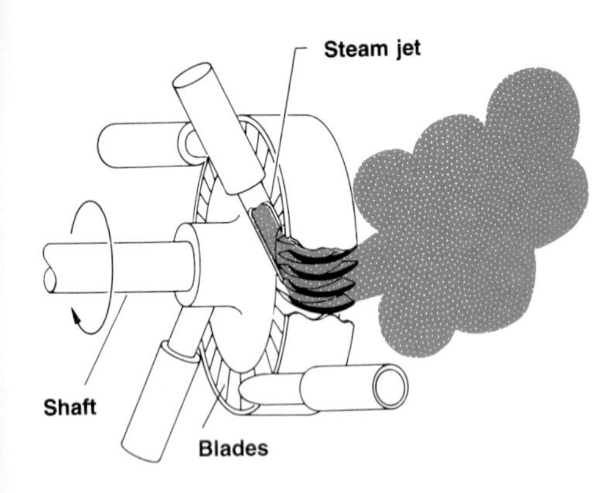

Figure 20-3 How is the steam turbine different from the reciprocating engine? Which one has fewer parts?

The steam turbine works like a high-speed windmill. As you can see in Figure 20-3, steam jets direct the steam at an angle against the blades of the turbine. The steam moves the blades much as air moves a pinwheel. Many electric generators are run by steam turbines that provide a power output great enough to take care of the needs of a city of 50,000 people. Yet, such a turbine could fit into an average classroom. The steam for modern turbines may have a pressure as high as 1,300 N/cm^2 and be as hot as 600° C.

Steam turbines work well only when the blades turn at a high speed. In making electric power, the turbine shaft is joined directly to the shaft of a high-speed electric generator. In a ship, however, where the propeller speed is much slower than the turbine speed, the turbine is attached to the propeller by gears that reduce the speed. On some ships the turbine runs an electric generator. Electricity from the generator powers electric motors used to drive the ship.

20.2 POWER FOR CARS

Gasoline engines and diesel engines are internal combustion engines. Steam engines and steam turbines are classified as external combustion engines. In an **external combustion engine,** the fuel burns outside the engine itself. The heat is transferred by steam to a cylinder or turbine in the engine. Gasoline and diesel engines are two examples of **internal combustion engines,** in which the fuel burns inside the cylinders of the engine.

The most common gasoline engine is the **four-stroke engine,** in which the piston makes four movements during one full cycle. Engines of this type are used in most cars.

Figure 20-4 shows the steps in the operation of the pistons in a four-stroke engine. The first is the intake stroke. The piston slides down during the intake stroke, lowering the pressure in the upper part of the cylinder. A mixture of air and gasoline vapor is drawn into the cylinder through the intake valve. The exhaust valve is closed at this time. The mixture of air and gasoline vapor that enters the cylinder comes from the carburetor, a device that mixes air and gasoline in the correct amounts. In some engines, a fuel-injection system is used instead of a carburetor to force a measured amount of fuel into the cylinder at the proper time.

The compression stroke, during which the piston travels upward, compresses the air-gasoline mixture and heats it up. This stroke follows the intake stroke. At this stage, both the intake and the exhaust valves are closed.

Just as the piston reaches the top of the stroke, an electric spark jumps across the gap of the spark plug to

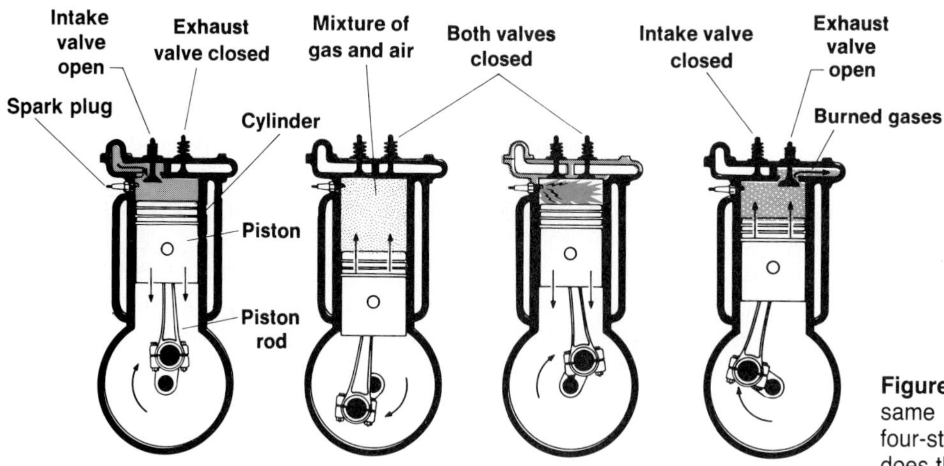

Figure 20-4 The diagram shows the same cylinder during the stages of a four-stroke cycle. How many times does the piston move down during one cycle?

explode the air-gasoline mixture. In the power stroke, the hot gases force the piston down in the cylinder. The power stroke is the only stage during which the burning gasoline does work by moving the piston.

During the exhaust stroke, the piston travels upward again, pushing the burned gases out through the open exhaust valve. This step is the fourth in the cycle and is followed by another intake stroke.

Since power is provided during only one of the four strokes, there must be some way of keeping the piston moving during the other three strokes. This is done in several ways. Notice in Figure 20-5 that the pistons are attached to a

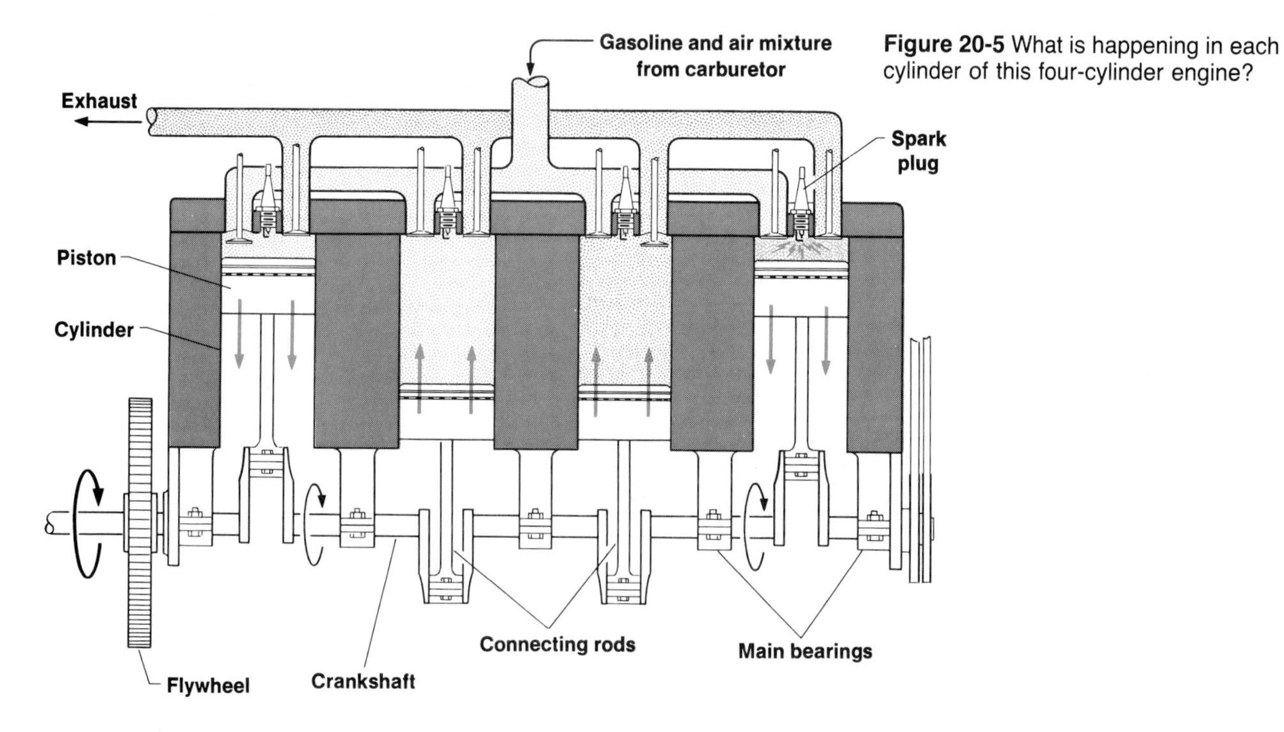

Figure 20-5 What is happening in each cylinder of this four-cylinder engine?

crankshaft. A heavy flywheel is connected to the crankshaft of the engine. The flywheel's inertia keeps the pistons moving during the other strokes. Another way to keep the pistons moving is to connect several cylinders to the same crankshaft. While the intake stroke takes place in one cylinder, the exhaust stroke is taking place in another, and the crankshaft is kept turning by the power stroke in still another cylinder.

If car engines had only one cylinder, the car would not ride smoothly. The more cylinders an engine has, the smoother it runs. Today, most cars have four or six cylinders to provide fuel economy.

Gasoline engines have positive and negative features. One very useful feature of a gasoline engine is that it can start or stop on a moment's notice. In contrast, a steam engine cannot be started until the steam pressure has built up. Also, gasoline engines develop a great deal of power with a fairly small engine.

Gasoline engines are moderately efficient. The best gasoline engines change only about 20 percent to 25 percent of the energy of their fuel into useful work. Friction, the heat of the exhaust gases, and running the cooling system, use up the rest of the energy. Early steam engines used only five percent to eight percent of their fuel's energy. Large high-speed steam turbines used in modern power plants are about 30 percent to 40 percent efficient.

Gasoline engines have some negative features. For instance, four-stroke engines can run in only one direction. They cannot be made to run backward. This problem is solved by having a reverse gear. Reciprocating steam engines can be run either forward or backward simply by shifting a sliding valve.

Diesel engines are more efficient than gasoline engines. There are three main differences between diesel engines and gasoline engines. Diesel engines burn a kerosene-like fuel, use no spark plugs, and have much higher compression ratios.

In the diesel engine, air enters the cylinder during the intake stroke. The rising piston then squeezes the air to about one eighteenth of its former volume. When it is so highly compressed, the air becomes hot enough to ignite the fuel. At the top of the compression stroke, an injector sprays fuel into the cylinder. Because the air is so hot, the fuel explodes at once. The hot gases push the piston down in the power stroke. This stroke is followed by the exhaust stroke.

The **compression ratio** of an engine is the ratio of the volume within the cylinder when the piston is at the bottom of its stroke to the volume within the cylinder when the piston is at the top of its stroke. Notice in Figure 20-6 that the compression ratio of a diesel engine may be as great as 18 to 1. In contrast, the compression ratio of the gasoline engine of a car is commonly about 8 to 1. A compression ratio of 8 to 1 is too low to ignite fuel without a spark plug.

In recent years, car engines have been built with lower compression ratios to reduce air pollution. However, the lower ratios are not without fault. Lower ratios use more fuel, which speeds up the reduction of the world's reserves of fossil fuel.

Diesel fuel is much like kerosene. It provides a little more energy per unit of volume and is slightly cheaper than gasoline. Also, diesel engines are more efficient than gasoline engines, largely because of their higher compression ratios. However, diesel engines are hard to start when cold, tend to be noisy, and need heavy engine castings to withstand the high pressures formed inside the engine. These problems are costly to overcome. As a result, diesel engines are used in large units like tractors or trucks, where the cost of the engine is only a small part of the total cost.

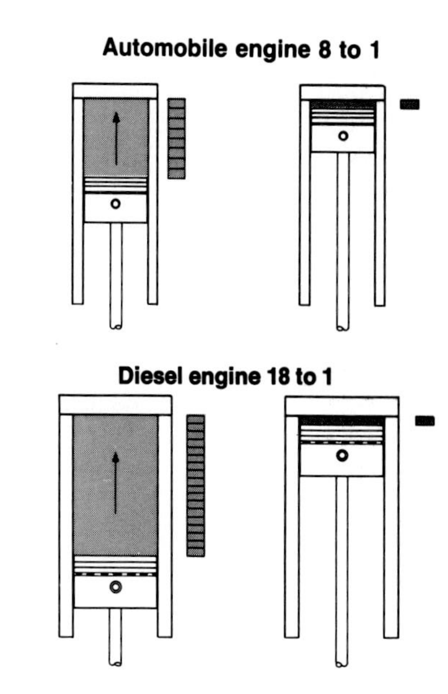

Figure 20-6 The cylinders on the left show the pistons at the bottom of their strokes and the cylinders on the right show the pistons at the top of their strokes. A diesel engine has a compression ratio of 18 to 1. What are some advantages of the diesel's high compression ratio?

Power is sent from the engine to the car's wheels. Remember that in Figure 20-5 each piston is attached to the crankshaft. Thus, if the piston moves, the crankshaft must also move. However, there is an important change in the motion. The pistons move up and down, but the crankshaft turns in a smooth rotary motion. As you can see in Figure 20-7, this change in motion is similar to the change in motion that takes place when you ride your bike. Your knees and upper legs move up and down, but the pedals go in a

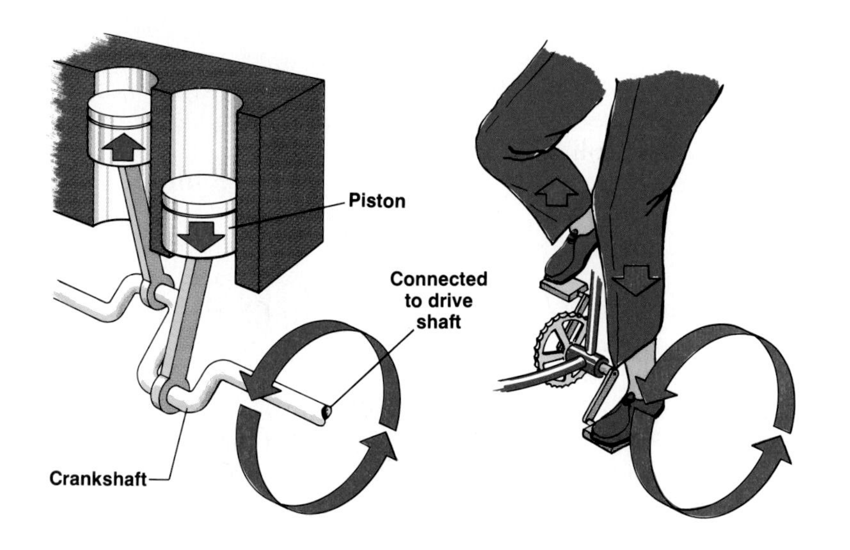

Figure 20-7 The up-and-down motion of the pistons is changed to circular motion in the crankshaft. How does the same change in motion occur when you ride a bicycle?

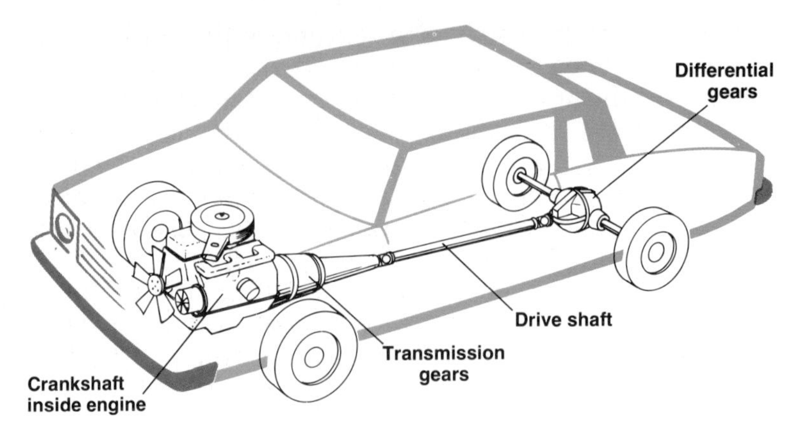

Figure 20-8 In a rear-wheel drive car, the power from the engine goes to the drive wheels through the transmission gears, the drive shaft, and the differential gears.

circle. In the engine, the pistons act like your upper legs, the connecting rods like your lower legs, and the crankshaft like the pedals.

The cylinders are timed so that first one, then another, then a third fires, and so on. A steady series of power strokes is sent to the crankshaft, which is connected to the transmission. The **transmission** is made up of gears that change the speed and direction of the car. The drive shaft transmits power from the engine to the wheels.

Figure 20-8 shows a car with the engine in the front and the drive wheels, or those wheels directly powered by the engine, in the back. Many cars have front-wheel drive (powered wheels in front), thus doing away with the long drive shaft down the middle of the car.

The transmission permits the driver to shift gears to get the best gear ratio, or mechanical advantage. In a car with manual transmission, the driver must depress a clutch to shift gears. With an automatic transmission, the driver does not need to shift gears; shifting is done automatically.

The drive shaft is attached to the rear axles through a set of gears known as a **differential** (DIF-*uh*-REN-*chul*) (Figure 20-8). The gears in the differential permit the rear wheels to turn at different speeds. Without the differential, a car could not make a turn without slipping. This is due to the fact that when a car turns, the outside wheels must turn faster than the inner ones.

Figure 20-9 What type of engine is used to power modern military and commercial aircraft?

20.3 ENGINES FOR AIR AND SPACE FLIGHT

Large aircraft use jet engines. Although gasoline engines are used in small aircraft, almost all large aircraft are powered by jet engines. Military and commercial aircraft, like the one shown in Figure 20-9, use jet engines.

Jet engines work on the action-reaction principle defined in Newton's third law of motion. The force that blows the high-speed gases out of the rear of the engine represents the action force. The force directed back on the engine by the escaping gas is the reaction force. This reaction force is the thrust that drives the plane forward. The jet plane uses a large amount of fuel, but it produces a thrust that can push a plane faster than the speed of sound.

The **ramjet** shown in Figure 20-10A is one type of jet engine used only after another engine has given the aircraft a high speed. Rocket engines, which are discussed in the next section, are usually necessary to boost the jet to a speed sufficient to compress the air for proper ramjet operation. The ramjet is the simplest of jet engines. It has no moving parts and can deliver great amounts of power at high efficiency when moving at high speed.

Another type of jet engine is the **turbojet,** which has air-compressor blades mounted on the front end of a long, central shaft, and turbine blades in the rear (Figure 20-10B). The purpose of the compressor is to pull in air and force

Figure 20-10 In what way are the ramjet, turbojet, and bypass jet engines alike?

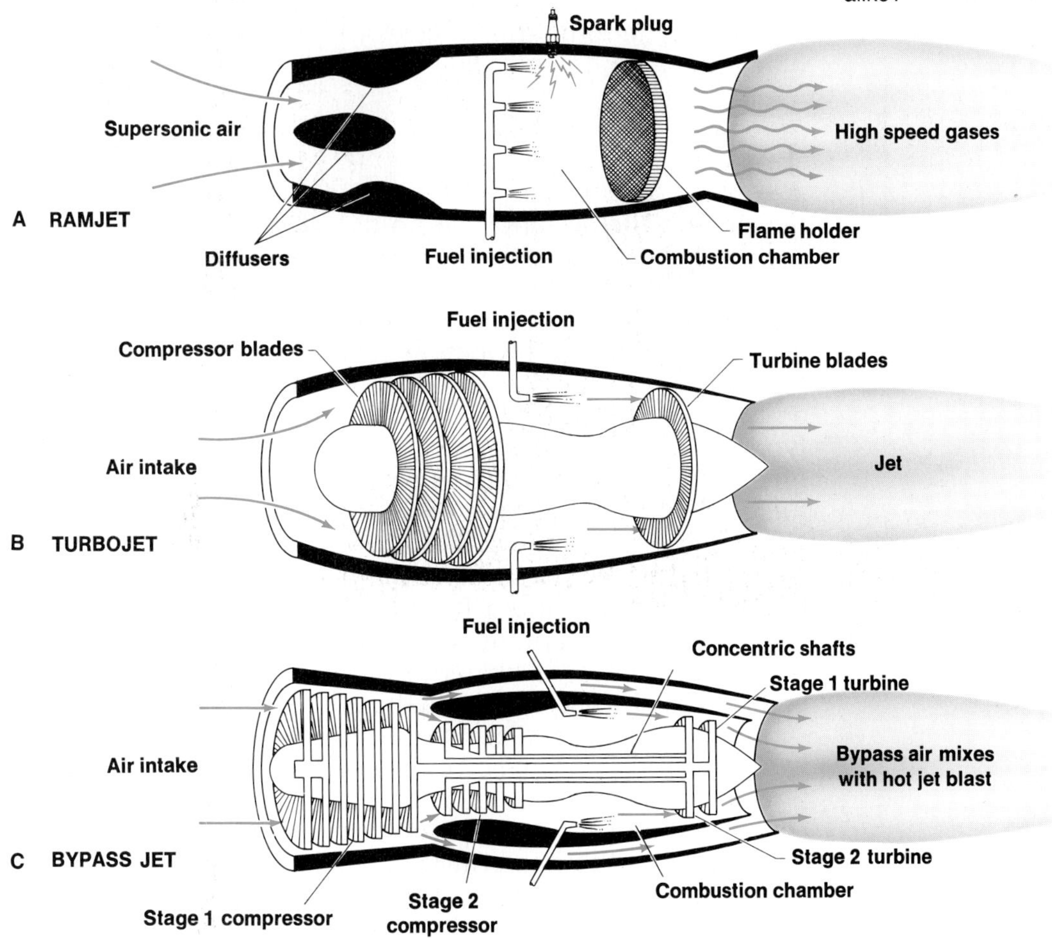

it into the combustion chamber. Here a fuel, somewhat like kerosene, is sprayed in and ignited.

The hot exhaust gases striking the turbine blades make the turbine turn at high speed. Then the hot gases rush out of the jet at the rear of the engine. The escaping gases give the engine its forward thrust. The spinning turbine powers the compressor at the front.

Another type of jet is the bypass jet, or fanjet. In a **bypass jet,** a large stream of cool air bypasses the combustion chamber and goes directly into the exhaust gases. See Figure 20-10C.

The noise of jet engines comes mostly from the high-speed exhaust gases. In the bypass jet, the noise is reduced by the bypass exhaust layer, which acts as a buffer between the exhaust gases and ordinary air. This jet engine is more efficient at low speeds than other types of jet engines.

Some jet engines, mostly on high-speed military aircraft, also have afterburners. The afterburner is a small, modified ramjet that is attached to the tail pipe of a turbojet. Recall that the ramjet works only when high-speed gases enter the front of its combustion chamber. When a ramjet is attached to the tail pipe of a turbojet, the exhaust gases of the turbojet supply the high-speed gases that make the ramjet work efficiently.

Rockets are space engines. Rocket engines, like jets, operate on the action-reaction principle. They differ, however, in that the jet takes oxygen from the air to burn its fuel, and the rocket does not. The rocket carries an **oxidizer,** which combines with the fuel to supply energy that propels the rocket. The oxidizer may or may not contain oxygen. With its own oxidizer on board to burn fuel, the rocket engine can operate outside the earth's atmosphere. Of all types of engines, rockets produce the greatest thrust for a given amount of weight. In rockets that use solid fuel, the fuel and oxidizer are formed to act like a combustion chamber. One of the main assets of solid fuel is that it can be stored in the rocket. Thus, the rocket is always ready for a quick launching. Liquid-fueled rockets, on the other hand, must have the fuel pumped in just before launch. However, solid fuel cannot be turned off after ignition in case of an emergency, but the liquid fuel can.

Figure 20-11 Spacecraft have no use for wings and other aerodynamic control surfaces. What is the purpose of these surfaces on the space shuttle? Why is the shuttle different from previous spacecraft?

In 1981 the first space shuttle was launched. The shuttle is similar to other space vehicles except that the main section of the shuttle can be recovered and used again. Most rockets cannot be reused. You can see in Figure 20-11 that the space shuttle has wings and tail surfaces designed to permit it to maneuver in the air. These control surfaces allow a pilot to land the shuttle like an airplane.

MECHANICS AND REPAIR

servicing engines and equipment

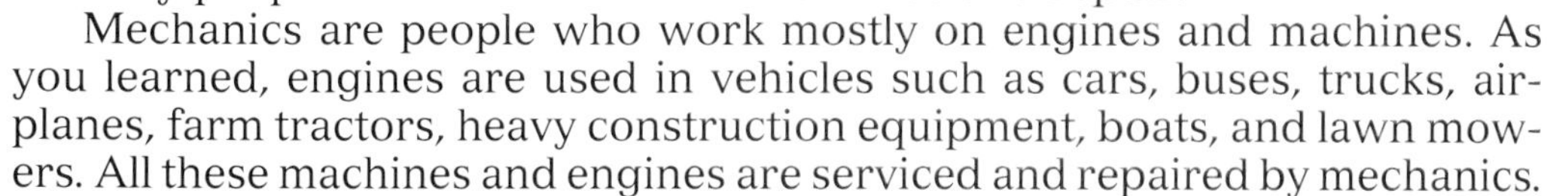

Where does your family take your car for a tune-up? Whom do you call when your furnace or refrigerator does not work? These jobs and many others are done by people who are skilled in mechanics and repair.

Mechanics are people who work mostly on engines and machines. As you learned, engines are used in vehicles such as cars, buses, trucks, airplanes, farm tractors, heavy construction equipment, boats, and lawn mowers. All these machines and engines are serviced and repaired by mechanics.

Some mechanics do light-duty repair such as tune-ups and adjustments. Others do complete engine overhauls. There are so many cars and other vehicles with engines that there are many jobs available for people who are qualified to do the work.

The repair field encompasses a wide range of additional jobs, including the repair of telephones, bicycles, industrial machinery, television sets, vending machines, locks, watches, and cameras.

Many jobs in mechanics and repair exist in the heating and cooling industries. In Chapter 19, you learned some of the principles of heat and refrigeration. These principles are put to good use by people who install, service, and repair furnaces, freezers, and air conditioners.

How does someone become a mechanic? Some high schools offer courses in car repair. Many industries have apprenticeship programs that give on-the-job training. These programs provide the technical skills necessary for repairing the products developed by a particular company.

Most apprenticeship programs are open to high school graduates who have good grades. Many of the industries voluntarily register their programs with the Bureau of Apprenticeship and Training of the U.S. Department of Labor. The bureau sees to it that the apprentices in these programs receive a fair wage and thorough on-the-job training.

For more information on a career in mechanics and repair, contact

U.S. Department of Labor
Employee and Training Administration
Bureau of Apprenticeship and Training
Frances Perkins Building
200 Constitution Ave. NW
Washington, DC 20210

CHAPTER REVIEW

SUMMARY

1. Engines do work that helps provide transportation, food, and clothing.
2. The steam engine marked the beginning of the Machine Age.
3. Steam engines and steam turbines are external combustion engines.
4. Diesel and gasoline engines are internal combustion engines.
5. The four-stroke engine is a typical gasoline engine.
6. Diesel engines are more efficient than gasoline engines and have a higher compression ratio.
7. Jet engines, which are based on the action-reaction principle, are used in aircraft.
8. Rocket engines can be used in space because they carry their own oxidizer and need no oxygen from air.

VOCABULARY

Match the item in the left column with the best answer in the right column. Do not write in this book.

1. bypass jet
2. compression ratio
3. differential
4. engine
5. external combustion engine
6. four-stroke engine
7. internal combustion engine
8. piston
9. ramjet
10. reciprocating engine
11. steam turbine
12. transmission
13. turbojet

a. engine in which the fuel burns inside the cylinders of the engine
b. most common gasoline engine
c. jet engine that is used only after another engine has given the aircraft a high speed
d. jet engine in which the noise is reduced by a layer of bypass exhaust gases
e. sliding device inside a cylinder
f. engine in which the piston moves back and forth
g. gears that change the speed and direction of a car
h. any machine that uses heat energy to do work
i. engine in which the fuel burns outside the engine itself
j. set of gears that permits the rear wheels to turn at different speeds
k. comparison of the volumes in the cylinder when the piston is at the bottom and top of its stroke
l. engine containing rotating blades
m. jet with air-compressor blades in front and turbine blades in the back of a central shaft

REVIEW QUESTIONS

1. What is an engine?
2. How does a steam turbine work?
3. How does the efficiency of a steam turbine compare with that of a reciprocating steam engine?
4. What is the difference between an internal and an external combustion engine?
5. Describe the operation of the four-stroke car engine.
6. List two advantages that a gasoline engine has over a steam engine.
7. Explain how a diesel engine operates.
8. State three differences between diesel and gasoline engines.
9. Explain the function of each of the following parts of the automobile: differential, transmission, and drive shaft.
10. What is the function of the rocket-fuel oxidizer?
11. Why is the ramjet rarely used even though it is the simplest and most efficient of all jet engines?
12. In what major way are jets and rocket engines alike? How are they different?

CRITICAL THINKING

13. How do engines help reduce the cost of food, clothing, and shelter?
14. Why are steam engines not used to power airplanes?
15. Why do economy cars have fewer cylinders than high-performance cars?
16. Sometimes cars need a tune-up because the timing of the cylinders is off and they do not fire at exactly the right time. How would this affect the car's gas mileage? Explain.
17. Why must an air filter in a car be changed regularly?
18. Explain why an engine does not create energy.

FURTHER READING

Keen, Martin. *How It Works*. New York: Grosset & Dunlap, 1976.

McCosh, D. "New Generation Turbos." *Popular Science* 232 (May 1988): 68–71.

McHugh, Mary. *Careers in Engineering and Engineering Technology*. New York: Franklin Watts, Inc., 1978.

Ross, F. *The Space Shuttle, Its Story and How to Make A Flying Paper Model*. New York: Lothrop, Lee and Shepard Books, 1979.

Webb, Robert. *James Watt: Inventor of the Steam Engine*. New York: Franklin Watts, Inc., 1970.

Weiss, Harvey. *How to Be An Inventor*. New York: Thomas Y. Crowell, 1980.

How the Sciences Work Together

The Process in Physical Science: Friction

For hundreds of years, inventors, and tinkerers, have tried to make perpetual motion machines, machines that would run themselves forever. Thousands of designs were developed. None worked. One reason is friction.

An overbalanced wheel has been proposed as a perpetual motion machine.

We now know that friction robs moving bodies of energy, bringing them to eventual standstill. Without friction, a pendulum would never stop swinging. Cars could go forever on a drop of fuel. A child standing on a San Francisco pier could send an ocean liner to Japan with a gentle push.

But friction not only impedes motion, it helps us control it. Without friction between shoes and pavement, our feet would go out from under us, as if on ice. Brakes would neither slow nor stop a car. All furniture would have to be bolted down or the slightest nudge would send it sliding across the room. Much of our everyday life hinges on either taking advantage of friction or trying to overcome it.

Ice on the ground reduces friction between shoes and pavement making it difficult to walk.

The Connection to Physics

There are several categories of friction: *sliding* friction, *rolling* friction, and *fluid* friction (the friction within moving fluids, called *viscosity*, or between fluids and solids).

Friction converts mechanical energy into heat energy. Wherever two surfaces rub against each other, heat is produced. Heat is often a problem for machinery. For instance, overheated pistons inside a car engine can swell so much they stick inside the cylinders, damaging the engine beyond repair. To prevent this, we lubricate engines with oil.

Adding oil to a car engine lessens the harsh effects of sliding friction with fluid friction.

The harsh effects of sliding friction can be lessened by substituting fluid friction, or lubrication. Another method is to substitute rolling friction. A prime example is the wheel. It is much easier to move a bed on rolling casters than one with fixed legs—no matter how smooth the tips.

The Connection to Chemistry

Until about 1940, the main cause of friction was thought to be surface roughness. Modern theory now attributes friction mainly to *adhesion* between surface atoms. This explains why very smooth surfaces, such as cleaved mica (which is smooth to within one atomic diameter), give friction at least as great as that of ordinary surfaces.

According to modern theory, lubricants work because they interpose a layer of lesser adhesion—or bonding—between two surfaces. Liquids and soft solids form relatively weak bonds; consequently they are effective lubricants.

The Connection to Biology

Friction can be a problem inside nature's machines, too. Like the moving parts of an engine, our bones can rub together. The body combats this by lubricating the joints with fluids and an elastic tissue called cartilage. When this protection breaks down, due to diseases such as arthritis, the friction can cause pain and sometimes cripple its victims.

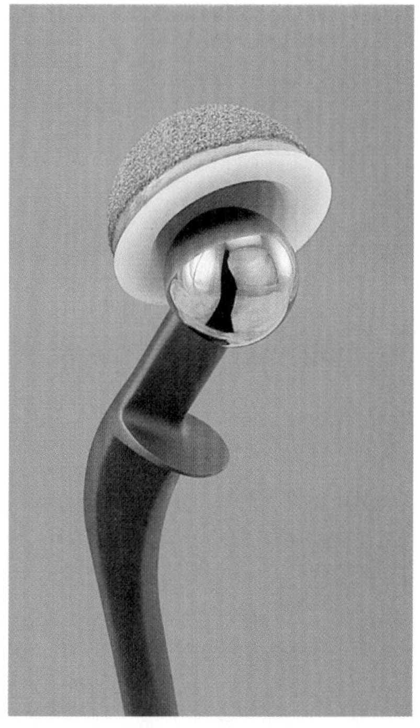

Artificial joints such as this hip joint use special metal alloys and high density polyethylene plastic to overcome friction.

Artificial joints are one way of regaining the advantage over friction once the natural joint has been damaged. Artificial joints use a chromium-cobalt-molybdenum steel on one side and a high density polyethylene plastic on the other. Each is polished to a mirror finish. With the help of the body's natural fluids as lubrication, these low-friction artificial joints have restored pain-free movement to many sufferers of severe joint disease.

WAVE MOTION AND ENERGY

UNIT 5

Did you know that when you "see" a rainbow, you are actually responding to energy waves that have traveled millions of kilometers from the sun to your brain? Your vision of that rainbow began as tiny particles of energy radiated by excited atoms of the sun. The energy traveled as light waves through the vacuum of space to earth's atmosphere. When the waves reached raindrops in the lower atmosphere, they were slowed down, reflected, and bent in such a way that they spread into a band of colors. As you look up at the sky, the waves pass into your eyes where they are slowed and bent again. Once the waves reach the nerves of your eyes, their energy stimulates the visual center of your brain, causing you to "see" a range of colors.

Another form of wave energy that your brain responds to is sound. When a herd of antelope, like those pictured, run across a plain, each animal's hooves pound against the ground producing vibrations that your brain detects as hoofbeats. How do sound waves travel from the source of the vibrations to your ears? How are they like light waves?

In this unit, you will learn more about how energy, in the form of sound and light, is transferred by means of wave motion. You will evaluate sound waves and light waves in detail, and learn how they enable you to hear and see. You will discover how light can be manipulated using screens, mirrors, lenses, prisms, and filters. Drawing upon your knowledge of light, you will then investigate the process of separating visible light into colors.

CHAPTERS

SOUND

Whether in the form of a dolphin's cry, a rattlesnake's hiss, or a child's laughter, sound is one of the most important tools higher lifeforms use in order to communicate. What is sound? How is it created and detected?

CHAPTER 21

SECTIONS

OBJECTIVES

- ☐ Identify the properties of wave motion.
- ☐ Distinguish between two types of wave motion (longitudinal and transverse).
- ☐ Demonstrate that sound is caused by vibrations.
- ☐ State a rule showing the relationship between frequency and pitch.
- ☐ Explain the concept of loudness as a measure of wave amplitude.
- ☐ Interpret the relationship between sound and music.
- ☐ Describe how sounds are made in musical instruments.

21.1 TRANSMITTING ENERGY BY WAVE MOTION

All waves are alike in many ways. A **wave** is a disturbance that propagates through a medium or space. There are many kinds of waves, most of which cannot be seen. You cannot see radio, heat, light, or sound waves. It is easier to understand these invisible waves if you first study a wave that can be seen: the water wave.

Figure 21-1 What form of energy transfer is shown in this photograph?

If you tap a pool of water with your finger, you send out a regular series of water waves, such as those shown in Figure 21-1. The highest point of the wave in Figure 21-2 is the **crest**, and the lowest part of a wave is the **trough**. You can see that the distance measured from the crest of one wave to the crest of the next wave (or from trough to trough) is the **wavelength**. The number of waves that pass a given point in one second is the **frequency**. The shorter the wavelength, the higher the frequency.

The material that carries the wave or transfers the energy is called the **medium**. In this case, the medium is water. The distance that a particle of the medium moves up or down from its rest position to the top of a crest or to the bottom of a trough is called the **amplitude**.

The energy of a wave depends upon its size, or amplitude. If you tapped the water gently, the energy and amplitude of the wave would be very small. However, if you dropped a large boulder into the water, you would produce a wave of large amplitude.

As the water wave travels, it loses some of its energy through friction. The energy also spreads out over a wider area. Therefore, the amplitude becomes smaller and smaller until finally the wave dies out. The energy of the wave has been changed to heat.

Figure 21-2 A side view of water waves. What is the distance from one trough to another called?

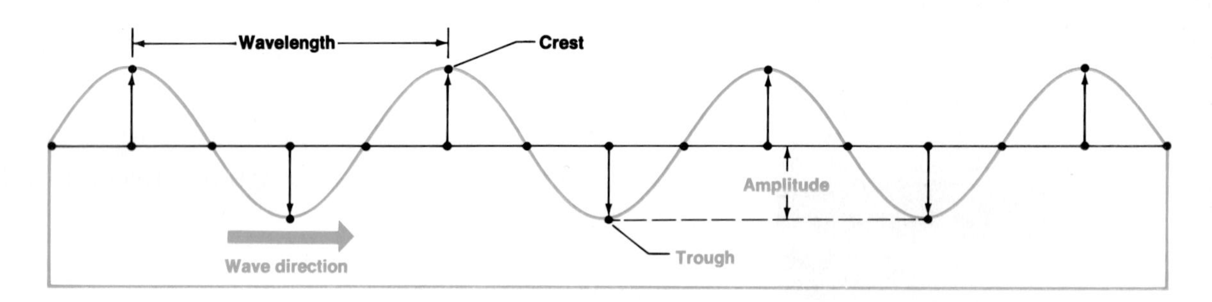

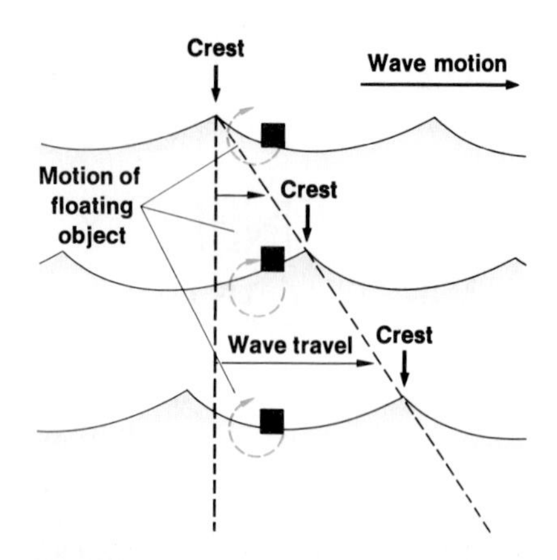

Figure 21-3 What is the difference between the motion of the water and the motion of the wave?

Again, suppose you tap the surface of a pool of water, sending out circular waves. A cork floating on the surface bobs up and down as the waves pass it. The cork is not carried along with the outward motion of the wave. The cork's motion on the surface of the water is demonstrated in Figure 21-3. As the wave passes, the water itself simply rises and falls a little, and returns to its starting point. Only the wave pattern of crests and troughs moves outward. The water itself and any object floating on the water do not move forward.

An exception to the rule that wave motion and the movement of water are unrelated occurs when waves reach a shoreline. At that point the waves "touch bottom" and turn into breakers. Then the water actually moves along with the wave up onto the beach.

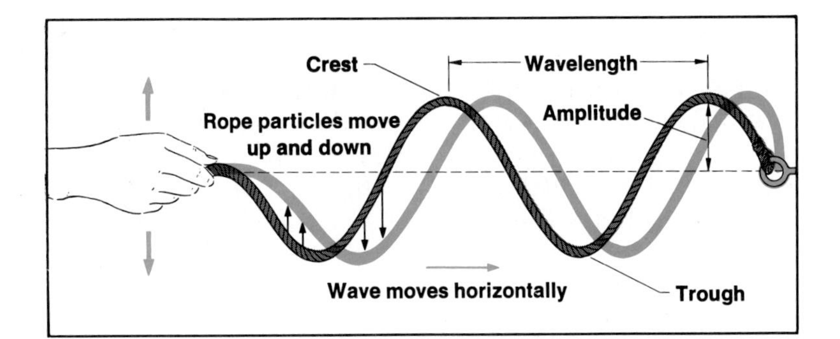

Figure 21-4 If you shake one end of a rope attached to a door, waves will begin to move along the rope toward the door. The gray rope represents the wave position at time 1. The blue line represents the wave position an instant later at time 2 when the wave crest has moved closer to the door. Did the wave amplitude change between time 1 and time 2?

Two types of wave motion can be identified. The motion of a wave can be seen in another way. Tie one end of a rope to a door knob and shake the other end sharply. Figure 21-4 shows how the waves move along the rope toward the door. If you shake the rope hard enough, the waves reflect back from the door before they die out.

The parts of the rope (like the particles of water) move up and down while the wave itself moves ahead. The type of wave produced by water and the rope is called a transverse wave. A **transverse wave** is a wave in which the particles of the medium move at right angles to the direction of the motion of the wave.

Another type of wave motion can be demonstrated by using a long, coiled spring on a smooth table top. Attach the spring at one end and give the free end a quick push-pull motion. This type of wave is called a longitudinal wave and is illustrated in Figure 21-5. A **longitudinal wave** is identified as one in which the particles of the medium move back

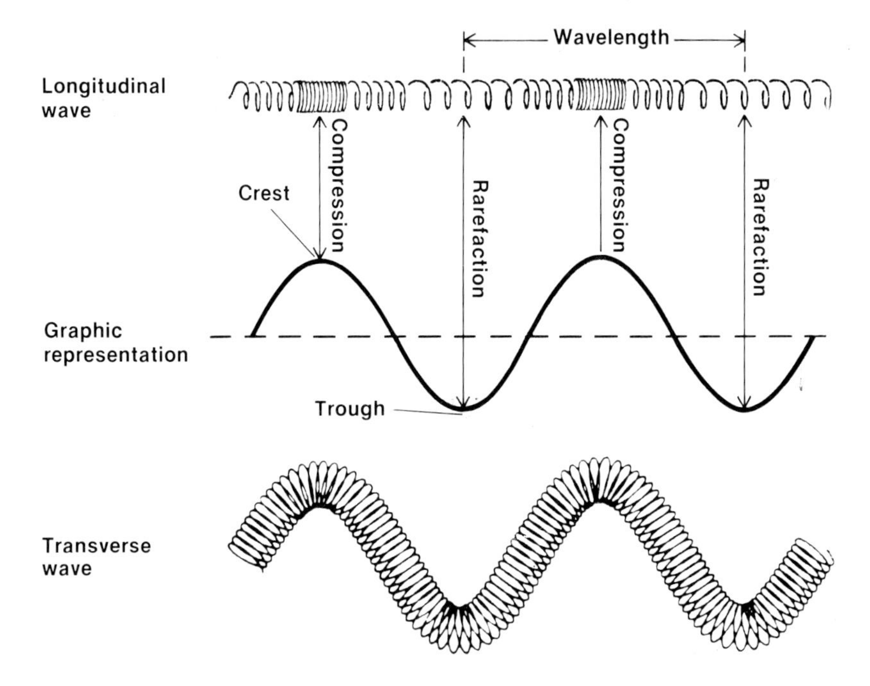

Figure 21-5 How does the movement of the particles in a coil spring compare in a longitudinal wave versus a transverse wave?

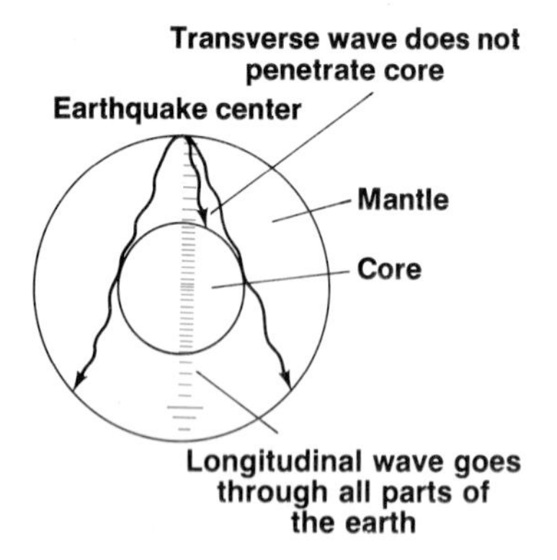

Figure 21-6 Cross section of the earth showing earthquake wave transmissions. The quake originates at the earthquake center and generates longitudinal and transverse waves.

and forth parallel to the direction of the motion of the wave. Compare the longitudinal and transverse waves shown in Figure 21-5. Whereas there is up and down displacement in a transverse wave, the particles in the medium of a longitudinal wave are compressed and then spread out. The compressed part of the wave is called a **compression**, and the spread-out part of the wave is called a **rarefaction** (RAR-*uh*-FAK-*shun*).

Observe that as the longitudinal wave moves lengthwise along the spring, no turn of wire moves along with the wave. Any given turn of wire moves back and forth only a short distance, but the energy of the waves is passed along the entire length of the spring.

The wavelength of a longitudinal wave is the distance from one compression to the next, or from one rarefaction to the next. The amplitude is the maximum distance forward or backward from its rest position that any particle (in this case a single turn of wire) of the medium moves. Sound waves are longitudinal waves.

Earthquakes produce both types of waves. The longitudinal earthquake wave is sometimes called a primary wave. This primary wave exerts a push-pull type of force, passing through all parts of the earth, including the central core. This central core consists of a solid inner section and a liquid outer layer. A primary wave travels about 8 km/sec, passing through the earth's diameter in about 26 min.

The transverse earthquake wave, also called a shear wave, has a snake-like motion and travels at about 4 km/sec through earth's crust and mantle. The transverse wave causes much of the damage produced by an earthquake.

A transverse earthquake wave, however, travels only through solids, not liquids. This fact supports the theory that the outer layer of the earth's core is liquid. The transverse earthquake waves do not pass through the liquid core to reach those parts of the earth that lie opposite the earthquake's center (Figure 21-6).

21.2 SOUND: A FORM OF WAVE MOTION

Sound is caused by vibrations. Sound waves are produced only by vibrating matter. If you place tiny paper riders on a piano string and strike the proper key, the paper dances as the string vibrates. Place some sand or paper clips on a drumhead and watch them move as you beat the drum. The sound lasts only as long as the vibrations last.

Sound can also be produced by colliding objects. Dropping a book on the floor or letting a drop of water fall into an empty pan produces sound. The collisions cause vibrations. **Sound** is a form of energy produced by the vibration of matter.

You can see how sound waves travel by using the coiled spring again. A quick shove against the free end of the coiled spring represents sound energy. The coils represent molecules of matter. The shove causes the coils to compress and spread apart just as sound causes the molecules in matter to compress and spread apart. Thus, a sound wave is a longitudinal wave. Air is a very common medium for sound waves, and it can act like the coils of a spring. Figure 21-7 illustrates the way air molecules bunch together to form a compression and spread apart to form a rarefaction.

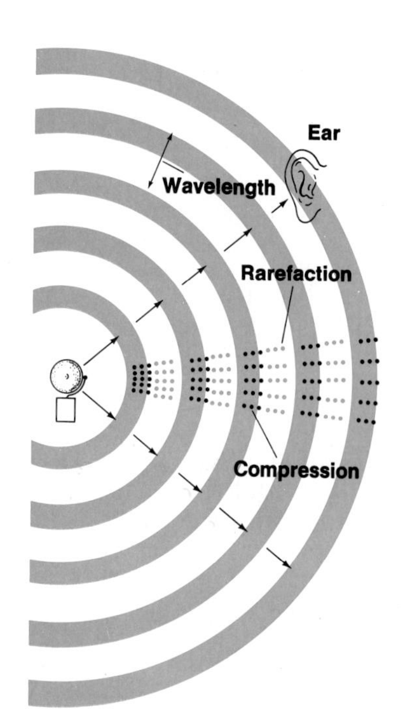

Figure 21-7 The vibrations of a school bell trigger longitudinal waves in the air. When the waves reach your ear, you hear ringing. How do the molecules in the air move in response?

Sound travels in solids, liquids, and gases. Some substances are good conductors of sound and others are not, but sound cannot be transmitted at all without a medium. The simple test illustrated in Figure 21-8 proves that sound energy does not travel through a vacuum. As air is pumped from the bell jar, the sound of an alarm clock grows weaker and weaker and finally disappears.

Now consider the problem of sound transmission on the moon or in space. Since there is no air, how can astronauts talk to one another? They use radios, because radio waves can move through a vacuum, whereas sound waves require a form of matter such as a solid, liquid, or gas in which to move from place to place.

You know that sounds travel through the air because you hear other people speak. You also know that sounds travel through solids because you can hear noises through walls and ceilings in your home and school. If you have ever had a tooth filled, you know that the sound of the dentist's drill carries very well through the solid bones of your skull.

You can do a few simple activities to see how much better sound travels through some solids than through air. Compare the sound of a stretched rubber band when you pluck it in air, and then while you hold one end in your teeth. Hold the base of a vibrating tuning fork firmly against one end of a meter stick. Press the other end of the stick against a table top or blackboard. Note how much better the sound travels through the solid stick than through air.

Liquids are also good conductors of sound. While swimming underwater in a pool, you may have noticed that you can hear the filter motor and other sounds. Certain animals, such as porpoises and whales, signal each other by making sounds. You will learn more about underwater sounds later in this chapter.

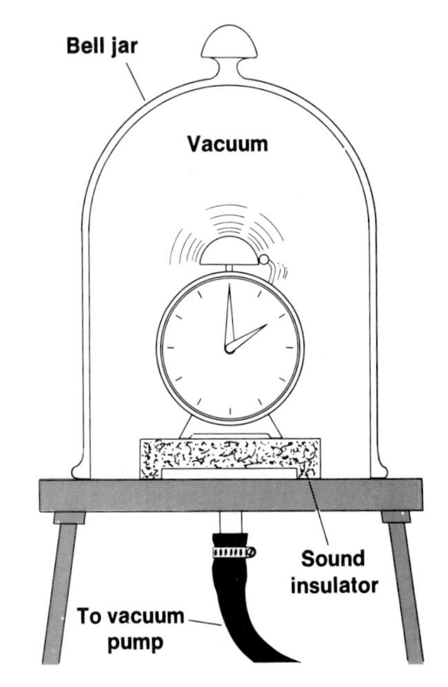

Figure 21-8 What happens to the sound of the alarm as air is pumped out of the bell jar, producing a vacuum?

PRODUCING SOUNDS WITH VIBRATIONS

activity

OBJECTIVE: Perform experiments with tuning forks to observe the effects of vibrations on various media.

PROCESS SKILLS

In this activity, you will produce sound waves in order to *observe* their effect on various materials and *draw conclusions* about your observations.

MATERIALS

several tuning forks of different frequencies
rubber mallet
container of water
ping-pong ball
string
tape
sheet of paper

PROCEDURE

1. Strike a tuning fork with a rubber mallet. If you do not have a mallet, tap the tuning fork on the rubber heel of your shoe. CAUTION: *Do not strike a tuning fork against furniture or walls.* Dip the tips of the tuning fork into a container of water. Record what happens.
2. Using string and tape, suspend a ping-pong ball from a table. Strike a tuning fork and hold one of the tips lightly against the hanging ping-pong ball. Record what happens.
3. Using tape, hang a sheet of paper from the side of a table. Hold a vibrating tuning fork lightly against the paper. Record what happens.
4. Repeat Steps 1, 2, and 3 with tuning forks of different frequencies. Record your observations.

OBSERVATIONS AND CONCLUSIONS

1. What happened when the vibrating tuning fork was placed in water?
2. How did the ping-pong ball and the paper react to the vibrating tuning fork?
3. How do you explain your observations?
4. What happened when different-size tuning forks were used?

What causes sound to travel so well through some substances but not through others? You may think the density of the substance is a factor. Steel is quite dense and carries sound very well. For instance, railroad workers often listen to the rails to find out if a train is coming. Through the rails, they can hear the sound of the train while it is still a long distance away.

Yet density is not the answer. Concrete and brick are also dense substances, but they are poor sound conductors. In fact, concrete blocks are often used in the walls of schools to block out noise. If all other factors are equal, the denser a substance is, the worse it conducts sound.

The best sound conductors are elastic substances. In these substances, the molecules are easily moved by sound energy. They transmit the sound energy from one molecule to the next throughout the length of the substance. The movement of the molecules is extremely small. For most sounds, the molecules move less than a millionth of a millimeter. The speed of sound is proportional to the elasticity of the medium and inversely proportional to the medium's density. In other words, the greater the elasticity, the greater the speed; the greater the density, the slower the speed.

Careful measurements of the speed of sound show that sound travels faster on warm days than on cold days. Sound travels through air at about 332 m/sec when the temperature is 0° C. As the air temperature gets warmer, sound travels about 0.6 m/sec faster for each rise of 1° C. This means that if the air temperature is 30° C, sound will travel 18 m/sec faster (0.6 × 30), or at a speed of 350 m/sec (332 + 18).

You may wonder why the speed of sound increases as the air gets warmer. Although air is no more elastic when it is warmer, its density decreases. The net result is that sound travels somewhat faster. Table 21-1 lists the speed of sound in various media.

TABLE 21-1

Medium	Speed of Sound at 20° C
Air	344 m/sec
Water	1,450 m/sec
Steel	5,000 m/sec

You can estimate how far away a lightning flash is by noting how long it takes before you hear thunder. Assume that sound travels about 350 m/sec. If the sound takes 3 sec to reach you, the lightning is about 1 km or 1,050 m away (350 m/sec × 3 sec = 1,050 m).

SAMPLE PROBLEM

A hunter fires a rifle and 2.5 sec later hears an echo of his firing bounced from a cliff. How far away is the cliff if the air temperature is 10° C?

SOLUTION

Step 1: Analyze

You are given that the air temperature is 10° C and the echo is heard 2.5 sec after the rifle is fired. The distance from the hunter to the cliff is the unknown.

Step 2: Plan

You know that the speed of sound in air at 0° C is 332 m/sec. Since you also know that sound travels 0.6 m/sec faster for each rise in temperature of 1° C, you can derive the speed of sound at 10° C as follows:

$$\frac{0.6 \text{ m/sec}}{1° \text{ C}} \times 10° \text{ C} = 6 \text{ m/sec change in speed}$$

$$332 \text{ m/sec} + 6 \text{ m/sec} = 338 \text{ m/sec}$$

Since you know the speed of sound and the time it took the sound to travel from the cliff to the hunter (2.5 sec/2 = 1.25 sec), you can calculate the distance using the following equation:

$$\text{distance} = \text{speed} \times \text{time}$$

Step 3: Compute and Check

Substitute all values in the equation.

$$\text{distance} = 338 \text{ m/sec} \times 1.25 \text{ sec}$$
$$= 422.5 \text{ m}$$

Since the distance of 422.5 m in 1.25 sec is a little more than the 338 m the sound would travel in 1 sec, your solution is complete.

Our ears detect sound waves. Use Figure 21-9 to follow the movement of sound in the ear. Sound waves enter the ear through the outer ear, which is shaped to collect the waves. It then travels down the ear canal to the eardrum. Sound waves striking the eardrum cause it to vibrate. As the eardrum vibrates, three small, hinged bones in the middle ear pass the vibration along to the inner ear.

The inner ear is filled with liquid in which are located many end-fibers of the auditory nerves. As the sound waves spread through the liquid, they vibrate these nerve endings.

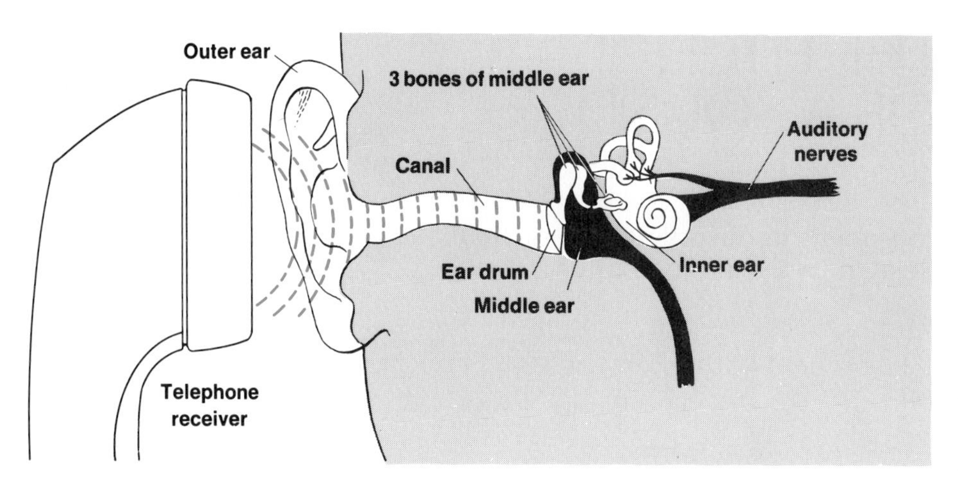

Figure 21-9 Trace the path of the sound from the telephone receiver to the auditory nerve.

The auditory nerves pick up the vibrations and send electric pulses to the auditory center of the brain. The brain, in turn, produces the sensation you know as sound.

Pitch is one measure of a sound wave. You learned earlier that sounds are caused by vibrations. The number of vibrations per second, or the frequency of the sound waves, is interpreted by the ear as **pitch.** The greater the frequency of waves that strike the eardrums, the higher the pitch of the sound. What is the frequency of a low-pitched sound?

The common unit for expressing wave frequency is vps, meaning vibrations per second. Humans are able to hear sounds with a frequency between 20 vps and 20,000 vps. The graph in Figure 21-10 represents the range of some audible sounds. These limits of hearing vary greatly in different people and can change with age. The upper limit of the hearing range is reduced sharply as a person grows older. Many animals, such as dogs and bats, can hear sounds of much higher frequency than those audible to the human ear. Dog whistles, which seem "silent" to humans, actually emit high-frequency signals beyond the range of human detection.

Figure 21-10 How much higher is the frequency of a bird's song than the frequency of a high soprano's aria?

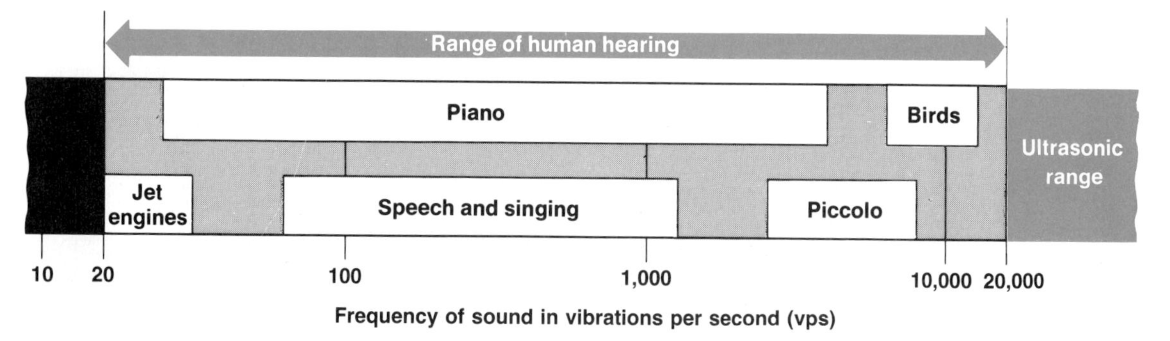

CHANGING THE PITCH OF A SOUND

activity

OBJECTIVE: Perform an experiment to observe contradictory yet explainable results related to a change in pitch.

PROCESS SKILLS
In this activity, you will *experiment* with pitch and *draw conclusions* about any changes noted.

MATERIALS
2 glass bottles
water
pencil

PROCEDURE
1. Fill two glass bottles with different amounts of water. Blow across the top of each bottle to determine which bottle produces the higher pitch.
2. Tap the bottles with a pencil and again determine which bottle produces the higher pitch.

OBSERVATIONS AND CONCLUSIONS
1. Which bottle produced a higher pitch when you blew across its top?
2. Which bottle produced a higher pitch when you tapped it?
3. How would you explain any difference that you noted?

Low-frequency waves, from 20 to 200 vps, produce deep bass (BASE) pitches. High-frequency waves, those above 8,000 vps, produce high-pitched sounds. An oscilloscope is a device that compares the frequency and intensity of sounds. Figure 21-11 shows wave patterns with different frequencies on an oscilloscope screen.

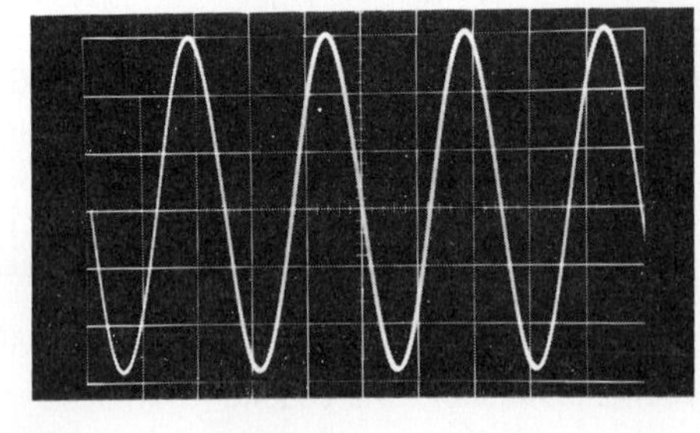

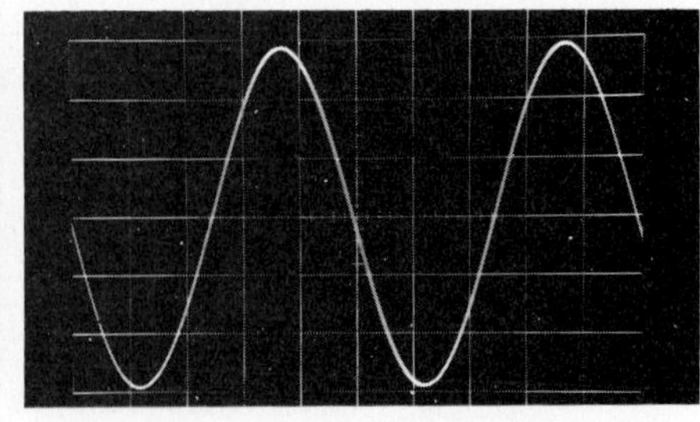

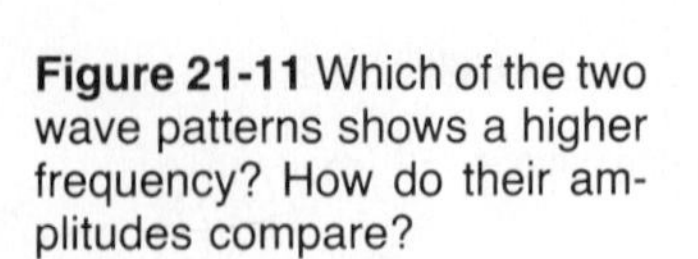

Figure 21-11 Which of the two wave patterns shows a higher frequency? How do their amplitudes compare?

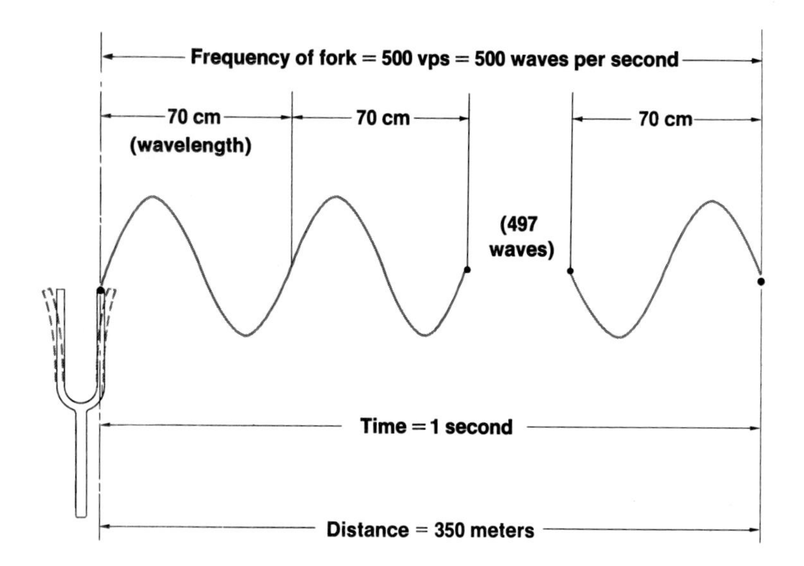

Figure 21-12 Five hundred waves per second, each 70 cm long, placed end-to-end, extend 350 m. How can you determine the speed of the sound from the data given in the diagram?

Velocity and frequency are related. Consider what happens when a tuning fork vibrates, sending out waves in all directions. Suppose the fork has a frequency of 500 vps. In other words, the fork sends out 500 complete waves in 1 sec. The waves are each 0.7 m (70 cm) long and are placed end-to-end, as in Figure 21-12. The first wave is 350 m away at the end of 1 sec (500 waves, each 0.7 m long, have left the fork). Therefore, the wave must have traveled at a speed of 350 m/sec. Notice that this speed is obtained by multiplying the frequency (500) by the wavelength (0.7). This relationship can be expressed as

$$v = fl$$
$$500/\text{sec} \times 0.7 \text{ m} = 350 \text{ m/sec}$$

where v is the velocity, f is the frequency of vibration, and l is the wavelength. This equation is very useful in science because it works for all waves, including sound waves.

For a wave in any one medium at a given temperature, the velocity v does not change. If the frequency is doubled to 1,000 vps, the wavelength is reduced by one half, or 0.35 m. High-pitched, high-frequency sounds have short wavelengths. Wavelengths of audible sounds range from 25 mm to about 17 m.

High-frequency sounds are used in industry. Sounds with frequencies above 20,000 vps are called ultrasonic. These sounds cannot be heard by humans.

Industry makes wide use of ultrasonics for cleaning and testing various objects, such as watches. Objects to be cleaned are dipped into a liquid. The ultrasonic vibrations are then directed into the liquid. The high-frequency vibrations cause a scrubbing action that cleans the objects.

SAMPLE PROBLEM

The musical tone called middle A is produced by a sound wave having a frequency of 440 vps. What would be the wavelength of sound waves of this frequency if the temperature of the air were 33° C?

SOLUTION

Step 1: Analyze

You are given the frequency of the sound and the temperature of the air. The wavelength is the unknown.

Step 2: Plan

Since you know the velocity of sound at 0° C, you can derive the velocity of sound at 33° C as follows:

$$332 \text{ m/sec} + (0.6 \text{ m/sec} \times 33° \text{ C}) = 352 \text{ m/sec}$$

Since you know the frequency and velocity, you can find the wavelength by solving for l in the following equation:

$$v = fl$$

$$l = \frac{v}{f}$$

Step 3: Compute and Check

Substitute all values in the equation.

$$l = \frac{352 \text{ m/sec}}{440/\text{sec}}$$
$$= 0.8 \text{ m}$$

Since a value of 0.8 m is within the range of a reasonable wavelength, your solution is complete.

To test for cracks or flaws, objects are subjected to ultrasonic waves. Any cracks in an object will scatter the waves and cause an echo. Ultrasonic waves are also used by some dentists in cleaning teeth and have a variety of other medical applications.

Sound waves can be reflected. When a sound wave strikes the surface of a material different from the one in which it is traveling and some of the wave is bounced back, **reflection** has occurred. Reflected sound waves are used in sonar technology. Sonar is an electronic method of using ultrasonic waves sent through the water and reflected back to the receiver. It can detect and determine the distance of a reflecting object in water. Radar uses radio waves in air in much the same way. The term sonar is derived from sound navigation and ranging. Sonar is used to survey the ocean floor, to detect objects, and to navigate. Submarine crews use sonar to locate other ships. Fishing boats carry sonar equipment to help find schools of fish. Certain animals such as bats and porpoises have natural systems similar to sonar.

Sound waves can also be refracted. When a sound wave moves from one medium into another, the sound wave is bent. The bending of waves as they pass from one medium to another is called **refraction.** You may have noticed that on a calm summer night, you can hear distant voices or sounds not usually heard. This is due to refraction. Sound waves travel faster through warm air. At night, cool air is closest to the earth. When sound waves hit the warm air above the cool air layer, they are bent downward and can be heard by someone far from the source of the sound.

Sound travels around corners. Even though you cannot see someone around the corner, you can hear him or her speaking. The spread of sound waves beyond the edge of a barrier is **diffraction.** This effect also occurs when there is an opening in the barrier. The amount of diffraction depends on the wavelength of the wave and the size of the barrier opening. Sound waves have wavelengths with a range of a few centimeters to a few meters, so they can be diffracted or bent around objects such as trees and people. As the wavelength decreases, the amount of diffraction also decreases. Light wavelengths are much shorter than sound wavelengths. Thus, light is not diffracted through an opening that diffracts sound.

The pitch of sound changes when the sound or listener moves. You have probably heard the pitch of the horn on an approaching car drop suddenly as the car passed you. Why should this occur? The horn itself does not change its frequency. Yet, the pitch that strikes your ears does change.

The number of vibrations per second that strikes your eardrum is changed if the source of the sound moves, or if you move. If the source is moving toward you, or you are moving toward it, the pitch is higher. More waves reach your ear per second than would be the case if you or the source were not moving.

On the other hand, the pitch is lowered if the vibrating object is moving away from you. The rise or fall of pitch that is due to relative motion between the observer and the source of sound is called the **Doppler effect**. The Doppler

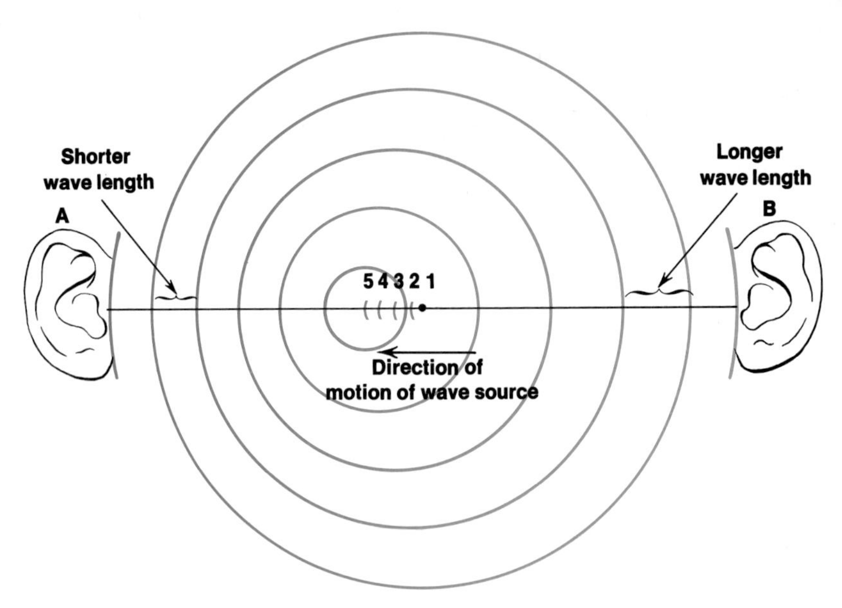

Figure 21-13 As the source of the sound moves from position 1 to position 5, what happens to the pitch heard by ear A? by ear B?

effect applies to all types of waves, not only to sound waves. Study Figure 21-13. Which of the two ears, A or B, hears the higher pitched sound?

Loudness is a measure of wave amplitude. The more energy a wave contains, the greater its size or amplitude. The effect of this energy on your ears is called loudness.

Whenever the speed of a source exceeds the speed of sound, a shock wave is formed. A jet plane compresses the air it moves into. At supersonic speeds, it catches up to and overtakes the pressure waves generated. This results in a loud sound called a sonic boom.

A sound becomes fainter as you move farther away from the source because the sound spreads out to cover a larger area. When you are twice as far away, the sound is only one fourth as loud. When you are three times as far away from the source, it sounds one-ninth as loud. The loudness of the sound varies inversely with the square of the distance. This is the same inverse-square law that applies to gravity, heat, light, and radio waves.

The inverse-square law applies only in "free space," when the waves are not reflected or focused. It does not hold true in your classroom. Here sound is reflected from walls and ceiling, keeping the sound from weakening as fast as it does in free space.

You can measure the loudness of sound by using a decibel meter. The sound waves that strike a microphone in the meter produce an electric current. The strength of the current depends on the loudness of the sound. The current is amplified and registered on a meter marked in decibels (db), the units of sound intensity (see Chapter 14).

A sound that can barely be heard by a person with good hearing is given a value of 0 decibels. This value is called the threshold of hearing. A sound is rated as 1 db louder than

TABLE 21-2

Decibel Ratings of Some Common Sounds	
Sound	**Decibels**
Threshold of hearing	0
Whisper	10–20
Quiet office	20–40
Automobile	40–50
Conversation	60–70
Heavy traffic	70–80
Air drills, riveters	90–100
Thunder	110
Threshold of pain	120
Jet airplane engine, 30 m away	140

another sound if a slight difference in loudness can be detected. Ratings of common sounds are given in Table 21-2.

Exposure to sounds that are too loud may cause a hearing loss. The U.S. Department of Labor has set limits on the time workers can be exposed to certain noise levels. Table 21-3 lists some of these limits. People exposed to noise levels greater than the values in the table are required to wear sound-muffling devices. Why must members of the ground crew who work around jet aircraft wear sound mufflers? Check the decibel rating of a rock band given in Chapter 14. How long can you listen to it safely?

Acoustics is the study of sound quality. The **acoustics** (*uh*-KOO-*stiks*) of a building is the sum of the qualities it has that provide distinct hearing. If the design of a theater or concert hall is such that the audience can hear well, it is said to have good acoustics. Acoustics also refers to the branch

TABLE 21-3

Safety Limits for Noise Exposure	
Sound Level (decibels)	**Time (per day)**
90	8 hr
95	4 hr
100	2 hr
105	1 hr
115	15 min

of science that deals with the study of sound. The concert hall in Figure 21-14 was designed using the principles of acoustics to improve the sound of music for all listeners.

One problem dealt with in acoustics is sound reflections. An echo is one kind of reflected sound. However, you cannot hear an echo if it is reflected back to you in less than about 0.10 sec. Since sound can travel a little over 30 m in this time, it is clear that the reflecting surface must be more than 15 m away if you are to hear the echo.

In some places, such as a concert hall, sound waves may be reflected back and forth several times. This gives rise to reverberations (*rih*-VUR-*buh*-RAY-*shunz*), or multiple echoes. If the walls of the hall strongly reflect waves, the echoes remain audible for a long time, producing poor acoustics. To prevent this, the walls and ceilings may be covered with materials that absorb sounds.

Sound waves can interfere with one another. Two sound waves traveling through the same medium simultaneously move independently, but also have a combined effect. When the displacements these two waves produce are combined, the resultant wave is their algebraic sum. This is known as the **principle of superposition.** The effects caused by two or more waves is called **interference.** Constructive interference occurs when the compression of one wave is added to the compression of another wave with the same frequency

Figure 21-14 The ceiling of the concert hall at Pike's Peak Center in Colorado Springs can be raised or lowered to adjust the acoustics.

and direction to produce reinforcement, or a louder sound. When a compression meets a rarefaction, a soft sound, or silence results. When tones of different frequencies occur at the same time, loudness varies at regular intervals. These variations in loudness due to interference are called **beats.**

21.3 THE SCIENCE OF MUSICAL SOUNDS

Some sounds are pleasing to the ear. If you strike your desk with a ruler or let a drinking glass fall to the floor, a jumble of irregular vibrations is produced. The resulting sound, which is often unpleasant to the ear, is termed noise. All noise has an irregular pattern of vibration.

Musical tones, on the other hand, have regular wave forms. Their wave patterns may be simple or they may be complex. However, a musical tone has a regular wave pattern that is repeated over and over again. For the most part, musical tones have a pleasing effect on the ear, whereas noise has an unpleasant effect.

A musical tone has the properties of pitch, loudness, and quality. The first two of these, pitch and loudness, have already been discussed. You learned that pitch depends upon the frequency and that loudness depends upon the energy, or the amplitude, of the sound wave. But what is meant by tone quality?

Suppose someone is playing a piano and someone else, a trumpet, in the next room. If they each take turns playing a note of the same pitch and loudness, are you able to tell the notes apart? How can you tell which instrument is being played even if you cannot see them? The property called tone quality helps you to identify a musical sound. The quality of a musical tone depends on the shape of its wave pattern. Each kind of musical instrument has its own unique wave pattern. In an orchestra, these waves combine to produce music.

Musical instruments are classified in three groups. Music is made in many different ways. Musical instruments are played by such methods as plucking, blowing, striking, rattling, or rubbing. These methods are used on instruments that can be placed into three groups: the wind instruments, string instruments, and percussion instruments. A fourth group of musical instruments, electronic equipment, is discussed at the end of this chapter.

Figure 21-15 How is the pitch changed on a trumpet? What is the source of the sound?

The wind instruments include the clarinet, saxophone, oboe, bugle, cornet, trumpet, tuba, trombone, French horn, flute, and piccolo. In these instruments, a column of air is made to vibrate by blowing. The musician in Figure 21-15 starts movement in the air column by vibrating his lips.

The string instruments include those that are plucked, such as the banjo, ukelele, harp, and guitar; those that are struck, such as the piano; and those in which a bow is drawn back and forth over the strings, such as the violin, viola, string bass, and cello.

Finally, the percussion instruments are those that are played by striking an object with some kind of stick or mallet. The percussion instruments include vibrating membranes, such as drums, and vibrating solids, such as the xylophone and the bell.

The pitch of wind instruments can be controlled by changing the length of air columns. When a musician moves the slide of a trombone in and out, the slide makes the air column shorter and longer. A short column along with the more rapid vibrations of the musician's lips produce a tone of higher frequency.

In some instruments, such as a trumpet, clarinet, or saxophone, the player changes the length of the air column by using keys or valves to block off parts of the air column. A flute player controls the air column by using the fingertips to open and close holes along the side of the flute.

Three factors affect the pitch of string instruments. The three factors that affect the pitch of a vibrating string are its length, tension, and mass (Figure 21-16).

For a given mass and tension, the longer a string, the slower it vibrates, and the lower its frequency. The deep bass notes of the piano or harp are produced by the longest strings. A violinist can raise the frequency of a string just by pressing down on it. This process prevents the entire string from vibrating. Tests have shown that the frequency of a vibrating string is inversely proportional to its length. For example, if you reduce a string to half its original length, its frequency doubles.

Increasing the tension of a string also makes it vibrate more rapidly and raises the frequency. To tune a string that sounds flat, or too low in pitch, the violinist turns a peg to tighten it. This increases the tension and frequency.

You have probably noticed the bass strings of a piano are not only longer and under less tension than the other strings, but they are also heavier. The heavy wire has more inertia and, therefore, vibrates more slowly and has a lower frequency.

Figure 21-16 What factors affect the pitch of this cello? Does it have a low or high pitch?

Instruments produce fundamental tones and overtones. If you pluck a stretched string lightly in the middle, the string vibrates as a whole, as shown in Figure 21-17A. The tone you hear when this is done is the lowest note that can be made by a vibrating string, called the **fundamental**.

However, if you pluck the string one quarter of the way from the end, it not only vibrates as a whole but also in segments, as shown in Figure 21-17B. The tone you hear has a different quality. Besides the fundamental, the first overtone is now present. **Overtones** are higher pitched tones produced by an object as it vibrates in segments. Overtones are tones the frequencies of which are whole-number multiples of the fundamental frequency. For example, if the frequency of the fundamental is 100 vps, the frequency of the first overtone is 200 vps. The frequency of the second overtone is 300 vps.

The number and strength of the overtones determine the quality of a musical tone. A string can vibrate not only as a whole but also in segments to produce overtones. Most of these overtones blend well with each other and with the fundamental, producing a pleasing effect. Overtones are often referred to as harmonics. Why is it important that a tuning fork produce no overtones?

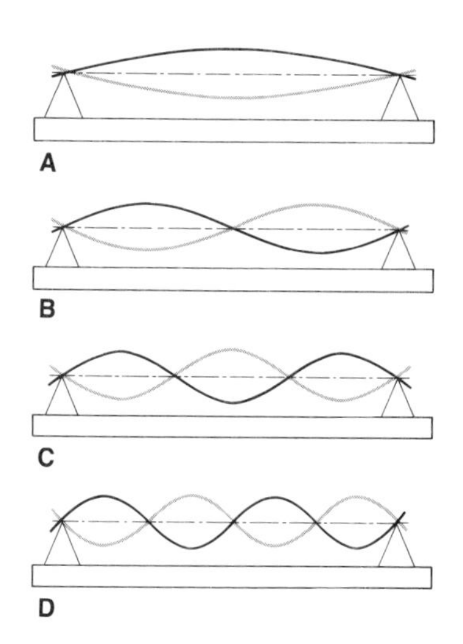

Figure 21-17 If the frequency of the fundamental tone (A) is 100 vps, what are the frequencies of the first (B), second (C), and third (D) overtones?

Rhythm, melody, and harmony are factors in musical composition. To the composer, the performer, and the listener, music is an art. But music is also a science.

The principal factors of a musical composition are rhythm, melody, and harmony. Rhythm is the timing pattern of music. Melody is the effect of single tones following each other in succession. Melody is often called the tune.

Harmony is the effect of two or more tones that are sounded together. Musical tones sounded together in harmony form a **chord.**

From experience, you know that certain tones harmonize well. Other tones produce an unpleasant effect, or discord, when you hear them together. Whether there is harmony or discord depends almost entirely upon how one frequency blends with another.

PRODUCING OVERTONES

activity

OBJECTIVE: Make a model of a musical instrument with which you can produce overtones.

PROCESS SKILLS

In this activity, you will *model* a musical instrument and *experiment* with it in a way that will help you to *draw conclusions* about how overtones are produced.

MATERIALS

2 boards, each with 2 triangular blocks securely attached
1 unattached triangular block
2 pieces of wire
2 spool-like fasteners
2 wire fasteners
metric ruler
2 equal weights

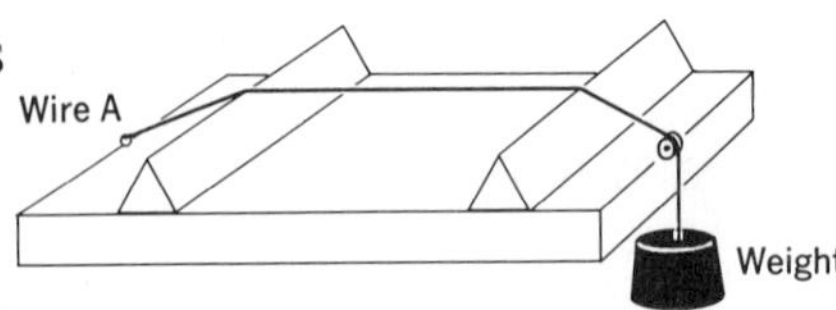

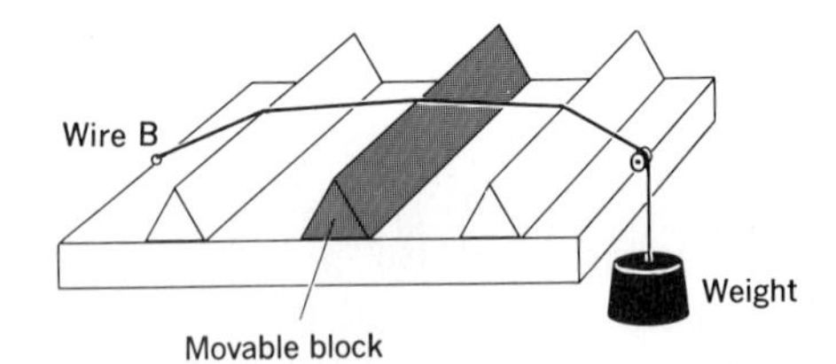

PROCEDURE

1. Set up two identical strings and place them under equal tension as shown.
2. Pluck the first string to produce a fundamental tone. Use a movable block of wood to slowly shorten the second string while plucking both strings at the same time. Record the length of the second string when you produce the first overtone, which occurs when the two strings vibrate together to produce a pleasant sound.
3. Shorten the second string further to see if there are other places where the two strings vibrate together to produce a pleasant sound. Determine the lengths of the two strings where overtones are produced.

OBSERVATIONS AND CONCLUSIONS

1. How short must the second string be to produce a note at the first overtone?
2. How did the lengths of the two strings compare at the point where the second overtone was produced?
3. What other tones can you produce?

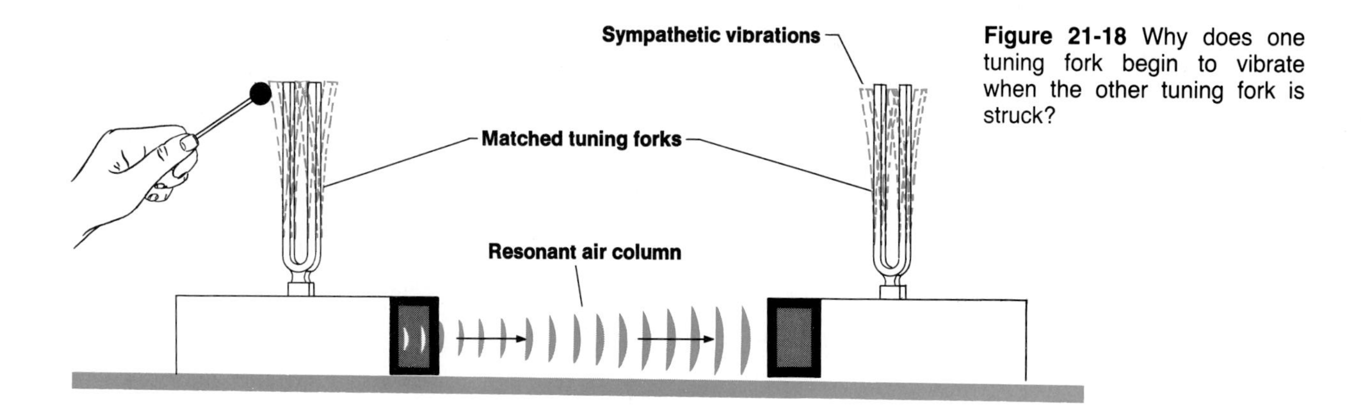

Figure 21-18 Why does one tuning fork begin to vibrate when the other tuning fork is struck?

Sounds are made louder by forced vibrations and by resonance. Sounds made by instruments are made louder in one of two ways: forced vibrations or resonance.

Sound waves can be transferred from one object to another. A vibrating tuning fork placed on a table causes the table to vibrate. This transfer of wave energy by contact between objects is known as **forced vibrations**. String instruments rely on forced vibrations to increase the loudness of the sound. That is, the string is attached to a surface that is forced to vibrate when the string vibrates. Since the vibrating surface is larger than the string alone, the sound becomes louder.

The strings of a piano create louder tones than those of a zither or harp for this reason. Piano strings are mounted on a sounding board, in which they produce forced vibrations. In much the same way, forced vibrations in the wooden body of a violin increase the sounds produced by the strings.

Another way in which sounds are increased is by resonance. **Resonance** occurs when an object vibrates at its own natural frequency due to a similar vibration of another object. The natural frequency is the one fundamental frequency at which an object will vibrate when disturbed. Moreover, an object will normally absorb energy only at its own natural frequency.

You can show how resonance works by placing two mounted tuning forks of the same natural frequency close together, as shown in Figure 21-18. If one tuning fork is struck, the other also begins to vibrate. However, wrapping a heavy rubber band around the end of a tine of the struck tuning fork changes the natural frequency. The second tuning fork no longer resonates with sympathetic vibrations.

Columns of air also have natural frequencies. This can be demonstrated by holding a vibrating tuning fork over a cylinder, as illustrated in Figure 21-19. As the tines of the fork vibrate over the air column, they send down waves that

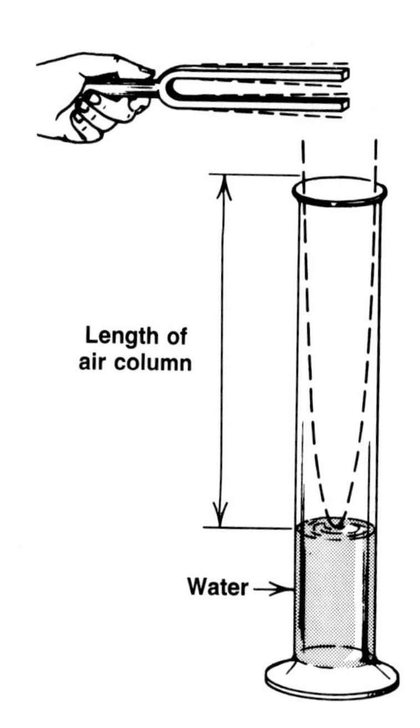

Figure 21-19 What is the relationship between the length of the resonating air column and the length of waves produced by the vibrating tuning fork?

PRODUCING FORCED VIBRATIONS

activity

OBJECTIVE: Perform an experiment to produce forced vibrations in various objects.

PROCESS SKILLS

In this activity, you will *experiment* with tuning forks vibrating at different frequencies in a way that will help you to *draw conclusions* about forced vibrations.

MATERIALS

several tuning forks of different frequencies

PROCEDURE

1. Strike a tuning fork and press its stem against the top of the table. Notice the sound.
2. Repeat Step 1 using tuning forks of different frequencies. Also experiment by placing the stem of each of the tuning forks against different surfaces such as windows, walls, and cabinet doors.

OBSERVATIONS AND CONCLUSIONS

1. What happens to the sound when the stem of each vibrating tuning fork of a different frequency is placed against the top of the table? Explain your observations.
2. What happens when the vibrating tuning forks are placed against other surfaces?

reflect back up from the water surface. CAUTION: *Do not attempt this without your teacher's permission and supervision. The glass may shatter if touched by the tuning fork.*

As water is slowly added, the air column inside the cylinder is shortened. At a certain point, the tone of the tuning fork suddenly sounds quite loud. At this point the air column has the same natural frequency as the tuning fork.

The sound grows louder because the sound waves of the tuning fork are in step with the sound waves inside the cylinder. How does this happen? First, a sound wave from the tuning fork goes down the cylinder. The wave is reflected by the water and goes back up the cylinder. At this point, it meets the next vibration of the tuning fork and is reflected back down the cylinder. Now the sound wave has not only the energy of the tuning fork but also the energy of the reflected wave. As long as the new waves from the tuning fork are in step with previous waves in the air column, the amplitude of the sound increases sharply. In other words,

SAMPLE PROBLEM

If the length of the resonant air column in Figure 21-19 is 50 cm, what is the frequency of the tuning fork? Assume the temperature of the air to be 0° C.

SOLUTION

Step 1: Analyze

You are given that the length of the resonant air column is 50 cm and the air temperature is 0° C. The frequency of the tuning fork is the unknown.

Step 2: Plan

Since you know that the fundamental wavelength is four times the length of the resonant air column, you can derive the wavelength using the column length as follows:

$l = 4 \times 50 \text{ cm} = 200 \text{ cm or } 2 \text{ m}$

Since you know the velocity of sound at 0° C (332 m/sec) and the wavelength, you can use the following equation to solve for frequency:

$$v = fl$$

$$f = \frac{v}{l}$$

Step 3: Compute and Check

Substitute all values in the equation.

$$f = \frac{332 \text{ m/sec}}{2 \text{ m}}$$
$$= 166\text{/sec or } 166 \text{ vps}$$

Since the value of 166 vps is within an acceptable range for sound wave frequency, your solution is complete.

the column of air in the cylinder resonates with the vibrating fork.

There is a relationship between the wavelength of the sound and the length of the resonating air column. The length of a resonating air column that is closed at one end is one fourth the fundamental wavelength of the sound it reinforces. In an open pipe, the length of the air column is one half the wavelength of the main frequency it reinforces.

Resonance is of great importance in many other branches of science. For example, when you tune a radio, you are adjusting its frequency to match that of the radio waves you want to receive. Energy of this frequency is then amplified while other radio frequencies are rejected.

BIOGRAPHY

ROBERT A. MOOG

the electronic musician

Many modern composers search for totally different sounds with which to create new music for the electronic age. Instead of working with the traditional instruments that use wood, strings, and valves, they depend on electronic equipment to produce sounds.

After the invention of simple recorders and amplifiers, musicians began to experiment with sounds from nature, such as winds, storms, ocean waves, and animal sounds. Soon they were able to mix one sound with another. Finally, an instrument was developed that could control how sound signals were generated and could modify sounds from a central console. This instrument, called the synthesizer, was the beginning of electronic music. The synthesizer enabled musicians to experiment with mixing instrumental music with sounds of nature, and even with the noises of traffic and industry.

The first synthesizer was built by RCA in 1955. It was able to imitate the sounds and styles of a harpsichord, an organ, a hillbilly band, and a dance band. However, it was not commercially successful because inexpensive semiconductors had not yet been developed. With semiconductors, a synthesizer would use less power, last longer, and be more compact.

In 1964, Robert Moog, an amateur musician with a PhD in engineering and physics, built the first commercially successful electronic music synthesizer. Moog's synthesizer can produce the sounds of all the common musical instruments. It also emits an odd collection of beeps, thumps, buzzes, and squawks that are formed into music by a system of electronic controls. In the late 1970s, Moog invented a digital synthesizer, which uses a computer to create and control sounds. The Moog synthesizer is used to play everything from rock to the classics.

CHAPTER REVIEW

SUMMARY

1. All waves have amplitude, frequency, wavelength, and shape.
2. In a transverse wave, the particles of the medium move at right angles to the direction of wave motion; in a longitudinal wave, the particles move back and forth parallel to the wave motion.
3. Sound is caused by vibrations.
4. Sound travels in solids, liquids, and gases, but not in a vacuum.
5. Pitch refers to the effect on the ear of the frequency of sound vibrations.
6. The velocity of a sound equals the frequency times the wavelength or $v = fl$.
7. Acoustics is the sum of the qualities that provide distinct hearing.
8. Noise has an irregular wave form whereas music has regular wave forms.
9. Musical instruments can be divided into wind, string, percussion, and electronic.
10. The pitch of a string instrument depends on the length, tension, and mass of the string.
11. Sounds are made louder by increasing the vibrating surface or by causing objects to vibrate in resonance.
12. The factors that describe a musical composition are its rhythm, melody, and harmony.

PROBLEMS

1. What is the wavelength of a sound with a frequency of 20 vps at a temperature of 20° C?
2. An ultrasonic generator sends 500,000 waves per second through the ocean. If each wave is 0.3 cm in length, what is the speed of sound in ocean water?
3. Find the wavelength of water waves that have a frequency of 0.75 wave per second and a velocity of 6.75 m/sec.
4. Assume that you first hear a police siren 240 m away. By the time it is only 60 m away, how much louder is the sound?
5. You hear an echo 2 sec after firing a shot. How far away is the reflecting surface if the temperature is 20° C?
6. (a) What is the wavelength of a tone resonating with a closed pipe 0.4 m long? Assume a temperature of 0° C. (b) What is the wavelength of a sound that resonates with an open pipe 0.4 m long? (c) Which pipe sounds a note of the higher frequency?

VOCABULARY

Match the item in the left column with the best answer in the right column. Do not write in this book.

1. acoustics
2. amplitude
3. chord
4. compression
5. crest
6. diffraction
7. Doppler effect
8. frequency
9. fundamental
10. interference
11. longitudinal wave
12. medium
13. overtones
14. pitch
15. rarefaction
16. refraction
17. resonance
18. sound
19. transverse wave
20. trough
21. wavelength

a. distance measured from the crest of one wave to the crest of the next
b. characteristic of sound caused by the frequency of the vibrations
c. bending of sound waves around corners
d. lowest part of a wave
e. bending of sound waves as they pass from one medium to another
f. particles of the medium move at right angles to the wave motion
g. effects produced by two or more waves moving in the same medium at the same time
h. vibration at an object's own natural frequency due to similar vibration of another object
i. distance a particle of the medium moves from its rest position
j. number of waves that pass a given point in one second
k. highest point of a wave
l. compressed part of a longitudinal wave
m. particles of the medium move back and forth in the direction of the wave
n. lowest note that can be made by a vibrating string
o. material that carries a wave
p. spread-out part of a longitudinal wave
q. rise or fall of pitch due to relative motion between the observer and the source of the sound
r. tones the frequencies of which are whole-number multiples of the fundamental frequency
s. sum of the qualities that provide distinct hearing
t. musical tones sounded in harmony
u. form of energy produced by vibrations of matter

REVIEW QUESTIONS

1. What is sound? What causes sound?
2. What is the amplitude of a sound wave?
3. What are compression and rarefaction? Do these terms refer to transverse or longitudinal waves?

4. Describe the difference between transverse and longitudinal waves.
5. Through what phase of matter does sound travel fastest? slowest?
6. Explain the meaning of ultrasonics. List several of its uses.
7. How many times greater than the speed of sound in air is the speed of sound in steel?
8. How much faster does sound travel through air at 30° C than at 0° C?
9. What is the Doppler effect?
10. What is meant by the acoustics of an auditorium?
11. How does noise differ from a musical tone?
12. How long must a closed pipe be to reinforce a particular fundamental tone?
13. How are overtones produced in vibrating strings? How are they related to the fundamental?
14. What are the three principal factors of a musical composition?
15. Give two examples of forced vibrations.
16. How is the pitch of a violin string raised?
17. What is meant by resonance?
18. Why does sound travel more rapidly through steel than through air?
19. Why does sound travel more rapidly through warm air than through cold air?
20. Explain three properties of a musical tone.
21. What is meant by the fundamental frequency of a string? How is it produced?
22. The vibrating lips of a trumpet player start the sound, but what produces the musical tones that you hear?
23. What is the relationship between pitch and the size of musical instruments?

CRITICAL THINKING

24. Why is it more common to hear the bass part of music through walls than the higher parts?
25. If you shouted at a wall 20 m away, could you hear an echo? Explain.
26. If all gases are equally elastic, do you think sound waves travel more rapidly through air or through hydrogen? through air or carbon dioxide? Explain.
27. What would be the effect if you doubled a 10 db noise? a 60 db noise?

FURTHER READING

Aldous, Donald. *Sound Systems.* New York: Franklin Watts, Inc., 1984.

Ardley, Neil. *Sound and Music.* New York: Franklin Watts, Inc., 1984.

Buckley, R. *Oscillations and Waves.* New York: Taylor and Frances, 1985.

Free, J. "Noise Zapper." *Popular Science* 230 (January 1987): 76–7.

Heuer, Kenneth. *Thunder, Singing Sands, and Other Wonders.* New York: Dodd, Mead and Co., 1981.

Kettelkamp, Larry. *Magic of Sound.* New York: William Morrow and Co., Inc., 1982.

Rossing, T. D. *Science of Sound: Musical, Electronic, Environmental.* Reading, Mass.: Addison-Wesley, 1982.

LIGHT

Using light means more than switching on a lamp. For example, when light carried by optical fibers is flashed on and off, coded messages can be transmitted over great distances with very little amplification. How is light used to magnify images? to create motion pictures?

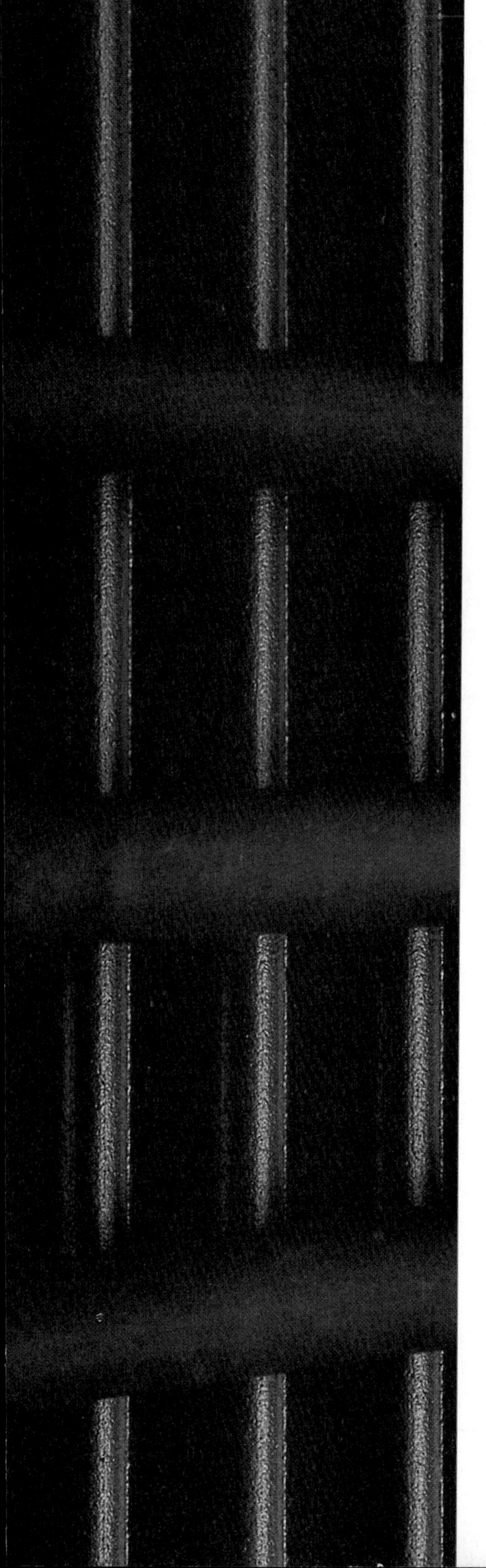

CHAPTER 22

SECTIONS

OBJECTIVES

- ☐ Describe how light is produced and transmitted.
- ☐ Demonstrate how light energy can be measured.
- ☐ Describe the properties and uses of lasers.
- ☐ Demonstrate the law of reflection.
- ☐ Identify the types of images formed by plane, convex, and concave mirrors.
- ☐ Describe how changes in the speed of light can explain refraction.
- ☐ Identify the types of images formed by concave and convex lenses.
- ☐ Describe the operation of optical systems such as magnifiers, projectors, telescopes, and the human eye.

22.1 PROPERTIES OF LIGHT

Light has the properties of both waves and particles. The first major studies on the nature of light were carried out late in the seventeenth century by Isaac Newton and Christian Huygens (HIE-*gunz*). Newton believed that a beam of light is made up of a stream of particles. Huygens formed a different theory, that light travels in waves.

Both scientists tried to find evidence that explained the properties of light by observing how it formed shadows and how images were formed by mirrors and lenses. Newton thought that if light travels in waves, then the wave action should cause the edges of shadows and images to be blurred. Huygens, however, felt that the blurred edges are too small to be seen by the human eye.

In the centuries that followed, the debate continued. It was not until the early part of this century that Max Planck and Albert Einstein discovered that light has the properties of both particles and waves. Prior to these findings, it was believed that the wave and particle theories were opposites that could not exist together.

In 1900, Plank proposed that light consists of tiny particles, which he called quanta. Einstein later applied Planck's theory to explain the phenomenon known as the **photoelectric effect**, in which a beam of light hitting a metal surface causes electrons to be knocked from that surface. He demonstrated that when the quanta of light struck the metal atoms, they forced the atoms to release electrons.

The photoelectric effect cannot be explained by the wave theory of light. The wave theory suggests that the higher the intensity of the incident light shining on the metal surface, the greater should be the energy distributed over the light's wave front and the greater should be the velocity of the electrons emitted by the metal. In contradiction to the wave theory, it was found that the velocity of the electrons did not increase with the intensity of the light on the surface. This phenomenon can be explained by assuming that light is absorbed as particles, in which case the kinetic energy and velocity of the emitted electrons are independent of the intensity of the incident light.

A particle of light, or quantum, is now referred to as a **photon**. Each photon consists of a tiny packet of energy. Photon energy depends upon the frequency of the light wave. A photon of violet light, which has twice the frequency of red light, has twice the energy. Photons of higher frequency ultraviolet light and X-rays have even more energy.

The photoelectric effect can be explained only by the particle theory of light. However, the wave theory has been supported by many experiments. For example, the effects produced by diffraction and interference, discussed in Chapter 21, can be explained by the wave character of light but not by the particle theory. Therefore, this dual nature of light is the theory that scientists accept today.

The "electric eye" doors that open when you pass through a beam of light operate on the photoelectric effect. Electric eyes are also referred to as photocells, or devices that produce an electric current (in the form of emitted electrons) when light shines on them. Another use of photocells is the counting device illustrated in Figure 22-1.

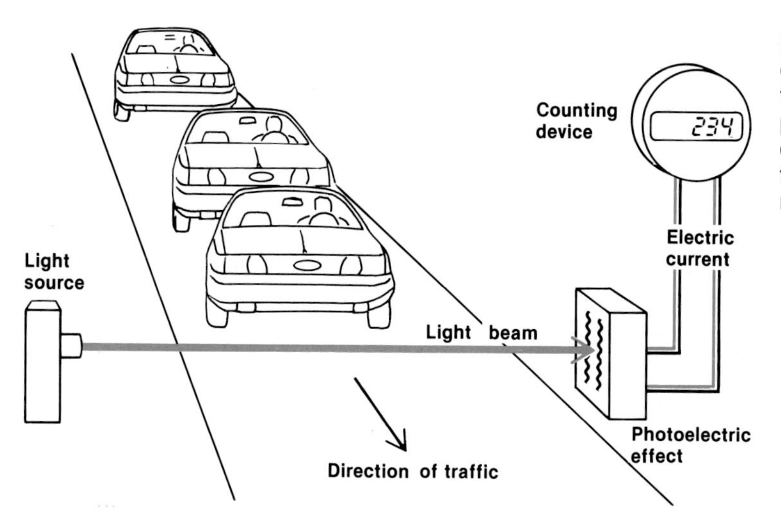

Figure 22-1 The photoelectric effect can be used in traffic studies to count the cars on a highway. When a car passes through a beam of light, the circuit is broken, and a counter adds one to the total. What happens when the car moves on?

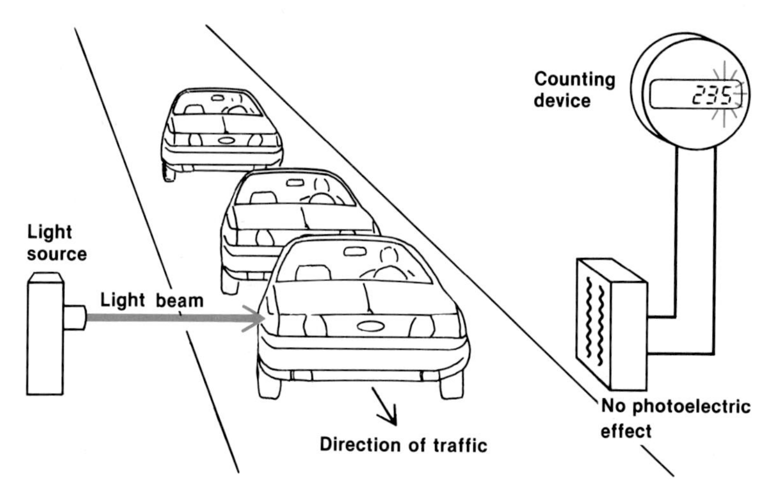

Light can be polarized. The discovery that light can be polarized, or separated into planes of vibration, seemed to support the wave theory of light. Figure 22-2 illustrates how light is polarized.

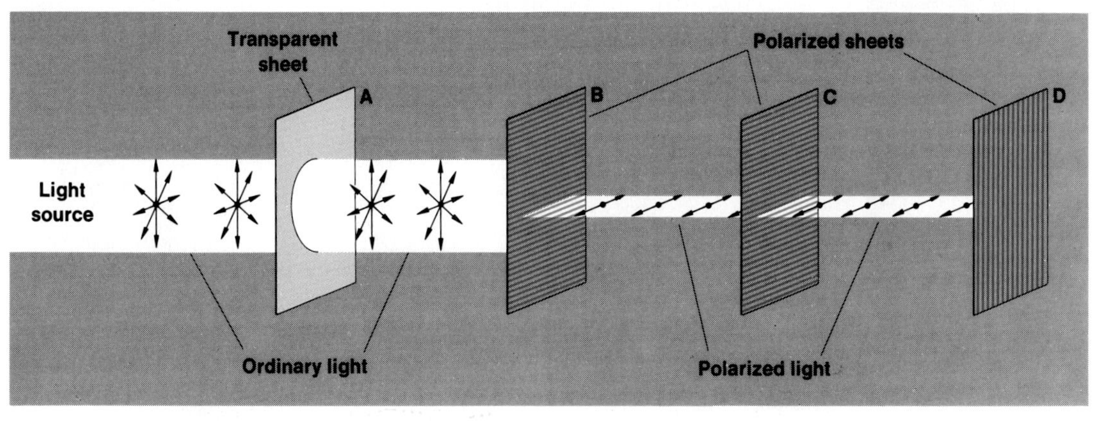

Figure 22-2 What happens to ordinary light as it passes through the transparent sheet? How is it affected when it passes through a polarized sheet? (Note that the arrows represent amplitude vectors drawn from the rest position of a transverse wave to the wave's crest or trough. These vectors indicate the direction of wave vibration. See also Figure 21-2.)

Figure 22-3 What is the effect of using a polarized camera lens (bottom) to photograph this scene?

When ordinary, unpolarized light, vibrating in multiple directions, passes through the transparent sheet, it does not change. However, as it passes through sheet B, which is polarized, only those light waves vibrating in one direction pass through. The rest of the vibrations are blocked. It is as if the light were "combed" through parallel slots.

The light that remains after going through sheet B also goes through sheet C, which is polarized in the same direction as sheet B. Polarizing-sheet D, however, is turned at right angles to sheet C. Therefore, the light waves that pass through sheet C are not lined up with the slots of sheet D and all light is blocked.

Certain crystals act like polarizing sheets because they block certain light waves. When the atomic planes of the crystals are lined up in a given direction, they permit the passage of light waves vibrating only in that same plane. This screening process that permits the passage of light waves vibrating only in one plane is called **polarization.** Light in which the wave vibrations are confined to a single plane is said to be polarized.

The fact that light can be polarized proves that light travels in transverse waves. Remember that in transverse waves, displacement is perpendicular to the direction of wave movement. If light traveled in longitudinal waves, such barriers would have no effect on the light. Why?

Light is not the only type of wave that can be polarized. All transverse waves can be polarized.

Figure 22-4 Irregular patterns of color revealed by polarized light indicate the points of greatest strain on this plastic drawing curve.

Polarized light can be useful. Have you ever used polarizing sunglasses? Much of the annoying glare you may experience in driving comes from sunlight that is reflected off various surfaces. Figure 22-3 shows that this glare is largely blocked out by polarized glasses or camera lenses.

Polarized light is a help to engineers in designing objects. A transparent plastic model of the object is stressed by twisting, stretching, and compressing. When viewed through polarized light, colored patterns in the plastic models show where the strains in the object are greatest. Where is the object under stress pictured in Figure 22-4 most likely to break?

Light is produced by excited atoms. Light can be produced by any of the following methods:

1. heating something until it becomes so hot that it gives off light. Any object heated until it emits light is said to be **incandescent.** An electric light bulb, a burning

match, and a candle are good examples of incandescence. The light is given off by particles of these objects heated to the point where they glow.
2. passing an electric current through a gas, as in neon lighting.
3. fluorescence. In this process, demonstrated by fluorescent bulbs, certain substances glow when exposed to radiation.

Whatever method is used, light is produced whenever electrons fall from one energy level in an atom to a lower level. To understand this transformation of energy, compare electrons in orbit around the nucleus of an atom with satellites in orbit around the earth.

Suppose you want to lift a satellite to an orbit 600 km above the earth. A great deal of chemical energy from fuel is needed to give the rocket enough speed to overcome gravity. Kinetic energy of the rocket is transferred to the satellite as potential energy. While the satellite is in orbit, the energy it has absorbed stays with it. But when the satellite falls back to earth, it releases this stored potential energy in the form of heat energy.

If you were to raise the satellite to an orbit only 300 km above the earth, less energy would be needed. As a result, the amount of energy released when the satellite fell to earth would be less than if it fell from a higher orbit.

Electrons in atoms behave in a similar way. As you learned in Chapter 3, the nucleus of the atom can be pictured as having "satellites," called electrons, in orbit at certain energy levels. An electron can absorb energy and move to a higher energy level in an orbit farther away from the nucleus. When this electron falls back to the starting energy level, it gives off this extra energy in the form of radiation, just as the earth satellite gives up its extra energy in the form of heat. The higher the energy level to which an electron is lifted, the greater the energy it gives off when it falls back to a lower level. In Chapter 23, you will learn how the amount of energy given off is related to the color you see.

The methods of making light are all just ways of exciting atoms by raising their electrons to higher energy levels. In each case, some other form of energy, such as heat or electric current, is used to excite the atoms.

Natural light on earth comes from stars. Daylight is caused by the sun, which is the nearest star. The intense light and heat of the sun and other stars is due to nuclear reactions. Moonlight is the sun's light reflected off the moon's surface.

Light travels from distant stars through space and needs no solid, liquid, or gas to carry its waves. In this way, light, which is electomagnetic waves, differs from sound, which cannot pass through a vacuum. **Electromagnetic waves** are transverse waves that have both electric and magnetic components. In Chapter 27 you will learn about other forms of electromagnetic waves, such as microwaves, radio waves, and X-rays. The different forms of electromagnetic waves are grouped according to wavelength in what is known as the electromagnetic spectrum. Visible light is only a tiny portion of the electromagnetic spectrum.

Light makes objects visible. You can see only when light strikes your eyes. The light may come directly from a source or it may be reflected off some object. You see most things by reflected light. In either case, if you are to see an object, light must come from the object and then it enters your eyes.

When light waves strike a substance, they may be reflected, absorbed, transmitted, or scattered as in Figure 22-5. If the substance transmits none of the light waves, it is described as **opaque.** You cannot see through opaque objects. An opaque object in the path of light casts a shadow because light is not transmitted through the object.

If light waves pass through a substance so that objects can clearly be seen through it, that substance is said to be **transparent.** Glass, water, and air are common examples of transparent matter.

Many substances are neither opaque nor transparent, but are somewhere in between. These substances, such as frosted glass and waxed paper, are said to be **translucent.** Light does pass through translucent matter, but is said to be scattered (Figure 22-5). As a result, things cannot be seen clearly through translucent substances.

Any light that is absorbed is changed into heat, warming up the object that absorbed it.

Light can be measured. The brightness, or intensity, of a light source can be measured by comparing it with the brightness of some standard source of light of known brightness. Originally, the light from a certain size candle made from the wax of sperm whales was used as the standard source of light. This was not a very reliable standard because brightness varied to some degree from one of these candles to another. In recent years, the standard of light brightness has been revised by international agreement to improve ease and precision of measurement. Rather than basing the standard on an object, such as a candle or other source of illumination, a standard value was established. Other sources are then compared to this chosen value. The **candela** (cd) is the standard unit of brightness. In practice, it is common to use incandescent light bulbs that have been compared with the standard. An ordinary 40-W light bulb has an intensity of about 35 cd and a 100-W light bulb will give about 130 cd.

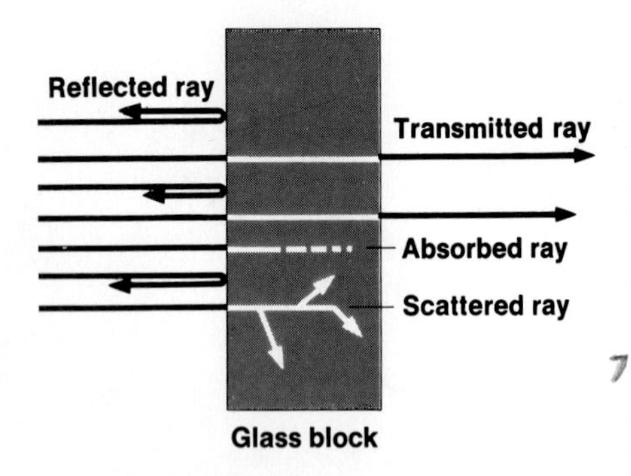

Figure 22-5 What can happen to light when it strikes an object?

The rate at which light energy is given off from a source is measured in lumens (lm). One candela is equal to about 12.5 lm. When new, a modern 100-W light bulb (incandescent type) has a brightness of about 1,600 lm.

People are often more interested in the amount of light falling on a surface than in the brightness of the source itself. The light on the page you are reading depends not only upon the brightness of the bulb, but also upon how far away the page is from the bulb. Light brightness follows the inverse square law. Suppose you measure the light from a lamp at a given distance from you. If you put the lamp twice as far away, you get only one fourth as much light.

The brightness of light can be measured by a light meter. A light meter is often used on (or in) a camera. Such a meter measures the light with a photocell. Many of the photocells depend on the photoelectric effect and produce a small electric current when light hits them. The greater the light, the stronger the electric current.

Home lighting can be measured by light meters and adjusted to save energy and insure adequate levels of light for different purposes. Another factor that affects the brightness of a room is the color of its walls. Lamps that might give enough light in a room with white walls could be too dim in a room with red or blue walls.

Some light sources are more efficient than others. Not all sources give the same amount of light for a given amount of input. One common source, the incandescent light bulb, has a low efficiency. A 25-W bulb uses 25 W of electrical power but changes only about 1.5 percent of this power into visible light. The rest is lost as heat.

A far more efficient light source is a white fluorescent bulb. Light from such a bulb is often called "cool" light because it gives off so little heat.

You can see from Table 22-1 that the efficiency of incandescent bulbs varies with size. How many 25-W bulbs

TABLE 22-1

Efficiencies of Light Sources

Bulb Type	Power Input (W)	Brightness (lm)	Efficiency (lm/W)
Incandescent	25	250	10.0
Incandescent	100	1,600	16.0
Incandescent	1,000	20,500	20.5
Fluorescent	40	2,320	58.0
Ideal	1	680	680.0

MEASURING THE BRIGHTNESS OF LIGHT

activity

OBJECTIVE: Perform an experiment with an oil-spot light meter.

PROCESS SKILLS

In this activity, you will *analyze data* collected while using an oil-spot light meter in order to *draw conclusions* about the effect of light on oil and about the application of the light meter.

MATERIALS

oil, grease, butter, or fat
paper
paper stand
2 light bulbs in sockets
2 electric wires with plugs

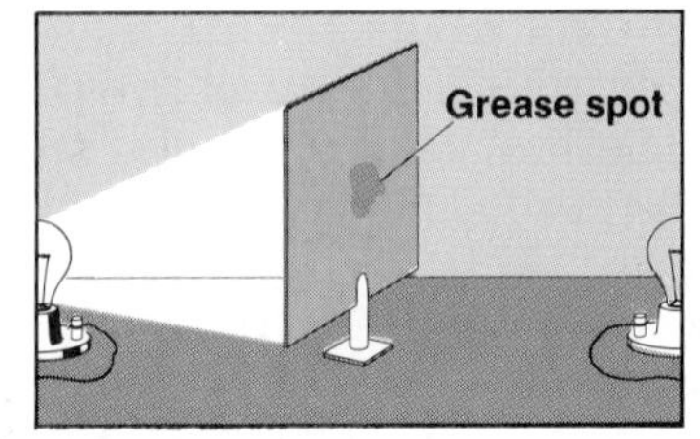

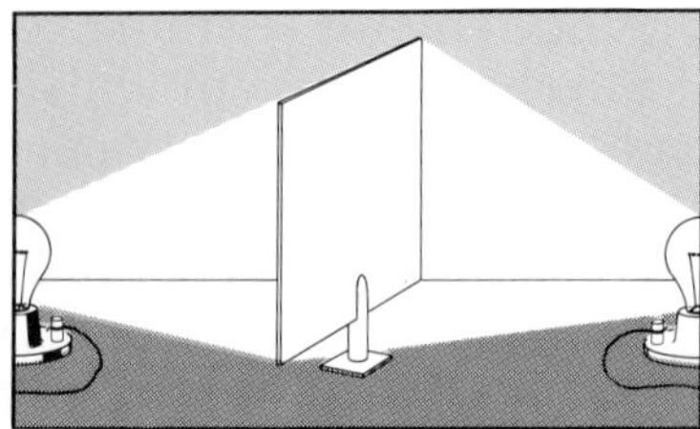

PROCEDURE

1. Set up the lights and paper stand as shown in the diagram. Rub a drop of oil, grease, butter, or fat onto a sheet of paper and put it in the paper stand. Turn on one light and look at the paper from the opposite side, as shown in the diagram on the left. Notice the grease spot.
2. Turn off the first light behind the paper and turn on the second light in front of the paper. Observe the grease spot again.
3. Turn on the first light so that both lights are on, as in the diagram on the right. Again look at the grease spot.

OBSERVATIONS AND CONCLUSIONS

1. Is the oil on the paper a good transmitter of light? Explain.
2. Is the oil a good reflector of light? Explain.
3. When the light was equally bright in front and behind the paper, what happened to the spot?
4. How can the oil-spot light meter be used?

would be needed to equal the total lumens emitted by one 1,000-W bulb? How many watts would be needed to light all of these 25-W bulbs?

The laser produces coherent light. In 1960, scientists developed the first laser, which produced a kind of light that never before existed in nature. The special properties of a laser light make it a useful tool in surveying, surgery, communication, and many other fields. For instance, you may have seen lasers used to read price codes at the supermarket checkout counter. What makes laser light so useful? The light from a laser has the following three properties:

1. It is monochromatic. White light consists of a wide range of wavelengths. Even light of one color contains a wide range of wavelengths. The light from a laser, however, is all of the same wavelength or frequency. Figure 22-6 contrasts the wavelengths of white light and monochromatic light.
2. It is coherent. Normally, light waves travel in random phase; that is, the crests and troughs of one wave are not lined up with the crests and troughs of another. In a laser, however, the crests and troughs of waves are in phase, as illustrated in Figure 22-7.
3. It travels in a plane wave front. Light normally tends to spread out in all directions in a curved wave front, the way water waves travel when you drop a pebble in a quiet pond. At distances over 1 km, even focused searchlight beams spread out. Light from a laser, however, travels in almost parallel lines with very little spreading, even at great distances (Figure 22-8). A laser beam may spread as little as 1 cm in 2 km. An early laser was aimed at the moon 384,000 km away, yet its beam spread to a diameter of only 3 km.

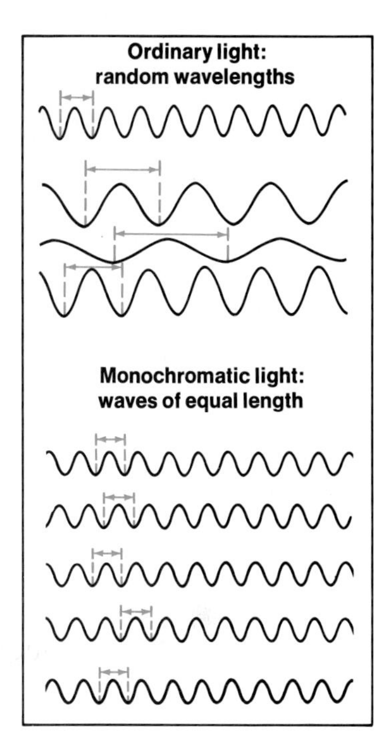

Figure 22-6 What is the major difference in the wavelengths of common light and laser light?

A **laser** is a device that amplifies light as it passes through energized material. The word "laser" is taken from the first letters of *l*ight *a*mplification by *s*timulated *e*mission of *r*adiation.

The frequency and wavelength of light produced when electrons drop from one energy level to another depend upon the difference in these energy levels. In a laser, only a single change in energy level is produced. As a result, light of only one frequency is emitted.

Before an atom can emit a photon of light, one of its electrons must absorb energy and move to a higher energy level. In the laser, the needed energy is "pumped in." Sometimes this energy comes from an outside light source. An electric discharge in the gas within the laser tube can also provide the needed energy. At this point, the atoms are said to be excited.

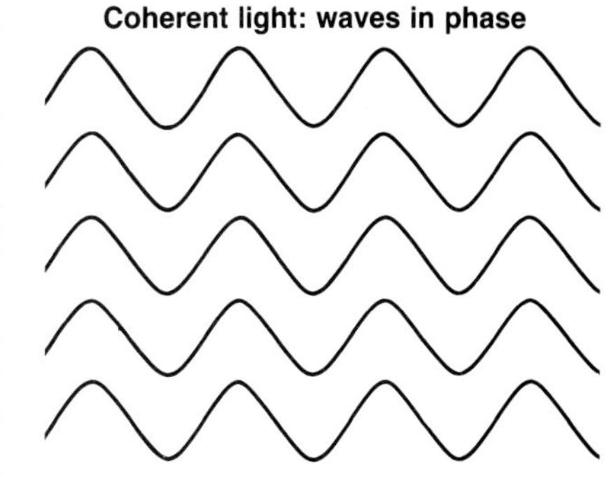

Figure 22-7 What is meant by the statement that laser light is coherent?

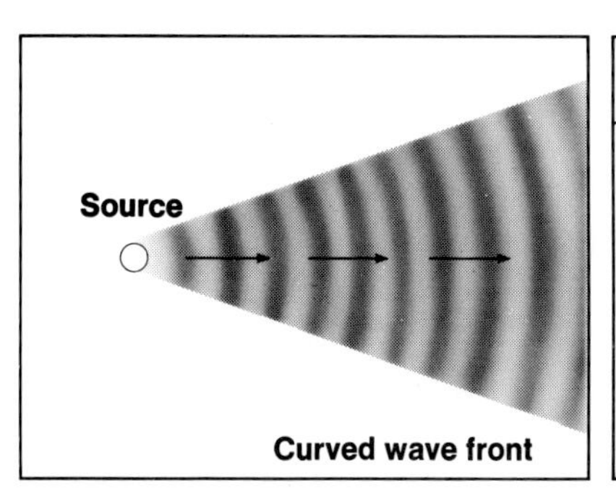

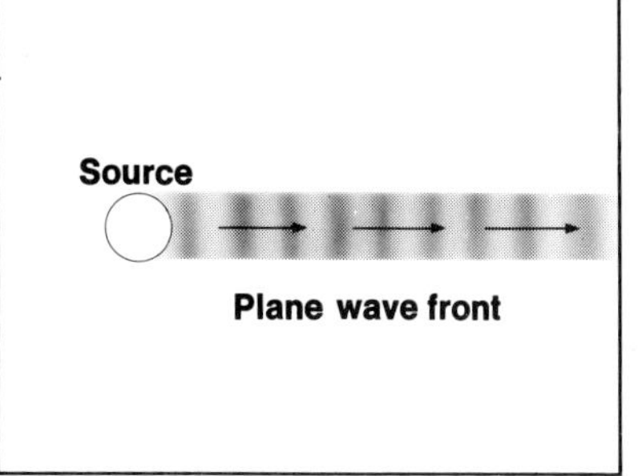

Figure 22-8 What is the effect of the curved wave front and the plane wave front on the light's energy?

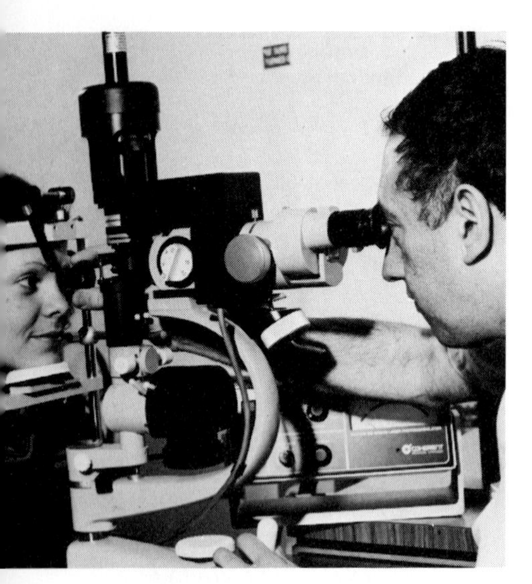

Figure 22-9 Why are lasers particularly suited for use in eye surgery?

Since excited atoms are unstable, they easily release their energy. Normally, the energy is released at random. In a laser, however, the emissions are said to be stimulated or activated. The first photon in the highly excited substance to be emitted will be reflected back and forth within the laser. In the process, the photon will strike billions of other excited atoms in a fraction of a second. This, in turn, triggers (or causes) the emission of more photons. These photons are all in phase, or coherent. A chain reaction is set up, and a great deal of energy stored in the atoms is released.

The triggering action is referred to as amplification. Because one end of the laser is only partially mirrored, a small fraction of the radiated energy leaks through. However, this small fraction still contains a great amount of energy.

The first laser devices were able to produce only pulses of light. Now, however, continuous wave lasers emit a constant beam of light. The substances used in lasers include gases, liquids, and solids.

Lasers have many uses. Scientists have found a wide range of uses for the laser. For example, it can be used as a range finder. The time interval between sending of the laser pulse and receiving the reflection can be recorded and the distance computed. Using the laser reflector placed on the moon by astronauts Armstrong and Aldrin, scientists were able to measure the distance to the moon to within a few centimeters and were able to locate the exact spot of each lunar-landing site.

The laser can amplify heat as well as light to melt extremely hard materials. Laser heat is used to drill tiny holes in diamonds. Drilling, which formerly took three days, is done in two minutes by a laser beam.

The laser is also used in medicine. The highly focused beam is used for delicate surgery. In Figure 22-9, a laser is being used to "weld" a detached retina to the back of the eye. The pulse of energy acts within a thousandth of a second. In fact, laser surgery takes place so fast that normal eye movements do not cause a blurring of the selected "welding" site. Lasers are also used in brain surgery and in the treatment of skin cancers. Scientists may one day use lasers to study molecular bonds. A laser can be directed to break one chemical bond without damaging others around it. X-ray lasers are being developed. With these, scientists can produce three-dimensional views of microscopic structures.

Lasers are used to make very lifelike, three-dimensional images called **holograms.** The laser light is split into two beams. One beam hits an object, and then is reflected to a photographic plate. The other beam is reflected from a set of mirrors directly onto the same plate. The difference in distance each beam traveled puts them out of phase and creates an interference pattern on the plate. This pattern gives the effect of a three-dimensional picture when light shines through the hologram. Holography is being used to find defects in machinery and diseases in animal tissue.

Fiber optics is an application of the physics of light. Thin fibers of pure glass or plastic are used to transmit information in the form of light. These fibers, as small as 0.025 mm in diameter, consist of a core surrounded by a cover called the cladding. Light entering the optical fiber at a certain angle can travel through bends in the fiber because light rays hitting the cladding are bent back into the core. A laser can be the source of light. The laser flashes at a very high rate, sending a coded message through the optical fiber to a device that changes the code back to the original signal.

Telecommunications companies have found that optical fibers are far superior to copper cables for transmitting information. Fiber-optic cables are not subject to electrical interference, need less amplification at the receiving end, and have greater information-carrying capacity in a less bulky medium. Whereas a copper cable is limited to 24 telephone conversations at one time, a fiber-optic cable is capable of transmitting over 8,000 conversations.

22.2 REFLECTION OF LIGHT

Images are formed by even reflection. In Chapter 21, you learned that sound waves can be reflected when they encounter a surface. Light waves that strike an object also can be reflected, in two basic ways. The reflected light is either even or uneven.

If light strikes a smooth surface, the light is reflected in an even pattern (Figure 22-10). If the surface is smooth

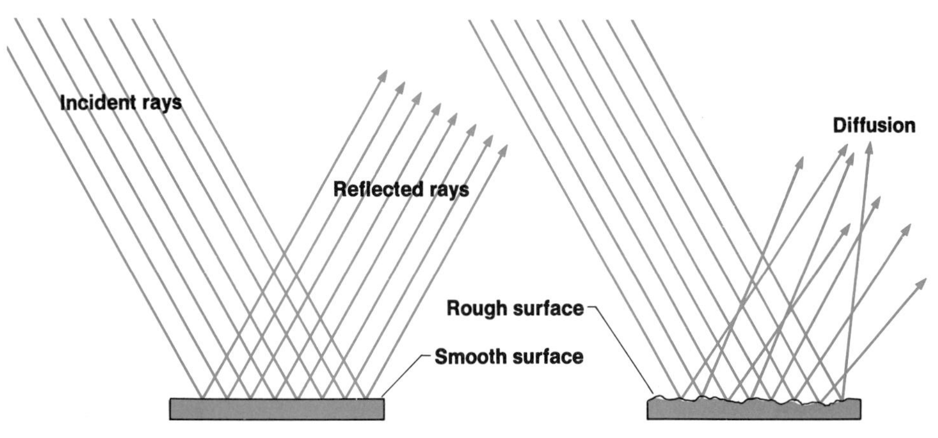

Figure 22-10 In which of the two surfaces can you see a mirror image?

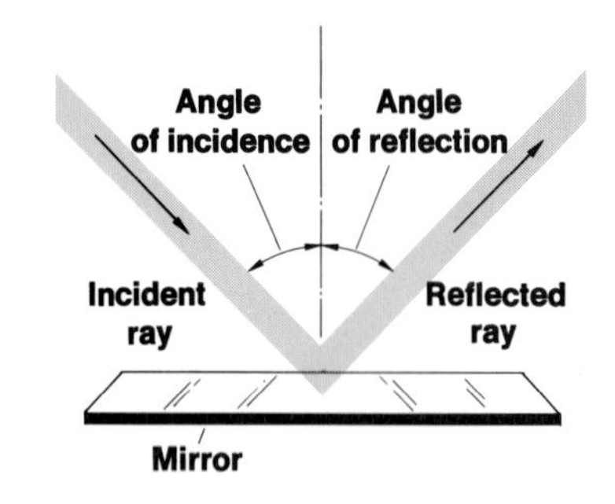

Figure 22-11 How are the angle of incidence and angle of reflection related according to the law of reflection?

enough, as in a mirror or highly polished metal plate, you can see an image in that surface. If the surface is rough, as in a plastered wall or a sheet of paper, the reflected light is scattered or diffused. The reflected light is uneven and cannot form an image. Instead, diffusion makes the surface itself visible.

A perfect plate-glass wall mirror would not be visible. You could only see images of other things reflected in it, not the glass itself. On the other hand, when you look at a plastered wall, you see no images, only the wall itself.

Plane mirrors obey the law of reflection. A flat, polished surface that reflects light without diffusing it is called a plane mirror. Notice in Figure 22-11 that the angle at which light is reflected from a plane mirror is the same as the angle at which it strikes the mirror. The angle between the perpendicular to a surface and the light beam striking the surface is the angle of incidence. The **law of reflection** states that the angle of reflection equals the angle of incidence. A ball bouncing on the sidewalk obeys the same law.

When light is reflected from a plane mirror, the pattern of the light rays is even or undisturbed. Figure 22-12 shows how an image is formed by a plane mirror. Note the path of the rays as they travel from the object P, to the mirror, and then to the eyes. Your eye has no way of telling that the light rays have been reflected. Instead, the light rays seem to come from a point behind the mirror, P′. You "see" the object at that point, called the image.

You have seen that images formed by plane mirrors are life-sized, equal in size to the object. The image is located as far behind the mirror as the object is in front of it. Images formed by plane mirrors are called **virtual images** because they do not exist where they appear to be, behind the mirror. Virtual images are always upright, or erect. They are also "left-handed," or mirror, images. Stand in front of a mirror and reach out with your right hand to shake hands with your image. Which hand does the image extend?

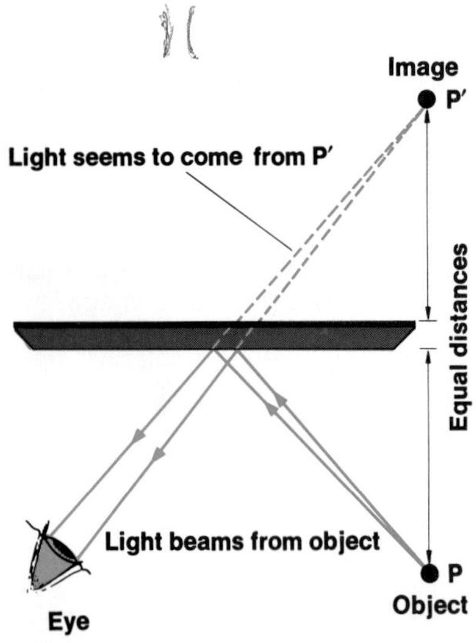

Figure 22-12 Compare the distance from the object to the mirror and from the image to the mirror. How do they compare? Where does the object appear to be?

Curved mirrors also obey the law of reflection. There are two basic kinds of curved mirrors: concave or convex. A mirror, lens, or any object the surface of which curves in or recesses is said to be **concave.** An object the surface of which curves out is **convex.** The inside of an orange peel has a concave shape and the outside is convex. Curved mirrors are often spherical, or shaped like a round section cut from a sphere. However, mirrors may have any kind of curve whatsoever.

REFLECTIONS FROM A PLANE MIRROR

activity

OBJECTIVE: Perform an experiment to observe images in a plane mirror.

PROCESS SKILLS

In this activity, you will *observe* images in a plane mirror that will help you to *draw conclusions* about how images are formed.

MATERIALS

3 plane mirrors
masking tape
clock

PROCEDURE

1. Place a clock in front of a plane mirror. Observe the image of the clock face in the mirror.
2. Hinge two square plane mirrors together with masking tape. Stand them on the table at a 90° angle from each other. Place the clock in front of the mirrors and observe the images of the clock.

OBSERVATIONS AND CONCLUSIONS

1. How do the numbers on the clock appear in the single plane mirror?
2. How do these numbers appear in the hinged mirrors?
3. How do you account for the difference in the images?

All mirrors follow the law of reflection discussed earlier. At whatever point on the mirror a ray of light falls, it will be reflected in such a way that the angle of reflection equals the angle of incidence. When the surface is curved, however, the images that result are quite different from those formed by a plane mirror.

You may have seen the odd images produced by the curved mirrors in an amusement park fun house. You also may have seen the images in a silver bowl or spoon, in soap bubbles, or in the curved chrome on cars. These images are reflected from surfaces that act like curved mirrors.

Curved mirrors have some important uses. In the world's largest telescopes, curved mirrors focus the light from distant stars. Curved mirrors are also used in searchlights, auto headlights, and flashlights to direct light to a certain spot.

In Figure 22-13 a spherical concave mirror is shown. The center of the sphere, C, of which the mirror is a section, is called the center of curvature. Line PO is called the principal axis and point F is the principal focus. All light rays that reach the mirror parallel to the principal axis will be reflected through the principal focus. By tracing the path of several rays from the object, you can locate its image.

An object, such as the arrow AB, is placed in front of the mirror at a point beyond C. A light ray from the tip A, parallel to the principal axis, hits the mirror and then reflects back through F. Another ray from A through C hits the mirror "head on" and reflects back upon itself along the same path. The image of the portion of the object that is at point A is located where these two reflected rays cross, at A′. Points B and B′ are located in the same way.

Notice that the image in Figure 22-13 is inverted. When an object is placed at a distance greater than the focal length, concave mirrors form what are known as real images, rather than virtual images. A **real image** is an inverted image formed by rays that converge at one location. In Figure 22-13, that location is line A′B′. If you held a piece of white paper along line A′B′, the paper would act as a screen on which the reflected light rays would converge. A small image of arrow AB would be projected on the paper screen. A "real" observable image actually exists at that location. In contrast to a real image, a virtual image does not actually exist where it appears to. The rays reflected from a plane mirror do not actually converge behind the mirror to form an image. Because the rays of a virtual image do not converge, a virtual image cannot be projected onto a screen.

Figure 22-13 Is the image formed by this curved mirror real or virtual? Is the image right side up or upside down?

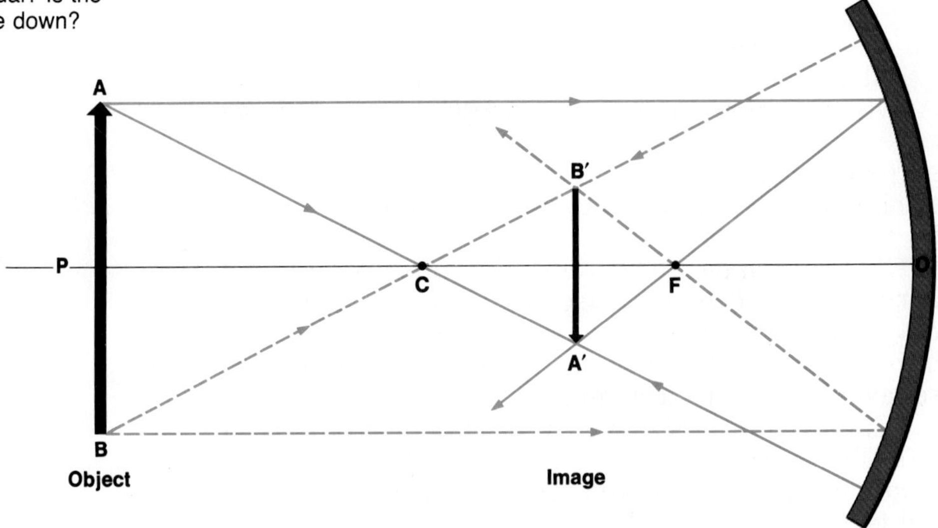

Concave mirrors are used in **reflecting telescopes**. The reflecting surface is commonly coated with aluminum. A telescope of this type is diagrammed in Figure 22-14. One of the largest reflecting telescopes at Mt. Palomar Observatory in California has a mirror more than 5 m in diameter.

Concave mirrors make good magnifiers. If you put an object inside the principal focus, F, of a concave mirror, the virtual image formed in the mirror is larger than the object. A shaving or makeup mirror utilizes a concave surface to magnify the image of your face.

The reflecting surface of a convex mirror bulges outward. The silvered balls used to decorate Christmas trees are good examples of convex mirrors. Convex mirrors are sometimes used as rear-view mirrors for cars and trucks. Although the image formed by a convex mirror is smaller than the object, it does provide a wide field of view. Convex mirrors always form a virtual image that cannot be projected onto a screen.

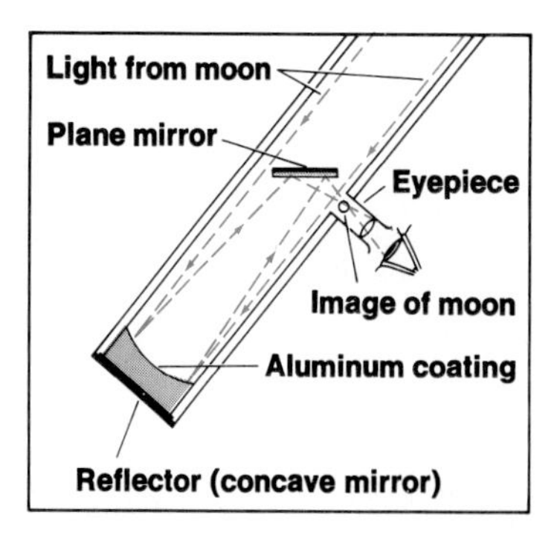

Figure 22-14 What kind of mirrors are used in this reflecting telescope? How many mirrors are there?

IMAGES IN A CONCAVE MIRROR

activity

OBJECTIVE: Perform an experiment to verify that concave mirrors produce real images.

PROCESS SKILLS

In this activity, you will *experiment* to find the image of a light source, which will help you to *draw conclusions* about the kind of image formed by a concave mirror.

MATERIALS

concave mirror in a holder
index card
movable light source such as a light bulb in a socket

PROCEDURE

1. In a darkened room, place a lighted bulb several feet in front of a concave mirror.
2. Move a small card back and forth between the light bulb and mirror until an image of the bulb can be seen on the card.
3. Describe the image of the light bulb.

OBSERVATIONS AND CONCLUSIONS

1. Is the image larger or smaller than the light bulb?
2. Is the image erect or inverted?
3. Is the image virtual or real? How can you tell?

22.3 REFRACTION OF LIGHT

Figure 22-15 Why does the coin become visible when water is added to the cup?

Seeing is not always believing. Have you ever ridden in a car on a hot, dry afternoon and seen that at times the highway ahead looked wet? The road looked as if there were reflections in it, as in a pool of water. But when you reached the spot, the water was gone.

This illusion is an example of a mirage. A **mirage** is an optical illusion caused by the bending of light rays as they pass through air layers of different densities. As you learned in Chapter 21, the bending of any waves, such as light or sound waves, as they pass from one medium into another is called refraction. As long as light passes through a medium of constant density, it travels in a straight line. When it enters a medium of different density, the light is refracted.

Refraction of light can be demonstrated, as shown in Figure 22-15, by putting a penny in the bottom of an empty cup and standing back so that the cup's rim just barely hides the penny from view. Without changing your position, pour water into the cup, being careful not to disturb the penny. What happens? The penny seems to rise in the cup until you can see it clearly because the light traveling from the coin to your eye is refracted as it passes from the water to the less dense air.

If you dip a straight stick at an angle into water, the stick appears to be sharply bent at the water line. This effect is also due to the refraction of light rays as they pass from one medium into another.

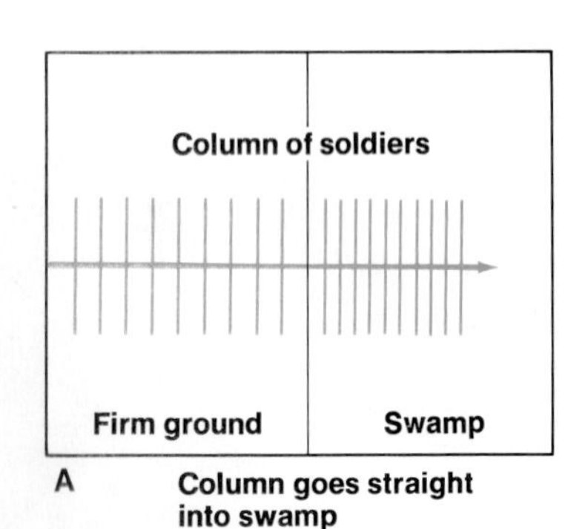

A Column goes straight into swamp

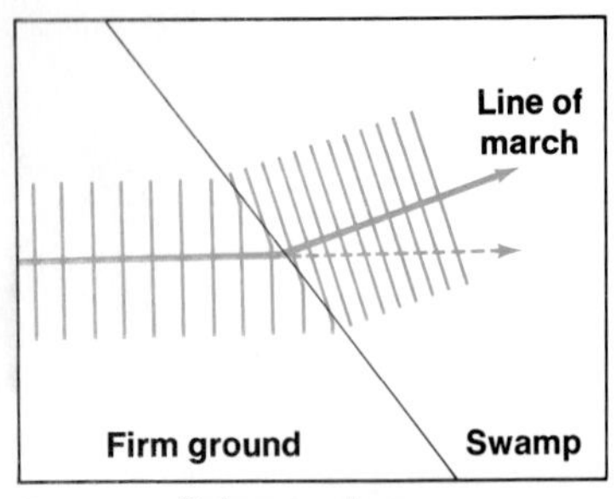

B Column enters swamp at an angle

Figure 22-16 What happens when the soldiers go directly into the swamp? when they enter the swamp at an angle? How does the column of soldiers compare to light beams?

Light refracts because it changes speed. When light goes from air into glass, it slows down from 300,000 km/sec to 200,000 km/sec. If it enters glass at an angle, light is bent into a new angle.

The bending of light can be compared to a column of soldiers marching from firm ground into a soggy swamp. As the first row of soldiers steps into the swamp, they slow down. The second and following rows keep coming and bunch up as they too enter the swamp. The soldiers, as shown in Figure 22-16A, continue to move through the swamp, traveling more slowly and closer together than they did on firm ground.

Much the same thing happens when the column enters the swamp at an angle. The soldiers in the front row do not all step into the swamp at the same time. Instead, they enter it one at a time. As they enter, each soldier falls slightly behind the soldier to the right. Thus, the line of soldiers who

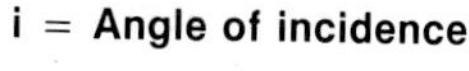

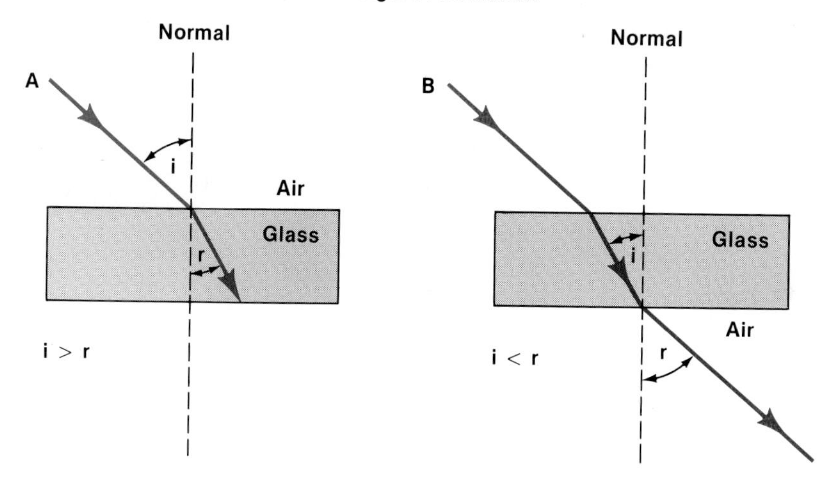

Figure 22-17 Light bends toward the normal as it passes from the less dense air into the denser glass (A) and away from the normal as it passes back into the air (B).

were formerly side-by-side is now turned to a new angle of travel. The following rows of soldiers bunch up behind (Figure 22-16B).

Which way is light bent when it goes from one medium to another at an angle? If light passes at an angle from a substance like air into a denser substance like water or glass, it bends toward an imaginary line called the normal that is perpendicular to the surface between the two substances. This is shown in Figure 22-17A. When light passes at an angle from a dense substance into one that is less dense, it bends away from the normal (Figure 22-17B).

Light travels at different speeds in different substances. As you can see in the example of the marching soldiers, the more they slow down, the more the lines bend. So the degree to which a substance is able to bend light is related to the speed of light in that substance. This refracting or bending ability of a substance is related to its index of refraction.

You can find the index of refraction of a substance by dividing the speed of light in a vacuum by the speed of light in that substance. In most cases, the speed of light in air, rather than in a vacuum, is used because there is very little difference between the two speeds.

The index of refraction is a characteristic property of a substance and can be used to identify that substance. For instance, diamond has an index of 2.42, whereas glass has an index of about 1.5. A jeweler can easily tell a fake from a real diamond by measuring its index of refraction.

Not all glass has an index of refraction of 1.5. The exact value depends on the kind of glass. The index of refraction for several substances is shown in Table 22-2.

TABLE 22-2

Substance	Index of Refraction
Vacuum	1.0000
Air	1.0003
Ice	1.31
Water	1.33
Glass (crown)	1.52
Glass (flint)	1.61
Diamond	2.42

SAMPLE PROBLEM

What is the index of refraction of glass through which light travels at 200,000 km/sec?

SOLUTION:

Step 1: Analyze

You are given that the speed of light in glass is 200,000 km/sec. The index of refraction for the glass is the unknown.

Step 2: Plan

You can derive the following equation from the definition of the index of refraction:

$$\text{index of refraction} = \frac{\text{speed of light in vacuum or air}}{\text{speed of light in substance}}$$

Since you know the speed of light in air and in the substance (glass), you can use the equation to calculate the index of refraction.

Step 3: Compute and Check

Substitute all values in the equation.

$$\text{index of refraction} = \frac{300{,}000 \text{ km/sec}}{200{,}000 \text{ km/sec}}$$
$$= 1.5$$

Since your answer is approximately equal to the index of refraction given for crown glass in Table 22-2, your solution is complete.

Air can also cause refraction. Most often you can disregard the difference between the speed of light in air and its speed in a vacuum. However, this difference does result in a slight bending of light as it enters the atmosphere. This bending allows you to see the sun before it rises and after it sets. Figure 22-18 shows how this happens.

Suppose that you are at C, looking toward the setting sun. As the ray AB enters the air, it bends as shown. This refraction occurs because the ray is traveling into denser air and, as a result, slows down. As you can see, the effect of this refraction is to "lift" the sun above the horizon. Refraction lengthens the hours of daylight because the earth receives the sun's rays before the sun actually rises and after it sets.

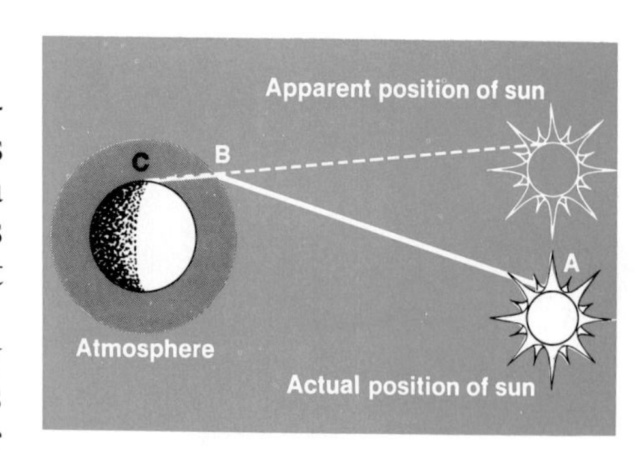

Figure 22-18 Why is the sun visible when it is still below the horizon?

You can see other ways in which air affects light. Look at a distant object using a line of sight that is directly above a heat source, such as a candle, gas burner, or hot plate. The dancing of heat rays on a hot metal roof, the shimmering of a hot roadway in the distance, and the twinkling of stars are all examples of distortions caused by the changing refraction of the light as it passes through air of varying densities. In working out star positions, astronomers must correct their readings to allow for the refraction of air.

Lenses refract light. A **lens** is a transparent substance having at least one curved surface. Although lenses differ in shape, size, and substance, the chief purpose of all lenses is to refract light. The amount of refraction produced by a lens depends both upon its shape and upon the index of refraction of the substance of which it is made. Lenses are used in telescopes, cameras, eyeglasses, microscopes, spectroscopes, movie projectors, and many other optical devices.

To understand how lenses affect light, first look at the path of light through a simple prism. Notice that in each prism in Figure 22-19A, the ray bends into the prism as it

Figure 22-19 In what ways are double convex and double concave lenses like prisms?

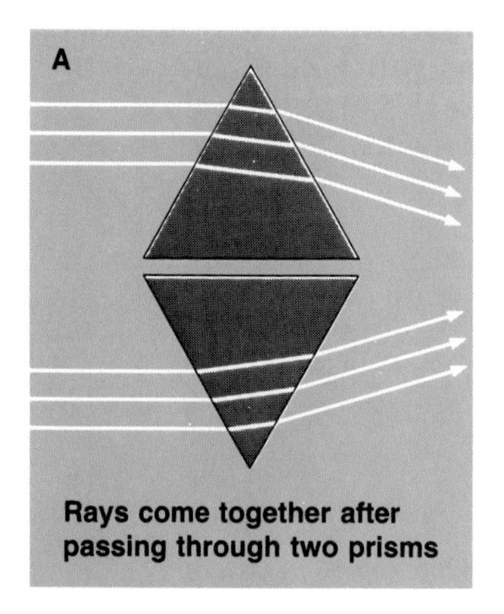

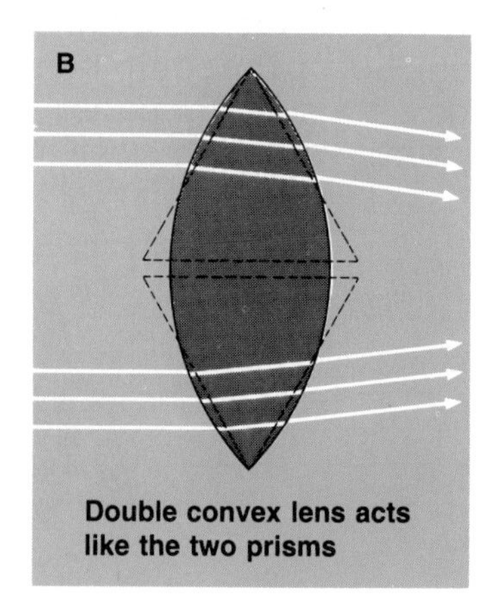

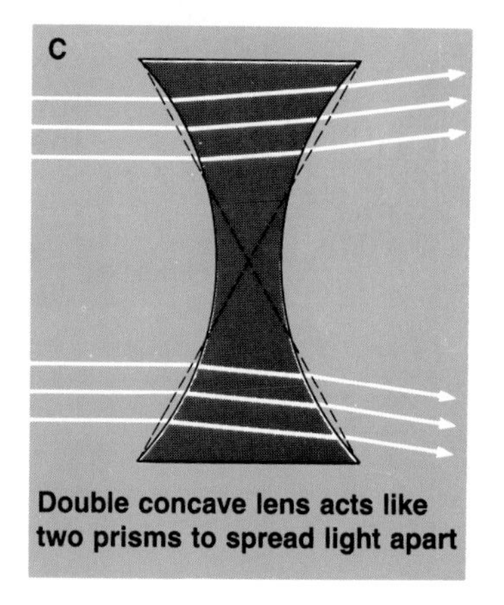

enters the glass and toward the thick part of the prism as it leaves. When two prisms are set base-to-base, light rays come together, or converge, after passing through the two prisms. This occurs because the light is refracted toward the thicker part of each prism.

The double convex lens in Figure 22-19B is very much like two prisms joined together with their side points smoothed down. Note that a convex lens is thicker at the center than at the edges. Since it always brings parallel light rays together, it is called a converging lens.

If two prisms are placed together point-to-point, refraction causes the light rays to spread apart, or diverge. As shown in Figure 22-19C, a double concave lens is like two prisms joined together, point-to-point. A concave lens is thinner at the center than at the edges and tends to spread light rays apart. Therefore, it is called a diverging lens.

A lens has a focal length. A common magnifying glass is a good example of a convex lens. With it, you can do some tests that will help you see how lenses affect light. With your teacher's supervision, focus the light from the sun on a sheet of paper by moving the lens back and forth until a dazzling spot appears on the paper. The spot appears where the light from the sun has converged after passing through the lens (Figure 22-20). This spot is so hot that the paper may char or even burn. CAUTION: *Do not perform this experiment without your teacher's supervision. Do not look directly at the sun since this can cause blindness.*

The light of the sun comes together at what is called the principal focus, or focal point, of the lens. The **principal focus** is the point at which the rays parallel to the principal axis converge after passing through the convex lens. The distance from the center of the lens to the principal focus is called the **focal length** of the lens. To find the focal length of a lens, simply measure the distance between the magnifying glass and the bright spot on the paper.

To see how lenses form images, use a convex lens and a luminous object, such as a lighted candle or bulb, in a darkened room. Set up the equipment as in Figure 22-21, focusing the image on a sheet of paper. Move the lens slowly along the stick until you get a clear image on the paper screen.

Figure 22-20 How can you find the focal length of a convex lens?

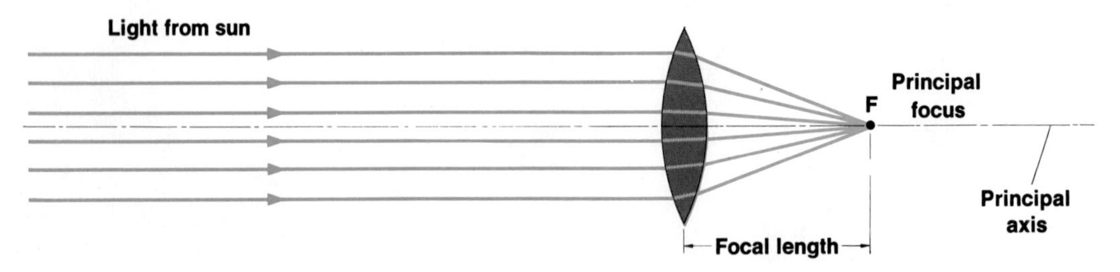

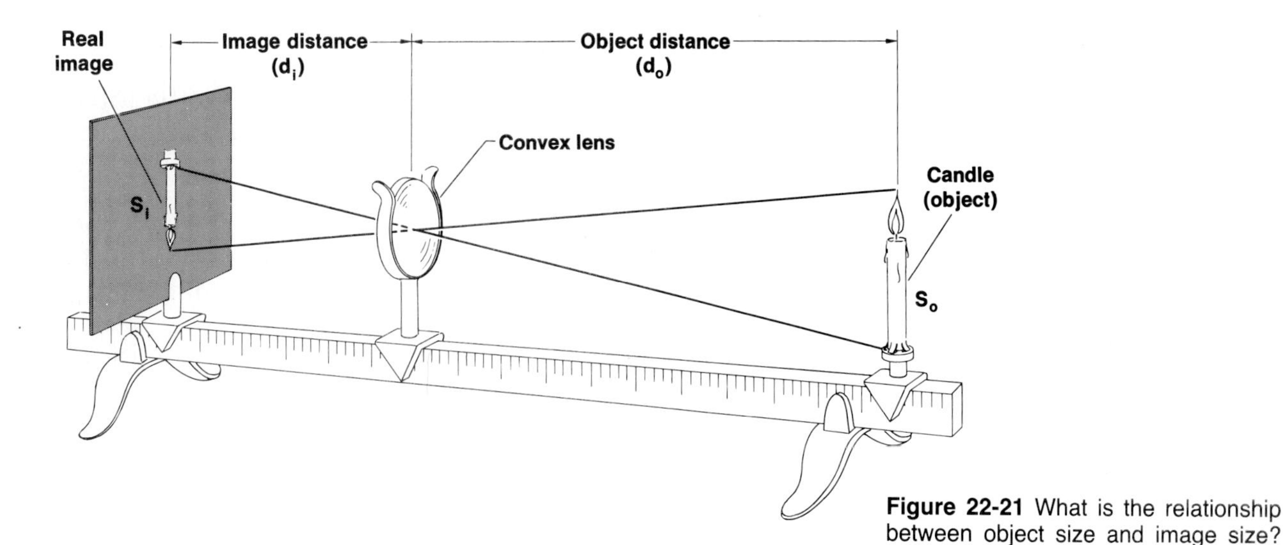

Figure 22-21 What is the relationship between object size and image size? object distance and image distance? Why is the image "real?" **CAUTION: Take care when using lighted candles to avoid burns or fire.**

The image you see is a real image. Notice that it is in full color, inverted, and, in this case, smaller than the object. The image is located very close to the principal focus of the lens.

The image is smaller than the object because the object is so far away. The closer the object is to the lens, the larger the image will be. The following formula shows how the distances and sizes of objects and their images are related to each other:

$$\frac{\text{size of image}}{\text{size of object}} = \frac{\text{distance of image}}{\text{distance of object}}$$

or

$$\frac{s_i}{s_o} = \frac{d_i}{d_o}$$

For example, if the image were 10 times as far away from the lens as the object, the image would be 10 times larger than the object. The magnification of the lens would be 10, or 10X.

Suppose you moved the candle from its location in Figure 22-21 to another position farther from the lens. What happens to the image? From Figure 22-22 you can see that the image moves closer to the lens and becomes smaller.

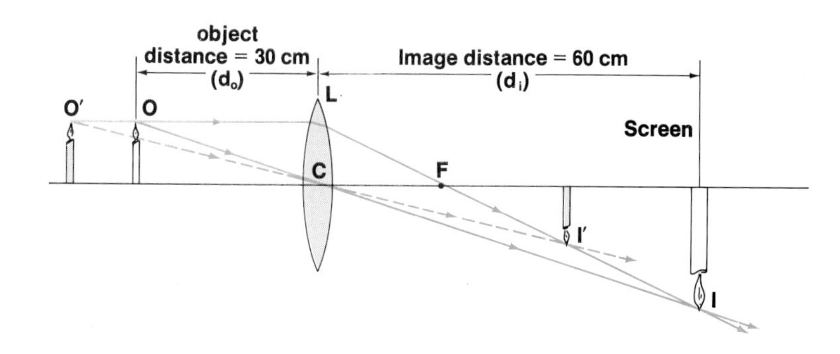

Figure 22-22 What happens to the image size and distance if the candle is moved from point O to O'?

If the lens had a focal length of 20 cm and the candle was 30 cm from the center of the lens, where would the image be? As an example, to find the image of point O (the top of the candle), you must find where rays of light from that point come together again after passing through the lens. You know that the ray parallel to the principal axis OL will refract through F, the principal focus. Another ray, OC, drawn through the center of the lens, will pass straight through without refracting. At the point I where both rays intersect is the image of point O.

Try following the same procedure for the base of the candle. A ray of light from the base, parallel to OL, goes straight through C and will not be refracted.

If you repeat this process for other points on the object candle, you will be able to draw the whole image of the candle. The image will be at the point shown in the diagram that is 60 cm from the lens. Since the image is twice as far away as the object, it is also twice as large. Like all real images, it is inverted.

Images can be located. As you move the candle nearer and farther away from the lens, you will find that you have to move the screen in order to focus the image. The closer the candle is to the focal point, the farther away is the image. This relationship between object and image location is shown by the following formula:

$$\frac{1}{f} = \frac{1}{d_o} + \frac{1}{d_i}$$

As shown in Figure 22-22, d_o is the object distance, d_i is the image distance, and f is the focal length. Instead of locating the image by drawing a diagram, you can find its location by using the formula. If the object is 30 cm from the lens and the focal length is 20 cm, you can solve for d_i:

$$\frac{1}{20} = \frac{1}{30} + \frac{1}{d_i}$$

$$d_i = 60 \text{ cm}$$

Lenses have many uses. In Figure 22-22, you can see that as the object is moved closer to the lens, the image moves farther away, becoming larger. If the object is placed just outside the focal point, the image is very large and far away. This is how slides and movie films are projected on a screen. The object is the lighted slide or film. It is placed just outside the focal length of the projector lens. Thus, a large, inverted image is formed on the distant screen. Why is the slide placed into the projector upside down?

SAMPLE PROBLEM

A 6-cm stick stands upright 12 cm from a convex lens with a focal length of 3 cm. (a) How far away from the lens is the image? (b) How large is the image?

SOLUTION TO (a):

Step 1: Analyze

You are given that the size of the object is 6 cm, the object distance is 12 cm, and the focal distance is 3 cm. The image distance is the unknown.

Step 2: Plan

Since you know f and d_o, you can solve for d_i using the equation

$$\frac{1}{f} = \frac{1}{d_o} + \frac{1}{d_i} \qquad d_i = \frac{1}{\left(\frac{1}{f} - \frac{1}{d_o}\right)}$$

Step 3: Compute and Check

Substitute all values in the equation.

$$d_i = \frac{1}{\left(\frac{1}{3\text{ cm}} - \frac{1}{12\text{ cm}}\right)}$$

$$= 4\text{ cm}$$

You can check your answer by substituting all values in the original equation. Since 1/3 = 1/12 + 1/4, your solution is complete.

SOLUTION TO (b):

Step 1: Analyze

You are given s_o, *f, and* d_o. You have derived d_i. The image size (s_i) is the unknown.

Step 2: Plan

You can solve for s_i using the equation

$$\frac{s_i}{s_o} = \frac{d_i}{d_o} \qquad s_i = \frac{d_i}{d_o} \times s_o$$

Step 3: Compute and Check

Substitute all values in the equation.

$$s_i = \frac{4\text{ cm}}{12\text{ cm}} \times 6\text{ cm}$$

$$= 2\text{ cm}$$

You know that the object is 3 times farther from the lens than is the image, therefore the object must be 3 times larger than the image. Since 6 cm is 3 times larger than 2 cm, your solution is complete.

If the object is moved closer to the lens, until it is at the focal point, no clear image appears on the screen. Instead, the rays from the object are parallel after passing through the lens. If a point source of light is placed at the focal point, a parallel beam of light can be projected.

If the object is moved still closer to the lens, inside the focal point of Figure 22-22, the rays will go through the lens and diverge on the other side. Remember, real images are formed only when rays coming from the object converge after passing through the lens. When the object is located inside the focal point, no real image can be formed. However, a virtual image can be seen by looking through the lens. As shown in Figure 22-23, the image is enlarged, and right side up. For instance, when you read with a magnifier, you hold it closer to the page than the focal length of the lens and keep your eye close to the lens.

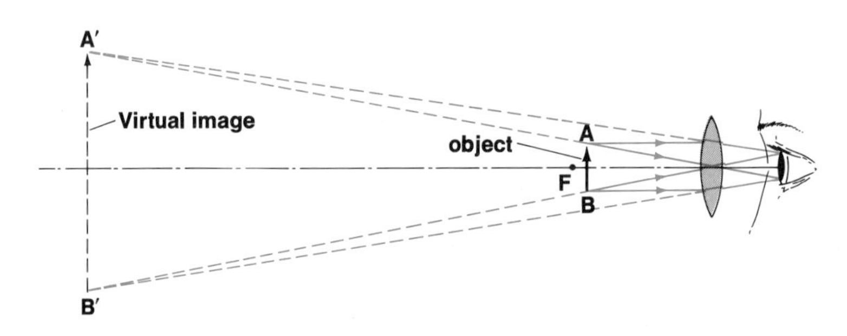

Figure 22-23 Why is the image upright in this diagram of a microscope?

Have you ever wondered how a camera works? A convex lens focuses an image on the film. In taking pictures of distant objects, the distance between the lens and the film is almost equal to the focal length of the lens. Why is this distance necessary?

To get sharp pictures of nearby objects, the distance between the film and the lens must be increased. Most cameras are equipped with an adjustment for changing the film–lens distance. A good lens, with a large opening, can give sharp pictures with a short exposure time. When the film is exposed, any movement becomes a blur. The shorter the exposure, the less blurring occurs.

The eye is like a camera. In many ways, your eye acts like a camera. However, it does not record an image on film. Instead, your eye is linked to your brain somewhat like a television camera is connected to its transmitter.

Figure 22-24 shows the parts of the eye that work like a camera. Light enters your eye through the tough transparent layer called the **cornea**, where it is refracted. It then goes through a flexible convex lens, where it is bent even more. This refraction causes an image to form on the **retina** (RET-*uh-nuh*), which is the light sensitive, innermost layer

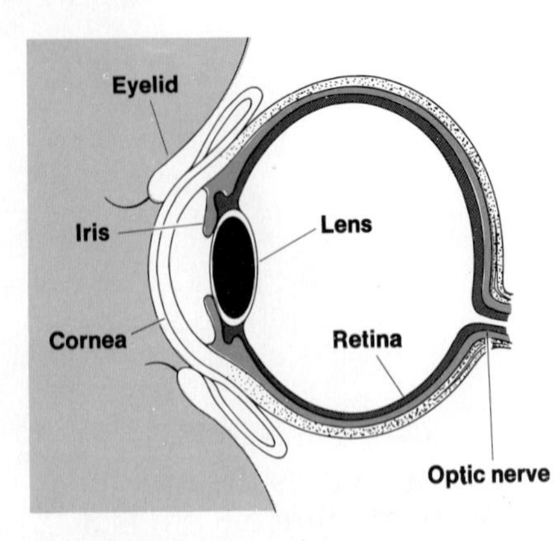

Figure 22-24 How does an eye's iris act like a camera's diaphragm?

of the eyeball. Light striking the retina produces nerve impulses that travel through the optic nerve to the visual center of the brain. It is in the brain that the sensation of "seeing" occurs.

The amount of light that can enter your eye is controlled by the iris, a colored ring that opens up in the dark and closes down to a tiny opening in bright light. The iris, which is just outside the lens, acts like the diaphragm, an adjustable opening of the camera. Your eyelids correspond to the shutter of the camera. They both open and close to expose the "film."

The lens of your eye is flexible. It can bulge out or flatten to focus on objects that are near or far away. To focus on nearby objects, muscles in your eye cause the lens to bulge out, shortening the focal length of the lens.

When focusing on objects over 6 m away, the lens is at its flattest and the muscles are relaxed. When the lens is relaxed, its focal length is almost equal to the distance from the lens to the retina.

Sometimes, however, images do not form clearly on the retina. **Nearsightedness** is a condition that results if a person has eyeballs that are too long, as shown in Figure 22-25, or lenses that are too thick. In either case, the image is brought to focus in front of the retina instead of on it. A nearsighted person sees a fuzzy image of distant objects.

Nearsightedness can be corrected by using concave lenses in front of the eye. These lenses cause light rays to diverge or spread out. As a result, the converging effect of the person's own eye lens is reduced.

Farsightedness is the condition that results if a person has eyeballs that are too short, or lenses that are too flat (Figure 22-26). This person sees a fuzzy image of nearby objects. A convex lens can correct this fault. Middle-aged persons often need to wear reading glasses after their ability to focus on nearby objects is lessened as their lenses become stiff with time. Some people need to wear bifocals (lenses with two different focal lengths) or trifocals (lenses with three different focal lengths) to be able to focus on objects at different distances.

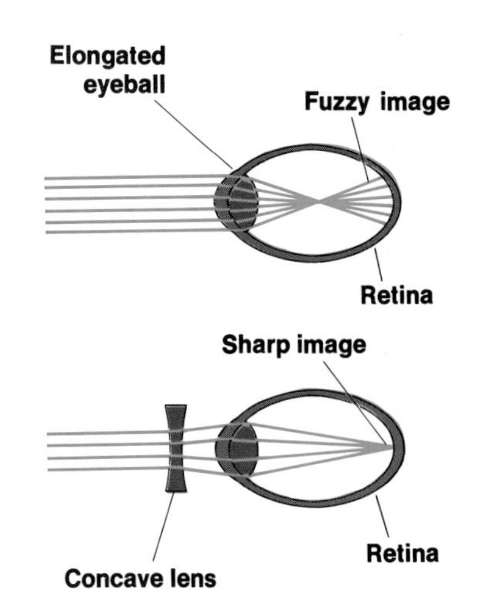

Figure 22-25 How does a concave lens correct nearsightedness?

Movies are illusions. You are familiar with how a movie projector throws an enlarged image of the film on a screen. Movies are an illusion of moving people and objects, formed by a rapid succession of projected images. What makes these images seem to move?

When you look at a lighted lamp, a real image of the lamp forms on your retina, producing the sensation of vision in your brain. When the lamp is off, the image disappears, but you still "see" it for about 0.10 sec. This effect is called persistence of vision.

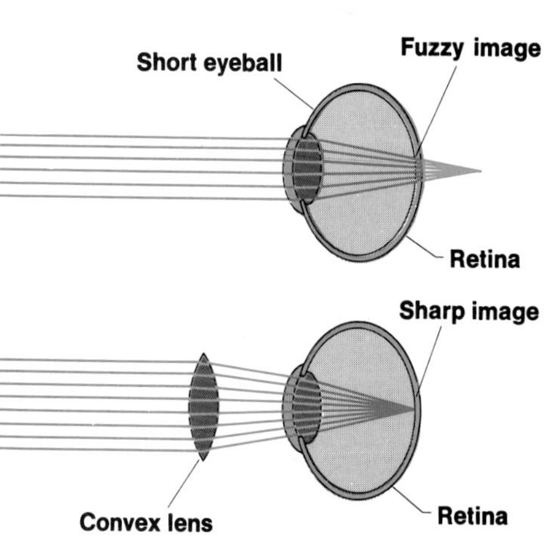

Figure 22-26 How does a convex lens correct farsightedness?

Persistence of vision makes movies possible. The movie projector throws a still picture on the screen for just a fraction of a second. Then, the picture is replaced by a slightly different one. This process is repeated 24 times per second in commercial movies, 16 times per second in home movies, and 30 times per second on television.

Because of the persistence of vision, you retain the image of one picture while the next one is shown, and the two pictures blend together. Since each picture differs only slightly from the previous picture and merges smoothly with it, you see what appears to be constant motion on the screen.

Some optical devices use more than one lens. The eye and the optical devices discussed up to now all had only a single lens to refract light. To correct a vision problem, one lens is added in front of the eye's own lens. Many optical devices use two lenses.

MAKING A MOTION PICTURE

activity

OBJECTIVE: Make a motion picture with a flip pad.

PROCESS SKILLS

In this activity, you will *model* movie film with a flip pad and *draw conclusions* about how your vision supports the illusion of motion.

MATERIALS

small note pad
pencil

PROCEDURE

1. Using a small note pad and pencil, start with the last page and make a simple line drawing, such as a stick figure of a person who is walking.
2. Turn to the preceding page and draw the same figure in the same place on the page with very slight changes in stride.
3. Keep repeating Step 2 until you have enough pages to flip the pictures rapidly with your thumb.

OBSERVATIONS AND CONCLUSIONS

1. What happened when you flipped the pictures?
2. How would you explain the result?

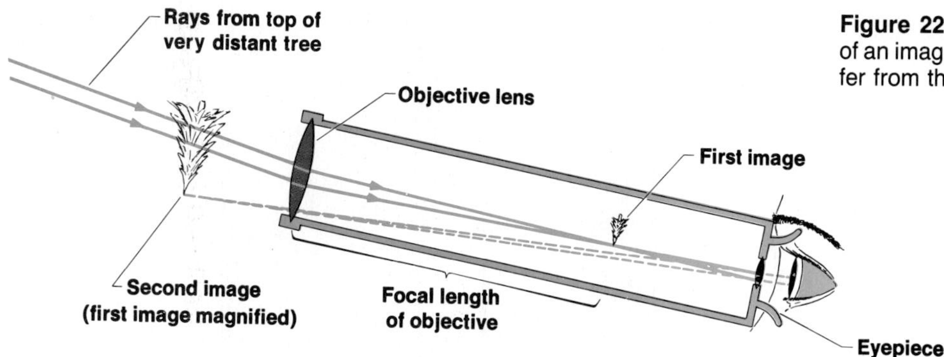

Figure 22-27 How does the formation of an image in a refracting telescope differ from that in a reflecting telescope?

One device that uses two lenses is a refracting telescope, shown in Figure 22-27. Recall that earlier in the chapter you learned about a reflecting telescope that uses a mirror to focus the light (Figure 22-14). In a refracting telescope, a lens instead of a mirror is used to focus the light.

A large convex lens, called the **objective**, is placed at the outer end of the long telescope tube. The objective lens of a telescope has a long focal length. A real inverted image of a distant object is formed by the lens near the lower end of the tube, very close to the principal focus of the lens. The image is viewed through an eyepiece, containing a small convex lens that magnifies the image.

The magnification (M) of this kind of telescope is found by dividing the focal length of the objective (f_o) by the focal length of the eyepiece (f_e):

$$M = \frac{f_o}{f_e}$$

Although the image is inverted, this is not a problem when viewing the moon or stars. (Why?) When the refracting telescope is used to view objects on the earth, another lens is added to turn the image right side up.

Whereas the telescope permits people to see objects that are far away, the microscope, shown in Figure 22-28, lets them view objects that are very near and very small.

The microscope has two convex lenses with the objective lens at the lower end of the tube. The objective lens has a very short focal length. The object to be viewed is placed just beyond the principal focus of this lens. A real, inverted, and enlarged image is formed near the upper end of the tube. The eyepiece is used to magnify the image still further.

The overall magnifying power of a microscope is found by multiplying the separate powers of each lens. If a microscope has an objective lens that enlarges 44X and an eyepiece that enlarges 10X, the final enlargement is 440X. The highest practical power possible with the optical microscope is about 2,000X. The electron and field-ion microscopes give magnifications that are much higher than this, but they do not depend on the refraction of light.

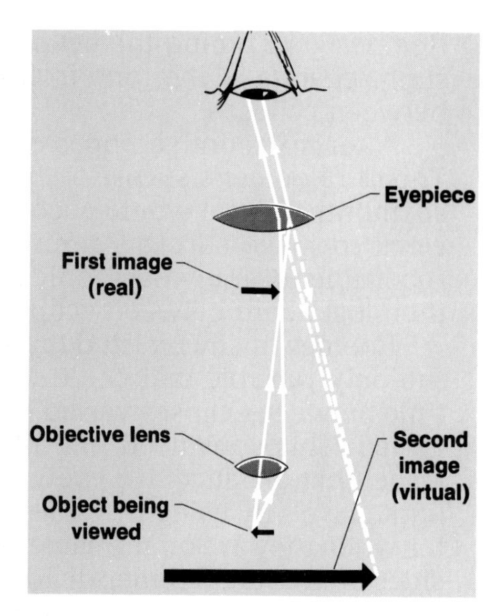

Figure 22-28 Compare this diagram of a compound microscope with that of a simple microscope in Figure 22-23. What are the major differences?

BIOGRAPHY

OLAUS ROEMER

finding the speed of light

How can you measure the speed of something as fast as light? Early scientists believed that light traveled instantly from one place to another. In the seventeenth century, however, a Danish astronomer, Olaus Roemer, found that light had a definite speed that could be measured.

The discovery came through Roemer's study of the moons of Jupiter. Roemer timed the eclipses of the inner moon. He noted that the moon went behind the planet every 42 hr 28 min. Most people would have been content to observe and record this information perhaps several dozen times, but Roemer kept timing the eclipses for several months. He began to notice a slight change in the time from one eclipse to another. More time elapsed between eclipses.

After six months, approximately 16 min were added to each rotation. Then to Roemer's surprise, the additional 16 min were lost during the next six months. What would account for these changes? Roemer built a special clock to make sure his timings were accurate. He kept on sighting and recording for four years, only to find that each year 16 min were gained and then lost again between eclipses.

Roemer finally realized that the distance between Earth and Jupiter was the only variable that could affect his timings. It occurred to him that the time between eclipses varied just as the distance between Jupiter and Earth varied. This convinced him that it took light a certain amount of time to travel that distance. He knew then that he could find the speed with which light travels by using the difference in the distance between Earth and Jupiter when they are on the same side of the sun and when they are on opposite sides of the sun. Roemer divided the difference in distance by the difference in time he had recorded (approximately 16 min, or 1,000 sec). He established the speed of light at 226,000 km/sec. We now know that Roemer's calculation is within 25 percent of the actual speed of light, which is approximately 299,800 km/sec.

CHAPTER REVIEW 22

SUMMARY

1. Light has properties of both waves and particles.
2. Light is produced by excited atoms, when the electrons move from a higher energy level to their original lower level.
3. Light energy can be measured.
4. The laser produces monochromatic coherent light.
5. Images are formed by even reflections.
6. The law of reflection states that the angle of incidence is equal to the angle of reflection.
7. Plane and curved mirrors obey the law of reflection.
8. Light is refracted when it passes at an angle from one medium to another.
9. The index of refraction is a ratio of the speed of light in air to that in another medium.
10. Convex lenses converge or bring light rays together.
11. Concave lenses diverge or spread light rays apart.
12. Your eyes enable you to see when light enters the cornea, is refracted to form an image on the retina, and is transmitted by nerve impulses to the brain.

REVIEW QUESTIONS

1. How do you know that light can pass through a vacuum?
2. When does an atom radiate light energy?
3. List three different ways of producing light.
4. Why is a candle flame visible? Why is a tree visible?
5. What can happen to light when it strikes an object?
6. Describe some uses of lasers.
7. What is the unit of light brightness?
8. What is the law of reflection?
9. How fast does light travel in a vacuum?
10. Why does a stick appear bent when it is dipped into water at an angle?
11. How do convex and concave lenses affect parallel light rays?
12. Where do rays from a distant object converge after passing through a convex lens?
13. If the object is a great distance away from a convex lens, where is the image located?
14. Where would you hold a magnifier to look at an object?
15. What is persistence of vision?
16. What causes nearsightedness? How can it be corrected?
17. What causes farsightedness? How can it be corrected?
18. Name three parts of a camera that correspond to parts of a human eye.

19. In what part of a refracting telescope is the objective lens found?
20. What are coherent light waves?
21. How is normal light different from monochromatic light?
22. What type of surface produces an uneven reflection? an even reflection?
23. What evidence is there to show that light is a stream of particles? that light travels in waves?
24. Does polarization show that light waves are transverse or longitudinal? Explain.
25. Distinguish between opaque, transparent, and translucent.
26. How is the index of refraction found?
27. When refraction occurs, in which direction is the light ray bent?
28. What is the relationship between the distances and sizes of objects and their images?
29. Where would you put a convex lens to throw an enlarged image of a lighted bulb on a screen?

CRITICAL THINKING

30. How could you use the photoelectric effect to make a burglar alarm?
31. If incoherent light is compared to a milling crowd at rush hour, to what would you compare coherent light?
32. How would you explain that, for certain substances, light with a long wavelength does not cause the photoelectric effect? Hint: Within certain limits, the energy of a photon is directly proportional to the frequency of the light-radiating source.

PROBLEMS

1. What happens to its apparent brightness as a torch moves from a point 10 m away from you to a point 40 m away?
2. How many 100-watt bulbs are needed to equal the total lumens of one 1,000-watt bulb?
3. How many watts of input are needed to light all the 100-watt bulbs needed in Problem 2?
4. When you stand 3 m in front of a plane mirror, exactly where is your image?
5. Find the speed of light in water if the index of refraction for water is 1.33.
6. If the image of a 5-cm candle is located three times as far from the lens as the object, what is the size of the image?
7. If an object is 60 cm away from a convex lens that has a focal length of 20 cm, where is the image located? How large is the image in relation to the size of the object?
8. Find the magnifying power of a refracting telescope if the focal length of the object lens is 10 m and of the eyepiece is 10 cm.
9. What is the total magnification of a microscope if the objective lens magnifies the object 36X and the eyepiece magnifies the image 10X?
10. If the light from the sun takes 500 sec to reach the earth, how far away is the sun?

VOCABULARY

Match the item in the left column with the best answer in the right column. Do not write in this book.

1. candela
2. concave
3. convex
4. cornea
5. farsightedness
6. focal length
7. hologram
8. incandescent
9. laser
10. law of reflection
11. lens
12. mirage
13. nearsightedness
14. objective
15. photoelectric effect
16. photon
17. polarization
18. principal focus
19. real image
20. retina
21. translucent
22. transparent
23. virtual image

a. angle of reflection equals angle of incidence
b. part of the eye on which images form
c. eye condition in which the eyeballs are too long or the lenses too thick
d. coherent light of only one wavelength that travels in almost parallel lines
e. distance from the center of a lens to the principal focus
f. object that glows after heating
g. limits the passage of light waves to those that vibrate in the same plane
h. light causes electrons to be ejected from metal
i. image that cannot be projected on a screen
j. mirror or lens that curves out
k. lifelike, three-dimensional image
l. mirror or lens that curves in
m. unit of brightness given off by a standard source of light
n. particle of light
o. eye condition in which the eyeballs are too short or the lenses too flat
p. where rays parallel to the principal axis converge after passing through the lens
q. substance through which light passes but objects on the other side are unclear
r. substance through which light passes and objects on the other side are clear
s. large convex lens with long focal length
t. optical illusion caused by light rays bending when passing through air layers of different densities
u. transparent substance with a curved surface
v. tough, transparent layer through which light enters the eye
w. inverted image formed by converging light

FURTHER READING

Asimov, Issac. *How Did We Find Out About the Speed of Light?* New York: Walker, 1986.

Sobel, Michael. *Light*. Chicago: University of Chicago Press, 1989.

Watson, Philip. *Light Fantastic*. New York: Lothrop, Lee, & Shepard Books, 1983.

COLOR

When you see a flamingo, the first thing you may notice is its bright pink plumage. How do you detect color? What makes pink different from blue?

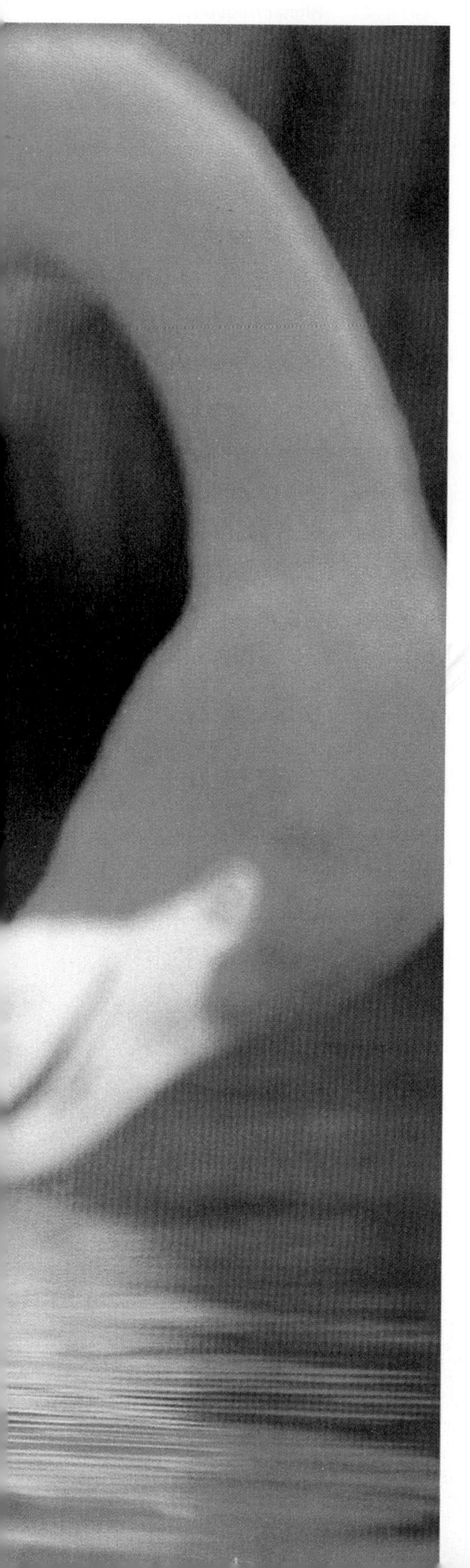

CHAPTER 23

SECTIONS

OBJECTIVES

- ☐ Identify the position of light in the electromagnetic spectrum.
- ☐ Describe how a prism is used to separate white light into colors.
- ☐ Explain how the primary colors of light interact.
- ☐ Discuss the result of combining pigments that have different primary colors.
- ☐ Describe how the spectroscope is used in research.
- ☐ Explain the Doppler effect with respect to light.

23.1 ELECTROMAGNETIC SPECTRUM

Light is part of the electromagnetic spectrum. You have learned from your study of sound waves that the shorter the wavelength, the higher the frequency of the sound. The same principle applies to the wavelengths of light. High-frequency light has a short wavelength. However, all wavelengths of light travel at the same speed, 300,000 km/sec in air. The shortest wavelength of light that can be seen by the

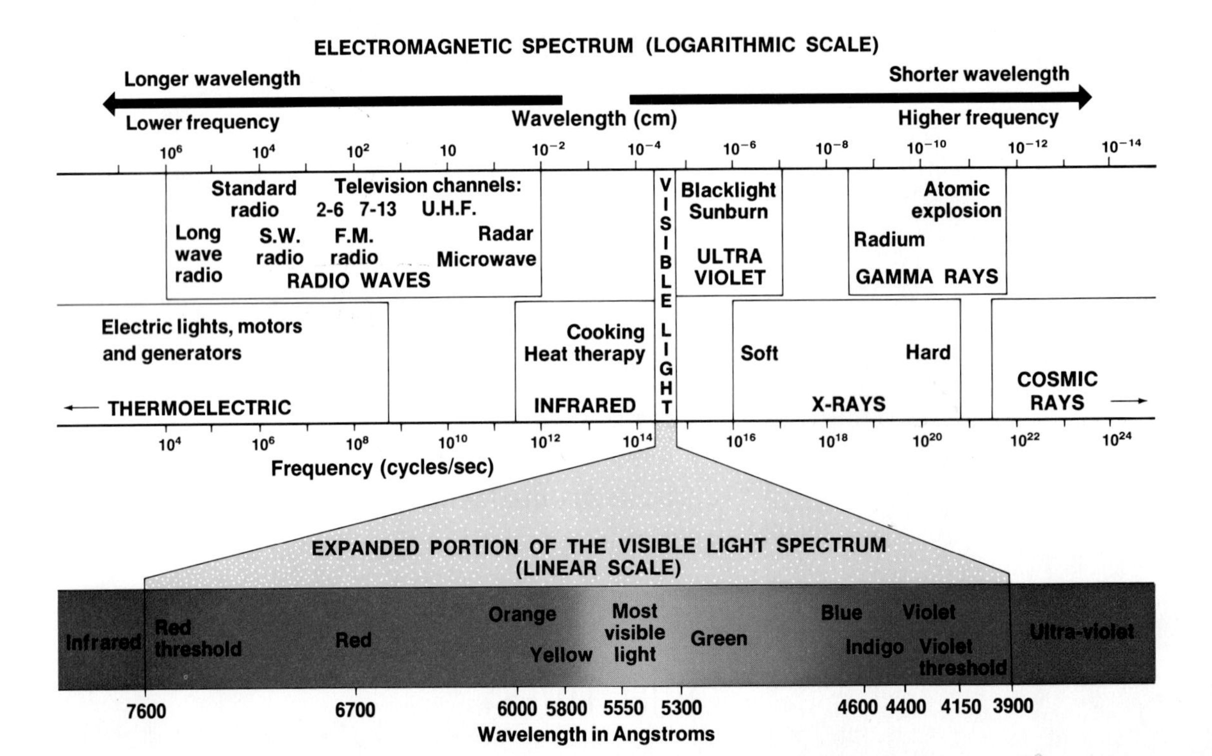

Figure 23-1 The electromagnetic spectrum. Which kind of waves have the shortest wavelengths? Which have the longest wavelengths?

human eye is violet light. Red light has the longest wavelength that can be seen by the human eye. As you can see in Figure 23-1, the wavelength of red light is nearly twice that of violet light.

There are many waves with frequencies higher and lower than those visible to the human eye. These waves are all alike except for frequency and wavelength. Frequency and wavelength account for all the different light that can be observed. The entire range of wavelengths from the shortest to the longest is the **electromagnetic spectrum.** The part of the electromagnetic spectrum that can be seen by the human eye is called **visible light.**

The top portion of Figure 23-1 uses a logarithmic scale. This means that for each division mark to the right, the frequency is ten times greater. Likewise, for each division mark to the right, the wavelength is ten times less. If this kind of compressed scale were not used, there would not be enough room on a page to show the entire electromagnetic spectrum.

You can see on Figure 23-1 that the visible part of the spectrum is only a small part of the entire spectrum. The wavelengths of visible light are given in units called angstroms (Å). Wavelengths of visible light are so short that scientists have adopted this small unit of length. Ten million angstroms are equal to 1 mm.

The human eye cannot see all the colors of the visible spectrum equally well. Yellow is the color that is most completely seen and colors with frequencies higher or lower than yellow are not seen as easily. Yellow appears brighter than other colors. Figure 23-2 shows the brightness of the colors to the average eye.

Infrared rays have frequencies below those of red light and are commonly referred to as radiant heat. These rays are given off by warm objects. Infrared rays give a feeling of warmth when they strike the skin. An infrared thermometer pointed at an object can measure an object's temperature by detecting the amount of radiant energy it gives off. One such thermometer, called a bolometer, is so sensitive that it can detect infrared rays from distant galaxies.

Light with frequencies slightly greater than those of violet light is called **ultraviolet light.** Since it has such a high frequency, ultraviolet light has more energy per photon than visible light.

Ultraviolet rays are sometimes called black light because they are invisible to the human eye. Camera film is very sensitive to ultraviolet light, but your eyes are not. This does not mean that ultraviolet light has no effect on your eyes. Quite the opposite is true. Because of its high energy per photon, ultraviolet light can harm your eyes. This is why you should not look at an ultraviolet lamp when it is on. Ultraviolet rays also tan and burn your skin, kill bacteria, and cause some substances to fluoresce, or give off visible light.

Beyond ultraviolet light are **X-rays,** with still higher frequencies and greater photon energy. They can pass through many solid objects. Denser objects, however, absorb more X-rays than do less dense objects.

Still higher in frequency are the **gamma rays,** which are produced in the nucleus of the atom. Most radioactive elements, such as radium, emit gamma rays (see Chapter 6).

Notice that there are no sharp dividing lines between the different sections of the electromagnetic spectrum but that they do overlap. For example, the low infrared rays can be used as radio waves; some electronic tubes produce X-rays with the same wavelength as gamma rays.

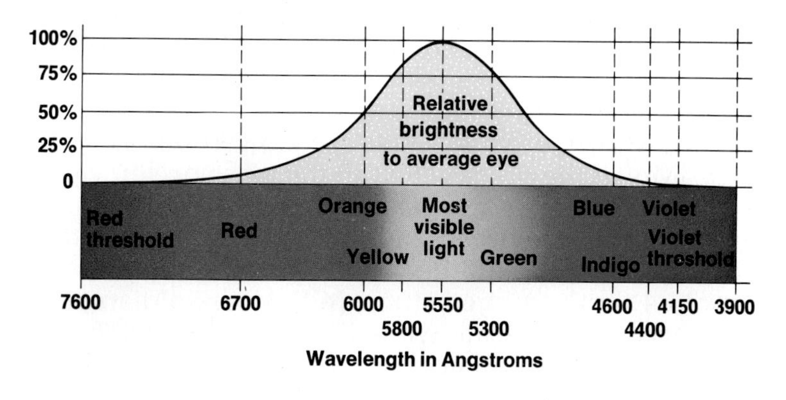

Figure 23-2 Study the diagram and explain why you think school buses are painted yellow.

Figure 23-3 When white light is separated into a spectrum by a prism, why does violet bend the most and red bend the least?

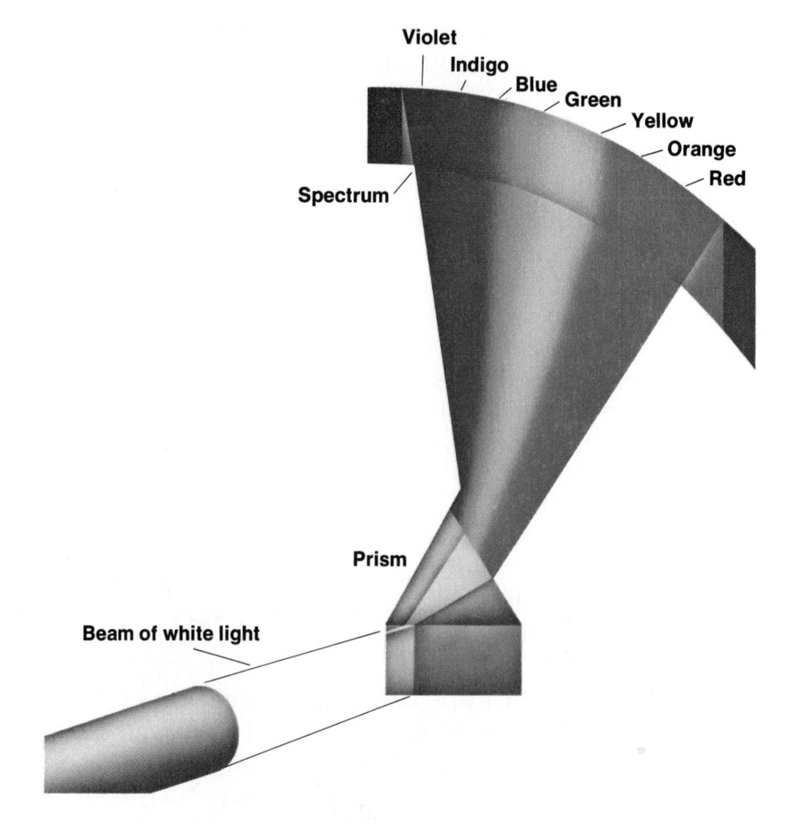

The colors of light can be separated. Sir Isaac Newton was the first to make a careful study of color. Newton allowed a beam of sunlight to pass through a glass prism. When he projected the beam onto a screen, Newton found that the light spread out into a band, or **visible spectrum,** of colors, as shown in Figure 23.3. By using a second prism, Newton was able to combine these colors again and produce a beam of white light.

How does the prism separate the colors of the spectrum? Recall that the speed of light changes as it passes from one medium to another. In Chapter 22, you learned that this change in speed causes light to bend, or refract. Each color bends at a different angle. Violet light is slowed down the most as it passes into a glass prism, and red is slowed down the least.

Since the waves of violet light are slowed down more than the waves of the other colors in the glass prism, violet light is bent the most. Red light is slowed down the least, so it is bent the least. The other colors, orange, yellow, green, blue, and indigo, are found between red and violet. Notice in Figure 23-4 that a cut gemstone, in this case a diamond, acts like a prism to separate white light into colors. That is why flashes of color can sometimes be seen in a diamond.

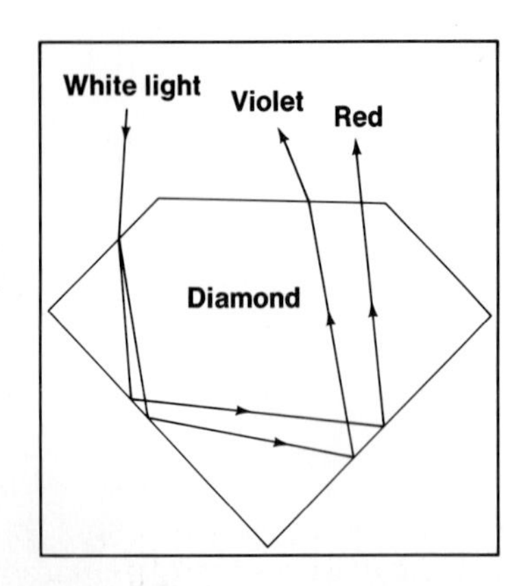

Figure 23-4 Why is it possible for a faceted diamond or other clear gem to emit flashes of color, even in white light?

You learned in the last chapter that a lens acts like a prism. Therefore, as light passes through a lens, it is not only refracted but separated into a spectrum of colors as well. How, then, does a camera lens focus a sharp image on film? Why are colors not separated?

Cameras produce sharp images by doing something similar to what Isaac Newton did with prisms. Newton used one prism to separate light and another prism to bring the light back together again. In much the same way, two or more lenses are used in cameras to produce clear pictures. The result is that light spread apart by one lens is combined again by another lens. In a good camera, seven or eight lenses are combined into two or three units to form a clear lens system. Such a color-corrected lens system will form a sharp image on film.

Water, dust, and air separate colors of light. Raindrops sometimes act like tiny prisms. They can refract sunlight to separate the colors. Figure 23-5 shows the path of two light rays, R (red) and V (violet), through a raindrop. Note that the light is refracted, reflected, and then refracted again.

The combined effect of thousands of raindrops, all found at about the same angle from the observer, makes a complete rainbow. A rainbow is usually seen when the sun comes out but while there is still moisture in the air. You can see a rainbow in the sky only if the sun is at your back and low in the sky.

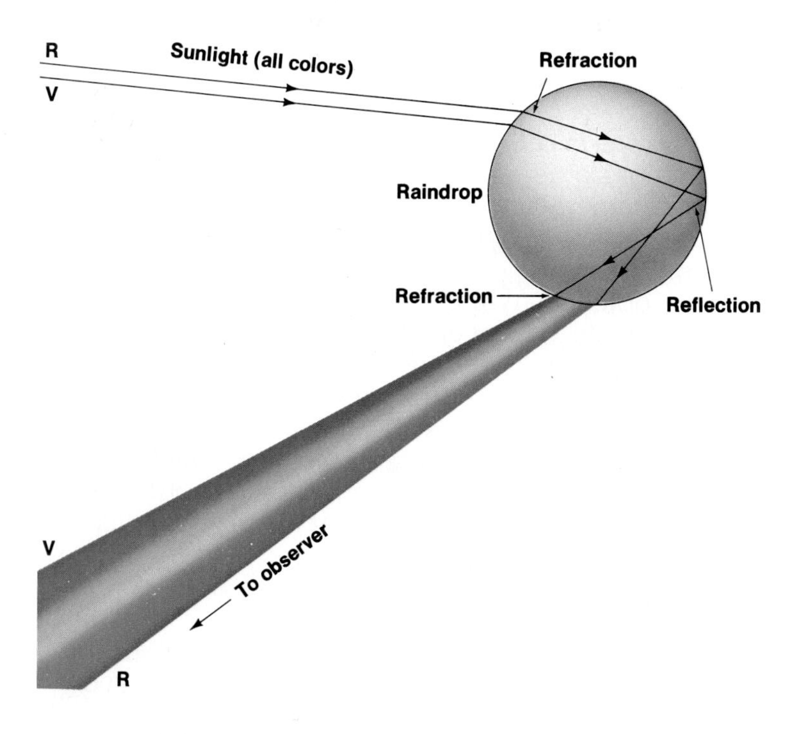

Figure 23-5 A rainbow is the result of refraction and reflection of light by raindrops. How many times is the light refracted? How many times is it reflected?

Have you ever wondered why the sky is blue during the day? The light of the sun must travel through several hundred kilometers of air on its path toward the surface of the earth. As sunlight passes through air, some of the light is scattered by tiny particles of air, water, and dust. Blue light, which has a relatively short wavelength, is scattered very easily, and seems to come to your eyes from all directions. Green, yellow, and red light, on the other hand, have longer wavelengths, and pass through the air with less scattering. This scattering of blue light gives the sky a blue color.

Ultraviolet light, which you cannot see but which can give you a sunburn, is scattered even more easily than visible blue light. Therefore, scattered ultraviolet light can give you a sunburn even in the shade or on a cloudy day.

Sometimes the sky turns a bright reddish-orange at sunset. This phenomenon can also be explained by the scattering of light. In the evening, the sunlight you see passes through the atmosphere at a sharp angle. Therefore, the sunlight must travel farther through the atmosphere to reach the earth. Over this long distance, the light is scattered even more than it is during the day. Moisture and dust in the air also increase the scattering of light. Red light, with its long wavelength, is scattered the least of all the colors, so it is able to pass through the atmosphere to your eye as the sun is setting.

23.2 COLORS OF OBJECTS

The color you see is reflected color. Why do some objects look white, others blue, and still others yellow when they are all lit by the same white light? The color you see depends on how these objects absorb, reflect, or transmit the light that strikes on them.

You see an object as white if it reflects light of all wavelengths equally well. However, if an object reflects mostly red light, the object looks red. If an object does not reflect light at all, it looks black. The amount of light reflected from different colored paper is shown in Table 23-1.

TABLE 23-1

Reflecting Ability of Paper

Color of Paper	Percent of Light Reflected
White	85
Ivory	67
Bright yellow	50–70
Dark red	14
Dark green	9
Dark blue	8
Flat black	2–4

Black is the absence of light. A piece of black velvet absorbs light of all wavelengths and reflects none. Light raises the temperature of the object that absorbs it, making the object warmer. This is why, in a hot, bright climate, light-colored clothing is worn instead of dark-colored clothing.

So far, only the color of objects seen in white light has been discussed. What happens when an object is viewed in filtered light? Hold a sheet of white paper in sunlight or in the beam of light from a projector. Now place a piece of red glass or plastic between the light source and the piece of paper. As you can see in Figure 23-6, the paper will look red. The red filter absorbs light of all wavelengths except red. Since only red light passes through, only red light can be reflected from the white paper. What if you replaced the red filter with a green one? What color would be reflected from the white paper? How would you explain this?

Suppose you go one step further. Keep the red filter, but use red paper instead of white. The red paper receives only red light. Since red paper reflects only red light, the paper looks red. If, instead, you hold a green filter in the beam of light, only green light strikes the red paper. What color will the paper reflect? Look again at Figure 23-6. Since the red paper absorbs green light, the paper cannot reflect any of the light coming from the green filter. As a result, the red paper looks black when green light shines on it.

Remember, an object's color depends on (1) the kind of light shining on it, (2) which wavelengths of this light it is able to transmit, if transparent, or (3) which wavelengths it reflects, if opaque.

Figure 23-6 Two beams of white light are projected, one through red glass onto white paper, and the other through green glass onto red paper. What effect do the glass filters have on the white light? What color light is reflected by each sheet of paper?

The primary colors of light are red, green, and blue. One way to study the effect of mixing colors is to look at a spinning disc painted various colors. Because of persistence of vision, which you learned about in the last chapter, you see the same effect as if all the colors were entering your eye at once.

A more practical way of mixing colors is to project several beams of colored light onto the same spot on a white screen. When all the colors of the spectrum are projected in this way, you see white. This result is like the second part of Newton's experiment, in which he used a second prism to join the colors that had been separated by the first prism.

As you can see in Figure 23-7, white light can also be obtained by projecting the right amounts of only three colors. These colors, red, green, and blue, are known as the **primary colors of light.** (Note: These colors differ from the primary pigments, which will be discussed later in this chapter.) By changing the relative amounts of the primary colors of light, all of the colors of the spectrum can be produced.

Any two colored lights that produce white when mixed together are called **complementary colors.** Blue and yellow are such a pair of complementary colors. If blue is removed from white light, the colors that remain will produce yellow. It would appear, then, that blue must contain all the colors except yellow. Therefore, if you mix blue with yellow, you will get white. Why do you think laundry bluing improves

Figure 23-7 What colors are produced when red and green light are mixed? when blue light is added?

CHANGING COLORS

activity

OBJECTIVE: Perform an experiment to observe how the color of an object is affected by different colored lights.

PROCESS SKILLS
In this activity you will *observe* color changes that will help you to *infer* the effect of colored light on the color of an object.

MATERIALS
projector
color filters
several colored objects of varying textures, such as a cloth, a dish, a piece of foil wrapping paper, and a piece of plain white paper

PROCEDURE
1. Darken the room and turn on a projector or other source of light. Cover the light with a colored filter.
2. Hold a series of objects of a variety of colors and textures in the light of the projector. Record the color of the light and the color of the object in this light.
3. Repeat Step 2 using different-colored filters.
4. View the objects again in white light. Compare their colors with the data that you recorded in Steps 2 and 3.

OBSERVATIONS AND CONCLUSIONS
1. What happened to the colors of the objects under different color lights?
2. Explain the differences you noted.

Figure 23-8 Stare at the center of the flag for 20 sec. Then stare at a sheet of white paper. What colors appear on the white paper? Why?

the appearance of white clothes that have become tinged with yellow? Red and blue-green are also complementary. When mixed together, they produce white.

There is a simple way to find the complement of any color. Stare steadily at a given color for about 20 sec. Then stare steadily at a sheet of white paper and you will see the complementary color appear before your eyes. For example, stare at a sheet of yellow paper. If you shift your gaze to a sheet of white paper, you will see a blue image appear. Why?

This effect is caused by retinal fatigue. That is, the cells on the retina of the eye become "tired" of viewing a given color (yellow in this example) for a long period of time.

Then, when you look at white paper, even though all colors enter the eye, the yellow does not register. As a result, you see blue, the complement of yellow.

To see this effect with several colors at the same time, stare at Figure 23-8. Then look at a sheet of white paper. Be sure that you hold your gaze steady both when looking at the figure and again when staring at the white paper.

The eye perceives color because the retina has three types of cells for registering color, called **cones.** Each type of cone is sensitive to one of the primary colors. If all three types of cones are equally stimulated, you see white. If only the cones sensitive to red are stimulated, you see red. If the cones sensitive to red and the cones sensitive to green are stimulated, you see yellow. Different combinations of the three primary colors make it possible for you to see all of the colors.

Figure 23-9 Each letter H is the same blue color and has the same brightness. The color and brightness of the disks are also exactly alike. How does the background affect what you see?

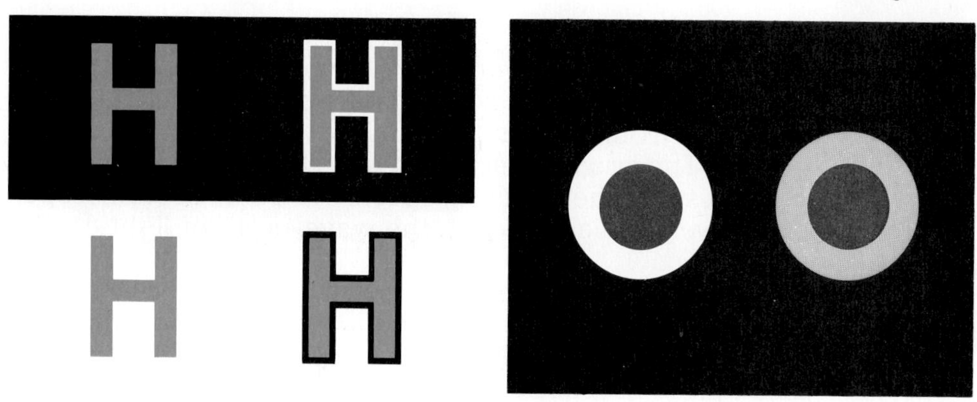

Other factors also seem to influence the colors you see. By looking at Figure 23-9, you can see how the brightness of colors seem to be affected by their background.

Some people are color blind. That is, they have defective color vision. The few people who are totally color blind see all objects in shades of gray. More often, however, a person is only partially color blind. Such a person cannot tell one color within certain color groups from another. For example, red, green, and yellow might all look like shades of yellow. Green, blue, and violet might all look like shades of blue. Figure 23-10 shows a test used to determine red-green color blindness.

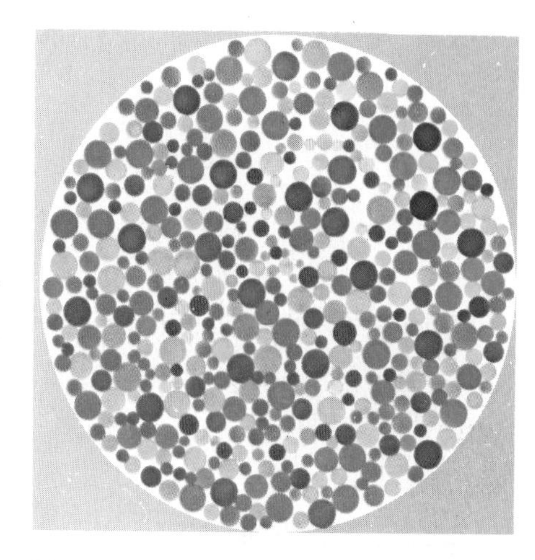

Figure 23-10 A typical test for red-green color blindness. What will a color-blind person fail to see?

The primary pigments are red, blue, and yellow. Perhaps you thought mixing yellow light with blue light would give you green, rather than white, light. In fact, when you mix pigments, or coloring matter, instead of lights, yellow and blue do produce green. Until now, the discussion has been about mixing light. Mixing pigments produces different results. Red, blue, and yellow are called the three **primary pigments.** Mixing the three primary pigments produces black, as shown in Figure 23-11.

Grind some yellow chalk and blue chalk together and mix the chalk with a little water. The mixture should look green. Yellow chalk appears yellow because it absorbs blue, indigo, and violet light. Blue chalk absorbs red, orange, and yellow light. Neither yellow nor blue chalk absorbs green light. Therefore, only green can be reflected from the mixture.

The four-color printing process is a good example of mixing pigments. A color picture is built up by printing the same picture on a piece of paper four times, one plate after the other. Each plate is inked with a different pigment, red, blue, yellow, and black. The black pigment is used to add sharpness to the picture.

Color plays an important role in research. A prism is part of a device called a **spectroscope** that is used to analyze light from objects, especially stars and other distant objects. The prism refracts the light and a system of lenses is used to study the resulting spectrum. Some spectroscopes also have a scale on which the wavelength of light can be read directly.

Some spectroscopes use a diffraction grating instead of a prism. The diffraction grating is a glass plate containing thousands of parallel lines per centimeter. As light passes through or is reflected from these very narrow lines, the light spreads out. The spreading bands of light interfere with one another, strengthening some waves and weakening others. The result is a spectrum of light.

The grooves on a record album act like a diffraction grating. Look at the reflection of a bare light bulb from the surface of a record to see the effect of diffraction.

The most common spectrum is the kind called a continuous spectrum, like the one in Figure 23-3. As the name indicates, all the colors of the rainbow are present in this type of spectrum. This kind of spectrum is produced when a solid, liquid, or very dense gas is heated until it glows.

Often, certain colors stand out more brightly than others. This indicates how hot the light source is. For example, the brightest part of the sun's spectrum is in the greenish-yellow region. Since that color is brightest at about 6,000° C, the temperature of the sun's surface must be approximately 6,000° C.

A second type of spectrum is produced by a gas under low pressure. When such a gas is excited by heat or electric current, it gives off light with only a narrow range of wavelengths. These different wavelengths form brightly colored

Figure 23-11 What color is formed when blue and yellow pigments are mixed? What color results when red is mixed with the blue and yellow?

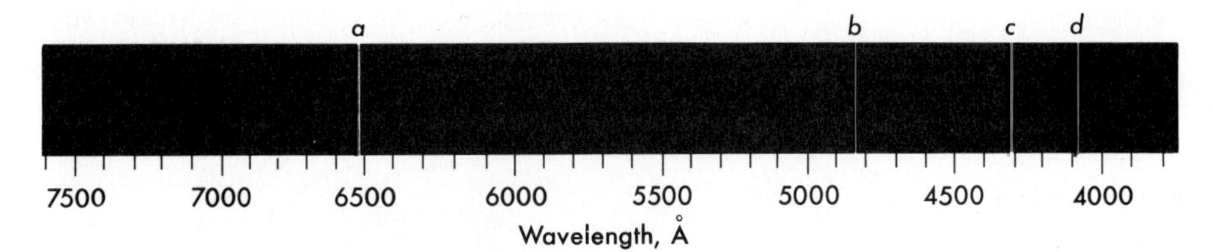

Figure 23-12 A bright-line spectrum of hydrogen. How is this bright-line spectrum formed?

lines on a dark background. Notice the bright lines in Figure 23-12. This type of spectrum is called a bright-line or emission spectrum. Each chemical element produces its own unique bright-line spectrum.

A third type of spectrum is similar to the continuous spectrum, except that it has dark bands cutting through it. This type of spectrum is produced when light is passed through a gas at low pressure. The gas filters out, or absorbs, specific colors. Thus, the spectrum appears continuous except for the dark bands where colors are filtered out. This type of spectrum is called a dark-line, or absorption, spectrum. A cool gas absorbs the same rays that it would give off if heated.

Most stars are dense masses of glowing gases. In viewing stars through a spectroscope, you would expect to see a continuous spectrum. Instead, a dark-line spectrum is seen. The dark-line spectrum is caused when some frequencies of the star's light are absorbed by the cooler gases above the star's surface. The dark lines in the spectrum indicate the elements that the star is made of.

How does a spectrum give scientists information about an element? When atoms of an element are excited, they emit light of specific wavelengths. The wavelengths of light produced by one element are different from those produced by any other element. Thus, the light from each element is the element's "fingerprint."

For instance, if you looked through a spectroscope at the light from a neon sign, you would see a pattern of bright lines. This pattern is produced only by the element neon. If mercury gas is present, it will form its own special pattern. The elements are "fingerprinted" by the spectra they produce. Scientists have used this method to show that neon and mercury, as well as over 50 other elements, exist on the sun. With the aid of the spectroscope, a chemist can detect an element even if only a tiny trace of it is present.

A shift in frequency causes the Doppler effect. In Chapter 21 (Figure 21-13), you learned that the Doppler effect is the shift in frequency that occurs whenever a wave source moves toward or away from you. In sound, the Doppler effect is heard as a higher or lower pitch. In light, the Doppler effect is seen as a shift in color towards either the blue or red end of the spectrum.

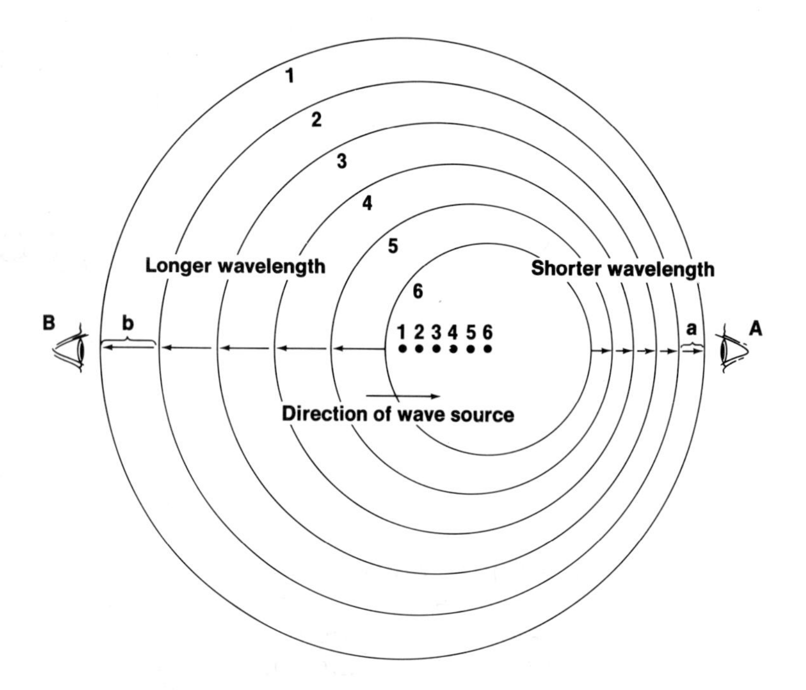

Figure 23-13 A diagram of the Doppler effect. Toward which end of the spectrum would the color shift when viewed from A? from B?

In studying starlight with a spectroscope, scientists can tell when the spectral lines of an element are shifted. If the lines are shifted toward the red end of the spectrum, this **red shift** indicates the star is moving away from earth.

Recent sightings have shown that the most distant galaxies within the range of the largest telescopes are moving away from earth at a speed of about 145,000 km/sec. This speed is almost half the speed of light. Other galaxies are moving away at speeds related to their distances from the earth. The discovery of the red shift has led to the theory that the universe is expanding rapidly.

Figure 23-13 shows how the Doppler effect causes a shift in color. Suppose that a light source, such as a star, is located at point 1. Light waves spread out in all directions with point 1 as the center. Each wave is separated from the next by one wavelength.

Now imagine the star moving rapidly toward the right. Each new wave starts from a new center, marked 2,3,4,5, and 6. What effect does this motion have on the wavelength of the light? The light received at A has a shorter wavelength (wavelength = a) and higher apparent frequency than that at B (wavelength = b).

If the star is green, its light would appear bluish to an observer at A, since blue light has a higher frequency, or shorter wavelength, than green. What about the person at B? To that person, the light would appear yellowish, since yellow has a lower frequency than green.

The Doppler effect can be observed in all parts of the electromagnetic spectrum. For example, radar uses the shift in frequency of radio waves to measure the speed of a car.

PHOTOGRAPHY

mastering light and color

Photography is the process of creating pictures using the properties of light. It is important for a professional photographer to have a knowledge of light and color. Both natural light and artificial light have unique characteristics that affect the quality of photographs.

Taking a photograph begins with using a camera to expose color or black-and-white film to light. Once the light is "captured" on film, developing the film and making a print completes the photo-making process. Career opportunities in photography may be related to any step in this process of generating a finished photograph. Positions include photographers, photo researchers, film and equipment manufacturers, film processors, and salespeople. Three specialized fields include scientific photographers, commercial photographers, and film processors.

Scientific photographers are responsible for documenting the visual information of many scientific disciplines. For example, medical photographers provide information used to diagnose and treat illnesses. They work with equipment such as X-ray machines, camera-equipped microscopes, and infrared scanning systems.

A commercial photographer takes pictures for illustrations and advertisements in magazines, books, and newspapers. These photographers are highly skilled in the use of photographic equipment. They are also talented in determining the proper composition, lighting, focus, and exposure.

Film processors and developers require a knowledge of the basic procedures of developing and printing photographs, as well as a knowledge of the basic principles of color. By using a variety of techniques and equipment, film processors can change the size, color, contrast, and other features of photographs.

Careers in photography generally require at least a high school education and technical knowledge of photography and photographic equipment. Training in these areas is available at many colleges, universities, art schools, and technical schools.

For more information on a career in photography, contact

Society for Photographic Education
P.O. Box 1651 F.D.R. Station
New York, NY 10150

CHAPTER REVIEW 23

SUMMARY

1. Visible light is the part of the electromagnetic spectrum that can be seen by the human eye.
2. Color is related to the frequency or wavelength of light waves.
3. White light can be separated into a spectrum of colors by a prism.
4. The color of an object depends on the light reflected from that object.
5. The primary colors of light are red, green, and blue.
6. The primary pigments are red, yellow, and blue.
7. A spectroscope can be used to identify elements in stars.
8. The Doppler effect is caused by a shift in the frequency of light to either the blue or red end of the spectrum as the wave source moves toward or away from you.

VOCABULARY

Match the item in the left column with the best answer in the right column. Do not write in this book.

1. complementary colors
2. cones
3. electromagnetic spectrum
4. gamma rays
5. infrared rays
6. primary colors of light
7. primary pigments
8. red shift
9. spectroscope
10. ultraviolet light
11. visible light
12. visible spectrum
13. X-rays

a. rays that can pass through solid objects
b. red, blue, and yellow
c. red, green, and blue
d. change in frequency of light from distant galaxies as they move away from the earth
e. two colors of light that produce white light when projected together
f. retina cells that perceive color
g. high energy rays produced in the nucleus of an atom
h. part of the electromagnetic spectrum that can be seen by the human eye
i. entire range of wavelengths from thermoelectric waves to cosmic rays
j. rays given off by warm objects
k. band of colors formed when light is spread out by a prism
l. rays that tan and burn skin
m. device used to analyze light from objects

REVIEW QUESTIONS

1. Name six kinds of rays found in the electromagnetic spectrum.
2. What determines the color of visible light?
3. How are X-rays and infrared rays alike? How are they different?
4. What is the effect of light energy when it is absorbed by an object?
5. Name the colors of the visible spectrum in order of decreasing frequency.
6. What two factors determine the color of an opaque object?
7. Explain why one object looks black and another looks white.
8. What happens to a beam of pure red light when it passes through a prism?
9. How does red light differ from blue light?
10. What color results when the three primary colors of light are mixed in equal proportions?
11. What color results when the three primary pigments are mixed in equal proportions?
12. What causes a rainbow?
13. What is the relationship between frequency and wavelength?
14. Why does a red object appear black when viewed through a green filter?
15. What is retinal fatigue? How can it affect your color vision?
16. What color do you get when you mix blue and yellow light? blue and yellow pigment? Explain your answers.
17. What is a spectroscope and how does it work?
18. Under what conditions would a dense gas produce a continuous spectrum when viewed through a spectroscope? Under what conditions does a gas produce a bright-line spectrum?
19. Describe the Doppler effect and give an application of this effect on light.
20. Explain how a dark-line, or absorption spectrum, is formed.

CRITICAL THINKING

21. You are told to write a message to your classmates on any color paper with any color ink. The paper will be illuminated in a dark classroom by a light source with a blue filter. What color combinations of ink and paper will allow your classmates to read the message?
22. How does sunscreen lotion prevent sunburn?
23. Our sun is a yellow star. What color would it appear to be to a viewer on a planet in a galaxy that our galaxy is rapidly approaching? Explain.

FURTHER READING

Babbitt, Edwin S. *The Principles of Light and Color*. Secaucus, N.J.: Lyle Stuart Inc., 1980.

Billmeyer, Fred W., Jr., and Max Saltzman. *Principles of Color Technology*, 2nd ed. New York: John Wiley and Sons, 1981.

Branley, Franklyn. *The Electromagnetic Spectrum*. New York: Thomas Y. Crowell, 1979.

Overheim, R. Daniel, and David L. Wagner. *Light and Color*. New York: John Wiley and Sons, 1982.

Simon, Hilda. *The Magic of Color*. New York: Lothrop, Lee, and Shepard Books, 1981.

How the Sciences Work Together

The Process in Physical Science: Viewing the Earth from Outer Space

Since 1960 an array of artificial satellites has used the vantage point of outer space to probe our planet. They have been used to monitor changing ocean and wind currents, measure the chemical composition of the atmosphere, and locate valuable mineral deposits. Recently satellites have given us a clear look at the enormous effect of human activities on the earth's environment.

A Nimbus satellite is adjusted prior to launch.

The Connection to Physics

Satellites use electromagnetic waves sent between the earth and the satellite to gather data and transmit information back to receiving stations on the ground.

Objects and materials on the earth's surface and in its atmosphere reflect electromagnetic radiation depending on their physical and chemical properties. Objects at different temperatures emit different wavelengths of infrared radiation and this information has been used to map shrinking rain forests. Areas covered with forest tend to be slightly cooler than areas that have been cleared of trees.

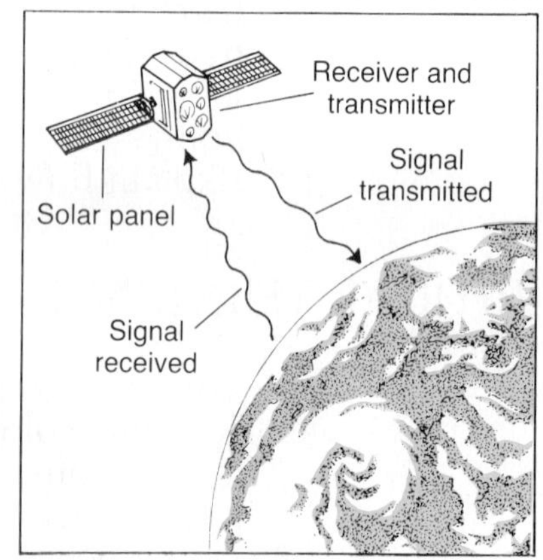

Satellites work by receiving and transmitting electromagnetic signals from the earth.

Weather satellites operated by the National Oceanic and Atmospheric Administration, (NOAA) monitoring infrared wavelengths, have measured a dramatic decline in rain forests in South America as trees are cleared for farms.

Computer enhanced satellite images of a rain forest in Western Brazil. Red areas indicate dense vegetation. Blue areas show deforestation.

The Connection to Biology

Biologists are interested in keeping track of tiny sea plants called phytoplankton, since all fish in the ocean

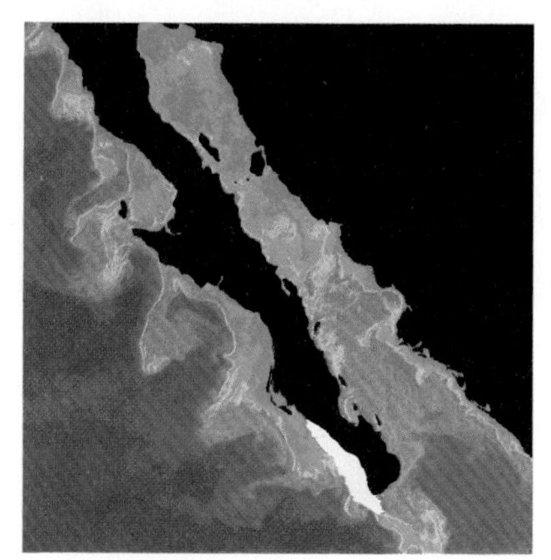

The burgundy and dark blue areas in this satellite image of the coast of Baja California indicates low levels of the sea plants phytoplankton.

either eat phytoplankton or eat other things that eat phytoplankton. The amount of phytoplankton in a region is a good indicator of the population of fish in that area. The amount of phytoplankton can be determined from the color of the ocean in a given area. Gathering this information by ship is impractical. Satellites, however, can provide readings for entire coastal regions and shed light on the status of huge populations of phytoplankton. Not only is this information useful to the fishing industry, but it provides basic information on the health of our oceans.

The Connection to Earth Science

Predicting weather and monitoring climates has been one of the most important uses of satellites so far. Weather satellites pick up visible and invisible radiation reflected from clouds. This information is relayed to receiving stations on the earth where computers are used to form images of moving clouds. These images are familiar sights on TV weather forecasts. They can tell you whether to carry an umbrella the next day or, more importantly, if and when to evacuate in the face of a life threatening hurricane.

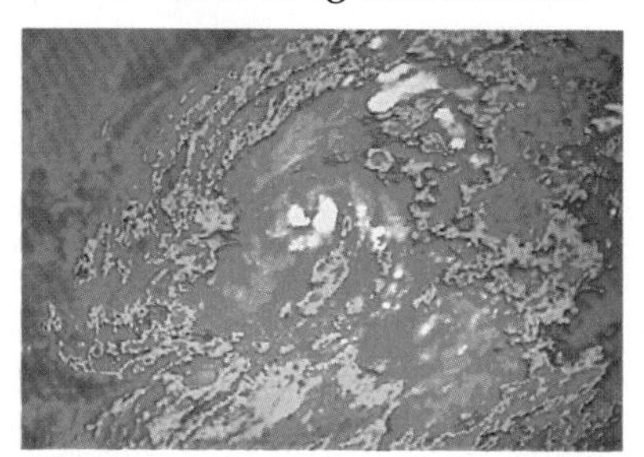

Computer enhanced satellite image of a hurricane over the Atlantic Ocean.

The National Aeronautics and Space Administration has launched five earth surveying satellites since 1972. These satellites, called Landsat 1-5, orbit the earth at an altitude of over 800 kilometers. They can detect details as small as 30 meters across on the earth's surface. Landsat takes 33,500-square-kilometer "snapshots" of the earth's surface at one time. Geologists use these images to quickly map the characteristics of vast portions of the earth saving

A Landsat image of New York City and New Jersey. Gray areas show roads and buildings. Red areas show vegetation.

weeks of airplane and ground survey time. From surface features geologists can tell whether or not there are useful mineral deposits underground. Also, geological faults can be studied more effectively leading to a better understanding of earthquakes.

ELECTRICITY AND MAGNETISM

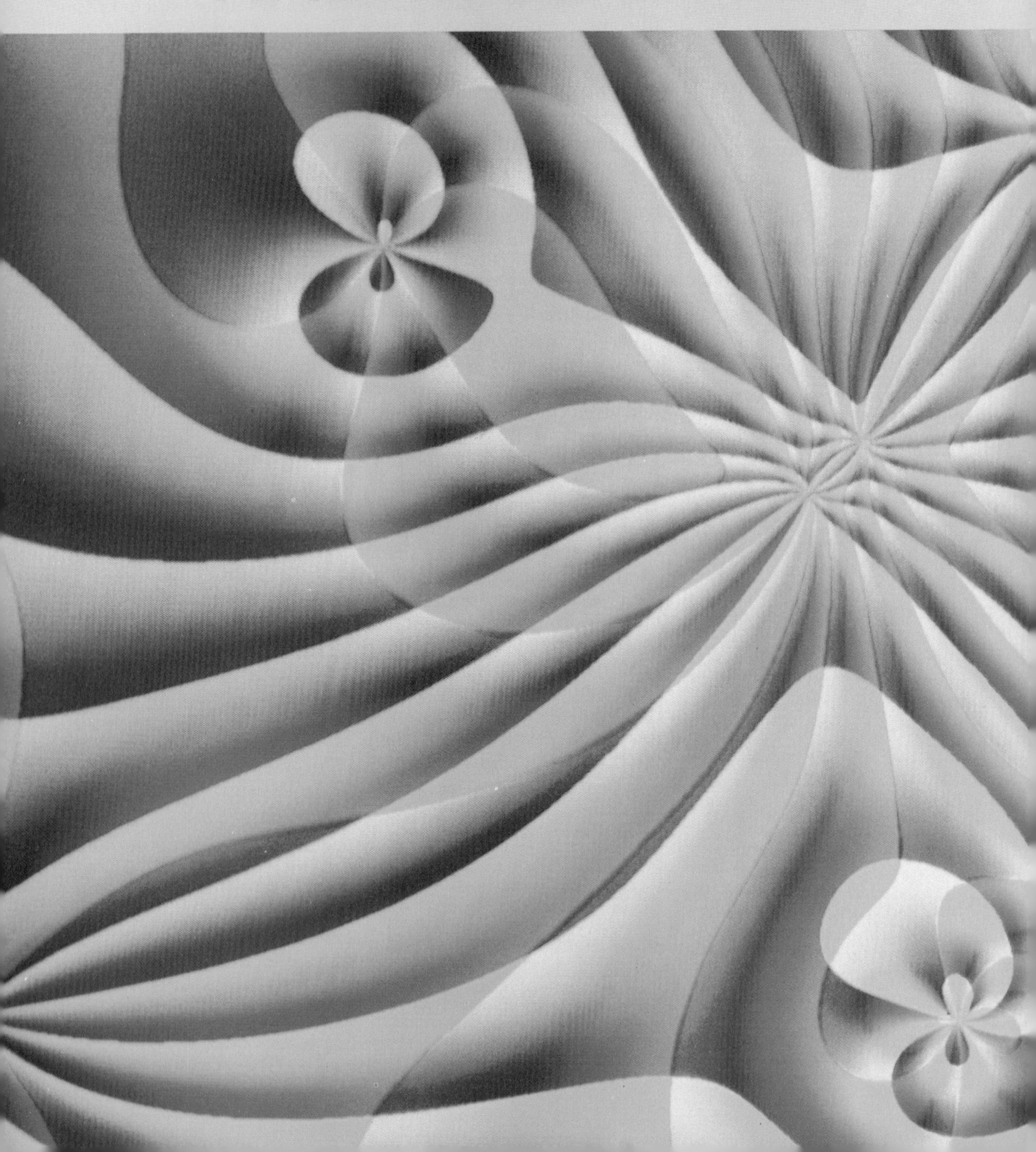

UNIT 6

For centuries scientists have tried to understand the forces of electricity and magnetism. Recently computers have greatly advanced our understanding of these forces. For example, this computer-generated image uses colors to represent the forces that exist between a group of electrical charges. The red areas represent electric forces acting in the horizontal direction and the yellow areas represent electric forces acting in the vertical direction. The blue areas represent the combined electric force.

You already may be familiar with the forces of electricity and magnetism. For instance, the earth generates a magnetic force that causes compass needles to point north. Compounds form when positively charged and negatively charged ions exert an electrical force of attraction on one another.

Our knowledge of the forces of electricity and magnetism has created an ever growing number of applications that have revolutionized the way we live. Almost all modern devices, from light bulbs, televisions, and toasters to computers, airplanes, and rockets apply the forces of electricity and magnetism.

In this unit, you will explore electricity, magnetism, and electromagnetism and the forces that exist among electric charges and magnetic materials. The ideas and concepts in this unit will help you expand your understanding of the world around you, particularly of the electrical devices you use every day.

CHAPTERS

ELECTROSTATICS

In less than a millisecond, a bolt of lightning unleashes the amount of electrical energy used by an average household in one year. How does the atmosphere produce electrical energy?

CHAPTER 24

SECTIONS

OBJECTIVES

- ☐ Describe how static charges are formed and how they interact.
- ☐ Tell how objects can be charged by contact and by induction.
- ☐ Identify where charges concentrate on objects.
- ☐ Explain why some static charges can be dangerous.
- ☐ Describe some of the uses of electrostatic devices.

24.1 ELECTRIC CHARGES AROUND YOU

Objects can receive a static charge. To understand how objects can be charged, you must first review the structure of the atom. Normally an atom contains equal numbers of positively charged protons (+) and negatively charged electrons (−). In such a state, the atom is neutral. When an atom loses an electron, however, the charges are no longer equal. Since one electron is missing, the atom has a positive charge. The lost or free electron has a negative charge.

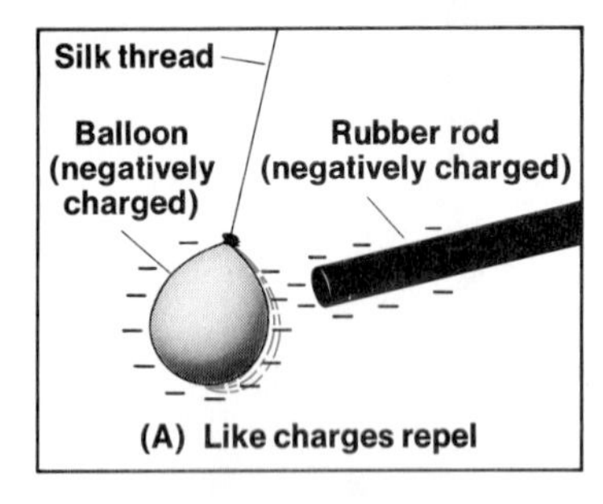

(A) Like charges repel

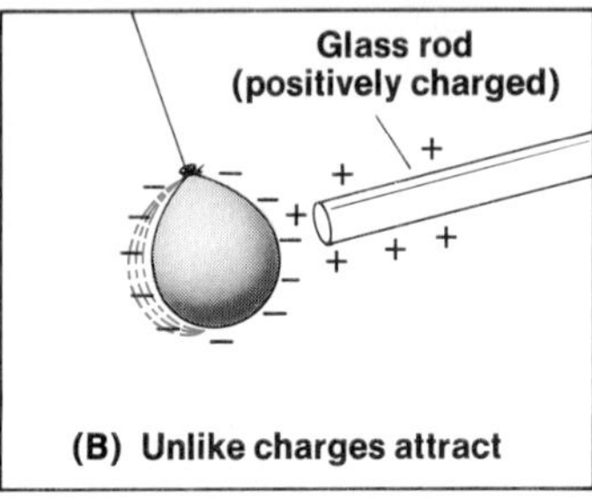

(B) Unlike charges attract

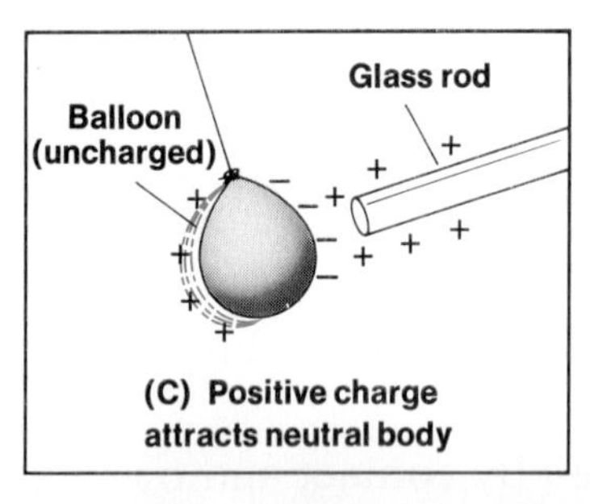

(C) Positive charge attracts neutral body

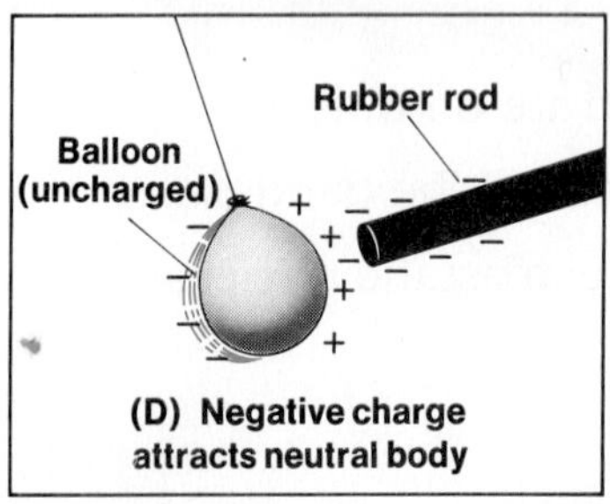

(D) Negative charge attracts neutral body

Figure 24-1 In which of the examples are objects attracted to each other? repelled?

When many atoms of an object gain or lose electrons, the entire object becomes charged. An object that receives extra electrons has a negative charge and an object that loses electrons has a positive charge. The accumulation of positive or negative charges on an object is referred to as **static electricity**.

If you rub two objects together, the contact causes electrons to be transferred from one object to the other. For example, if you rub a rubber rod (or almost any plastic) with wool or fur, the rubber picks up electrons from the wool. In much the same way, your comb gets a negative charge when you run it through your hair. On the other hand, if you rub a dry glass rod with silk, the silk removes electrons from the glass, leaving the glass with a positive charge. It is important to remember that it is only the electrons that move when an object receives a static charge.

When you walk across a rug, the contact between the rug and your shoes builds up a negative charge on your body. If you then touch a metal doorknob, the electrons travel from your hand to the doorknob. You experience a shock.

Charged and uncharged objects interact. To find out how static charges affect each other, hang a balloon from a thread, as illustrated in Figure 24-1A. Give the balloon a negative charge by rubbing it with wool or fur. Now charge a rubber rod in the same way and hold it near the balloon. The balloon will be repelled by the rod. If two objects both have a negative charge, they repel each other. Similarly, if two objects have a positive charge, they repel each other. In general, like charges repel each other.

Now give a glass rod a positive charge by rubbing it with silk. Then hold the rod near the negatively charged balloon, as in Figure 24-1B. Notice that the positive rod attracts the negative balloon. From this activity, you can see that unlike charges attract each other.

Further tests show that if the balloon is uncharged, it is attracted to either the positive or negative charges (Figures 24-1C and 24-1D). Thus, charged objects attract uncharged, or neutral, objects.

Static charges attract and repel by means of a field of force they create. A field of force is the region around a charged object within which a force can be detected. When an electron or other charged object is brought into this field, it is affected by the force.

What affects the strength of the force between two charged objects? First, the greater the charge, the greater the force. Second, the closer the objects, the greater the force.

FORCES OF ATTRACTION AND REPULSION

activity

OBJECTIVE: Learn what causes one object to be attracted to or repelled by another object.

PROCESS SKILLS

In this activity, you will *observe* the interaction of two objects and *infer* the cause of the attraction and repulsion between them.

MATERIALS

- tape
- thread
- polystyrene ball
- ring
- ringstand
- plastic comb
- piece of woolen cloth
- aluminum foil

PROCEDURE

1. Use tape to attach one end of a piece of thread to a polystyrene ball. Tie the other end to a ring on a ringstand.
2. Rub a plastic comb with a piece of woolen cloth. Bring the comb near the ball. Record what happens.
3. Touch the comb to the ball. Record what happens.
4. Wrap a single layer of aluminum foil around the ball. Repeat Steps 2 and 3. Record what happens.

OBSERVATIONS AND CONCLUSIONS

1. What happens to the ball in Step 2 when the comb is placed near it?
2. What happens to the ball in Step 3 when it is touched by the comb?
3. What happens to the foil-covered ball when the comb is placed near it? when it is touched by the comb? Explain.
4. Explain the principle demonstrated by this activity.

Suppose you measure the force between two charged objects a certain distance apart. When the objects are moved twice as far apart, the force between them becomes one fourth as much. If the distance between them is made three times as great, the force will be only one ninth as much. The force between charged objects varies inversely with the square of the distance between them. This relationship is known as the inverse-square rule. In Chapter 15, you learned that the inverse-square rule also applies to the force of gravity between two objects.

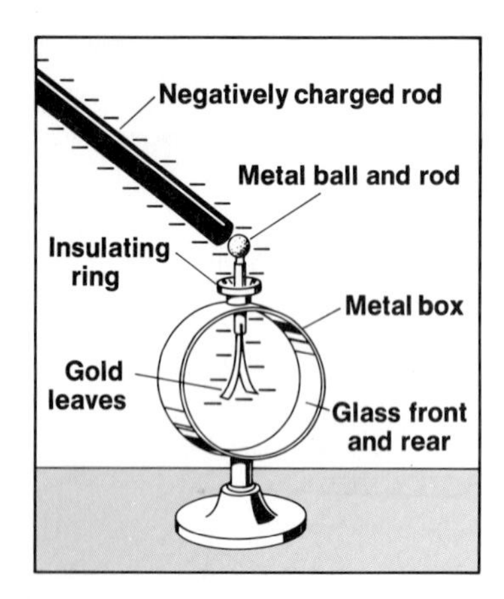

Figure 24-2 The knob of this electroscope is being touched by a negatively charged rod. Why do the leaves of an electroscope spread apart?

Electroscopes can detect static charges. A device used to detect an electric charge is called an **electroscope.** The balloon shown in Figure 24-1 detects charges. However, a gold-leaf electroscope like the one shown in Figure 24-2 is much more sensitive than a rubber balloon. In addition, the electroscope can be used to indicate how much charge is present.

The electroscope in Figure 24-2 becomes charged when the metal knob at the top is touched by a charged object. Notice the insulating ring at the top of the round metal box. The insulating ring keeps the charge on the knob from leaking to the metal box itself. As a result, the charge travels down to the gold leaves inside the metal box. The glass walls of the electroscope protect the leaves from being moved by air currents.

The leaves of an electroscope can be made of aluminum or any thin metal foil. Leaves of gold can be made very thin and light. Because their mass is very low, they can detect a very small charge.

When the metal knob of an electroscope is touched by a negatively charged rod, some of the electrons move from the rod to the knob. The electrons then travel down the metal rod. When the electrons reach the gold leaves, each leaf becomes negatively charged. Therefore, the like-charged leaves repel each other. The greater the charge, the farther the leaves are spread apart.

What happens when an electroscope is touched by a rod with a positive charge? When a positive rod touches the knob of the electroscope, electrons flow from the electroscope to the rod. As the electrons leave the electroscope, the gold leaves develop a positive charge. Once again, the gold leaves repel each other.

Figure 24-3 An uncharged object can become charged if it is touched by another charged object. Once both are charged, how do the objects behave?

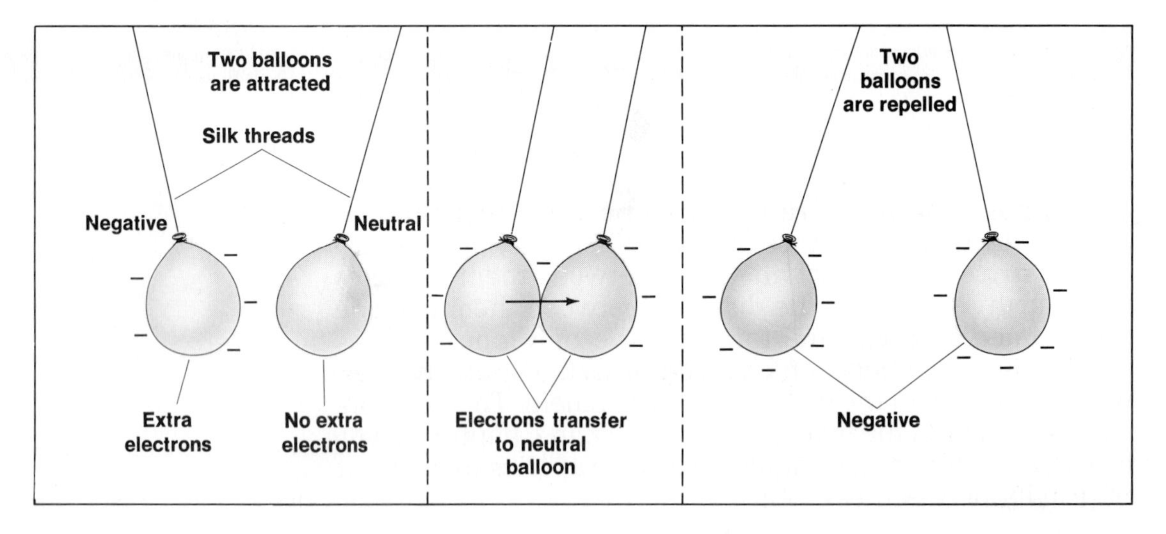

Objects can be charged by contact or by induction. You have learned that objects can be charged by touching them with other charged objects. This process, shown in Figure 24-3, is called charging by contact. Notice that after the charged balloon touches the uncharged balloon, both balloons have the same charge.

Objects can also be charged by a process called induction. **Induction** occurs when a charged object produces an opposite charge on a nearby object. You learned earlier that an uncharged object will be attracted to a charged object. The balloon in Figure 24-4 is attracted to the charged rod. This attraction occurs because some electrons on the balloon are repelled by the negative charge on the rod. These electrons move to the far side of the balloon, leaving the positive charge on the side of the balloon near the rod. The unlike charges on the balloon and the rod will then attract each other.

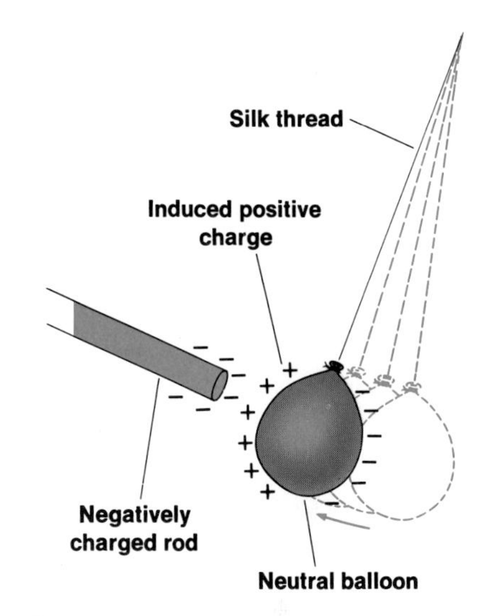

Figure 24-4 How is a positive charge induced on the neutral balloon?

This brief shifting of electrons results in induced charges. An induced charge is one that appears when a nearby charge briefly upsets the electron balance. When the charged rod is removed, the displaced electrons return to their normal positions, and the balloon again becomes neutral.

The shift of electrons caused by an induced charge can be prevented by grounding an object. **Grounding** occurs when the induced charge of an object is absorbed upon contact with a much larger object. The balloon in Figure 24-5, for example, is being grounded by the hand.

Most often, grounding means connecting with the earth. But in this case, just touching a charged balloon with your finger will ground the balloon because your body is large enough to drain off the excess electrons. A negatively charged rod placed near the balloon will repel electrons from the balloon into your body. If the ground (your finger) is removed, the electrons cannot get back to the balloon. The balloon is left with a shortage of electrons and thus has a positive charge. This process is referred to as charging by induction.

Notice that when an object is charged by induction, it always receives a charge opposite to that of the inducing charge. In this case, the negative charge on the rod induced a positive charge on the balloon.

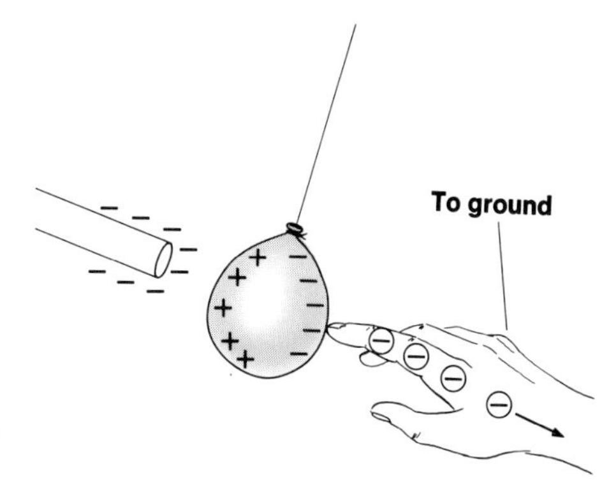

Figure 24-5 What kind of charge is being induced on the balloon? How does the hand affect this induced charge?

A charge can be placed on a conductor. You may have noticed that most of the charged objects in previous examples are nonmetals. Nonmetals are poor conductors of static charges. Charges do not move very well or very far on a poor conductor.

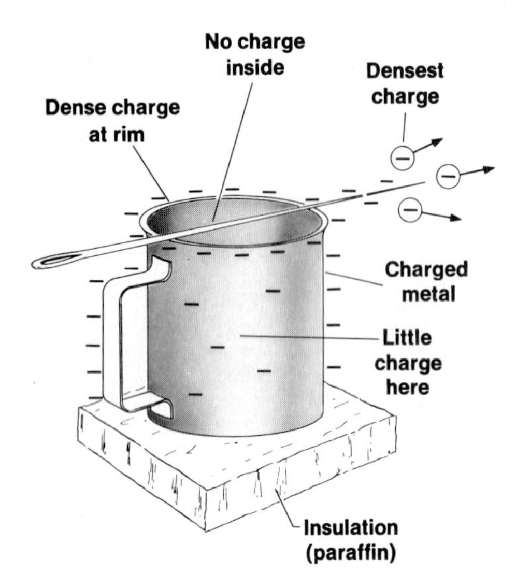

Figure 24-6 Where do charges build up on a conductor? Where are they most likely to leak off?

What happens when you put a charge on a good conductor? As in the electroscope, the charge moves throughout the entire conductor very quickly. Most metals are good carriers of charge.

A static charge does not spread evenly throughout all parts of a conductor. The charge will collect on outer surfaces and concentrate on sharp or pointed surfaces. If a charge is placed on a metal cup, as shown in Figure 24-6, most of the charge collects on the rim of the cup. The charge on the bottom and the sides of the cup is much weaker. Curiously enough, no charge is found on the inside of the cup. This is true even if the charge was first put on the inside of the cup. The charge moves to the outside surface at once. The reason for this movement is that like charges repel. Therefore, electrons move away from each other to the outermost surfaces of a conductor.

If a sharp needle is placed across the top of the cup, any charge placed on the cup will quickly drain away. The charge moves to the sharp point of the needle and leaks off into the air. This process is important in the operation of many electrostatic devices, including lightning rods.

Remember that there is no charge on the inside of the cup shown in Figure 24-6. All hollow conductors act in the same way. Because charges move to the outside surface of a conductor, the space inside is shielded from the effects of these outside charges. This means that you are probably safe from lightning inside a metal car or in a building with a steel frame.

24.2 HAZARDS OF STATIC ELECTRICITY

Static charges can produce sparks and shocks. Static charges can cause fires and explosions wherever there is flammable material. Static electricity has caused many disasters in the petroleum and dry-cleaning industries. Sparks from static charges have set fire to hydrogen balloons and to anesthetic gases used in surgery. Accidents such as these, however, can be prevented by following certain fire prevention measures.

Did you ever get a shock when you touched the door handle of a car? The shock is caused by a static charge. A car can build up a static charge in a number of ways. Friction between the tires and the road or between the car and the air, for example, may cause a transfer of electrons.

You may have noticed grounded wires sticking up from the road just before a highway toll station. These wires neutralize the static charge built up on the car and prevent both the driver and the station attendant from getting shocked as they exchange coins.

To neutralize a charge, the normal electron-proton balance must be restored. There are two ways of restoring this balance to an object: (1) The charged object can be given an equal and opposite charge. (2) The charged object can be grounded. The earth acts as a storehouse of free electrons, which can neutralize positive charges.

Charged objects also lose their charge simply by being exposed to air. The charge leaks off into the air. The amount of moisture in the air greatly affects the rate of discharge. On a humid summer day, discharge occurs quite rapidly. On a cold, dry winter day, discharge occurs quite slowly.

Lightning is caused by static charges. Prior to the early 1750s, lightning was a mystery that frightened most people. Benjamin Franklin, however, believed lightning was a form of static electricity, similar to sparks. In 1752, with his famous kite and key experiment, Franklin proved his theory.

How do moving electrons build up charges during a storm? Winds in a storm push around masses of air and clouds. The effect is like pushing a piece of fur on a plastic rod. These moving masses tend to gain or lose electrons, and build up positive or negative charges.

When a negatively charged cloud mass forms near the earth's surface, an opposite charge is induced on the objects directly below it. When the attraction between the two charges becomes great enough, the electrons jump from the moving cloud to earth. This discharge of electrons heats air and causes flashes of light called lightning, as shown in Figure 24-7. The heating of air also causes the air to expand rapidly. This expansion produces thunder.

Figure 24-7 How is lightning related to static electricity?

Franklin also noticed that a pointed object loses a static charge quickly. He reasoned that perhaps a building could be protected from lightning by attaching some grounded points to it. The lightning rod is based on this reasoning.

How does a lightning rod work? Suppose a house, similar to the one in Figure 24-8, has a number of pointed metal lightning rods connected to the ground. The positive charge built up in the clouds by a storm is slowly and quietly discharged by the lightning rods.

If a cloud has a negative charge, the lightning rod will drain extra electrons from the cloud to the ground. If the cloud has a positive charge, as in Figure 24-8, the earth supplies the electrons needed to help neutralize the cloud. Sometimes, so many electrons stream from (or to) the metal points of the rods that the rods glow.

If the charge is not neutralized fast enough, lightning will strike. In that event, lightning rods act as conductors. They carry the charge safely into the ground. However, lightning rods can be dangerous if they are not properly grounded. Why do you think this is so?

You may think that a television antenna can serve the same purpose as a lightning rod. However, this is not the case. If the antenna is not grounded properly, lightning can come into the house, ruining the TV set and perhaps causing a fire.

Figure 24-8 In what way is a lightning rod like the needle on the cup shown in Figure 24-6?

24.3 USING STATIC ELECTRICITY

Large static charges can be produced. By the seventeenth century, scientists knew that they could produce static charges by rubbing various substances together. In 1672, Otto von Guericke (*vuhn* GAY-*rih-kuh*), a German inventor, built a machine based on this principle. His machine looked like a grindstone. Instead of using a stone, however, he used a ball of sulfur. His machine resembled the one pictured in Figure 24-9 and produced static charges. Guericke's machine and others like it are called electric generators. An **electric generator** is a device that changes mechanical energy to electrical energy.

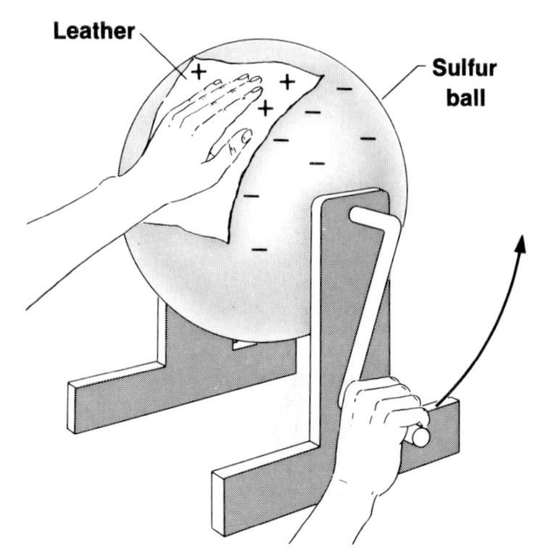

Figure 24-9 How can a static charge be produced by this early generator?

Large static charges can be formed in a generator called the Van de Graaff generator (Figure 24-10). The generator consists of a hollow metal sphere standing on insulating supports. Inside is a motor-driven rubber belt stretched between two pulleys. As the belt moves, an emitter comb transfers electrons to the belt so that it becomes charged. The belt moves into the metal sphere, where the charge is transferred from the belt to the sphere by a collector comb. The charge then moves to the outside of the sphere. How is this similar to the experiment in Figure 24-6? The Van de Graaff generator is used by nuclear physicists as a source of charged particles for atom bombardment.

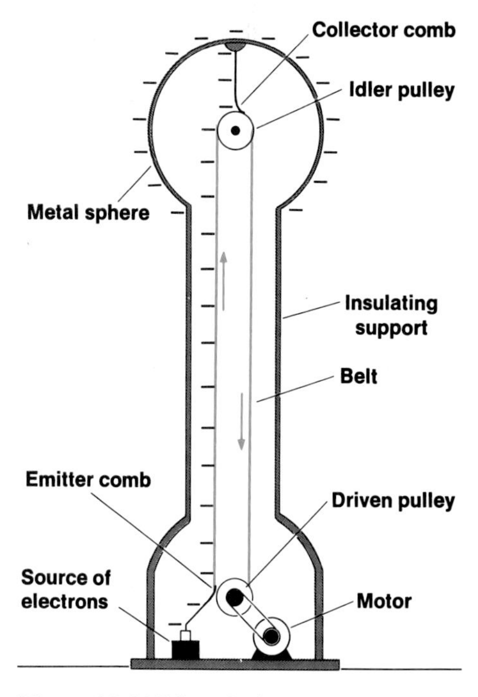

Figure 24-10 Where is the static charge stored in the Van de Graaff generator?

A capacitor can store a static charge. While Benjamin Franklin was working in America, two scientists working in Europe built a device in which static charges could be stored. The scientists were amazed at the violent shocks they got from a glass jar coated inside and out with metal foil. This jar became known as a Leyden jar, or a condenser. The term "condenser" was used because the jar was able to "condense" the so-called "electrical fluid" believed at that time to provide static electricity. Today this device is called a capacitor. A **capacitor** is a device that stores static charges. Capacitors consist of metal plates that are separated by an insulator.

To understand how a capacitor works, consider the following example. Suppose a small negative charge is put on metal plate A in Figure 24-11 on the next page. The electroscope leaves spread apart. Now suppose grounded metal plate B is brought close to plate A. As the figure shows, the negative charge on plate A induces a positive charge on plate B. This induced positive charge attracts more of the electrons on plate A to the side nearest plate B. The electroscope leaves come together as electrons move away from the leaves up toward plate A. This indicates plate A now holds a larger charge. The two plates and the insulating air space between them form a capacitor.

Without the induction of plate B, you could not put a larger charge on A. The charge would leak off as fast as you added it. The closer A is to B, the greater the charge the capacitor can hold. Also, the larger the plates, the more charge the capacitor can hold.

The capacitor is one of the most widely used devices in electronics. When you tune your radio, you are turning a variable capacitor. Such a capacitor has two sets of plates that mesh together without touching. As the plates open up, the capacity decreases. When this happens, the frequency of electrical waves received rises. If the plates are meshed together, the capacity increases and the frequency of electrical waves goes down. By this method, you can tune your radio to match the frequency of the desired station.

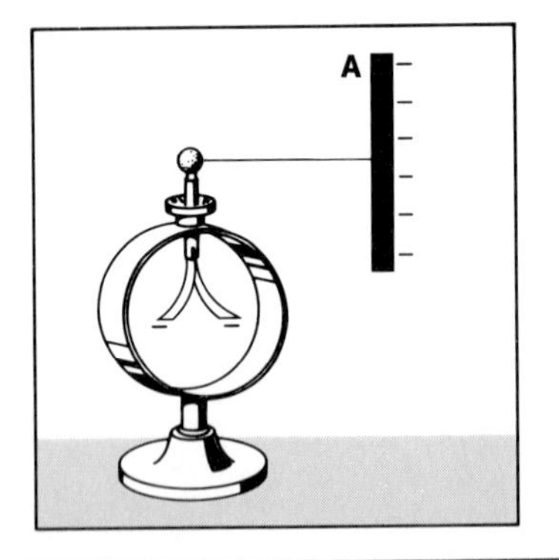

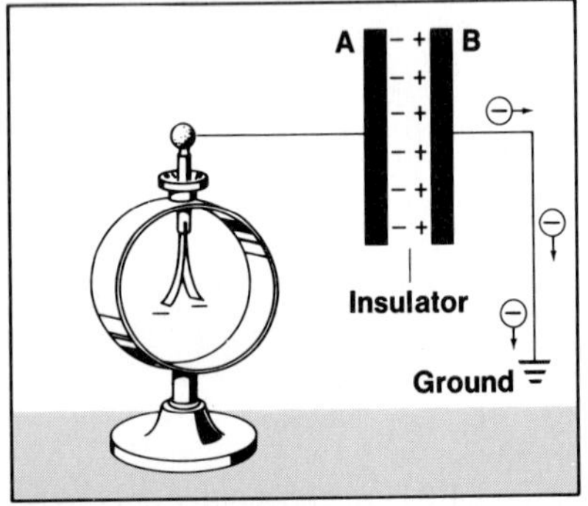

Figure 24-11 Explain why plate A can hold a larger charge if plate B is moved next to it.

There are many uses for static charges. In 1938, Chester Carlson invented a process of copying printed matter without the use of liquid inks or chemicals. Instead, Carlson's process used tiny static charges and dry powdered inks. The name he gave to the process, xerography (*zih*-RAHG-*ruh*-FEE), comes from the Greek words for dry writing.

In a photocopy machine, a beam of light transfers the image from a printed page onto a positively charged drum. The drum is coated with a photoconductor, a substance that conducts a charge only when exposed to light. When a strong light shines through the paper, it leaves a shadow of the printing on the drum. Where the light strikes the charged drum, the photoconductor carries away the positive charge. Where the light does not strike, however, the positive charge remains.

At this stage, a dry ink with a negative charge is dusted onto the drum. The negatively charged ink is attracted only to the positively charged area of the drum. The drum rotates and is pressed against a sheet of paper, which attracts the ink. The ink, in turn, is heated to fuse it to the paper.

Static charges can also be used to filter air. Static-charge filters can be small units used in home heating systems to remove dust and pollen. Similar huge filters are used to clean smoke from smokestacks. Static charges are used inside these filters to ionize, or charge, the particles in the air. The air is then passed over plates or wires that are also highly charged. Some of the plates are positive and some are negative. The ionized particles are attracted to the charged plates and removed from the air. One type of industrial filter is called the Cottrell device, illustrated in Figure 24-12. (See also Chapter 14 and Figure 14-10.)

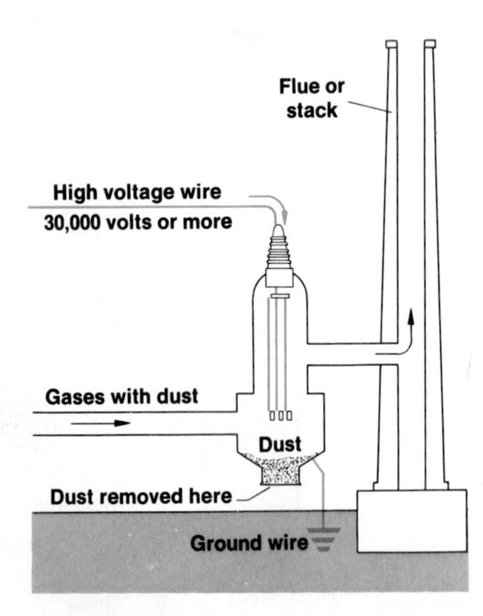

Figure 24-12 The Cottrell device removes dust and dirt from gases before the gases enter the flue and escape into the air. How does the Cottrell device work?

BIOGRAPHY

BENJAMIN FRANKLIN

lightning and static electricity

Benjamin Franklin was a gifted American scientist, scholar, and statesman. He excelled in such varied fields as writing, printing, government service, and many sciences, including optics and agriculture. Franklin was also an inventor. Some of his inventions included the Franklin stove, the bifocal spectacles, and the lightning rod.

Franklin was born in Boston, Massachusetts. He was tenth in a family of seventeen children. Franklin learned to read at an early age. However, after only two years of training, his formal education ended. It was Franklin's love of reading and his desire to learn that made him so successful.

Franklin's most well-known scientific work involved electricity. He believed that what he called "electrical fire" was a fluid that was attracted by other matter such as water and metals. He thought that when a body containing a large amount of electrical fire came near one with a smaller quantity, a discharge would equalize the electrical fire between the two bodies. To prove his theory, Franklin carried out his famous kite experiment.

In this experiment, Franklin tied a metal key to the lower end of a thin wire that was attached to a kite. He did not hold onto the wire or key, however. He knew that a charge of lightning could be carried down the wire. Instead, he attached a short piece of string to his end of the wire. When charges from the clouds flowed down the wire, a bright spark jumped from the key. This spark was the static charge Franklin had predicted. This experiment proved that lightning is a form of static charge.

Based on the results of his kite experiment, Franklin proposed that a building could be protected from lightning by attaching pointed iron rods to the sides of the building. He suggested that these iron rods would draw off the electrical fluid and carry it safely to the ground. Franklin's protective iron rods, now found on many buildings, are called lightning rods.

Franklin approached the study of science as he did all his other work. He felt that to be successful, a person has to work just a little harder than anyone else. Franklin's death at age 84 was mourned both in America and Europe.

CHAPTER REVIEW 24

SUMMARY

1. When an object gains electrons, it has a negative charge.
2. When an object loses electrons, it has a positive charge.
3. Like charges repel each other; unlike charges attract each other.
4. The force between two charged bodies depends on the size of each charge and the distance between the charged bodies.
5. Charged objects attract uncharged objects.
6. Electroscopes detect static charges.
7. Objects can be charged by contact and by induction.
8. Static charges move to the outer surfaces of a conductor.
9. Lightning is caused by static charges.
10. Lightning rods reduce the build-up of charges and provide an easy path to the ground if lightning strikes.
11. Static charges can be stored in capacitors.
12. Xerography and electronic air filters are some uses of static charges.

VOCABULARY

Match the item in the left column with the best answer in the right column. Do not write in this book.

1. capacitor
2. electric generator
3. electroscope
4. grounding
5. induction
6. static electricity

a. accumulation of a charge on an object
b. device used to detect an electric charge
c. device that changes mechanical energy to electrical energy
d. absorption of the induced charge on an object by a larger object or by the earth
e. device consisting of metal plates separated by an insulator
f. when a charged object produces an opposite charge on a nearby object

REVIEW QUESTIONS

1. What kind of charge does an electron have?
2. What charge does an object have if it loses some of its electrons? if it gains electrons?
3. How do like charges behave toward each other? How do unlike charges behave toward each other?
4. Explain why a neutral object is attracted by an object with either a positive or negative charge.
5. What is a field of force?
6. What is the purpose of the electroscope?

7. When you rub hard rubber with wool, what kind of charge does each substance pick up? Explain.
8. What happens when a conductor with a positive charge is grounded?
9. Explain how an insulated metal ball becomes charged when you touch it with a negatively charged rod.
10. In what two ways can a charge be neutralized?
11. What kind of charge is induced by a positive charge?
12. What are some dangers caused by a buildup of static charges?
13. What is lightning?
14. What is a capacitor used for? How is it made?
15. Explain how a capacitor stores up electric charges.
16. Why do you sometimes see a spark when you shuffle your shoes across a nylon rug and then touch a metal doorknob?
17. Explain why you are more likely to get an electric shock when touching a metal doorknob on a cold, dry day than on a hot, humid day.
18. In what two ways can lightning rods protect buildings from damage?
19. Explain how an image is transferred from printed material to the drum of an electrostatic copier.
20. Explain how an image is transferred from the drum of an electrostatic copier to a sheet of paper.

CRITICAL THINKING

21. Describe two ways of putting a positive charge on an electroscope.
22. What happens to the leaves of a charged electroscope when you touch the knob with your finger? Explain.
23. Describe how a negative charge can be put on an insulated conductor using a rod with a positive charge.
24. A gold-leaf electroscope has a positive charge. As you bring a charged rod toward it, the leaves spread farther apart. What kind of charge is on the rod? Explain.
25. Which has more need of lightning protection, a brick smokestack or a steel skyscraper? Explain.

FURTHER READING

Asimov, Isaac. *How Did We Find Out About Electricity?* New York: Walker, 1973.

Chapman, Phil. *Electricity*. (From Young Scientist Series). Tulsa, Okla.: EDC, 1976.

Epstein, S. *The First Book of Electricity*. New York: Franklin Watts, Inc., 1977.

Hockey, S.W. *Fundamentals of Electrostatics*. New York: Barnes and Noble Books, 1972.

CURRENT AND CIRCUITS

The night skyline of Paris blazes brightly with light, serving as a reminder that the energy of electric charges can be harnessed. How is this possible? What role do electric charges play in producing the electricity we use to run our appliances and light our homes?

CHAPTER 25

SECTIONS

OBJECTIVES

- ☐ Describe the nature of an electric current.
- ☐ Distinguish between conductors and insulators.
- ☐ State the relationships among current, potential difference, and resistance.
- ☐ Compare the functions of series and parallel circuits.
- ☐ Describe how an electric current can make heat and light.

25.1 ELECTRONS ON THE MOVE

An electric current is a stream of moving electrons. In Chapter 3, you learned that an electron is an atomic particle that has a negative electric charge. Electrons exist in all atoms and can be easily removed from the atoms of certain elements. These free electrons can be pushed along through substances called conductors. The movement of electrons within a conductor is called an **electric current.** The path along which an electric current travels is referred to as an **electric circuit.**

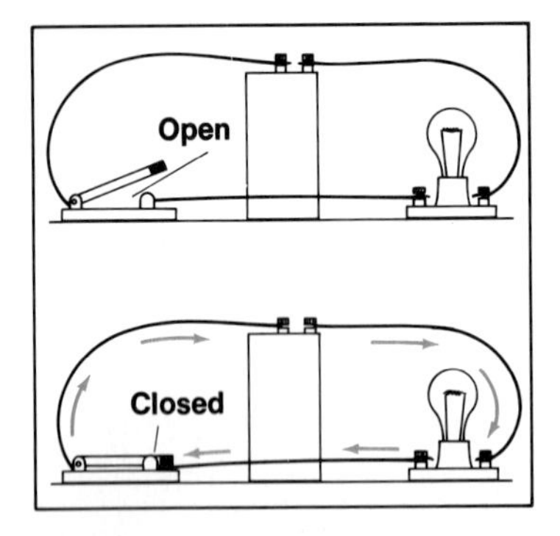

Figure 25-1 In which of the diagrams is the path of electrons broken? In which is it complete? What is the function of a switch in an electric circuit?

Suppose that a copper wire is used to make an electric circuit. If the wire is cut and the loose ends attached to a switch, the current in the circuit can be turned on or off. Look at Figure 25-1. Opening the switch stops the flow of electrons just as an open drawbridge stops the flow of cars across a river. When a switch is open, the circuit is called an open circuit. When the switch is closed, the circuit is complete and the current flows again. When the switch is closed, the circuit is called a closed circuit.

A flow of electrons can be compared with a flow of water. Just as a current of water could be measured in molecules per second, an electric current can be measured in electrons per second. However, the electron has such a tiny charge that vast numbers of them are needed to make even a small current. A more useful unit of charge is the coulomb. A **coulomb** is the charge on 6.25 billion billion electrons.

Just as water is more easily measured in liters, electrons are more easily measured in coulombs. One coulomb of electrons flowing through a wire per second is an **ampere** of current. A 100-watt light bulb requires a current of a little less than 1 ampere. The ampere is measured with an instrument called an ammeter.

Electrons are pushed along an electric circuit by a potential difference. To keep electrons flowing in a circuit, some sort of push or electric pressure is needed. This push can be better understood if you compare an electric circuit to a water fountain. Look at the fountains in Figure 25-2. In order to maintain a flow of water in the pipes, the water must be pushed. This push comes from the weight of the water that is at a level higher than the fountain.

Figure 25-2 How does pressure keep water flowing in a water fountain? Which fountain has greater potential energy?

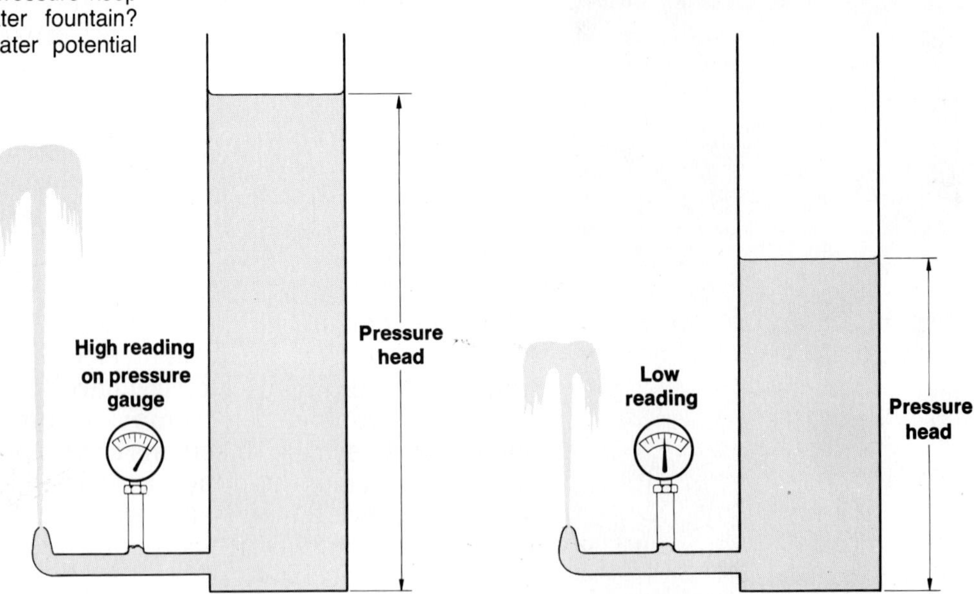

Recall that anything lifted to a height above its starting point has potential energy that can be released when the object falls. In a water system, the water is raised by a pump, and the stored energy is released when the water falls. There is a potential difference between the starting and ending levels of the water.

Another way in which water can receive a push is shown in Figure 25-3. The water in the pipe is pushed by a turning pump. This water system works like an electric circuit. In an electric circuit, a battery or generator acts as a pump, taking electrons from atoms. These electrons create a difference in the level of the charge on the two posts of the battery or the generator. Therefore, there is a potential difference between the two posts. In Figure 25-3 this difference is indicated by the plus and minus signs over the posts. The negative post of a cell is called the cathode and the positive post is called the anode.

Electrons will flow from an object of high electron potential (the cathode) to one of low potential (the anode). This electrical potential difference is measured in units called **volts.** An instrument called a voltmeter is used to make such measurements. An increase in voltage produces a larger current. A common cell used in a flashlight gives a potential difference, or voltage, of 1.5 volts. If two cells are used in a flashlight, the voltage is doubled, and the current produced is twice as large. You will learn about cells in Chapter 26.

When electrons flow steadily in one direction, the flow is called a direct current (DC). The dry cell used in a flashlight is a source of direct current. This current always travels in the same direction. When the direction of the current keeps changing in a circuit, it is called an alternating current (AC). Electric generators are sources of alternating current. Your home is supplied with alternating current by local energy plants. How the generators at an energy plant operate will be discussed in Chapter 27.

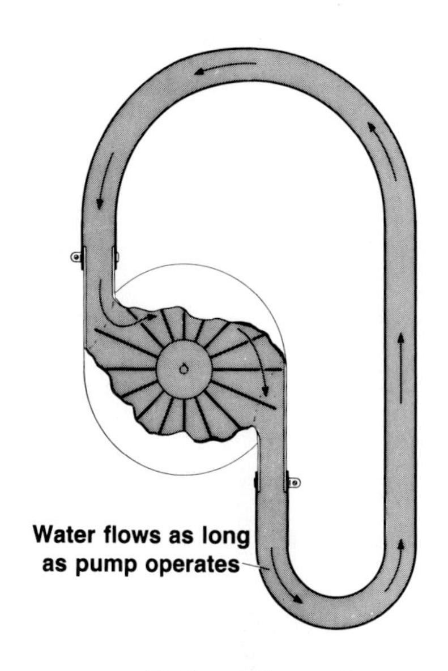

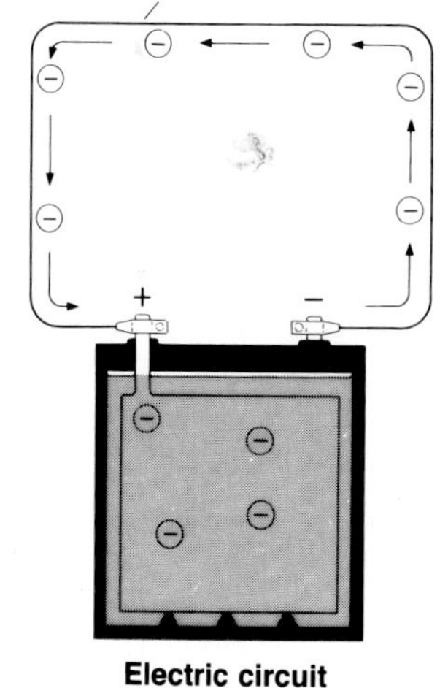

Figure 25-3 What is the source of potential difference in the electric circuit? How does it compare to the source of pressure in the water system?

Substances differ in how well they conduct currents. Given a source of electric current, how are the electrons carried from one place to another? In Chapter 24 you learned that a conductor is any substance that allows electrons to move through it easily. Therefore, conductors are used to carry electric currents from place to place. The best conductors are metals such as silver, copper, gold, and aluminum. Because of the high cost of silver and gold, only copper and aluminum are commonly used as conductors.

Some materials will not carry a current at all, even if the voltage is very high. Materials that do not readily carry a current are called **insulators,** or nonconductors. Most insulators are nonmetals such as plastics, rubber, porcelain, and glass. In normal conditions, air is also a nonconductor. When it contains moisture, however, air becomes a conductor.

Certain elements, such as germanium and silicon, are called semiconductors. A **semiconductor** is a solid material the ability of which to carry a current is between that of a conductor and an insulator. Most electronic equipment in use today contains semiconductors. Devices that use semiconductors are usually small and long lasting. They produce little heat and require little power to operate.

You know that water flows more freely through a clean water pipe than through one that is partly clogged. You could say that a clogged pipe offers resistance to the flow of water. In much the same way, conductors can resist the flow of a current. **Resistance** is a conductor's opposition to the flow of a current. An electric current is greater if the resistance of a conductor is reduced. The unit of electrical resistance is the **ohm.** The symbol for ohm is the Greek letter omega, Ω. The electrical resistance of the conductor in a circuit depends on the thickness, composition, length, and temperature of the conductor.

In Figure 25-4A, tap water is being forced through a short, thick pipe and a long, thin pipe. Even though the water pressure on both pipes is the same, the thick pipe allows more water to flow than does the thin pipe. Similarly, the same voltage sends a larger current through a thick wire

Figure 25-4 How are the water-carrying capacities of the pipes like the electrical-carrying capacities of the wires?

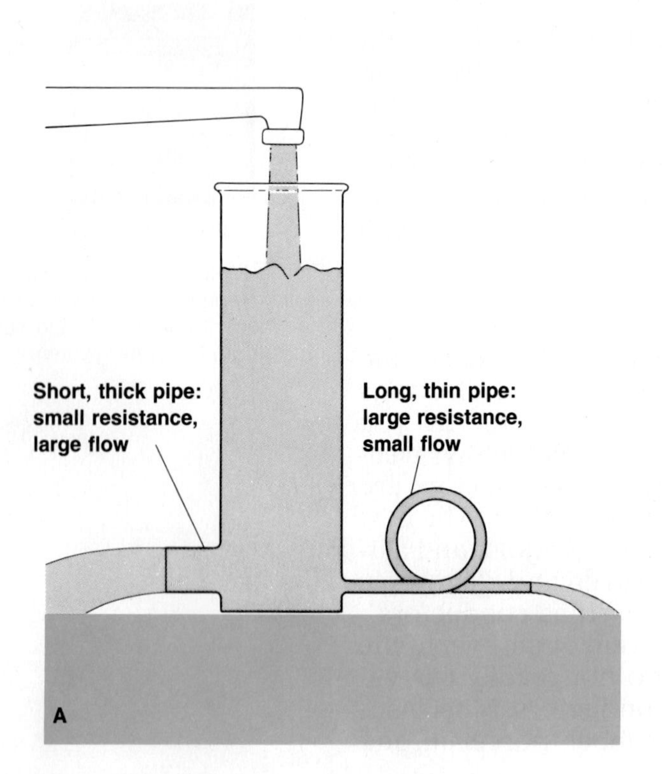

than it does through a thin one. Compare the ammeter readings in Figures 25-4B and 25-4C. Which setup allows for the larger flow of current? Assuming all other factors are equal, a thick wire will always have less resistance and thus allow a greater flow of electric current than a thin wire.

The relationship between resistance and wire thickness can be expressed in terms of a wire's cross-sectional area. For example, 300 m of copper wire that is 2.5 mm thick has a resistance of 1 ohm. If it were twice as thick, this wire would have four times the cross-sectional area. The resistance would be only 0.25 ohm, and the wire could carry four times as much current given the same voltage. Therefore, the resistance of a wire conductor varies inversely with its cross-sectional area.

The resistance of a substance also depends on its composition. Recall that most nonmetals are nonconductors. Their resistance is very high. Semiconductors have a resistance that is between those of nonmetals and metals. The best conductors, such as silver and copper, have very low resistance. A poor metallic conductor, nichrome, is made from an alloy of nickel and chromium. Nichrome offers about 50 times as much resistance as copper. However, even nichrome is a better conductor than most nonmetals.

Nichrome is used to make heating elements in toasters, irons, and other appliances. Other metals are used as resistors in light bulbs. When electrons are forced by a high voltage through such resistors, the electric energy is changed to heat. If the temperature is high enough, some of the electric energy is changed to light energy. This is what happens in the filament of a light bulb.

The longer a wire, the greater its resistance. If the resistance of 300 m of wire is 1 ohm, then 600 m of the same wire will have a resistance of 2 ohms. The resistance of a wire varies directly with its length. How would resistance affect the current in an appliance if several extension cords were joined to reach a distant source?

Temperature is another factor that affects resistance. The resistance increases as a conductor becomes hotter and it decreases as the conductor becomes cooler. Have you ever noticed the lights dim for an instant when an electric iron is plugged in? The lights dim because the wires inside the iron are still cool. These cool wires have a low resistance and therefore draw a large current. At that moment, less current is available for the lights. When the wires in the iron become hot, their resistance increases, and the current decreases. The room lights then regain their brightness.

In Chapter 19 you learned that certain substances called superconductors have no resistance to current when they are cooled to near absolute zero (−273° C). Once started, a current will continue to travel in a superconductor circuit even after the source of current is removed.

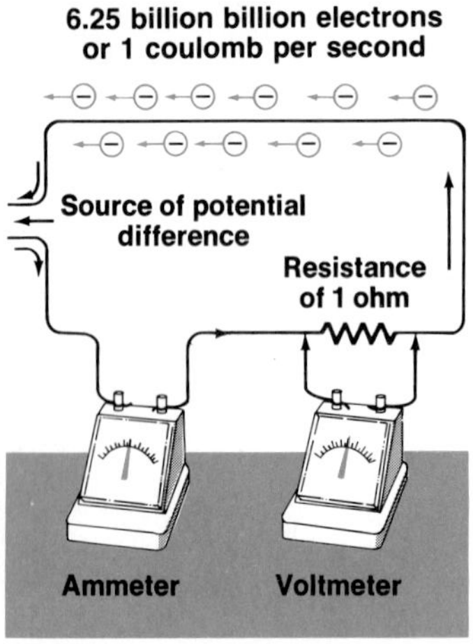

6.25 billion billion electrons
or 1 coulomb per second

Source of potential
difference

Resistance
of 1 ohm

Ammeter

Voltmeter

Figure 25-5 One volt of potential difference can push a current of one ampere (or one coulomb per second) through a resistance of one ohm.

Why should cold metals offer less resistance to currents than do hot metals? Scientists believe that electrons moving through a wire are slowed by the vibration of the atoms in the wire. The atomic movement increases when the metal is warmed and decreases when it is cooled. When a conductor is cooled, its atoms are less likely to get in the way of moving electrons. In superconductors, the atoms seem to be slowed to a point where they do not block electron flow at all. What problem prevents widespread use of superconductors?

The amount of resistance in an electric current can be controlled by a device known as a rheostat, or variable resistor. When connected to a circuit that contains a lamp, a rheostat can be used as a dimmer switch. When the resistance is raised, the amount of current is reduced and the lamp light becomes dim. When the resistance is lowered, the flow of current is increased and the lamp light becomes brighter.

There is a relationship among volts, amperes, and ohms. A potential difference of 1 volt causes a current of 1 ampere to move through a resistance of 1 ohm. This principle is illustrated in Figure 25-5. You have learned that two major factors, voltage and resistance, affect the amount of electrons that flow through a wire. Therefore, if there is a change in either voltage or resistance, there must also be a change in the current.

If 1 volt pushes 1 ampere through 1 ohm, 5 volts are needed to produce five times as much current (5 amperes). If the resistance were then doubled to 2 ohms, twice as much voltage would be needed. That is, it would now take 10 volts to push the 5-ampere current through a 2-ohm resistance.

The relationship among voltage, current, and resistance can be stated in a formula known as Ohm's law:

$$V = I\,R$$

where V is the voltage, I is the current, and R is the resistance. From this example, 10 volts = 5 amperes × 2 ohms.

Electric energy can be measured in kilowatt-hours. In Chapter 18 you learned that the units of power in the metric system are the watt (W) and the kilowatt (kW). Recall also that work = power × time. Thus energy can also be expressed in terms of power × time. The electric energy used in most households is measured in units called kilowatt-hours (kWh). As the name suggests, the kilowatt-hour represents one kilowatt of energy used for one hour.

SAMPLE PROBLEM

How large a current will 120 volts send through a resistance of 2 ohms?

SOLUTION

Step 1: Analyze

You are given that the voltage is 120 volts and the resistance is 2 ohms. The current is the unknown.

Step 2: Plan

Since you know that $V = I\,R$, you can use this equation to find current, by solving for I:

$$I = \frac{V}{R}$$

Step 3: Compute and Check

Substitute all values in the equation.

$$I = \frac{120 \text{ volts}}{2 \text{ ohms}}$$
$$= 60 \text{ amperes}$$

Since you can check your answer by multiplying 60 amperes × 2 ohms to get 120 volts, your solution is complete.

An electric meter like the one shown in Figure 25-6 records the number of kilowatt-hours used in your home. Your electric bill is computed by multiplying the number of kilowatt-hours by the cost of electricity per kilowatt-hour. If, for example, your family uses 1,200 kWh of electricity in one month at a cost of 5 cents per kWh, the total monthly bill would be 1,200 × \$0.05 = \$60.00

Another relationship exists between power and electricity. One watt of power is used when 1 volt pushes a current of 1 ampere through a conductor. The word equation for this relationship is stated as follows:

$$\text{power} = \text{voltage} \times \text{current}$$

or using symbols

$$p = V\,I$$

Figure 25-6 An electric meter typical of the models found in most buildings. What units are measured by the electric meter?

SAMPLE PROBLEM

Find the power, in watts, of a toaster that draws a current of 5 amperes on a 115-volt line.

SOLUTION

Step 1: Analyze

You are given that the current used is 5 amperes and the voltage is 115 volts. The power is the unknown.

Step 2: Plan

Since you know that power equals voltage times current, or

$p = V\,I$

and since you know two of the three values in this equation, you can solve for p.

Step 3: Compute and Check

Substitute all values in the equation.

$$p = 115 \text{ volts} \times 5 \text{ amperes}$$
$$= 575 \text{ watts}$$

Since you can check your answer by dividing 575 watts by 115 volts to get 5 amperes, your solution is complete.

25.2 PATHS FOR ELECTRONS

Electrons can flow in a series circuit. Figure 25-7 shows eight electric lamps connected in a single circuit. There is only one path for electrons to follow. Therefore, an electric current must pass through all eight lamps, one after the other. This type of circuit, in which the current has only one path to follow, is called a **series circuit.** What would happen if one light bulb in a series circuit burned out? The electrons could not flow through any of the bulbs because the circuit has been broken.

In a series circuit, the total resistance is found by adding the resistance of all the separate parts. For example, if the resistance of each light bulb is 70 ohms, the total resistance of the string of lights will be 8×70 ohms, or 560 ohms.

The current in a series circuit is the same everywhere along the circuit. If the current that goes through one lamp is 0.2 amperes, then 0.2 amperes is the current that goes through each lamp.

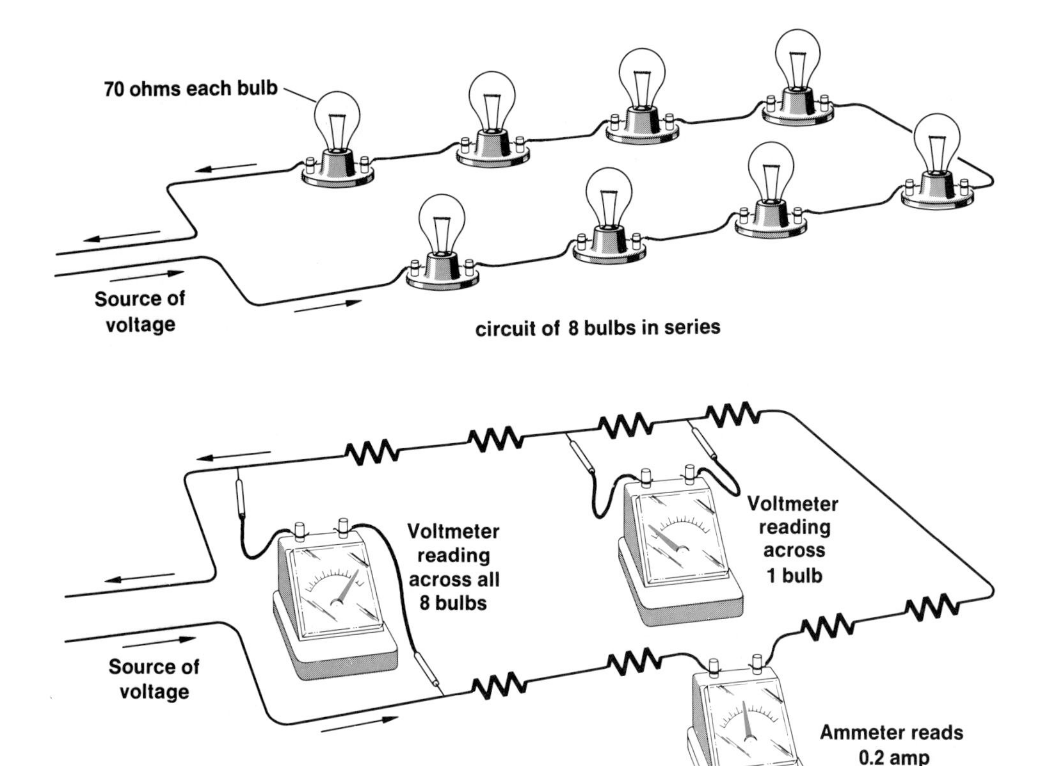

Figure 25-7 Is the voltage the same in all parts of a series circuit? Is the current the same?

The total voltage in a series circuit is the sum of the voltages across each resistance. As you can see in the wiring diagram in Figure 25-7, the voltage across all eight bulbs is higher than the voltage across one bulb.

Electrons can flow in a parallel circuit. Recall that a series circuit has only one path for the current. A **parallel circuit** contains two or more branches through which a current can flow. It is the circuit used in most households. In the parallel circuit shown in Figure 25-8, the current divides and flows along two branches. These branches join together again, and current flows out of the circuit. The following statements are true of all parallel circuits:

1. If the circuit is broken in any branch (by removing a bulb, for instance), electrons can still pass through the other branches because each branch has its own complete path for electrons.
2. The voltage across all of the branches is the same. The potential difference between the entrance and exit to this circuit is the same regardless of the path taken.

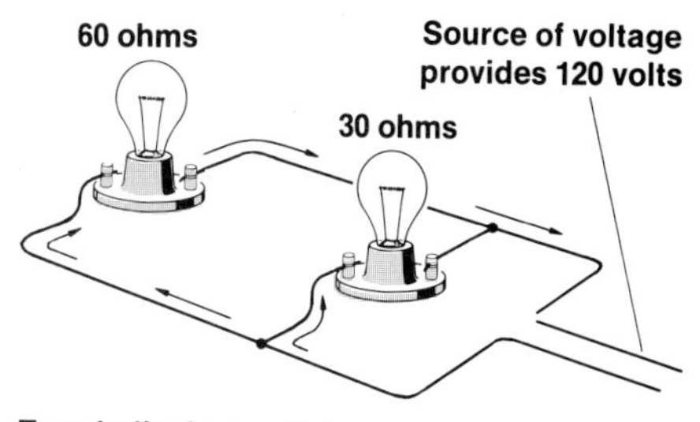

Two bulbs in parallel (wiring diagram shown below)

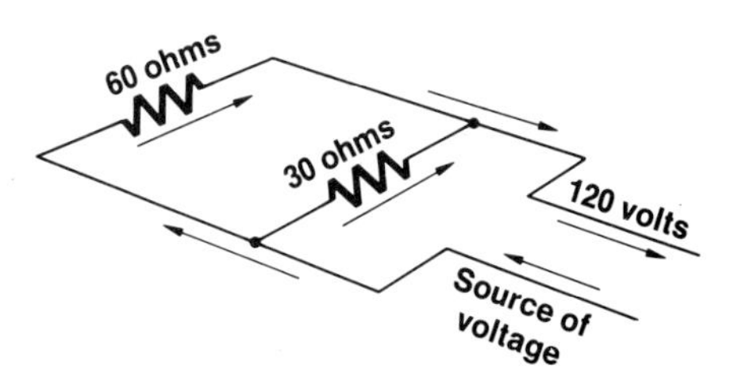

Figure 25-8 Through which bulb in the diagram will the greater amount of current flow?

3. The total current in the circuit equals the sum of the currents in each of the separate branches.
4. As more branches are added to the circuit, the total resistance decreases. The more pathways that are opened for the current, the less total resistance the current meets.

SAMPLE PROBLEM

How many volts are needed to pass a current of 0.2 amperes through eight bulbs in a series circuit if each bulb has a resistance of 70 ohms?

SOLUTION

Step 1: Analyze

You are given that there are eight bulbs, each with a resistance of 70 ohms. The current is 0.2 amperes. Voltage is the unknown.

Step 2: Plan

Because this is a series circuit, you know that the total resistance is the sum of the resistance of each part, or 560 ohms (8 × 70 ohms). Since you know Ohm's law,

$$V = I R$$

and since you know two of the three values in this equation, you can solve for V.

Step 3: Compute and Check

$$V = 0.2 \text{ amperes} \times 560 \text{ ohms}$$
$$= 112 \text{ volts}$$

Since you know that 112 volts equals the sum of the voltages across each resistance (8 × 0.2 × 70 = 112), your solution is complete.

A ***short circuit is dangerous.*** Have you ever been showered by sparks while using an electric appliance? What caused this to happen? To answer this question, consider the resistance, voltage, and current in the circuit at the time.

Suppose that a person is using an electric toaster in which the resistance is 23 ohms and the voltage at the plug is 115 volts. According to Ohm's law, the current flowing is 115 volts/23 ohms, or 5 amperes.

SERIES AND PARALLEL CIRCUITS

activity

OBJECTIVE: Demonstrate how the addition and the removal of light bulbs affect a series circuit and a parallel circuit.

PROCESS SKILLS

In this activity, you will *model* series and parallel circuits and *collect data* on their characteristics.

MATERIALS

8 6-volt light bulbs
8 sockets
2 6-volt batteries
2 switches
small screwdriver
insulated copper wire
wire cutters
paper
pencil

PROCEDURE

1. Arrange a switch and three 6-volt light bulbs in a series circuit. Be sure the switch is in the "off" position. Use a 6-volt battery to operate the circuit. Have your teacher check the circuit before you turn the switch to the "on" position. CAUTION: *Do not touch any part of the circuit, except the switch, when the power is on.*
2. Arrange three 6-volt light bulbs in a parallel circuit. Put a switch on the main line of the circuit. Be sure the switch is in the off, or open, position. Use a 6-volt battery to operate the circuit. Have your teacher check the circuit before you turn the switch to the on, or closed, position. Record which of the two circuits gives the brighter light.
3. Open each switch and remove one bulb from each circuit. Close the switches and record what happens in each case. Open the switches and replace the two bulbs.
4. With the switches open, add one bulb to each of the circuits. Close the switches and record what happens in each case.

OBSERVATIONS AND CONCLUSIONS

1. Which group of three bulbs gives the brighter light?
2. What happens when one of the bulbs is removed from its socket in the series circuit? in the parallel circuit?
3. How does the brightness of the four-bulb setup compare with that of the three-bulb setup in the series circuit? in the parallel circuit?

Now suppose the insulation on the cord to the toaster became so worn that the bare copper wires touched each other. As they touched, the wires "shorted," or formed a short circuit. The electrons would then flow from one copper wire to the other, instead of going through the toaster. A short circuit causes a sudden and large drop in resistance in the electric circuit.

If the resistance of the short circuit drops to 1 ohm, instead of 23 ohms, the current would become 23 times as great (115 amperes). This large current produces intense heat in the copper wires in the toaster cord and would quickly burn the covering of the cord. If the short circuit lasts more than an instant, it could spark and potentially start a fire.

Fuses and circuit breakers protect homes from short circuits. Is there any protection against short circuits caused by defective appliances? Protection can be built into a house by using fuses or circuit breakers in every circuit. Any current that goes through the circuit must first go through the fuse or circuit breaker.

A fuse contains a short wire, made of an alloy of lead, that melts easily. If the circuit carries a load that is too large, the fuse melts, breaking the circuit and thereby stopping the current. A fuse that has melted is commonly referred to as a "blown" fuse.

A 20-ampere fuse will carry up to 20 amperes safely all day without getting hot. If the current is much greater than 20 amperes, however, the fuse wire melts and breaks the circuit. A 115-ampere current would melt the 20-ampere fuse instantly.

Many homes and industries use circuit breakers instead of fuses. Circuit breakers act like switches to open a circuit if the current becomes too great. Temperature-sensitive devices similar to a thermostat inside the circuit breakers determine when the current has exceeded the limit. Circuit breakers are safer and easier to use than fuses. They are also less expensive in the long run, since they can be reset and used over again. Fuses cannot be reused. Once they melt, they have to be replaced.

Circuit breakers are also used in the headlight circuits of many cars. If a short circuit occurs while someone is driving at night, the lights blink on and off. In older cars, when a fuse burned out, the lights went out and did not relight, leaving the driver in danger.

If the cause of a melted fuse is a short circuit or a defective appliance, the defect must be repaired before the fuse is replaced or the circuit breaker is reset. Otherwise, the fuse will melt again.

BLOWING A FUSE

activity

OBJECTIVE: Demonstrate how a fuse works.

PROCESS SKILLS
In this activity, you will *model* a fuse and *observe* how it works.

MATERIALS
6-volt battery
insulated copper wire
wire cutters
aluminum foil
tape
cardboard
scissors

PROCEDURE
1. Use scissors to cut a 15 cm × 15 cm piece of cardboard.
2. Cut a piece of aluminum foil about 1 cm × 10 cm into the shape of a bow tie. The strand between the two sides of the bow should be only about 1 mm wide. Tape the bow to the cardboard.
3. Cut two 30-cm pieces of copper wire. Attach one wire to the positive battery terminal and the other wire to the negative terminal.
4. Touch the left side of the bow with one wire and the right side of the bow with the other wire. CAUTION: *Handle the wire only along the insulated part. Do not touch the bare ends of the wire.* Observe the strand between the sides of the bow.

OBSERVATIONS AND CONCLUSIONS
1. What happened to the bow strand when the sides of the bow were touched by the wires?
2. How does this setup compare with a fuse?

25.3 HEAT AND LIGHT FROM ELECTRICITY

Electricity produces heat. Earlier in this chapter, you learned that an electric current produces heat when passing through a conductor. If that conductor is a copper wire in an extension cord, carrying normal household current, very little heat is produced. However, a nichrome wire, as is used

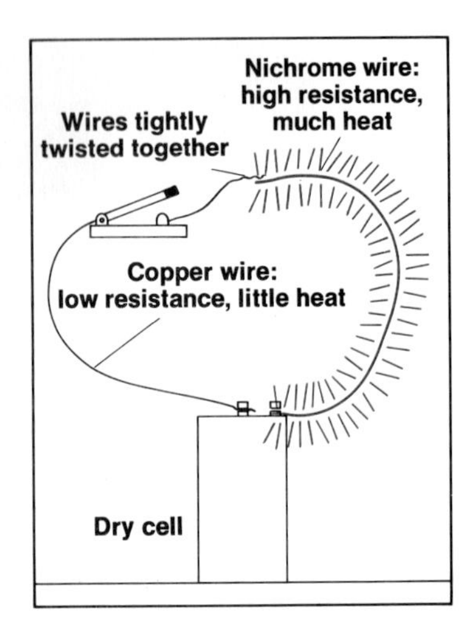

Figure 25-9 What happens to a high-resistance wire when a current flows in it?

inside a toaster, gets red hot when it carries normal household current. Three factors determine the amount of heat produced by an electric current: (1) the resistance of the conductor, (2) the amount of current, and (3) the length of time that the current is on.

You have learned that resistance is caused by collisions between atoms of the metal and the electrons of the current. You also learned that resistance is dependent upon wire thickness, length, and composition. Resistance causes heat. The more resistance the electrons meet in being pushed along, the more heat is produced.

The difference in resistance of two wire samples can be demonstrated by connecting lengths of copper and nichrome wire to each other and to a dry cell as illustrated in Figure 25-9. The same amount of current flows in each wire, but different amounts of heat are produced. The heat produced in each wire depends upon its resistance. The greater the resistance, the greater the heat for a given current.

The amount of heat produced also depends on the amount of the current. When more electrons flow through a conductor, the conductor gets hotter. To demonstrate this, the current from a single dry cell is passed through a thin nichrome wire as shown in Figure 25-10. An ammeter is used to measure the current. Now, instead of one cell, several cells in series are used. As you can see by looking at the ammeter, more current is being generated. In addition, the wire gets much hotter.

Laboratory tests indicate that the heat increases with the square of the current. For example, when the current is doubled, four times as much heat is produced for a given resistance. Heat is also proportional to the length of time the current flows. The longer a heater is turned on, the more heat it gives off.

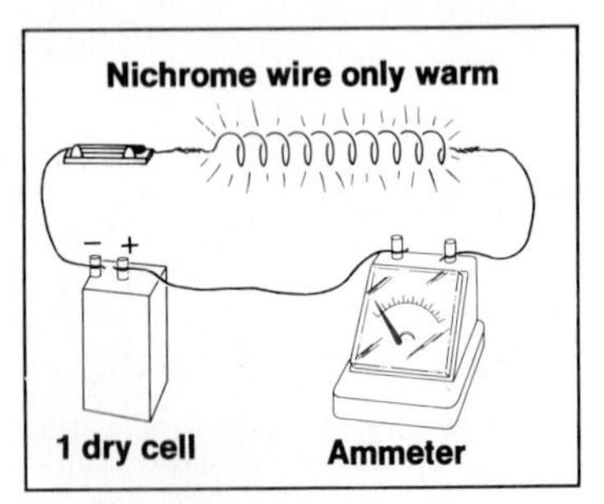

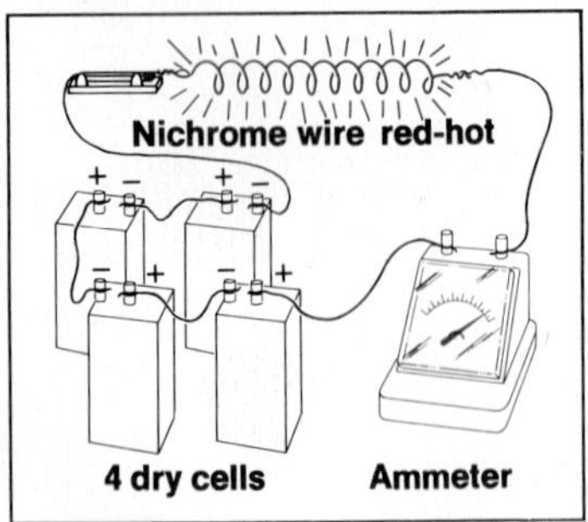

Figure 25-10 What is the relationship between the current and the amount of heat produced in a wire?

Electricity produces light. The first type of electric light was the electric arc lamp. An arc is a large spark produced when there is a small gap in an electric circuit. If there is enough voltage, the current will jump the gap, producing a blinding light. The electric arc lamp is used in some types of searchlights and movie projectors. However, the arc lamp is not practical for most lighting purposes.

During the 1880s, Thomas Edison and others developed the incandescent light bulb like the one shown in Figure 25-11. The idea behind this type of electric light is to heat a wire or filament to the point where it glows brightly. The filament used must be made of a substance that has a very high melting point. To keep the filament from burning, air, which contains oxygen, is removed from the bulb. Instead of air, a gas such as nitrogen or argon is used because it does not allow the wire to burn.

Even though it is widely used, the incandescent bulb is not an efficient source of light. Less than 2 percent of the energy used is released in the form of light waves. The rest of the energy is radiated as heat. Large incandescent bulbs are more efficient than small ones. For example, one 150-watt bulb gives as much light as 10 25-watt bulbs.

Incandescent bulbs would also be more efficient if a larger current were forced through them. The bulbs would glow more brightly, but they would not last as long. The added current would make the wire so hot that it would burn out in a short time.

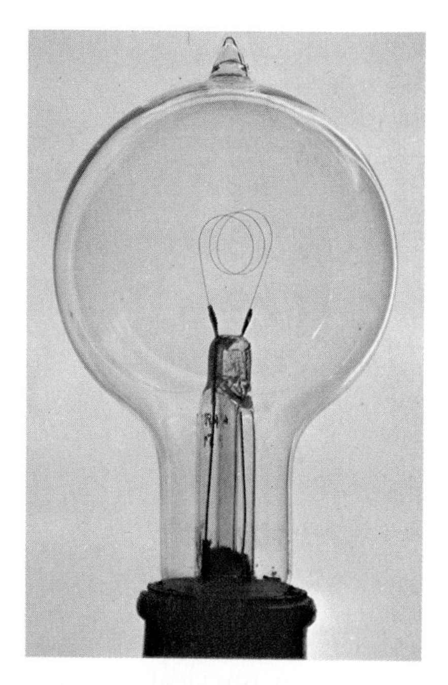

Figure 25-11 A thin filament is visible in this photo of an early light bulb. Why are most modern bulbs frosted so that the filament cannot be seen?

At night, cities are brightly lit with signs that contain many different colored lights. Most people refer to these lights as neon lights. However, the only lights with neon in them are the orange-red lights. The other colors are produced by different gases. Sodium, for example, produces a bright yellow light and mercury vapor produces a blue-green light. How do these colored lights work? Gases at normal pressures act as insulators, but at low pressures they become good conductors. When high voltage is applied under low pressure, electrons are stripped from the atoms of the gas and an electric discharge occurs, causing the gas to glow.

The mercury-vapor lamp has largely replaced all other types of lamps for lighting streets and highways. These lamps are also used to outline bridges like the one shown in Figure 25-12. This lamp gives more light with less current and less glare than many other types of lamps.

Fluorescent lamps work somewhat like mercury-vapor lamps. They use high voltage and a low-pressure mercury-vapor arc to produce ultraviolet light. In Chapter 23 you learned that ultraviolet light has frequencies beyond the range visible to you. The inside wall of the fluorescent light bulb is coated with powdered minerals called phosphors. These minerals convert the ultraviolet light to visible light. This process is known as fluorescence. The color of the light produced in this way depends on the kind of mineral powder used to coat the inside of the bulb.

Fluorescent lamps are even more efficient than neon lamps and are better for normal lighting purposes. Certain activities, such as photography, require lamps that give true color values to objects they light up. The light provided must be very similar to sunlight if the objects are to look natural. Some types of fluorescent lamps produce light of this quality.

Even the best types of lamps are still not efficient. There is much room for improvement, and research is being done in this field. New lamps with panes of luminescent glass are being developed. The luminescent panes absorb light energy during the day and release it at night when a small electric current is passed through them.

Figure 25-12 Mercury vapor lamps are replacing other types of highway and bridge lighting in many parts of the world. What do the lines of colored lights represent in this time-exposure photograph?

THOMAS EDISON

the master inventor

Thomas A. Edison was one of the truly great inventors. Born in Milan, Ohio, in 1847, Edison became interested in physical science at an early age. Because of his intense curiosity, he set up a small chemistry laboratory in the basement of his home when he was only 10 years old. Six years later, Edison had built his first invention. At that time, he was working as a telegrapher for a railroad company. His invention grew out of a need to tap out a routine signal on a railroad telegraph every hour. Edison thought this was a waste of time, so he invented a timing device hooked up to the telegraph that could send the hourly signal for him.

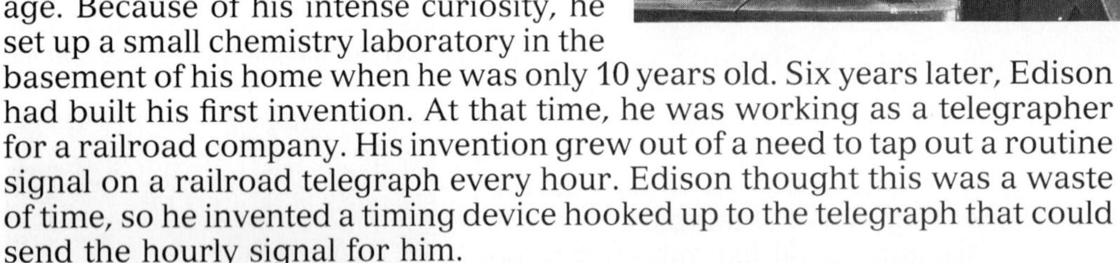

Edison had a variety of jobs over the next few years. While working at each job, he managed to invent a machine or to improve existing machinery to make his work easier. At the age of 23, Edison received $40,000 for the patent rights from one of these inventions. With this money, he set up a research laboratory in which he created most of his 1,100 inventions.

Edison soon became known as a genius. In 1877, he invented the phonograph. The next year, he and his co-workers developed the incandescent light bulb. Edison's other inventions included the motion picture projector, the typewriter, the electric generator, the telephone, and the diode.

One of Edison's discoveries left him, as well as most other scientists of his day, quite baffled. Edison had noticed a flow of current moving between a hot and a cold electrode inside a vacuum tube. He patented the tube without knowing exactly what was happening inside the tube or how it could be used. His simple discovery laid the groundwork for the electronics industry.

Once he had an idea for an invention, Edison seldom quit. When 10,000 tests failed to solve a problem in developing a storage battery, he was not discouraged. "After all," Edison said, "I have disposed of 10,000 ways that won't work." His efforts finally resulted in the widely used Edison storage battery.

Edison's intense creativity and ambition kept him working until his death in 1931, at age 84. Because of his numerous inventions, Edison will long be remembered as a man who had a great impact on the future.

CHAPTER REVIEW

SUMMARY

1. An electric current is a stream of moving electrons.
2. The path followed by an electric current is called an electric circuit.
3. An ampere is a flow through a wire of one coulomb per second.
4. A potential difference is the force behind the movement of electrons through conductors.
5. Electrons can travel in a conductor but not in an insulator.
6. The resistance of a conductor depends upon its thickness, composition, length, and temperature.
7. Ohm's law states that voltage is equal to current multiplied by resistance.
8. The electric energy used in the home is measured in kilowatt-hours.
9. Power in watts is found by multiplying voltage times current.
10. In a series circuit, the current has only one path to follow.
11. In a parallel circuit, there are two or more branches through which current can flow.
12. A short circuit causes a sudden drop in resistance in a circuit.
13. Fuses and circuit breakers protect buildings from the dangers of short circuits.
14. The amount of heat produced by a current in a wire depends on the resistance of the conductor, the amount of current, and the length of time the current is on.
15. Electric currents produce light by heating a wire or by passing through a low-pressure gas.

VOCABULARY

Match the item in the left column with the best answer in the right column. Do not write in this book.

1. ampere
2. coulomb
3. electric circuit
4. electric current
5. insulator
6. ohm
7. parallel circuit
8. resistance
9. semiconductor
10. series circuit
11. volt

a. stream of moving electrons
b. unit of electrical current
c. circuit with two or more branches
d. unit of electrical resistance
e. unit of electrical potential difference
f. path along which a current travels
g. material that does not readily carry an electric current
h. charge on 6.25 billion billion electrons
i. circuit with only one path for the current to follow
j. opposition of a substance to a current
k. material with current-carrying capabilities between that of a conductor and an insulator

REVIEW QUESTIONS

1. What is the difference between an open and a closed circuit?
2. What causes an electric current to flow?
3. In what way are voltage and water pressure alike?
4. What is an electric conductor? an insulator? Give examples of each.
5. What is the unit of electrical resistance? of electric current?
6. What is the difference between volt and ampere?
7. List four factors that affect the resistance of a conductor. Explain how the resistance is affected by each factor.
8. Explain why copper wire is used in extension cords and nichrome wire in electric toasters.
9. What two factors determine the size of the current in an electric circuit?
10. How is the resistance of a parallel circuit affected when more branches are added to the circuit? Explain.
11. How is a fuse like a switch?
12. What is a circuit breaker?
13. What three factors determine the total amount of heat that is produced when a current flows in an electric iron?
14. How much does the heat produced in a wire increase when the current through it is tripled?
15. How does an electric current produce light in an incandescent bulb?
16. How does an electric current produce light in a neon tube?
17. How does an electric current produce light in a fluorescent lamp?

CRITICAL THINKING

18. Compare the cost of using 100 W for 1 hr with using 50 W for 2 hr.
19. If the entire string of holiday lights goes out when one bulb burns out, are they connected in series or parallel? How do you know?
20. Suppose a 10-ampere and a 20-ampere fuse are both wired in series in an overloaded line. Which will be likely to blow first? Explain.
21. Can a 30-ampere fuse be used to replace a 20-ampere fuse to protect a home-wiring circuit? Explain.
22. Should a 10-ampere fuse be used to replace the fuse in Question 21? Explain.

PROBLEMS

1. How many volts are needed to push 7 amperes through an electric heater that has a resistance of 16 ohms?
2. How many amperes will a 6-volt battery send through a 3-ohm circuit?
3. What is the resistance of a toaster if 110 volts produce a 5-ampere current through it?

4. A light bulb draws 0.5 amperes on a 10-volt line. (a) What is the resistance of the bulb in ohms? (b) If you plugged the bulb into a 110-volt line, what would its current be? (c) How much extra resistance would you have to put in series with the bulb so that it would draw only 0.5 amperes on the 110-volt line?
5. A parallel circuit has four similar lamps, each on its own branch of the circuit. Each lamp draws 1 ampere on a 110-volt line. (a) What is the total current used by all four lamps? (b) What is the resistance of each lamp?
6. A piece of copper wire 300 m long has a resistance of 1 ohm. About how much current is sent through 900 m of copper wire of the same thickness by a 3-volt battery?
7. An electric iron draws 5.5 amperes on a 115-volt line. What is the iron's wattage?
8. A theater uses 50 light bulbs of 200 W each for 4 hr. Calculate the cost at $0.06 per kWh.
9. How much does it cost to light your school room for 5 hr at $0.06 per kWh? (Assume 10 fluorescent fixtures of 80 W each.)

FURTHER READING

Ardley, Neil, and Eric Laithwaite. *Discovering Electricity*. New York: Franklin Watts, Inc., 1984.

Cooper, Alan. *Electricity*. (From the *Visual Science Series*.) Morristown, N. J.: Silver Burdett, 1983.

Demurs, Ralph. *The Circuit*. New York: Viking Press, 1976.

Leon, George D. *The Electricity Story: 2,500 Years of Experiments and Discoveries*. New York: Arco Publishing, Inc., 1988.

Math, Irwin. *More Wires and Watts: Understanding and Using Electricity*. New York: Charles Scribner's and Sons, 1988.

Wade, Harlan. *Electricity*. Milwaukee, Wis.: Raintree Publishers, Ltd., 1977.

SOURCES OF ELECTRIC CURRENTS

You have learned that static charges can be released in a controlled manner to produce an electric current. What is the source of such a current? Can electric currents be produced by any other methods?

CHAPTER 26

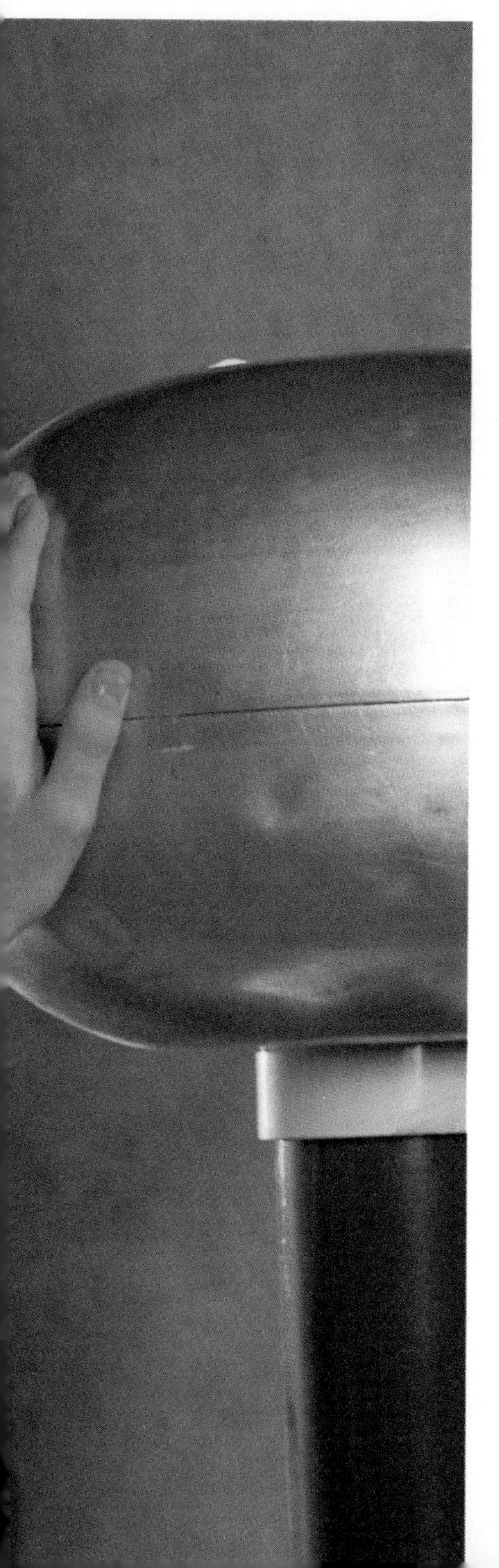

SECTIONS

OBJECTIVES

- ☐ Describe the action of chemical cells.
- ☐ Compare series and parallel wiring of cells.
- ☐ Describe the action and use of the storage battery.
- ☐ Identify five sources of electric current.

26.1 CELLS AND BATTERIES

Cells are made with different metals. Alessandro Volta made the first cell for producing electricity in 1798. An **electrochemical cell** is a device that changes chemical energy into electrical energy. Volta made his cell by placing two different metals into a liquid that was able to conduct a current. The two metals, or poles, are called **electrodes.** The liquid in a cell is called the electrolyte. An electrochemical cell consists of two electrodes and an electrolyte that acts chemically on at least one of the electrodes.

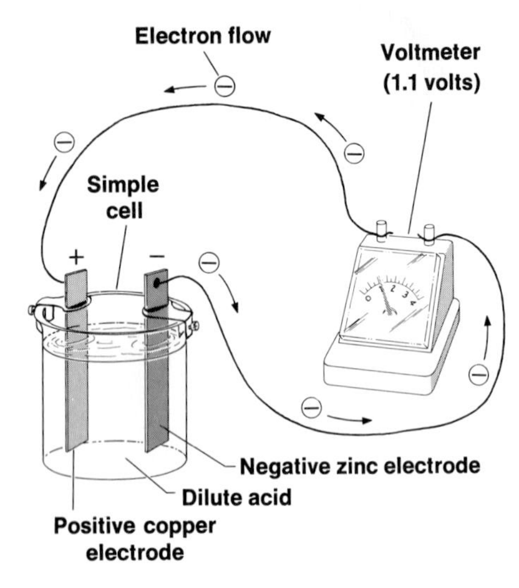

Figure 26-1 What controls the size of the voltage produced by the two metal plates?

To understand how an electrochemical cell works, consider what happens when zinc and copper are put into a solution of dilute sulfuric acid. This setup is illustrated in Figure 26-1. Atoms of zinc dissolve from the zinc plate. The zinc atoms enter the acid solution as positive ions. This reaction is indicated by the following equation:

$$Zn(s) \rightarrow Zn^{2+}(aq) + 2 \text{ free electrons}(aq)$$

The free electrons formed in this reaction collect on the zinc plate, giving it a negative charge. Therefore, the zinc becomes the negative electrode, or negative pole, of the cell. As you learned in Chapter 25, the negative pole is also referred to as the cathode.

Meanwhile, hydrogen ions from the acid remove electrons from the copper electrode. In this reaction, hydrogen ions form molecules of hydrogen gas as follows:

$$2H^{+}(aq) + 2 \text{ free electrons}(aq) \rightarrow H_2(g)$$

The loss of electrons from the copper leaves it with an excess of positive charges. Thus, the copper becomes the positive electrode, or positive pole, of the cell. Recall that the positive pole is called the anode.

When the positive and negative electrodes are connected by an outside conductor, such as a wire, electrons move along the wire from the zinc plate to the copper plate. Meanwhile, the chemical reactions within the cell supply the energy to keep the electrons flowing. The current goes from the negative electrode through the wire to the positive electrode. Since the current goes in only one direction, it is called a direct current (DC).

The electrochemical cells you are probably most familiar with are the ones found in flashlights. Flashlight cells are often referred to as dry cells. The electrolyte in a dry cell is a paste, whereas the electrolyte in a wet cell is a liquid solution. Most car batteries contain six wet cells. A collection of two or more electrochemical cells that are connected together form a **battery.** Although the dry cells used in flashlights are commonly called batteries, this is not correct. A flashlight dry cell is a single cell. A battery contains two or more cells.

Voltage depends on the elements used. Electrodes of different elements placed in an electrolyte transform chemical energy into electrical energy. Two electrodes made of the same element will not produce voltage. Electrodes must be made of two different materials in order to generate voltage. In Chapter 25, you learned that voltage is a difference in electrical potential. This potential difference occurs between two different electrodes. For example, zinc and carbon, the electrodes used in common dry cells, produce a potential difference of 1.5 volts.

Combinations of different elements produce different voltages. However, for a given combination of two elements, the voltage is always the same, even if the size of the cell varies. A large zinc-carbon cell produces electrons more rapidly than a small cell, but their voltages are the same. The relative activities of some elements are listed in the electrochemical series in Table 26-1. In this series, hydrogen is given a rating of zero, a standard value against which all other elements are compared.

The series of elements listed in the table is often called an activity series. The most active elements are at the top of the list, and the least active are at the bottom. Each element in the list replaces any element below it in a compound. For example, zinc replaces the hydrogen in sulfuric acid in the single replacement reaction in Figure 26-1.

$$Zn(s) + H_2SO_4(aq) \rightarrow ZnSO_4(aq) + H_2(g)$$

When any two of the elements in Table 26-1 are put into an electrolyte, the voltage shown on the voltmeter is the potential difference between them. For instance, copper with a rating of +0.34 and zinc with a rating of −0.76 produce a potential difference of 1.10 volts [0.34 − (−0.76) = 1.10]. To produce the greatest voltage, choose one electrode made of an active element near the top of the list and a second of a less active element near the bottom of the list. Only a few pairs of electrodes are commonly used in electrochemical cells. Cost, useful life, and other considerations greatly limit the choice of elements used.

TABLE 26-1

Electrochemical Series	
Element	**Activity Rating**
Lithium	−3.04
Potassium	−2.92
Calcium	−2.87
Sodium	−2.71
Magnesium	−2.37
Aluminum	−1.66
Manganese	−1.18
Zinc	−0.76
Iron	−0.44
Nickel	−0.25
Tin	−0.14
Lead	−0.13
Hydrogen	0.00
Copper	+0.34
Mercury	+0.79
Silver	+0.80
Gold	+1.50

Cells may be connected in series or in parallel. In Chapter 25, you learned that light bulbs can be wired in series or in parallel. Electrochemical cells can be wired together in much the same way. In a series hookup, the anode of one cell is wired to the cathode of the next cell (Figure 26-2A).

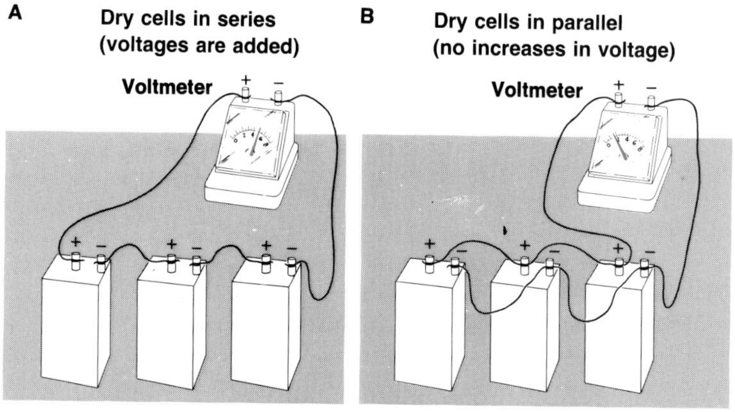

Figure 26-2 Which produces the greater voltage, cells in series (A) or cells in parallel (B)? What advantage is gained by connecting dry cells in parallel?

MAKING AN ELECTROCHEMICAL CELL

activity

OBJECTIVE: Demonstrate how galvanometer readings of an electrochemical cell vary with electrode composition.

PROCESS SKILLS

In this activity, you will *model* an electrochemical cell and *collect data* on the effect of using different electrode pairs.

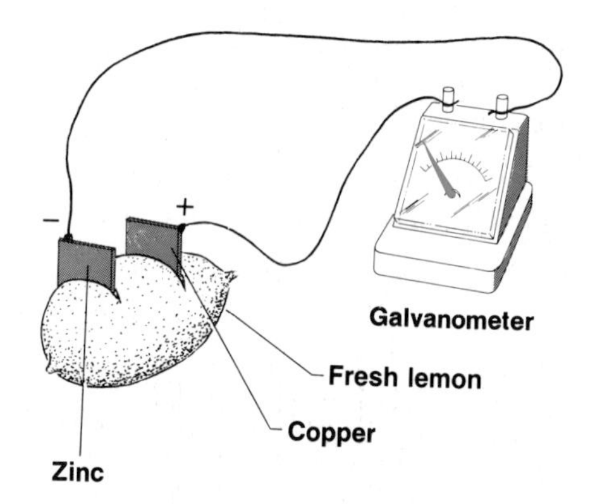

MATERIALS

insulated copper wire
wire cutters
galvanometer
lemon
strips of copper, zinc, and various other metals
knife
paper
pencil

PROCEDURE

1. Make two parallel cuts 6 cm apart along the middle of a juicy lemon. Insert a copper strip into one of the cuts and a zinc strip into the other.
2. Cut two equal lengths of insulated copper wire. Use wire cutters to remove the insulation from both ends of each wire. Connect one end of each wire to one of the terminals of a galvanometer.
3. Touch the free end of one wire to the copper strip in the lemon. Touch the free end of the other wire to the zinc strip as demonstrated in the diagram. CAUTION: *Handle the wires only where they are insulated.* Record the galvanometer reading for the copper-zinc cell.
4. Replace the strips of copper and zinc with strips of different metals. Record the galvanometer readings for each pair of electrodes.

OBSERVATIONS AND CONCLUSIONS

1. What was the function of the lemon in this activity?
2. How did the galvanometer reading that was produced by the copper-zinc cell compare to the expected value calculated by using the data in Table 26-1?
3. Compare the actual galvanometer reading to the expected values for each type of cell tested.
4. Which pair of electrodes resulted in the largest galvanometer reading?

When all of the cells are connected, the voltage is the sum of the individual voltages. For instance, three dry cells attached in series, each of 1.5 volts, form a 4.5-volt battery.

If the same three dry cells are connected in parallel, however, they form a battery that produces only 1.5 volts. In the parallel arrangement, the anode of one cell is connected to the anode of the next, and likewise for the cathodes (see Figure 26-2B). The parallel arrangement can be compared to making a cell with a larger positive pole and a larger negative pole. The advantage of this arrangement is that it lengthens the life of the battery. Since voltage depends only on the kinds of poles used and not on their size, the voltage is not changed in a parallel hookup.

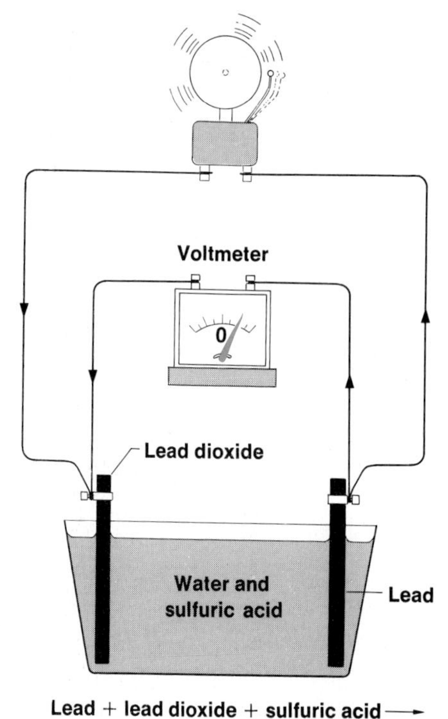

Figure 26-3 Is the lead storage battery being charged or discharged?

Batteries and cells have many uses. The lead storage battery has been a very useful source of electric current. Its special value is that it can be used more than once.

A storage battery can be made by placing two clean lead plates into a container of dilute sulfuric acid. Almost at once, a thin film of lead sulfate forms on each plate. If a current from two or three dry cells, arranged in series, is passed through the solution for a few minutes, bubbles of gas form around each plate. The brown coating that forms on the positive plate indicates that electrolysis of water is taking place. Oxygen, forming at the positive plate, reacts with the lead to form brown lead dioxide (PbO_2). At the negative plate, hydrogen gas is released. When these reactions take place, the cell is being charged.

The cell can be discharged by connecting it to a small flashlight bulb, or a doorbell (Figure 26-3). The cell can be recharged by connecting the positive pole of a battery of two dry cells to the positive plate of the storage cell. A voltmeter shows that the potential difference of this charged cell is about 2 volts.

A car battery is made up of six cells of 2 volts each, connected in series. The charging and discharging actions just described take place over and over again in a car's storage battery. The current that charges the battery comes from the car's generator or alternator. A gauge on the car's instrument panel indicates whether the battery is charging or discharging at any given time.

Each cell of a car battery has a set of plates made of spongy lead and a set of lead dioxide plates. The sulfuric acid acts on these plates while the cell is discharging, forming lead sulfate and water. The chemical equation for the reaction is

$$Pb(s) + PbO_2(s) + 2H_2SO_4(aq) \rightarrow 2PbSO_4(aq) + 2H_2O(l) + \text{energy}$$

lead + lead dioxide + sulfuric acid → lead sulfate + water

To charge the battery, electrons are sent through it in the opposite direction. This reverses the chemical reaction:

$$2PbSO_4(aq) + 2H_2O(l) + \text{energy} \rightarrow Pb(s) + PbO_2(s) + 2H_2SO_4(aq)$$

$$\underset{\text{sulfate}}{\text{lead}} + \text{water} \rightarrow \text{lead} + \underset{\text{dioxide}}{\text{lead}} + \underset{\text{acid}}{\text{sulfuric}}$$

Notice that energy must be put into the cell to charge it. Electric energy from a battery charger or the car's alternator is stored in the cell in the form of chemical energy, which is later released. These chemical reactions are reversible. As a result, cells of this kind may last for years.

How do you know when a storage battery is charged? Notice in the equation that water is formed during the discharge of the cell. Water is not as dense as sulfuric acid. Therefore, the specific gravity of the solution drops as the cell discharges. If the specific gravity of the cell is about 1.25 to 1.30, the cell is fully charged. If the value is less than 1.15, the cell is discharged.

Keep in mind that a high or low specific gravity reading indicates only whether the cell is charged, not the condition of the cell. A new cell in good condition could be discharged, or an old cell in poor condition could be charged.

There are many kinds of electrochemical cells. Most flashlight cells are made of carbon and zinc. A carbon-zinc cell produces 1.5 volts. This cell is useful for devices that operate on low voltages.

In recent years, cells using different substances have been made. One type, called the alkaline cell, has an anode of manganese and a cathode of zinc, with a highly alkaline electrolyte. The major advantage of this cell is that it works well when a high charge is needed for a long time. Under most conditions, the alkaline cell will last up to 10 times longer than a standard carbon-zinc cell of equal weight.

Another type of cell is the nickel-cadmium cell. It has an anode of nickel oxide and a cathode of cadmium. This cell produces about 1.2 volts. Not only is it an efficient source of energy, but it can be recharged. In this respect, the nickel-cadmium cell is like the lead-acid storage battery used in cars. These cells can be recharged many times without breakdown.

By examining Table 26-2, you can see that cells made of mercury or silver are not rechargeable. Also, they are made of expensive metals. The advantage of these cells is that they can be made very small. The size makes these cells useful in small devices such as hearing aids.

Cells that produce even more energy for their size are made with lithium and nickel fluoride. Storage batteries using these substances have 10 to 15 times more energy per gram than do the lead-acid batteries used in cars.

ELECTROLYTES AND CURRENT

activity

OBJECTIVE: Learn how the concentration of an electrolyte affects the amount of current generated by a cell.

PROCESS SKILLS

In this activity, you will *collect data* on the concentrations of electrolytes and their effect on current generated, *organize data* in table form, and *analyze data* by preparing a graph.

MATERIALS

insulated copper wire
wire cutters
ammeter
2 1.5-volt dry cells connected in series
600-mL beaker
2 metal clamps
2 strips of copper
250 mL distilled water
table salt
stirring rod
metric balance
paper
pencil
graph paper

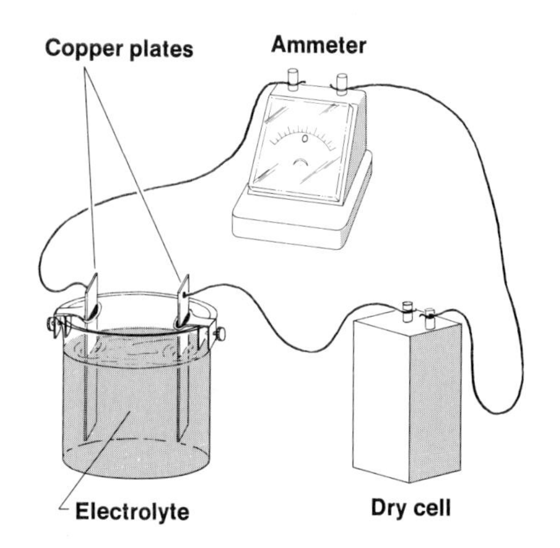

PROCEDURE

1. Prepare a table to record your data. Divide the table into two columns. Label one column "Ammeter Readings (amperes)." Label the other column "Concentration of Electrolytes (grams of salt)."
2. Cut three equal lengths of insulated copper wire. Use wire cutters to remove the insulation from the ends of each wire.
3. In a beaker, prepare an electrolyte by dissolving 0.25 g of table salt in 250 mL of warm distilled water. Set up the beaker with the rest of the equipment as illustrated in the diagram.
4. Once the wires are in place, read the ammeter. Record your results.
5. Dissolve an additional 0.25 g of table salt into the electrolyte solution. Read the ammeter. Record your results.
6. Repeat Step 5 until you have added a total of 1.50 g of salt to the solution and recorded the currents generated for each 0.25 g of added salt. Graph the ammeter readings versus the electrolyte concentrations.

OBSERVATIONS AND CONCLUSIONS

1. Describe the relationship between the electrolyte concentration and the amount of current generated in terms of your graph line.
2. How do varying concentrations of an electrolyte affect the amount of current generated?

The energy output of chemical cells is measured in watt-hours per kilogram. Table 26-2 indicates that the actual energy produced by each cell is still far below its potential output. Why is there a difference between the potential and the actual energy output?

TABLE 26-2

Output of Chemical Cells

Type of Cell	Typical Use	Actual Energy (watt-hours per kg)	Theoretical Energy (watt-hours per kg)	Rechargeable?
Lead-acid	car ignitions	20	180	yes
Zinc-carbon	flashlights	30	330	no
Nickel-cadmium	TVs, tools	40	230	yes
Mercury	hearing aids	100	250	no
Silver-oxide	hearing aids, watches	100	290	no
Lithium-nickel fluoride	electric cars	330	1,650	yes

One cell not listed in Table 26-2 is the fuel cell. The **fuel cell** is different from any other cell in that the fuel (generally hydrogen) is fed to the cell while it is in use. This cell consumes fuel in much the same way that cars use gasoline.

The fuel cell is very efficient. It converts up to 80 percent of its potential energy into electric current. Compare this value with an average of 20-percent efficiency for a car engine.

Fuel cells are now used only in spacecraft. With their efficient use of energy, fuel cells may someday be used to power cars and other devices. At the present time, such cells are far too costly for home use.

Cells have many uses. Today just about anything that runs on electricity can be made to run on an electrochemical cell or a battery. Batteries make it possible to use electric devices in places where there is no electric outlet. Also, batteries are safer because there is less danger of shock when batteries are used for power.

Shavers, power tools, movie cameras, electric typewriters, tape recorders, toothbrushes, TV sets, toys, fire alarms, and even some cars are now battery powered (Figure 26-4).

Figure 26-4 What are some of the benefits and limitations of an electric car?

Modern batteries are smaller, longer lasting, and more reliable than earlier models.

One of the most dramatic uses of electrochemical cells is for providing a small electric current to keep a defective heart beating. The pacemaker, a structure in the heart, stimulates the heart to beat. When the heart's own pacemaker breaks down, an artificial pacemaker can be implanted in the chest to stimulate the heart. The power for this device comes from a few tiny cells, each about as large as a fingernail. The pacemaker, such as the one in Figure 26-5, can keep a human heart beating for years before the cells need to be replaced.

The chemical reactions that take place inside electrochemical cells can sometimes occur outside cells. Underground or underwater steel structures, such as gas lines, piers, and ship hulls, may corrode because there is an electric current from the structure to the soil or water. The current is caused by a potential difference between the steel and the materials in the soil or water.

One effective way of stopping this form of corrosion is to reverse the current. It is possible to direct the flow of electrons to the steel from its surroundings. This reverse current is produced by placing blocks of zinc near the steel. Since zinc is higher in the electrochemical series, it acts like the negative pole of a cell. The current is then sent to the steel, which serves as the positive pole. In this process, the zinc corrodes. When it is used up, the zinc is simply replaced. The cost is low when compared to the cost of replacing the structure itself.

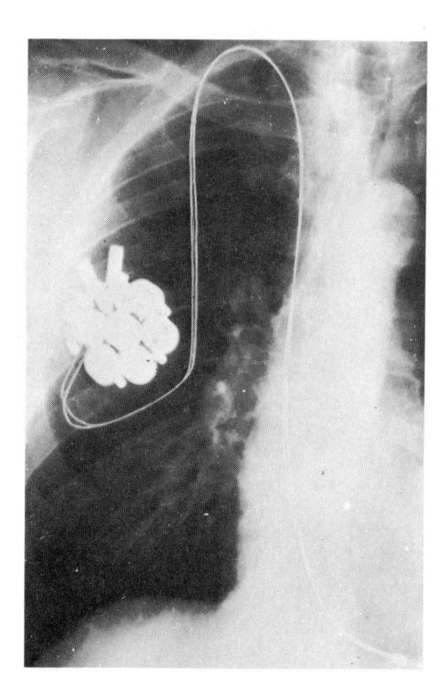

Figure 26-5 X-ray of a battery-powered pacemaker implanted in a human chest. What is the function of a pacemaker?

26.2 OTHER SOURCES OF ELECTRICITY

The thermoelectric effect produces an electric current. The changing of heat energy into electric energy is called the **thermoelectric effect.** When two different metals are joined to form a circuit, as shown in Figure 26-6, a sensitive galvanometer can detect an electric current when heat is applied at one junction of the metals. The amount of current generated is even greater if the other junction of the metals is cooled. A thermocouple is a device that includes such junctions in an electric circuit. When several of these devices are connected in series, the arrangement is called a thermopile.

In a thermocouple, the amount of current generated depends on the temperature difference at the hot and cold junctions of the wires. The greater the temperature difference, the greater the electric current. The current produced can be measured. Thus, the thermocouple can be used as a thermometer to measure heat energy. The range of such a device depends on the types of metals used. Iron and copper are useful up to 275° C, while platinum and rhodium can be used to measure temperatures up to 1,600° C.

A thermocouple is a very useful measuring device for several reasons. (1) It can be made highly sensitive, measuring temperature differences as small as 0.001° C. (2) The galvanometer can be placed far from the object being measured. For instance, a thermocouple can be placed in the wall of a jet engine and the temperature can be read on the pilot's instrument panel. (3) The thermocouple can cover a temperature range far greater than that of the common glass thermometer.

Thermocouples are also useful in many homes. A thermocouple can be used as a safety switch in a furnace. Most gas furnaces have a pilot light that ignites the main gas jets whenever the furnace is turned on. Have you ever wondered what might happen if the pilot light in a gas furnace went out? Does the house fill with explosive gas? No. A thermocouple, placed in the flame of the pilot light, keeps this from happening. It acts as a switch to shut down the main gas line if the flame goes out. The flame of the pilot light generates a small electric current in the thermocouple. This current keeps a valve in the main gas line open. When the flame goes out, the current stops, the valve closes, and gas is prevented from flowing into the house.

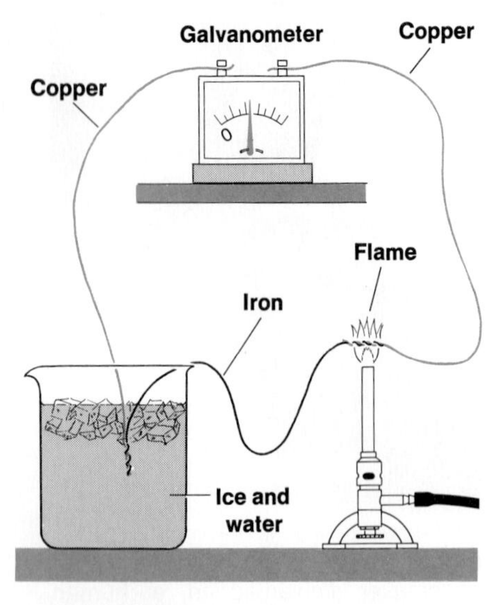

Figure 26-6 How is an electric current produced in a thermocouple?

The photoelectric effect produces an electric current. Recall that when a beam of light strikes certain metals, the

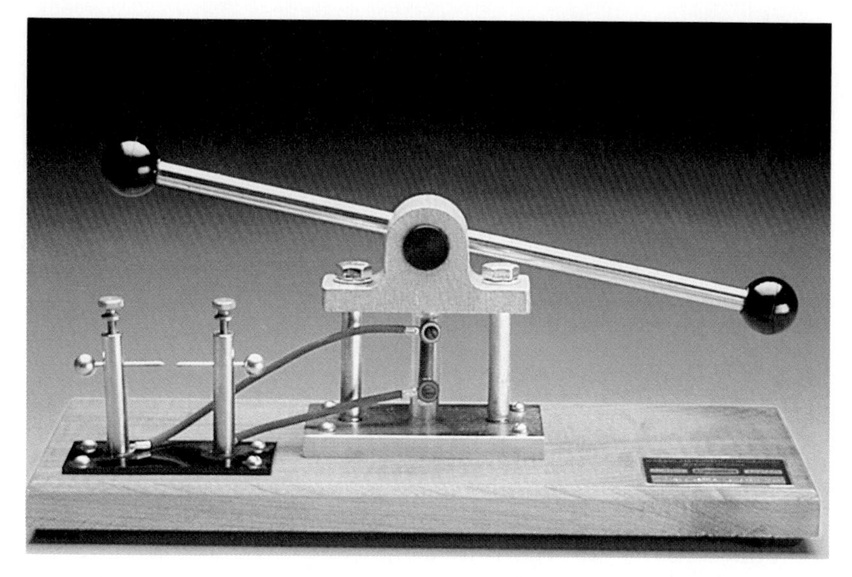

Figure 26-7 This piezoelectric voltage generator is used to demonstrate the piezoelectric effect. When a crystal is compressed by the screw-style pressure device (long metal handle), a spark several millimeters long is generated between the two vertical metal posts.

light energy is absorbed by the metal and transferred to the electrons. When these electrons gain enough energy to overcome the force that holds them to the surface of the metal, they are knocked out of the metal. If the metallic source of electrons, the emitter, is connected to a circuit, these electrons close the circuit and can produce an electric current. In Chapter 22, you learned that this action, in which light produces an electric current, is known as the photoelectric effect. In Figure 22-1, look again at the counting device, the operation of which is based on the photoelectric effect. The number of times light hits a photoelectric cell and generates an electric current is the quantity that is actually being counted by this device.

The piezoelectric effect produces an electric current. When certain crystals are subjected to a mechanical force, the opposite surfaces of the crystal acquire electric charges. If there is a potential difference between these surfaces tiny currents may result. This effect is demonstrated by the device shown in Figure 26-7. This change of mechanical energy into electric energy is called the **piezoelectric** (*pee-*AY-*zoe-ih-*LEK-trik) **effect.** For example, tourmaline and quartz crystals under mechanical stress produce currents and are used in some microphones and phonographs. In microphones, piezoelectric crystals change sound waves into electric signals, and in phonographs they convert the vibrations of the needle into an electric signal. In addition, crystals are used to produce the carrier wave of radio stations. These crystals can produce very stable frequencies so that the station signal does not drift from one place to another on the radio dial.

Electromagnetic induction can produce an electric current. In 1831, both Michael Faraday of England and Joseph Henry of the United States found that a magnet could be used to produce an electric current. Their results can be tested by performing the activity illustrated in Figure 26-8. Start by connecting each end of a coiled wire to a galvanometer. Then insert a magnet into the coil of the wire. What happens to the galvanometer when you do this? Now pull the magnet away from the coil. How is the galvanometer affected?

The galvanometer needle moves any time there is a current produced. The needle moves in one direction when the magnet is pushed into the copper wire coil and in the other direction when the magnet is pulled out. If the magnet does not move, the meter shows no current.

This activity proves that a magnet moving near a conductor produces a current in that conductor. The motion of a conductor through a magnetic field is employed by power plants to produce the electric current that is used in homes, schools, and factories.

Current is produced only when there is motion between the magnet and the coil of wire. However, it does not matter which moves, the magnet or the coil. Faraday called this process of producing a current by moving a magnet near a conductor or by moving a conductor near a magnet **electromagnetic induction**. The current produced in this way is called induced current.

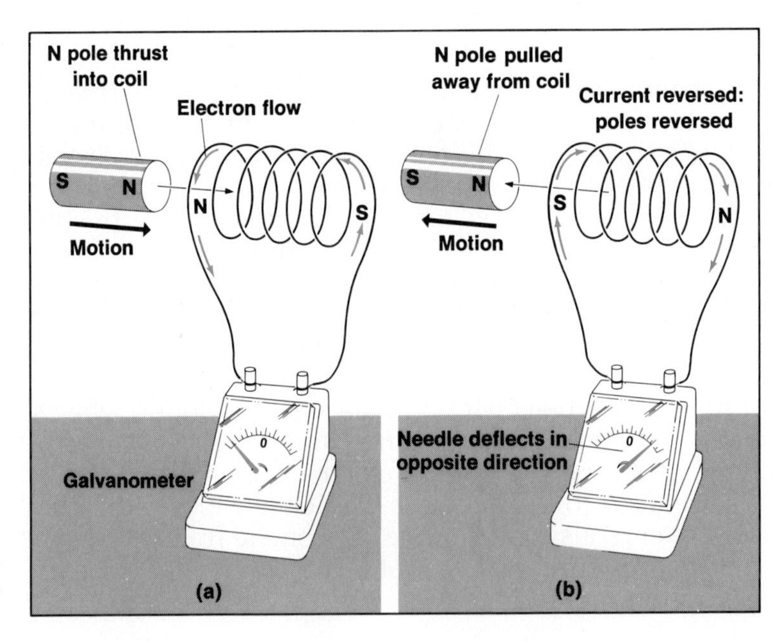

Figure 26-8 What happens when a magnet is pushed into or pulled out of a coil of wire?

BIOGRAPHY

ALESSANDRO VOLTA

the search for electric current

For several hundred years, scientists searched for the "invisible fluid" that they thought was an electric current. Scientists knew that an electric current could build up to form static charges. However, these charges could not be controlled. As soon as they accumulated, the static charges were released in a spark or flash. Until static charges could be released in a controlled manner, there seemed to be little hope of harnessing this energy and putting it to work.

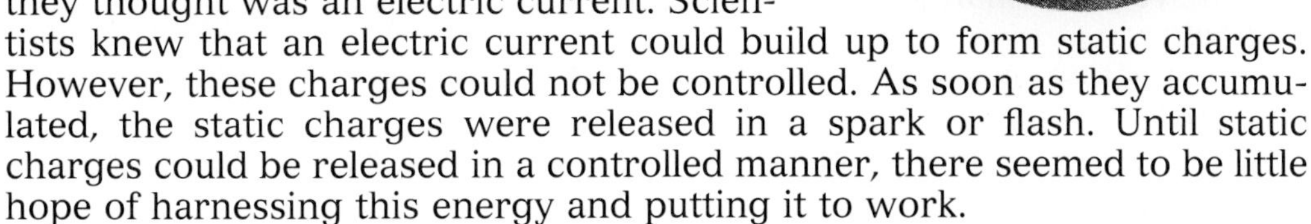

Alessandro Volta was an Italian physicist and chemist. He spent most of his life trying to find a way to harness electric energy. Volta was born in Como, Italy, in 1745. He became interested in electricity at an early age. His approach to solving the problem of electrical energy was quite logical. He first worked on ways to study static electricity. In 1775, Volta invented the electrophorus. An electrophorus is a device used to generate static electricity. Two years later, he invented the electroscope, a device used to detect static charges. Once Volta could produce and detect static charges, it was only a matter of time before he devised a way to produce an electric current. Volta's efforts were helped along by work being done by Luigi Galvani.

In 1791, Galvani published a paper that claimed electric currents could be produced by animal tissues. He stated that when a knife and probe were held in a certain way while dissecting an animal such as a frog, the frog's leg would twitch. Galvani reported that this movement was caused by something inside the tissues called "animal electricity." Volta claimed that the source of the electricity was not the frog, but the different metals used in the knife and probe. He referred to this source as "contact electricity."

Volta began to experiment with different kinds of metals. He produced an electric current by using two unlike metals and an acidic solution, creating the first source of a constant current. His electric cell contained two metals (electrodes) in a liquid that was able to conduct an electric current (electrolyte). Volta presented his work to the Royal Society of London. To show how his electric cell worked, he used it to decompose water by electrolysis, to electroplate precious metals, and to form an electromagnet.

Volta received many awards and a great deal of recognition. Probably most meaningful was the naming of the unit of electrical potential difference, the volt, in his honor. In 1827, Volta died at age 82 in Como, Italy.

CHAPTER REVIEW

SUMMARY

1. Electric currents can be produced from electrochemical cells.
2. Current that flows in only one direction is called direct current (DC).
3. The voltage of a cell depends on the elements used as electrodes.
4. Electrochemical cells can be connected in series or in parallel.
5. A storage battery can be recharged by reversing the current through its cells.
6. Electrochemical cells can be made of many different substances and have many different uses.
7. Heat can be changed into an electric current by the thermoelectric effect.
8. Light can be changed into an electric current by the photoelectric effect.
9. Forces acting on certain crystals can produce an electric current by the piezoelectric effect.
10. Magnets can induce an electric current in a wire by electromagnetic induction.

VOCABULARY

Match the item in the left column with the best answer in the right column. Do not write in this book.

1. battery
2. electrochemical cell
3. electrode
4. electromagnetic induction
5. fuel cell
6. piezoelectric effect
7. thermoelectric effect

a. process by which current is produced when a magnet is moved near a conductor
b. pole of an electrochemical cell
c. change of mechanical energy into electrical energy by certain crystals
d. fuel is fed to the cell while it is in use
e. device that changes chemical energy into electrical energy
f. change of heat to an electric current
g. two or more cells joined together

REVIEW QUESTIONS

1. What are the parts of a simple electrochemical cell?
2. Who made the first useful electrochemical cells?
3. What is the function of the lemon in a lemon cell?
4. How is voltage related to the size of the electrodes used in a cell?
5. Is a dry cell really dry inside? Explain.

6. What is the relationship between a cell and a battery?
7. Which are the negative and positive poles of a zinc-copper electric cell? Describe how they become charged.
8. What is the advantage of connecting several dry cells in series? What is the advantage of connecting them in parallel?
9. How is a storage battery charged?
10. Why does the fluid in a storage battery have a higher specific gravity when it is charged than when it is discharged?
11. Identify the electrodes and the electrolyte in a car's storage battery.
12. What is one advantage of an alkaline cell over a carbon-zinc cell?
13. In what way is a nickel-cadmium cell superior to an alkaline cell?
14. Compare the actual energy output of a lithium-nickel fluoride battery with that of a lead-acid battery.
15. How do zinc blocks slow down the corrosion of underground or underwater steel structures?
16. List five sources of voltage.
17. How is the thermoelectric effect used to produce a current?
18. What action produces the piezoelectric effect?
19. Describe the photoelectric effect.
20. Describe how an electric current is produced by induction.

CRITICAL THINKING

21. If lead and magnesium are used as the electrodes in an electrochemical cell, which one will be negative? Explain.
22. One kind of fuel cell uses hydrogen and oxygen bubbled into the cell through porous electrodes. What is the product of this reaction? Why would the electrodes last longer than those of other kinds of cells?
23. Which cell would have a greater voltage, one that uses aluminum and tin, or one that uses tin and copper? Explain your answer.
24. Four 2-volt cells are connected in series to form a battery. A second battery is formed by connecting four 2-volt cells in parallel. Which battery has the greater voltage? Explain.

FURTHER READING

Asimov, Isaac. *How Did We Find Out About Superconductivity?* New York: Walker, 1988.

Demurs, Ralph. *The Circuit.* New York: Viking Press, 1976.

Gutnik, Martin J. *Electricity: From Faraday to Solar Generators.* New York: Franklin Watts, Inc., 1986.

MAGNETISM AND ELECTROMAGNETISM

A superconductor (large disk) cooled to minus 290° F repels magnetic forces, keeping a magnet (small disk) suspended overhead. Scientists hope to apply this phenomenon in designing magnetically levitated trains that run without touching their rails. How would these trains save energy?

CHAPTER 27

SECTIONS

OBJECTIVES

- ☐ Describe the functions of magnets and magnetic fields.
- ☐ Summarize the actions of like and unlike magnetic poles.
- ☐ Discuss the link between electric currents and magnetic fields.
- ☐ Relate electromagnetic induction to the production of alternating and direct currents.
- ☐ Explain how induction is used in such devices as relays, motors, buzzers, and bells.
- ☐ Analyze how coils and transformers are used.

27.1 MAGNETISM

Like poles repel and unlike poles attract. More than 2,000 years ago, people were aware of a kind of stone that attracted bits of metal. They also discovered that an iron needle or bar would become magnetized when stroked with this stone. When this needle or bar was placed on a pivot so that it could turn freely, it would always swing around until one end pointed north. This kind of stone, which is pictured

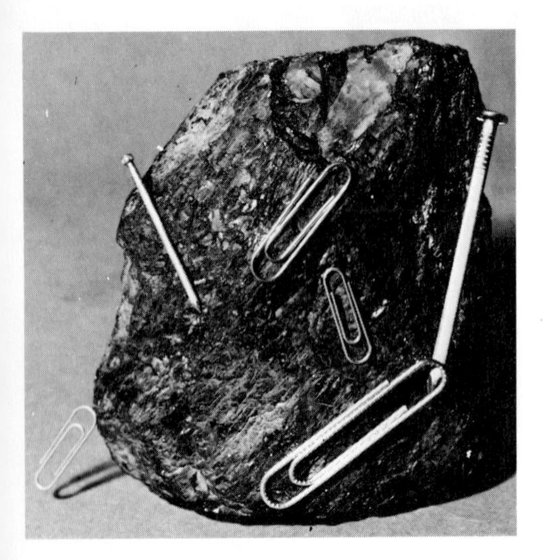

Figure 27-1 Lodestone commonly consists of natural iron oxide (Fe_3O_4). What is one of the properties of lodestone?

in Figure 27-1, is a natural magnet called **lodestone.** Although early scientists could not explain why lodestone worked the way it did, it was put to good use. As early as the twelfth century, sailors were using lodestone to magnetize their ship's sailing compass. A sailing compass was simply a magnetized needle on a pivot. The end of the needle that pointed north was called the north (N) pole. The other end of the needle was called the south (S) pole. A sailing compass and a lodestone were necessary equipment found on all ships sailing the seas.

The two ends of the magnetized needle of a compass are referred to as magnetic poles. A **magnetic pole** is the part of a magnet where the magnetic force is strongest. How do magnetic poles affect each other? To answer this question, look closely at Figure 27-2. Notice that when two N poles are close together, they repel each other. On the other hand, when the N pole of one magnet comes near the S pole of another, the two poles attract each other. In general, like magnetic poles repel each other and unlike magnetic poles attract each other. How does this behavior of magnetic poles compare with the action of static charges studied in Chapter 24?

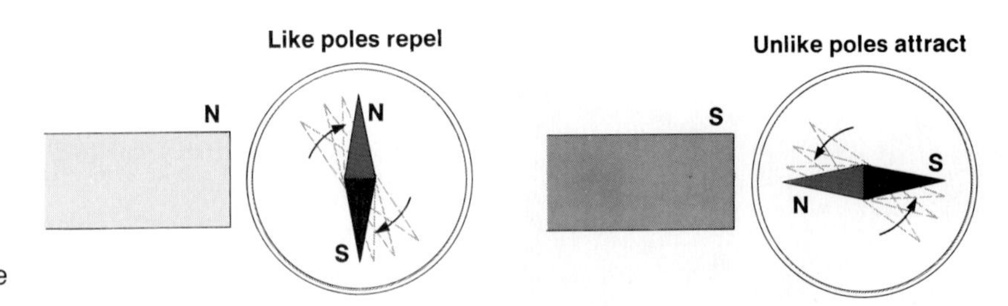

Figure 27-2 How do like and unlike magnetic poles react to each other?

In Chapter 24 (see page 493), you learned that the force between static charges follows the inverse-square law. The force between two magnetic poles also varies according to the inverse-square law. If the distance between the magnetic poles is doubled, the force becomes about one fourth as much. At three times the distance, the force between the poles is only one ninth, and so on. The force between the poles is also related to the magnetic strengths of the poles. The greater the strength of the poles, the greater the force between them.

All magnets have force fields around them. When iron filings are sprinkled on a pane of glass placed over a bar magnet, a pattern of lines like those in Figure 27-3 results.

This pattern indicates that the magnetic force extends out into the space around the magnet. The space in which the magnetic force is felt is described as a magnetic field.

Because of the way the iron filings line up around the magnet, magnetism is described as lines of force. These lines indicate the direction and strength of the magnetic field. The denser the lines, the stronger the magnetic field. Where along the bar magnet is the field strongest?

Figure 27-4A shows two magnets placed with their opposite poles near each other. The dense pattern of iron filings indicates that the lines of force go from the north pole of one magnet to the south pole of the other magnet. In Figure 27-4B, the magnets are placed with their north poles together. Notice how few filings there are between the two poles. The lines of force represented by this pattern of iron filings indicate that like poles repel.

Magnetic lines of force can penetrate many substances. Glass is an example of a material through which magnetic force can travel. However, other kinds of matter, such as steel, can be used to shield magnetism. A magnetic compass is of little use inside a submarine, for instance, because the interior is shielded from the earth's magnetic field by the steel walls of the vessel.

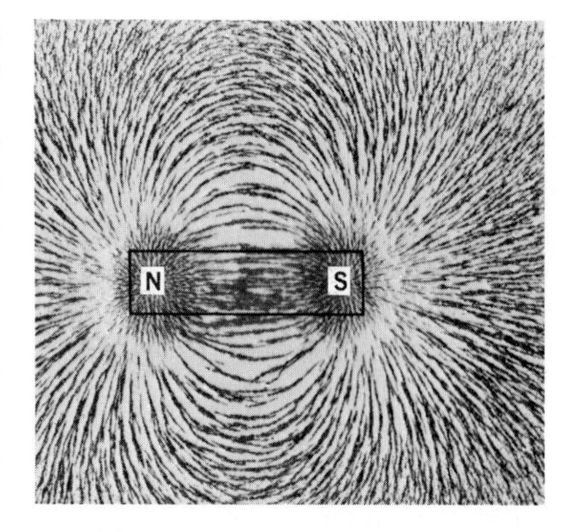

Figure 27-3 A bar magnet is placed beneath a pane of glass with iron filings sprinkled on it. What does the arrangement of the iron filings indicate about the force of the magnet?

The earth has a magnetic field. Even before Sir William Gilbert began experimenting with magnets in 1600, it was known that the earth was magnetized. Gilbert described the magnetic properties of the earth in one of his first books. He explained that the earth behaves as if a huge bar magnet were buried deep below its surface. Gilbert also suggested that this imaginary buried magnet is slightly tilted away from the earth's axis of rotation. That is, the magnetic poles are not in line with the poles of earth's axis of rotation, as illustrated in Figure 27-5.

Figure 27-4 What relationship between magnetic poles is indicated by the arrangements of the iron filings?

A

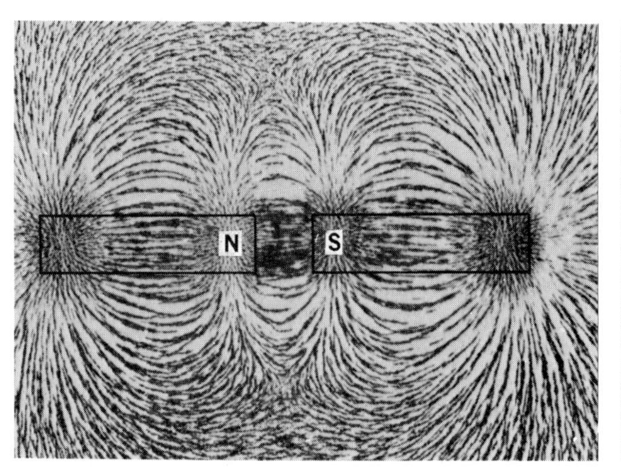

B

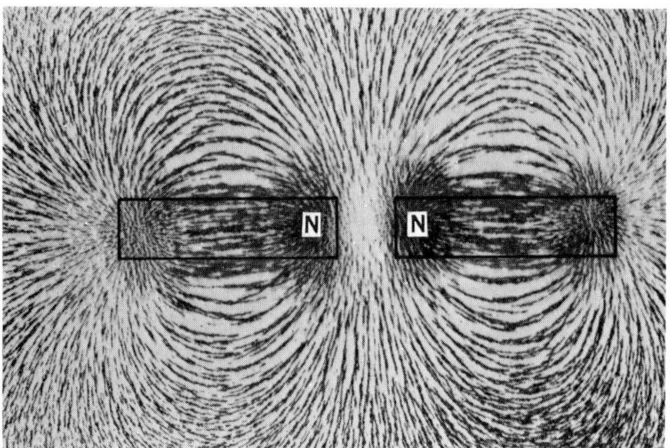

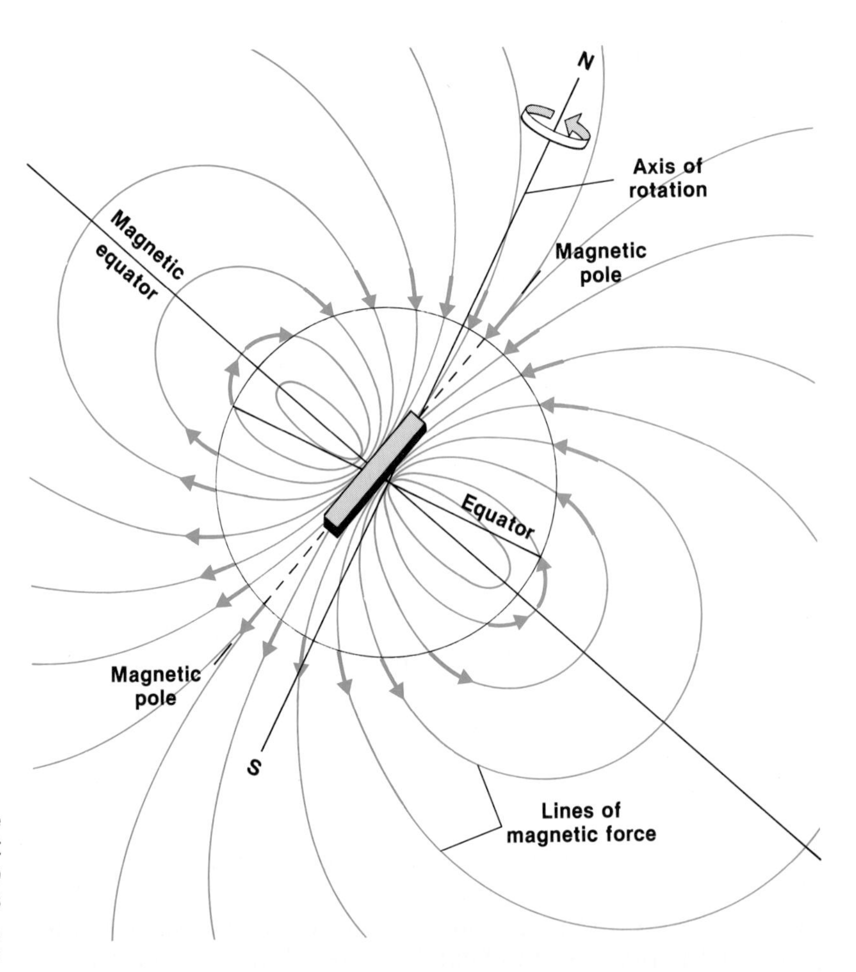

Figure 27-5 Gilbert suggested that the earth behaves as if a huge bar magnet were buried below its surface. He also realized that the earth's magnetic poles do not coincide with the axis of rotation, and thus do not coincide with true north and true south.

Gilbert's hypotheses proved to be correct. A compass needle points to the magnetic pole, not to the true, or geographic, pole. In San Diego, for instance, a compass points about 15 degrees east of true north, whereas in Boston it points about 15 degrees west of true north. This angle between the magnetic pole and the geographic pole is called the declination. Navigation charts must be constantly updated because the magnetic poles are slowly changing position.

There is a link between magnetic forces and electric currents. In 1820, a Danish teacher named Hans Christian Oersted noticed that a compass needle moved whenever a current passed through a nearby wire. After many tests, Oersted found that an electric current always produces a magnetic field. Why do you think this occurs?

activity

BLOCKING MAGNETIC FORCE

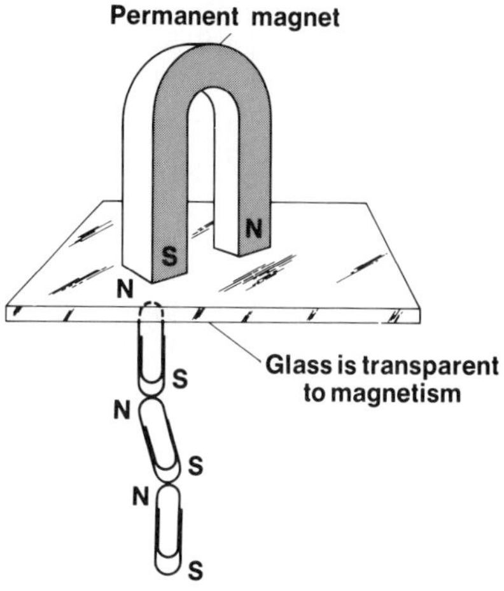

OBJECTIVE: Discover which materials are transparent to magnetic forces and which are opaque.

PROCESS SKILLS
In this activity, you will *observe* the effect of several materials on magnetic forces and *classify* those materials as to their ability to block magnetic forces.

MATERIALS
permanent magnet
box of paper clips
glass plate
pieces of cardboard, wood, plastic, copper, iron, steel, and lead
paper
pencil

PROCEDURE
1. Prepare a data table to record your observations. Divide a sheet of paper into two vertical columns. Write the heading "Transparent Material" at the top of one column and "Opaque Material" at the top of the other.
2. Place a permanent magnet on one side of a glass plate and several paper clips on the other side of the glass, as shown in the diagram. Notice what happens to the paper clips. Record your observations under the appropriate heading in the data table. Consider the glass or any material to be transparent to magnetism if it does not shield the magnet's magnetic force.
3. Repeat Step 2 using pieces of cardboard, wood, plastic, copper, iron, steel, and lead instead of glass. Under the appropriate table heading, record what happens to the paper clips in each case.

OBSERVATIONS AND CONCLUSIONS
1. Which of the materials tested are transparent to magnetism?
2. Which of the materials tested are opaque to magnetism?
3. How do materials attracted by a magnet affect magnetic lines of force?

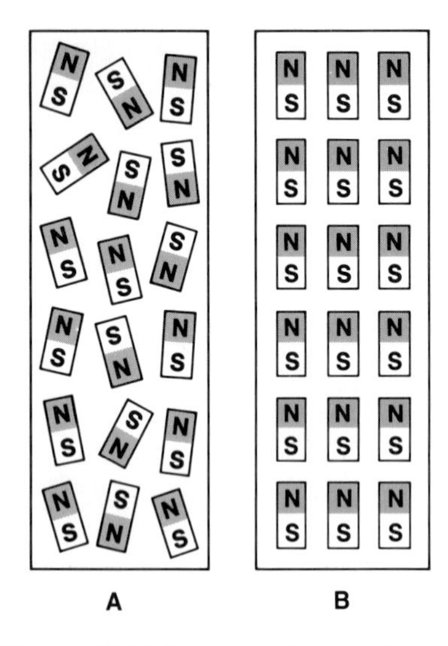

Figure 27-6 The magnetic domains are random in A and aligned in B. Which of the diagrams illustrates the pattern in a magnet?

When you recall that an electric current is made of moving electrons, you may begin to understand what causes matter to act like a magnet. All matter is made of atoms, with electrons revolving in orbit as well as spinning on their own axes. Therefore, each electron within an atom can be thought of as a tiny electric current. As is true of all electric currents, each of these electrons has a magnetic field. In most matter, however, about the same number of electrons spin in one direction as spin in the opposite direction. Thus, the magnetic fields cancel each other out. In iron, and in a few other substances, however, more electrons are spinning in one direction than in the opposite direction. As a result, not all of the magnetism in the atom is canceled out.

In a substance that can be made into a magnet, the atoms align themselves into regions called **domains.** Within the domains, the magnetic fields of the atoms are lined up so that they all point in one direction. The domains themselves, however, are not lined up. They are in a random arrangement so that their forces cancel each other. However, when a magnet is brought near, the domains all line up so that their north poles point one way and their south poles point the other. This arrangement is shown in Figure 27-6. Iron, nickel, cobalt, and a number of alloys can be made into magnets by making their domains line up.

The action of domains helps to explain an interesting behavior of magnets. The N pole cannot be separated from the S pole of a bar magnet by cutting the magnet in half. If you cut a magnet in half, you would find that each half is still a complete magnet, with N and S poles. Regardless of how many times a magnet is cut, and no matter how tiny the pieces become, each piece is still a complete magnet.

If a bar magnet is hammered sharply, the pattern of domains will be broken up and the bar will become demagnetized. Heating a magnet has the same effect. For iron, this temperature is 750° C; for nickel, 340° C; and for cobalt, 1,100° C.

Magnetizing some alloys can be difficult. However, once it is done, they are hard to demagnetize. These alloys make good permanent magnets. One of the best alloys is Alnico, a steel alloyed with aluminum, nickel, and cobalt. Some Alnico magnets can lift over 1,000 times their own weight.

Other substances, such as soft iron, are easily magnetized. However, they lose their magnetism readily. When a magnetic substance is described as soft, it means that it loses its magnetism easily. Soft metals make good temporary magnets and are used as cores for electromagnets.

Based on what you now know about magnetism, can you explain why the earth has a magnetic field? One theory states that slow movement of the liquid iron core in the earth's interior produces electric currents. These currents, in turn, produce the earth's magnetic field.

27.2 ELECTROMAGNETISM

Electricity can be used to produce magnetic fields. In Chapter 26, you learned that magnetism can be used to make an electric current. In this lesson you will learn how an electric current can be used to make a magnet. A straight wire carrying an electric current has a magnetic field that is represented by the concentric circles in Figure 27-7A. If this wire is twisted into a coil, as in Figure 27-7B, the magnetic field changes shape and becomes much stronger than that formed by the straight wire. Compare the shape of the magnetic field in Figure 27-7B with the shape of the field surrounding the bar magnet in Figure 27-3.

Figure 27-7 Compare the strengths of the magnetic fields produced in each of these setups.

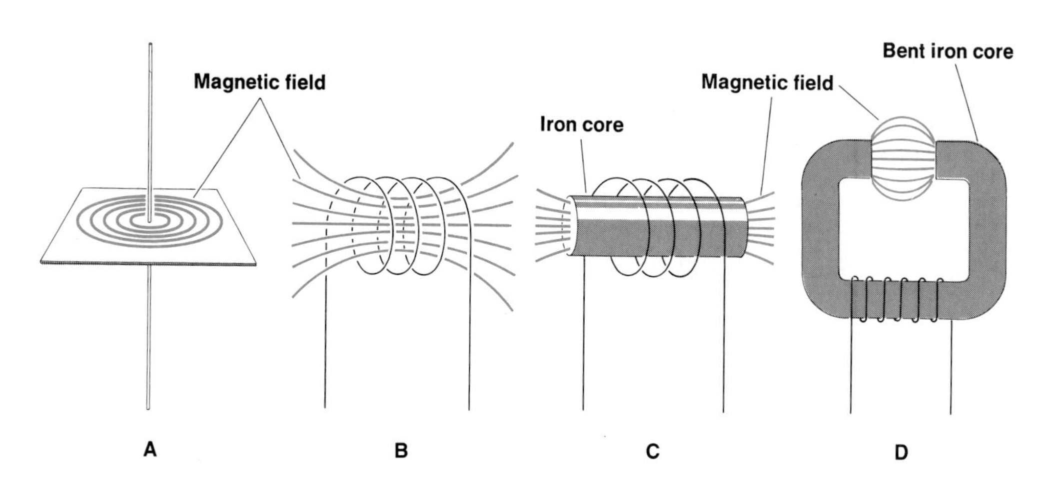

The magnetic field produced by a coiled wire becomes stronger if a soft iron core is placed inside the coil, as in Figure 27-7C. The magnetic field of the coiled wire makes the magnetic domains in the iron align. The iron core becomes a magnet. Such a setup, which combines electric current from the coil of wire and a magnetized iron core, is called an electromagnet. An **electromagnet** is a magnet produced by direct current.

If the core of an electromagnet is long enough, it can be bent so that the two opposite poles are brought close together. This situation is illustrated in Figure 27-7D. A very strong magnetic field is produced in the gap between the poles. (See Figure 27-4A.) The dense pattern of lines of force between opposite poles illustrates the strength of an electromagnet in the shape represented in Figure 27-7D.

A simple electromagnet is illustrated in Figure 27-8. Since the core is made of soft iron, the magnetic force is lost quickly when the electric current is stopped with the switch. Thus, the iron tacks being held by the magnet can be dropped simply by switching off the current.

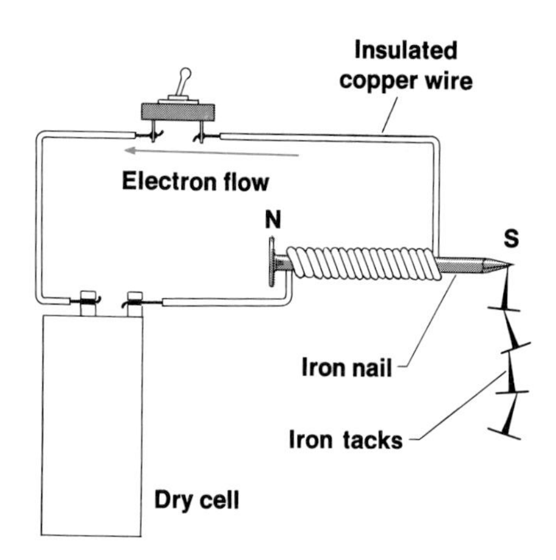

Figure 27-8 What are the main parts of this simple electromagnet?

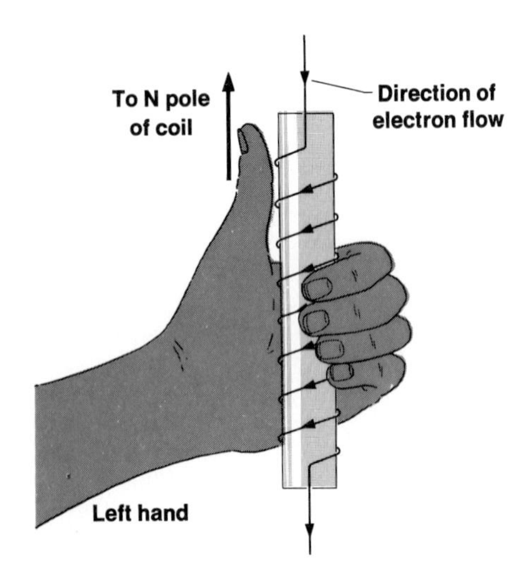

Figure 27-9 How is the left-hand rule used to determine the north pole of an electromagnet?

Three things can be done to increase the strength of an electromagnet: (1) the amount of electric current flowing in the coil can be stepped up; (2) the number of turns of wire forming the coil can be multiplied; and (3) a more permeable core, such as certain iron alloys, can be used. The term permeable refers to how well a substance can concentrate the magnetic lines of force. Soft iron is a highly permeable substance, whereas non-magnetic materials, such as wood or plastic, have low permeability.

The strength of a magnetic field is measured in a unit called the **tesla** (T). The tesla represents the density of the lines of force per unit area within the magnetic field. The magnetic pull of the earth is about 0.00005 T at the equator and twice as much (0.0001 T) at the poles.

In Chapter 19, you learned that when certain substances are cooled to near absolute zero (−273° C) they become superconductors. Having very little resistance, superconductors can carry huge currents. As a result, they can be used to make **supermagnets,** or magnets that are hundreds of times stronger than electromagnets at normal room temperatures. One of the first supermagnets, weighing only 4 N, was small enough to fit easily into a person's hand. Yet this magnet produced a magnetic field of 4.3 T. A common magnet having the same strength would have weighed 170,000 N.

The left-hand rule can be used to find the north pole. There are times when it is important to know which end of a magnet is the north pole. Scientists have found that the north pole can be determined quickly by using a simple guide called the **left-hand rule.** This rule is illustrated in Figure 27-9. If you wrap the fingers of your left hand around the electromagnet so they point in the direction in which the current flows (that is, from the negative to the positive terminals of a battery), your outstretched thumb will point toward the N pole of the coil. If you reverse the direction of the current, the poles of the coil will also reverse.

A common device in which an electromagnet is used is the **relay,** or the magnetic switch, which appears in Figure 27-10. When the switch at A is closed, the electromagnet attracts the piece of iron, B, called an armature. This opens the switch C in the second circuit. Notice that the armature is connected to a spring and a contact point. When switch A is opened, the magnetic force is turned off and the spring again closes contact C. This setup is a very simple electric relay. The relay can operate even with a very small current in the first circuit. A relay much like this one turns on the starting motor in a car when you turn the ignition key.

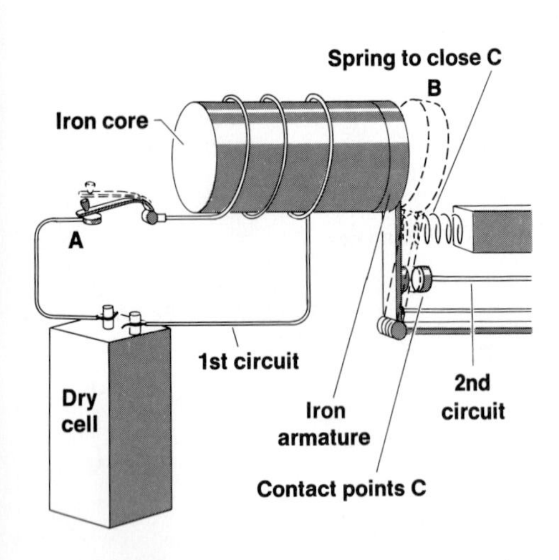

Figure 27-10 What happens to the second circuit when the first circuit is closed?

ELECTROMAGNETS

activity

OBJECTIVES: Learn how to set up an electromagnet and to determine what has the most effect on its strength.

PROCESS SKILLS

In this activity, you will *experiment* with several electromagnets in order to *draw conclusions* about the characteristics that affect their strength.

MATERIALS

insulated copper wire
wire cutter
3 soft iron cores
4 dry cells of low voltage
3 switches
pins
paper clips
paper
pencil

PROCEDURE

1. Set up three simple electromagnets similar to the one in Figure 27-8. Label the electromagnets A, B, and C. Electromagnet A should have 15 turns of wire and one cell. Electromagnet B should have 30 turns of wire and one cell. Electromagnet C should have 15 turns of wire and two cells attached in series.
2. Prepare a data table to record your results. Divide a sheet of paper into three columns. Write the letter A at the top of the first column, B at the top of the second, and C at the top of the third column.
3. Place the core of electromagnet A near a pile of pins and paper clips. Turn on the switch. Observe how many pins and clips are attracted to the core. Record your results in your data table.
4. Repeat Step 3 with electromagnets B and C. Record your results in the appropriate columns of your data table.

OBSERVATIONS AND CONCLUSIONS

1. Which electromagnet is the weakest?
2. How does increasing the number of turns of wire affect the strength of the magnet? What effect does increasing the voltage have?

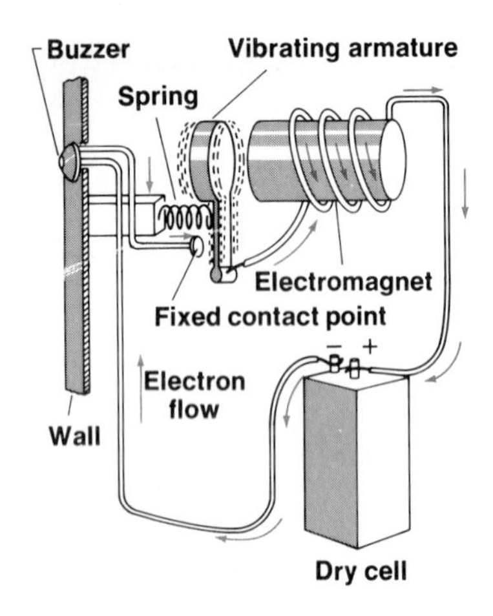

Figure 27-11 Why does the armature in this doorbell vibrate when the current is turned on?

Electric doorbells and buzzers are devices that use an electromagnet to break an electric current. Look at Figure 27-11. When a current flows through the circuit, the electromagnet attracts the armature. This action opens (or breaks) the circuit at the fixed contact point. Current stops flowing, and so the magnetic attraction stops. Then, the spring pulls the armature away from the electromagnet, closing the circuit again. The armature vibrates back and forth, opening and closing the circuit as you press the button, thus producing a buzzing or ringing sound.

Electric motors and generators are similar. An **electric motor** is a device that converts electric energy to mechanical energy. The magnetic field of an electric current is the force that operates an electric motor. An electric motor is really a magnetic motor; its armature, a movable electromagnet, is turned by magnetic forces. The armature spins because it is constantly attracted and repelled by fixed magnets, called field magnets. As illustrated in Figure 27-12A, the poles of the armature are first attracted to the opposite poles of the field magnets. Then, when the armature reaches the opposite poles, the current reverses its flow. This change in direction also changes the poles of the armature. Thus, the poles of the armature are repelled by the nearby poles of the field magnet (Figure 27-12B). The armature continues to turn, moving away from the poles of the field magnet. At each half-turn of the armature the current is reversed, resulting in a constant spinning motion.

The current is reversed by contacts called brushes, which slide over a split-ring commutator. One half of the commutator is attached to one end of the armature wire. The other half is attached to the other end of the armature wire. The commutator turns between the brushes that are

Figure 27-12 (A) The south pole of the armature is attracted to the north pole of the field magnet. (B) When it reaches the magnet, the current is reversed and the armature's pole becomes a N pole.

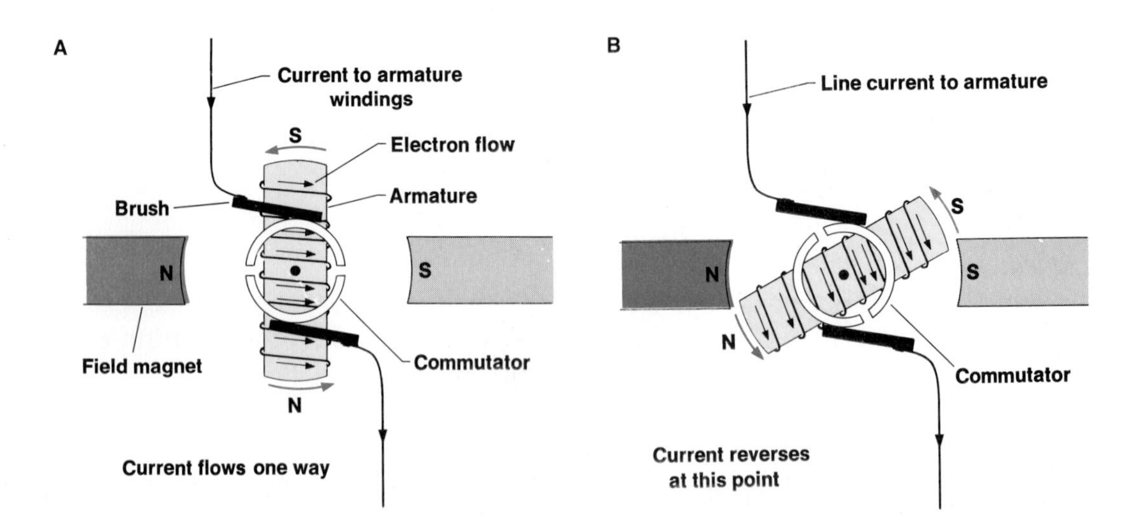

attached to the outside source of the electric current. As a result, with every half-turn, the commutator and the armature wire receive current in the opposite direction.

Although there are many types of electric motors designed for specific uses, they all have one important thing in common. All motors use a magnetic force to change electric energy into useful mechanical energy.

In many ways, an electric generator is the opposite of an electric motor. Recall that an electric generator is a device that changes mechanical energy into electric energy. In Chapter 24, you learned about electric generators, such as the Van de Graaff generator, that produce static charges. Another type of electric generator produces electric energy when coils of wire in the armature cut through magnetic lines of force. When this happens, an electric current is set up in the coils of wire. Recall from Chapter 26 that this process is called electromagnetic induction, and the currents produced are called induced currents.

A generator requires an outside source of energy to turn its armature and to force it through the magnetic field of the magnets. In an electric power station, the energy comes from falling water, from burning fuel, or from nuclear reactors. Thus, the input of a generator is mechanical energy and the output is electrical energy.

In an electric motor, the input and the output are reversed. In the generator, magnetic forces plus motion produce a current. In the motor, an electric current plus magnetic forces produce motion. Generators and motors are built in much the same way and can sometimes be used interchangeably.

Generators are commonly built to produce alternating current (AC). Look at the loop of wire turning in a magnetic field in Figure 27-13. Note that the armature wire ABDC in position 1 is moving parallel to the lines of force. In this position, the lines of force are not being cut, therefore no current is induced. The graph in Figure 27-13 shows that, at position 1, the current, and thus the voltage, is zero. When the armature loop makes a quarter turn to position 2, it is cutting directly across the magnetic lines of force. Now current is induced and there is a positive voltage of 60 volts as indicated by the graph.

Another quarter turn brings the armature wire again parallel to the lines of force, and again the voltage falls to zero. Now, as the loop rotates further, it cuts the magnetic field once more, but in the opposite direction. This change of direction produces a current in the opposite direction, thus inducing a negative voltage, as indicated by the curve on the graph.

Recall that current flowing first in one direction and then in the other is called alternating current, or AC. Much of the electrical energy used in this country is in the form of

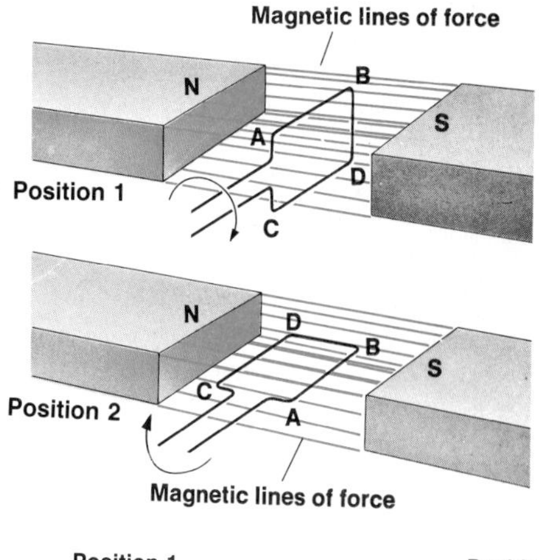

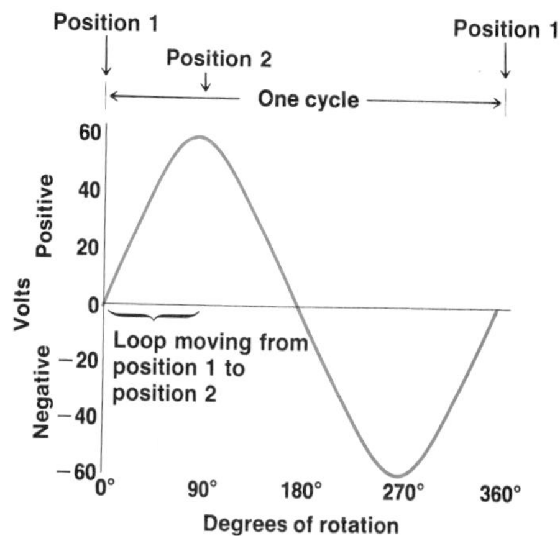

Figure 27-13 The use of a rotating loop to produce alternating current and a graph showing changes in voltage during one full rotation of the loop.

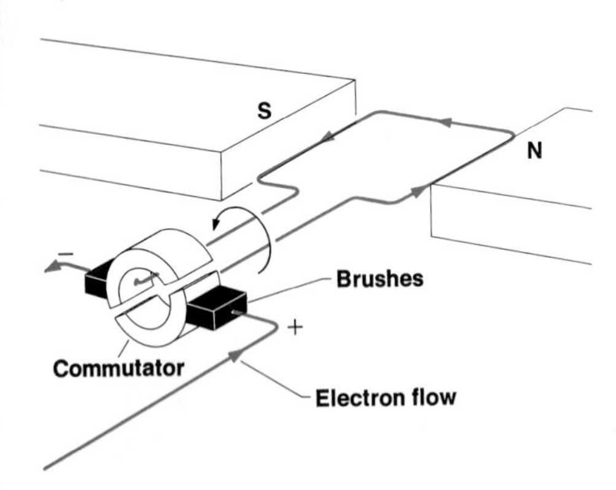

Figure 27-14 What effect does this split-ring commutator have on the alternating current produced by the coil?

AC. The graph in Figure 27-13 represents one full cycle of AC, induced when the armature wire makes a complete turn, passing both a N and a S pole. The frequency of the AC in the United States is 60 cycles per second. An armature having only one pair of poles would have to spin at a speed of 60 turns per second to produce this frequency.

Alternating current can be changed to direct current by a rectifier. Although AC is fine for most uses, direct current (DC) is sometimes needed for some circuits. For instance, DC is used for charging batteries and for electroplating. Some motors run only on direct current. A device that changes AC to DC is known as a **rectifier.** An AC generator can be adapted to supply direct current with a commutator. A commutator is a split ring, as illustrated in Figure 27-14. Each end of the commutator is connected to one end of a coil that is spinning in a magnetic field. The induced current in the coil is alternating, but as the coil and commutator turn, each of the stationary brushes is touching one of the half rings. The connections to the outside circuit are reversed at the same time the direction of the current in the coil alternates. The result is a pulsating direct current in the outside circuit.

Currents are induced in transformers and induction coils. An electric current is produced when a bar magnet is pushed into or pulled out of a coil of wire. Likewise, a current results when a conductor moves perpendicular to the lines of force in a magnetic field. Motion of the magnetic field relative to the wire coil resulting in an electric current is called electromagnetic induction. The same result can be obtained by using an electromagnet and switching the current on and off.

Suppose two coils of wire are wrapped around a common iron core as shown in Figure 27-15. Only one, called the primary coil, is attached to a dry cell. Recall that a dry cell is a source of current. When the switch in the primary circuit is closed, the iron core becomes magnetized. This action has the same effect on the secondary coil as if a magnet were pushed very rapidly into it. Therefore, a pulse of current flows briefly in the secondary circuit, as indicated on the meter. This current is called an induced current. The device that uses an iron core and a current in a primary coil in order to induce a current in a secondary coil is called an **induction coil**.

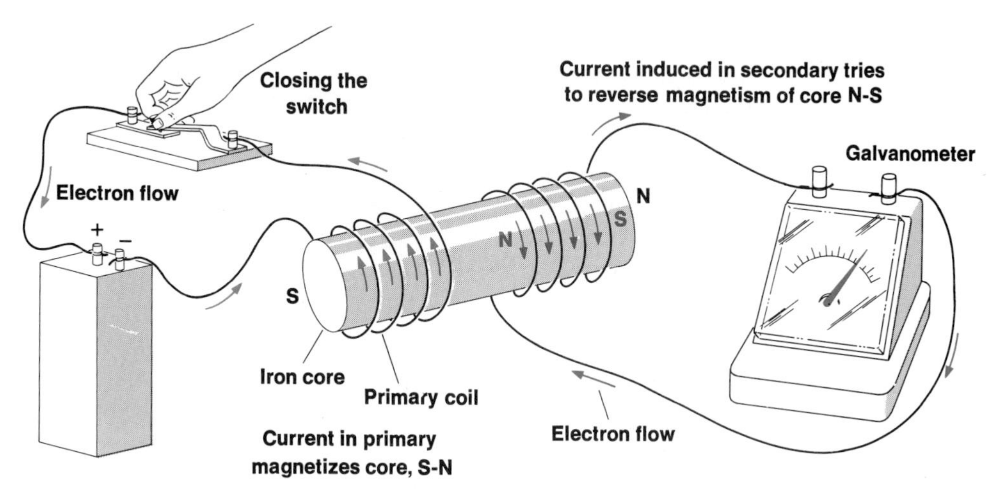

Figure 27-15 Why is it important to turn the switch on and off in an induction coil? What causes the current in the secondary coil?

When the switch in the primary circuit is opened, the meter needle again moves, but in the opposite direction. The effect is the same as if a magnet were being pulled out of the secondary coil. A current is induced in the secondary coil only when a change occurs in the magnetic field within the secondary coil. Opening and closing the primary circuit changes this magnetic field. The same effect can be achieved by connecting the primary circuit to an AC source. An alternating current reverses direction about 60 times per second in the United States. This change in direction produces the relative motion necessary for electromagnetic induction.

The induction coil can produce high voltages. The more turns of wire there are in the secondary coil, the higher the voltage of the induced current. In Figure 27-15, the primary and secondary coils have the same number of turns. However, if the secondary coil had twice as many turns as the primary coil, the voltage of the secondary would be twice that of the primary. The current in the secondary, however, would only be half of that in the primary.

The induced voltage depends upon the number of lines of force in the magnetic field cut per second by all the turns of wire in the secondary coil. Thus, a very high voltage is induced in the secondary coil when it has many turns relative to the number of turns in the primary coil.

A special kind of induction coil, called an ignition coil, is used in a car. The primary coil has only a few turns of insulated wire wrapped around an iron core while the secondary coil has thousands of turns of wire. The voltage produced in the secondary coil is high enough to produce a spark at the spark plugs. This spark ignites the fuel-air mixture in the engine's cylinders.

Figure 27-16 Why is the voltage doubled in the secondary coil? What happens to the current?

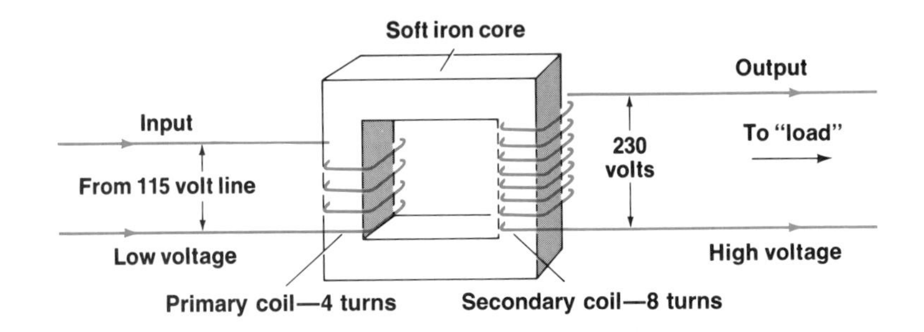

A **transformer** is a device that changes the voltage of an alternating current. A transformer is much like an induction coil. It contains two coils of insulated wire wound around a soft iron core. Alternating current is fed into the primary coil, which produces a changing magnetic field. As this field cuts through the secondary coil, AC is induced in the secondary coil.

If, as in Figure 27-16, there are twice as many turns in the secondary coil as in the primary coil, the voltage induced will be twice the input voltage. If such a transformer were plugged into a 115-volt line, it would have an output of 230 volts. This type of transformer is called a step-up transformer.

This description makes it seem as if transformers create power. However, you know this cannot be true. Power input must be essentially equal to power output. Recall that power = voltage × current. Although the voltage is doubled in the secondary coil, its current is cut in half. In fact, the current in the secondary conductor is less than half because some energy is lost as heat.

A step-down transformer has fewer turns in the secondary coil than in the primary coil and thus reduces the voltage. If the secondary coil has one tenth as many turns as the primary coil, the output voltage will be only one tenth as much. However, its current becomes nearly ten times as large as that in the primary coil.

Step-up transformers are used to increase the voltage in power lines. The cost of sending electric energy is lowest if the voltage is high and the current is low. This condition exists because energy loss in wires is proportional to the square of the current. High voltage results in low current, which, in turn, lowers energy losses in transmission lines such as those in Figure 27-17.

Step-down transformers can be used whenever low voltages are needed, as in electric bells and chimes. Step-down transformers may also be used whenever large currents are wanted. For example, the large current needed for electric welding is obtained from step-down transformers. Welding transformers use a very thick wire in the secondary coils. This wire can handle the high currents produced without losing too much heat.

Figure 27-17 Why is it necessary to step up the voltage in power lines?

SAMPLE PROBLEM

A step-up transformer is used on a 120-volt line to produce 3,000 volts across the secondary coil. (a) If the primary coil has 80 turns, how many turns must the secondary coil have? (b) If the current in the primary circuit is 20 amperes, what is the power produced?

SOLUTION TO (a)

Step 1: Analyze

You are given that the input voltage is 120 volts, that the induced voltage is 3,000 volts, and the primary coil has 80 turns. The number of turns in the secondary coil is the unknown.

Step 2: Plan

Since you know that the ratio of the induced voltage to the input voltage equals the ratio of the number of turns in the secondary (S) coil to the number of turns in the primary (P) coil, you can derive the following equation and solve for the unknown:

$$\frac{\text{induced voltage}}{\text{input voltage}} = \frac{\text{S turns}}{\text{P turns}}$$

$$\text{S turns} = \frac{\text{induced voltage} \times \text{P turns}}{\text{input voltage}}$$

Step 3: Compute and Check

Substitute all values in the equation.

$$\text{S turns} = \frac{3{,}000 \text{ volts} \times 80 \text{ turns}}{120 \text{ volts}}$$
$$= 2{,}000 \text{ turns}$$

SOLUTION TO (b)

Step 1: Analyze

You are given that the current in the primary circuit is 20 amperes. The power produced is the unknown.

Step 2: Plan

Since you know the voltage and the current, you can use the following equation to solve for power:

$$\text{power} = \text{voltage} \times \text{current}$$

Step 3: Compute and Check

Substitute all values in the equation.

$$\text{power} = 120 \text{ volts} \times 20 \text{ amperes}$$
$$= 2{,}400 \text{ watts}$$

Since a value in watts can be derived by multiplying volts times amperes, your solution is complete.

MICHAEL FARADAY

Davy's greatest discovery

Michael Faraday was an English physicist and chemist. Born in 1791, Faraday received almost no formal education during his early years. At the age of 14, he started working at a bookbinding company. While at work, he read many of the books brought into the shop for rebinding. After reading an article on electricity that appeared in one of the books, Faraday decided to learn more about this subject.

With a new interest in electricity, Faraday set up a laboratory in his home. It was there that he duplicated some of the experiments he had read about. He built a crude electrostatic generator and an electric cell, and used these two devices to perform simple experiments in electrochemistry.

Faraday's great opportunity came when he attended a lecture given by the chemist Sir Humphrey Davy. Faraday was so impressed with the lecture that he applied for work as an assistant to Davy. He was accepted and within a few years began making his own discoveries.

Convinced of the relationship between electricity and magnetism, Faraday discovered how to produce an electric current from a magnetic field. He found that he could generate an electric current by moving a permanent magnet into and out of coiled wire. This phenomenon, known as electromagnetic induction, was first demonstrated by Faraday in 1831.

Among his other accomplishments, Faraday conceived of the basic laws of electrolysis. Faraday stated that the amount of chemical change produced by a current was proportional to the amount of electricity used. Because of Faraday's work in this area, a unit of electricity was named for him. One faraday is equal to 96,400 coulombs of electricity.

Faraday's contributions to the fields of physics and chemistry far outweighed those of many other scientists of his day. He became so successful that his fame exceeded that of Davy, the man who discovered Faraday's talents. Many of Faraday's inventions, such as the electric generator, are still used today.

CHAPTER REVIEW 27

SUMMARY

1. Like poles of a magnet repel each other and unlike poles attract each other.
2. All magnets have force fields around them.
3. The earth has a magnetic field.
4. There is a relationship between magnetism and electric currents: an electric current always produces a magnetic field.
5. The left-hand rule states that if you wrap the fingers of your left hand around an electromagnet so that they follow the direction of the current, then your outstretched thumb will point toward the north pole.
6. When a coil of wire cuts through magnetic lines of force, an electric current is induced in the wire.
7. An electric motor changes electrical energy to mechanical energy.
8. A electric generator changes mechanical energy to electrical energy.
9. A rectifier changes alternating current to direct current.
10. Induction coils and transformers are electromagnetic induction devices.

VOCABULARY

Match the item in the left column with the best answer in the right column. Do not write in this book.

1. domain
2. electric motor
3. electromagnet
4. induction coil
5. left-hand rule
6. lodestone
7. magnetic pole
8. rectifier
9. relay
10. supermagnet
11. tesla
12. transformer

a. part of a magnet where magnetic force is strongest
b. magnet produced by direct current
c. device that changes electrical energy to mechanical energy
d. device that changes alternating current to direct current
e. device that changes the voltage of alternating current
f. magnetic switch
g. way of finding the N pole of an electromagnet
h. natural magnet
i. very strong magnet that works near absolute zero
j. uses a current in a primary coil to induce a current in a secondary coil
k. magnetic region of aligned atoms
l. unit of magnetic field strength

REVIEW QUESTIONS

1. State two laws that apply to magnetic poles.
2. Name three metals that can be strongly magnetized.
3. What is a magnetic field and how would you study it?
4. Describe the earth's magnetic field and suggest one of its causes.
5. How strong is the magnetic field of the earth at the equator? at the poles?
6. Why must compass readings be corrected when they are used for navigation?
7. What are magnetic domains and how do they affect the magnetism of matter?
8. Explain why not all elements can be strongly magnetized.
9. In what two ways can a permanent magnet be weakened?
10. What happens to the strength of the magnetic field around a current-carrying wire when that wire is coiled into a loop?
11. What three factors affect the strength of an electromagnet?
12. What is a unit of magnetic field strength called?
13. What is a supermagnet? Explain the advantages of using a supermagnet.
14. What is a relay? How is it used?
15. How does electromagnetic induction occur?
16. What factors determine the amount of voltage produced by electromagnetic induction?
17. How can an alternating current of an electric generator be converted to a direct current?
18. Discuss the construction and use of step-up and step-down transformers.
19. Explain the construction and action of an ignition coil.

CRITICAL THINKING

20. To make a permanent magnet, would you use a substance that is easy or difficult to magnetize? Support your choice.
21. Suppose you wanted to have a magnetic compass in your car. Where would be the best place to mount the compass? Why?
22. Why are certain materials said to be transparent to magnetic lines of force?
23. Why is the magnetic field greater at the poles of the earth than at the equator?
24. If atoms of an element have an excess number of electrons spinning in one direction, what can you predict about the magnetic property of the element?
25. Is an ignition coil more like a step-up or a step-down transformer? Explain.
26. Explain why an electric motor might also be called a magnetic motor.

PROBLEMS

1. The primary coil of a transformer has 200 turns of wire at 1,100 volts and 10 amperes. (a) What is the power produced in the primary coil? (b) What is the voltage across the secondary coil of 20 turns? (c) How many amperes are in the secondary circuit? (Assume 100-percent efficiency.)
2. Suppose you get a 220-volt output from a 2,640-volt line that carries 1 ampere of current. If there are 120 turns on the primary coil, how many turns of wire are there on the secondary? What is the current in the secondary?

FURTHER READING

Adler, David. *Amazing Magnets*. Mahwah, N.J.: Troll Publishers, 1983.

Ardley, Neil. *Exploring Magnetism*. New York: Franklin Watts, Inc., 1984.

Gutnik, Martin J. *Michael Faraday: Creative Scientist*. Chicago: Children's Press, 1986.

Rojansky, Vladimir. *Electromagnetic Fields and Waves*. New York: Dover Publications, Inc., 1980.

Vogt, Gregory. *Electricity and Magnetism*. New York: Franklin Watts, Inc., 1985.

ELECTRONICS

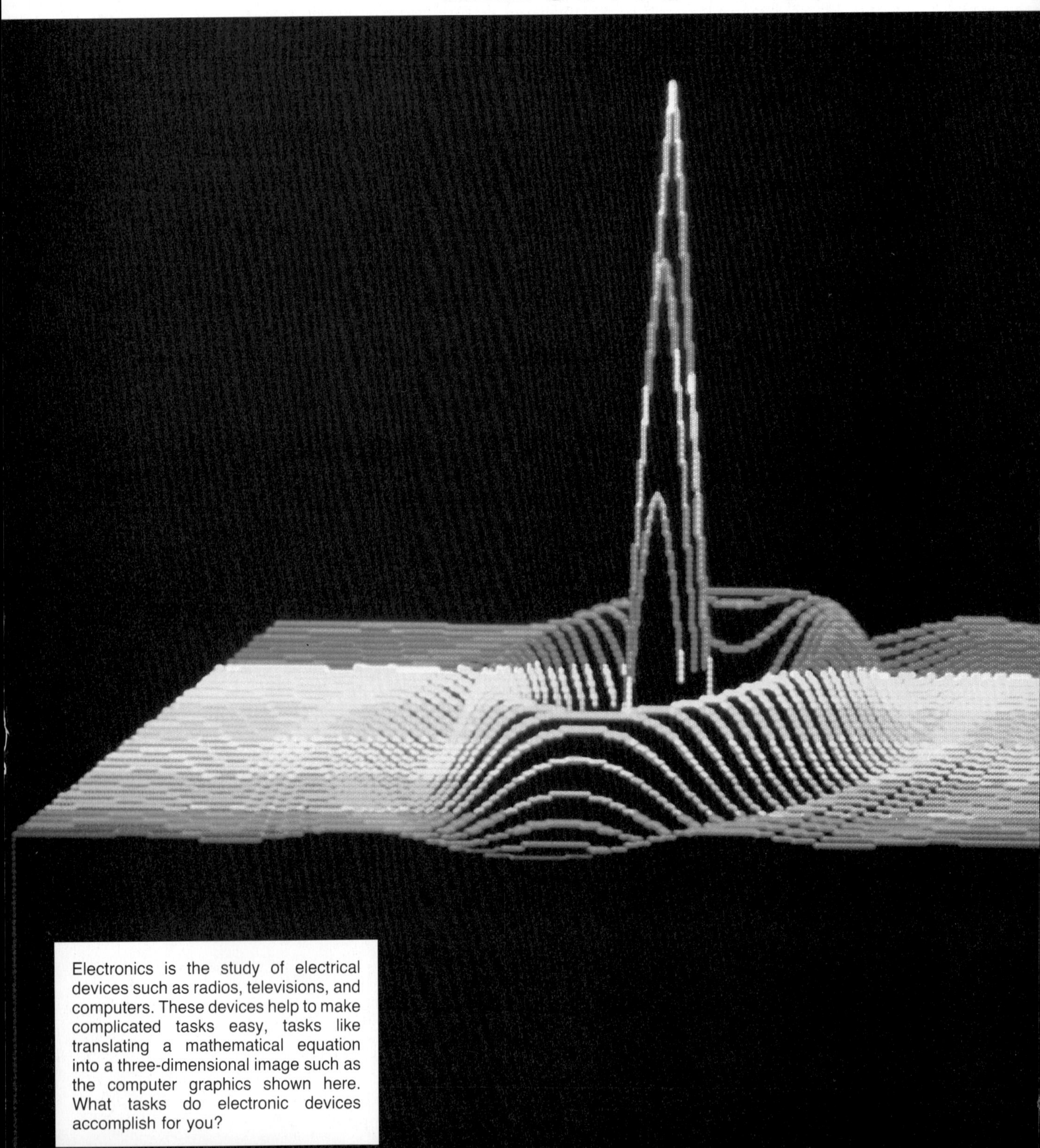

Electronics is the study of electrical devices such as radios, televisions, and computers. These devices help to make complicated tasks easy, tasks like translating a mathematical equation into a three-dimensional image such as the computer graphics shown here. What tasks do electronic devices accomplish for you?

CHAPTER 28

SECTIONS

28.1 Radio Waves

28.2 Tubes and Transistors

28.3 Cathode-Ray Tubes

28.4 Devices Using Phototubes

28.5 Computers

OBJECTIVES

- ☐ Describe how radio waves are broadcast.
- ☐ Relate the discovery of the vacuum tube and transistor to electronics today.
- ☐ Explain the function of the cathode-ray tube and its uses.
- ☐ Explain the photoelectric effect.
- ☐ Describe the basic functions of a computer.

28.1 RADIO WAVES

Radio stations produce carrier waves. Whenever you turn on your radio, you can tune in any given station just by turning a dial. Even though your radio antenna is receiving radio waves from many broadcasting stations simultaneously (from commercial broadcasts, "ham" operators, and many other short wave sources), these signals are not mixed together on your radio. How is it that you can tune a radio to so many different signals? This question can be answered by considering first the radio broadcasting station and then your radio receiver.

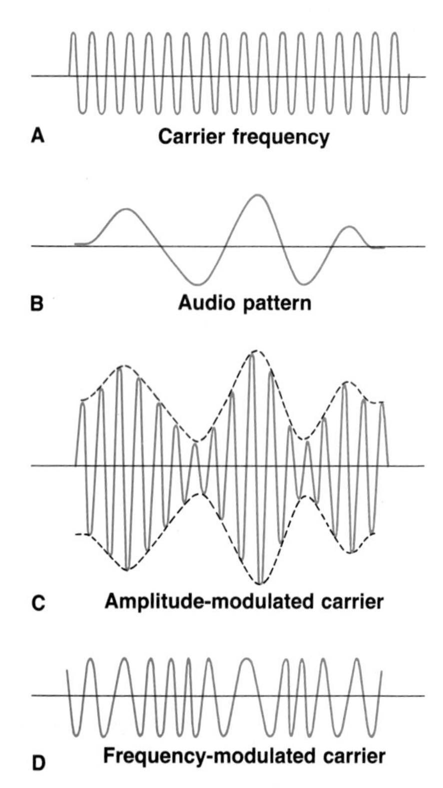

Figure 28-1 The amplitude of carrier waves (A) is changed to match the pattern of audio waves (B) for AM broadcasting (C). How are carrier waves changed for FM transmission (D)?

In order to transmit sound, the radio station changes audio waves produced in the studio to an electrical pattern. This pattern, in turn, goes through a modulator and is combined with the station's radio-frequency waves. **Modulation** is the process of shaping or coding the radio waves with the pattern of the audio waves. Because they carry an audio frequency, radio waves are called carrier waves.

One method of broadcasting is called amplitude modulation, or AM. In this method, the amplitude (height) of the carrier wave is changed to match the pattern of the audio wave, as illustrated in Figure 28-1. In this method, the frequency of the carrier wave remains constant.

Another method, called frequency modulation, or FM, is also used for broadcasting. In transmission by FM, the amplitude of the carrier wave is constant, but the frequency of the carrier wave is changed to carry the audio pattern, as illustrated in Figure 28-1D.

To keep stations from interfering with each other, each station must broadcast its carrier wave at its own assigned frequency. Also, ranges of frequencies called bands have been agreed upon worldwide for use in standard amateur, police, airline, military, and other broadcasts.

Radio waves are found in a broad band of the electromagnetic spectrum, a portion of which is illustrated in Figure 28-2. The band width assigned to FM is wider than that available for AM. The combined use of frequency modulation and high frequencies reduces interference from natural and man-made sources.

You have learned that each station broadcasts at its own assigned radio frequency. However, this still does not explain why the different signals do not interfere with each other in the radio receiver. It is because every electronic circuit has its own natural frequency. When the radio's circuit is adjusted with a tuner, the natural frequency of the radio circuit is changed. A radio can only pick up a broadcast that matches the adjusted frequency of the radio. The frequency of the radio circuit is easily changed, or tuned, by adjusting the radio's variable capacitor.

AM waves reflect off the ionosphere. The region of air from about 80 km to about 600 km above the surface of the earth can reflect standard AM radio waves. This region is called the ionosphere. Some of the air in this region is ionized by cosmic rays and ultraviolet light from the sun. The lower part of the ionosphere acts like a mirror, reflecting the longer wavelengths of AM radio broadcasts back to earth (Figure 28-3). As a result, the reflected signals can be received at very long distances from the station. FM radio and TV use carrier waves of very high frequencies that are not reflected back to earth. For this reason, FM and TV signals can be received only if the transmitter and receiver are in a straight line from one another and are not blocked by the earth.

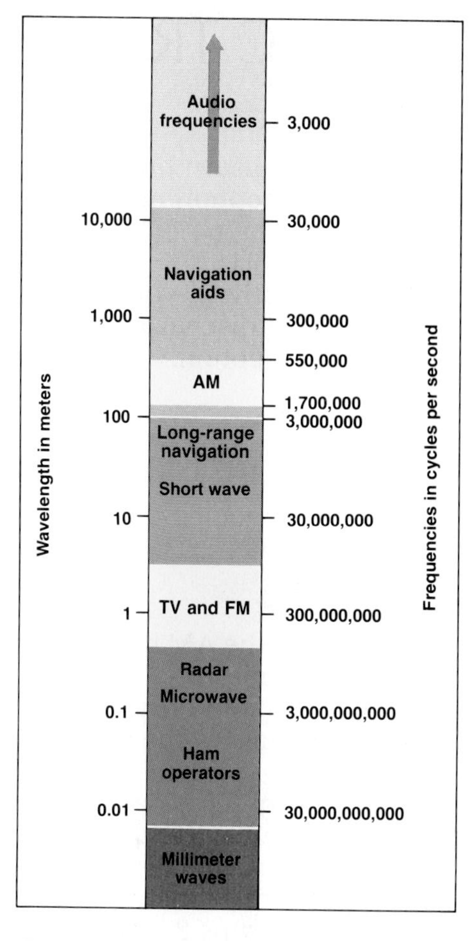

Figure 28-2 Does the standard AM broadcasting band have a higher or lower frequency range than that of TV and FM?

Radar sends and receives its own signals. The concept of radar originated in 1922. In that year, the U.S. Navy observed that high-frequency radio waves were cut off when ships passed through the line of transmission. In 1934, the Navy experimented with radar by bouncing high-frequency pulses off airplanes. However, radar was not widely used until the early years of World War II, when it was further developed by British scientists.

Radar is an electronic instrument used to locate moving or fixed objects. The word radar comes from the first letters of the phrase radio detecting and ranging. Radar operates quite simply. A powerful radio transmitter sends out pulses of high-frequency waves. A receiver picks up echoes, or pulses, reflected from the objects hit by these waves. The distance to an object is found by measuring the time it takes for the pulse to reach the object and reflect back to the receiver. The radar equipment automatically converts the time recorded for the round trip into the distance to the object (half the round-trip time × velocity of the waves). The direction of the object is determined by the direction in which the antenna is pointed when the signal is received.

A radar screen resembles the picture tube of a TV set. Constant streams of reflected signals are converted into glowing lights on a radar screen.

Today, radar affects many areas of our lives. It is used to analyze weather conditions, to track space flights, and to guide ships and planes through clouds and fog. Radar is even used to study migrating birds.

28.2 TUBES AND TRANSISTORS

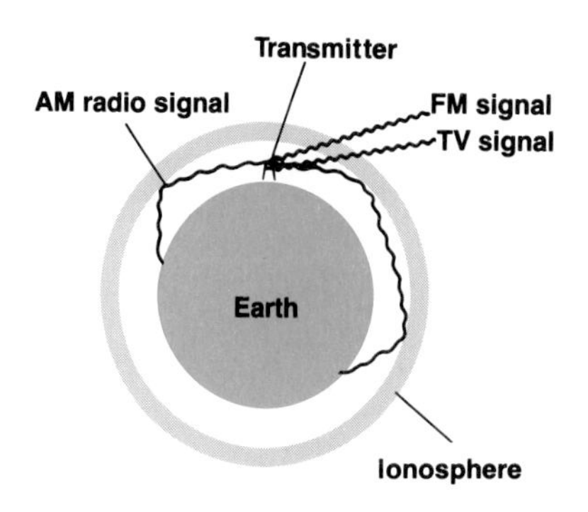

Figure 28-3 Why is it possible to hear AM radio broadcasts at longer distances than FM or television broadcasts?

Edison helped to develop the electron device. Electron devices are devices that control the flow of electrons. The earliest form of these devices, the vacuum tube, resulted from a discovery made but not utilized by Thomas Edison in 1883. At that time, Edison was working on ways to improve his newly invented light bulb. He was trying to find a way to keep a black deposit from forming on the inside surface of

STATIC TRANSMITTERS

activity

OBJECTIVE: Learn how electrical discharges such as sparks affect reception on various radio bands.

PROCESS SKILLS

In this activity, you will *experiment* with various producers of electrical discharges and *communicate* how these discharges affect radio reception.

MATERIALS

portable radio	insulated copper wire
plastic rod	wire cutters
piece of woolen cloth	paper
dry cell	pencil

PROCEDURE

1. Turn on the AM band of a radio. Set the dial between stations and turn up the volume.
2. Turn on and off any electric switch in the room. Record the effect this action has on the radio.
3. Cause a spark near the radio by rubbing a plastic rod with wool or by shuffling across a rug and touching your hand to a grounded object. Record your results.
4. Short-circuit a dry cell by attaching a wire to one post and scraping the other end of the wire across the other post. Record your results.
5. Repeat Steps 2, 3, and 4 with the radio tuned to the FM band.

OBSERVATIONS AND CONCLUSIONS

1. How was the radio affected by the actions in Steps 2, 3, and 4?
2. How was the radio affected when Steps 2, 3, and 4 were repeated with the radio tuned to the FM band?
3. How would distance between the static transmitter and the radio affect your results?

the bulb. This deposit was produced as metal evaporated from the hot metal filament in the bulb. Edison tried various ways to keep the deposit from forming, one of which is modeled in Figure 28-4. He put a charged metal plate inside the bulb and connected it to the positive post of a battery and included a sensitive ammeter in the circuit.

At one point, Edison noticed that when the metal plate had a positive charge, there was a deflection in the milliammeter. Today, scientists know that the milliammeter measured electrons given off by the hot filament. The electrons were attracted to the positive plate, and thus produced a current. A current produced in this manner is called the Edison effect, even though Edison never made any use of his discovery.

Some years later, an English scientist, John Fleming, used the Edison effect to detect radio signals. He evacuated the air from a tube similar to the one made by Edison. However, instead of a hot filament, Fleming used a wire with a negative charge (cathode). When radio waves hit the cathode, the electrons of the cathode absorbed some of the waves' energy, causing the cathode to give off electrons just as Edison's hot filament did. The electrons were attracted to the positive plate. Thus, the first crude vacuum tube was built.

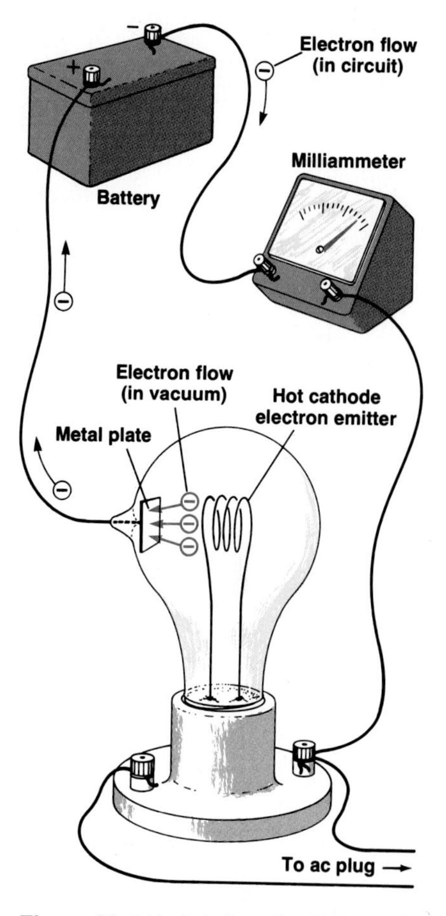

Figure 28-4 Explain how the Edison effect applies to this diagram. What is the purpose of the battery in the diagram? Why is an AC current needed here?

The diode and triode are basic electron devices. A **diode** (DIE-*ode*) is an electron device that contains two electrodes—a cathode and a plate. A diode allows current to pass in one direction only. Therefore, it acts as a rectifier, which changes back-and-forth alternating current (AC) into a pulsing direct current (DC) going in one direction.

At one time, the equipment used to send and receive radio signals contained vacuum tubes. When a radio wave struck an antenna, it produced a small alternating current that flowed to the plate in a diode. Since the current flowed only in one direction (while the plate was positive), the AC current was changed to a pulsing direct current in the diode. None of the coding was lost when the AC was changed to DC, so the pulsing direct current had the same modulated pattern sent out by the broadcasting station.

In 1906, an American inventor named Lee De Forest put a third electrode, called a grid, in a vacuum tube. This new electron device containing three electrodes was called the **triode**, or three-element tube. As shown in Figure 28-5, the grid is often a wire mesh between the cathode and the plate.

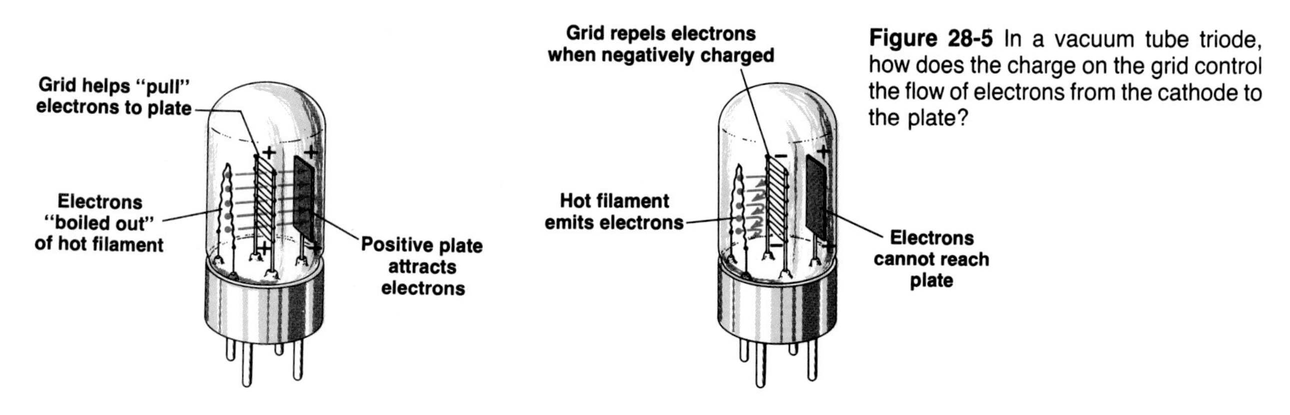

Figure 28-5 In a vacuum tube triode, how does the charge on the grid control the flow of electrons from the cathode to the plate?

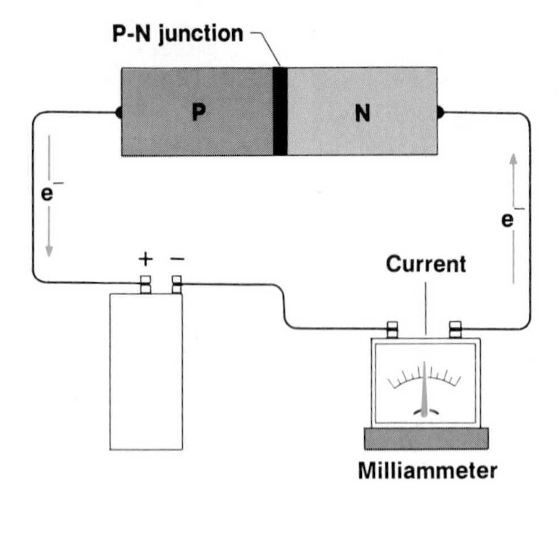

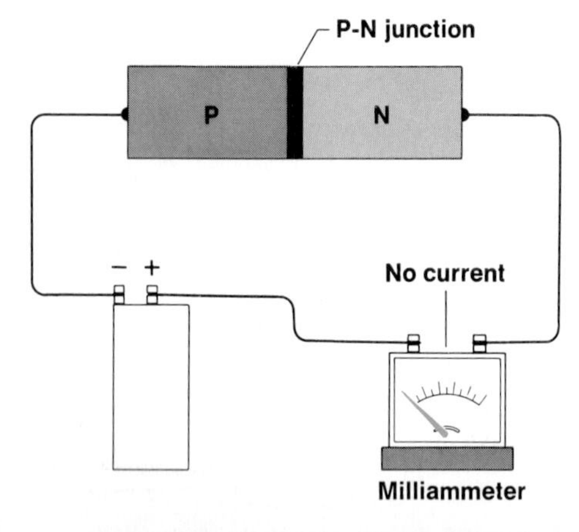

Figure 28-6 Why is there a flow of electrons in the top circuit but not in the bottom circuit?

When the grid is given a negative charge, it repels electrons coming from the cathode, keeping many of them from reaching the positive plate. If the grid is given a high enough negative charge, none of the electrons will reach the plate. However, if the grid is made positive, a larger flow of electrons will be attracted from the hot filament to the plate. The grid acts like a valve controlling the amount of current passing through the tube.

Transistors make circuits smaller. In 1947, physicists at Bell Laboratories developed the **transistor**, a small, solid-state electron device that functions like a vacuum tube. In solid-state devices, electrons flow through solid materials rather than through a vacuum. Transistors are usually made of thin wafers of the semiconductors silicon and germanium. (Recall that semiconductors are substances with a limited ability to conduct current.) Although made of solid wafers, these devices are still referred to as electron devices.

Vacuum tube diodes and triodes have been replaced with semiconductor diodes and triodes, which are more commonly referred to as transistors. Products that once used vacuum tubes are made much smaller with transistors in their circuits. Also, since they do not use a hot filament, transistors operate on much less energy than vacuum tubes and need no warm-up time.

In one type of wafer used in transistors, certain impurities are added so that free electrons are provided within its crystal structure. This type of wafer is called a negative or N-type semiconductor. If other types of impurities are added to the wafer, it becomes a collector of electrons and is called a positive or P-type semiconductor.

A transistor is made by placing one P-type and one N-type semiconductor together to form what is called a P-N junction. A P-N junction can be used in place of a vacuum tube diode. Like a diode, a P-N junction permits a charge to flow in only one direction (Figure 28-6).

Transistors with N-P-N or P-N-P junctions use three semiconductor wafers and are used in place of vacuum tube triodes. The emitter is the semiconductor wafer that gives off free electrons. It can be compared to the cathode of a vacuum tube. The collector is the wafer that receives electrons, similar to a vacuum tube's plate. The third wafer, called the base, acts like the grid of a vacuum tube triode. Transistor radios use simple circuits with P-N-P triodes.

Like vacuum-tube triodes, transistors can both rectify and amplify, or strengthen, electric currents. Their ability to amplify is particularly useful. For example, when radio waves from a broadcasting station strike an antenna, a weak current is induced in the antenna. Although this current fluctuates to match the radio signal, it is far too weak to operate the speakers of a radio or television set containing transistors. However, the current is strong enough to change the charge on the base of a transistor. Small changes in the base-emitter current cause larger variations in the collector-emitter current. The current reaching the collector is amplified, being many times greater than the strength of the signal originally picked up by the antenna.

The first transistors were much smaller than the vacuum tubes they replaced. Today, transistors are often engraved, by the thousands, onto a material such as silicon. These small assemblies of transistors and other components are called integrated circuits, or microchips. Integrated circuits, usually smaller than a penny, make it possible to build electronic circuits that are amazingly small and compact. A standard home computer, for example, made with vacuum tubes, would be too large to fit into a typical classroom.

An integrated circuit that can execute a set of instructions (a program) is often called a microprocessor, the central processing unit of modern computers. The microprocessor is used to control a wide range of products such as microwave ovens, video games, and car engines.

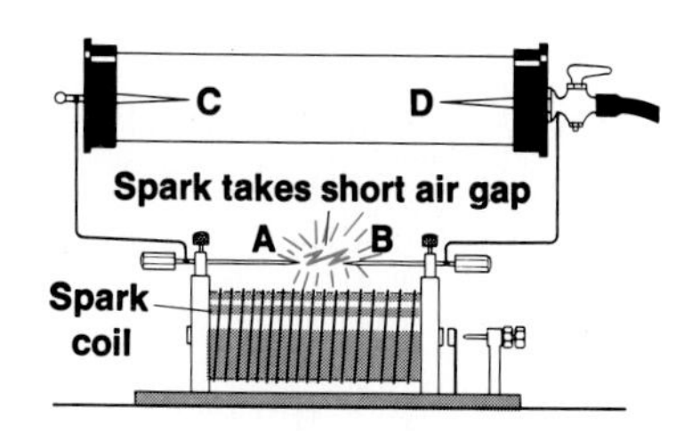

Figure 28-7 In this cathode-ray tube, does it require a greater potential difference to send a current through the high-pressure air between A and B or through the low-pressure air between C and D? Why?

28.3 CATHODE-RAY TUBES

Cathode rays flow in low-pressure gases. Dry air at normal pressure is not a good conductor of electricity. To send a current through the short gap between A and B in Figure 28-7, for instance, requires a potential difference of about 3,000 to 4,000 volts. When most of the air is pumped out of the tube, the electrons travel across the long gap from C to D more easily than from A to B.

The discovery that low-pressure gases are good conductors was made by Sir William Crookes, a British scientist, in about 1875. Crookes sealed a wire electrode in each end of a glass tube. A pump was attached to the tube so that air could be removed. As Crookes slowly pumped the air from the tube, the current increased. At first, a thin, crackling arc jumped from one electrode to the other. As more air was removed, however, the arc spread until the whole tube glowed with a pale pink light. When the pressure was decreased even more, the glow became fainter and disappeared. The end of the tube opposite the cathode (the negative electrode) began to glow. This glow meant that some kind of unseen rays were coming from the cathode. Crookes called these cathode rays. **Cathode rays** are streams of high-speed electrons leaving the negative electrode.

A cathode-ray tube is shown in Figure 28-8. Electrons boil off the hot cathode and are pulled through ring-shaped

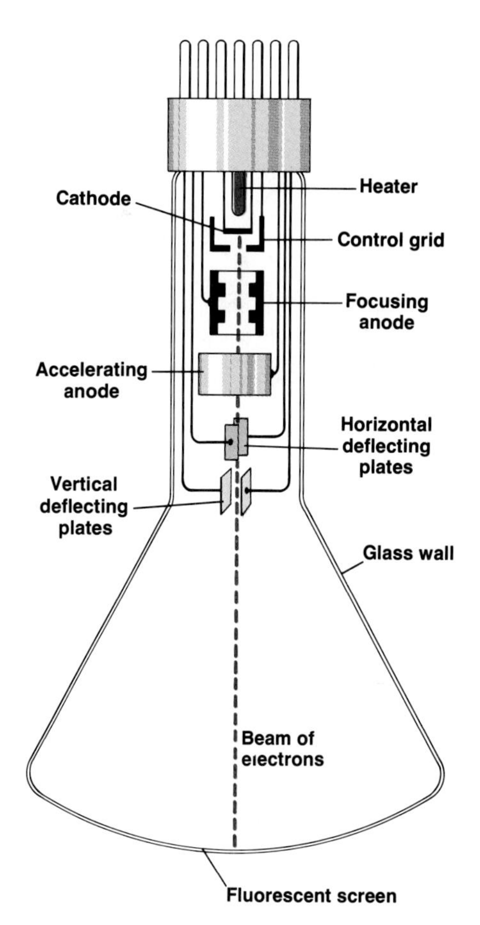

Figure 28-8 How is the beam of electrons made to move from side to side or up and down in this cathode-ray tube?

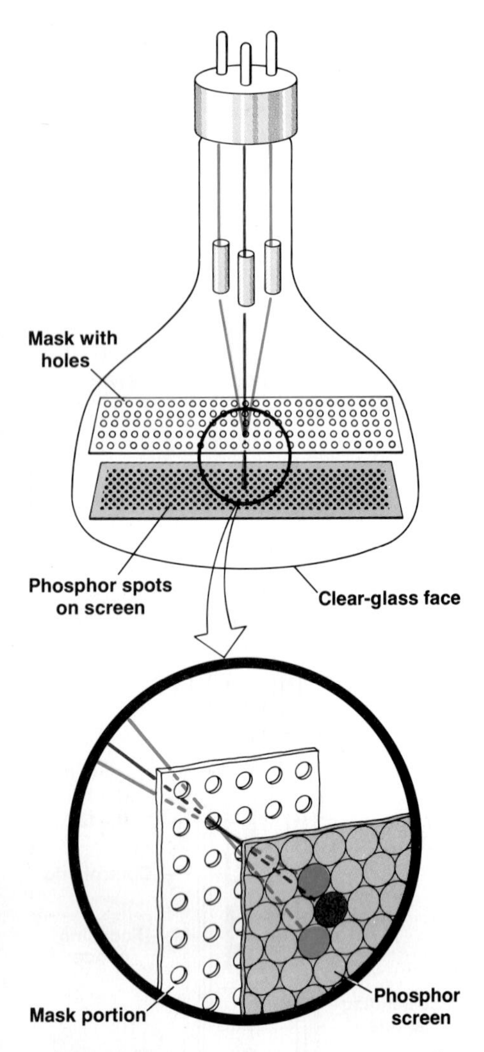

Figure 28-9 What is the purpose of the three color beams in this television picture tube?

positive electrodes, or anodes. This arrangement of electrodes that produces a focused beam of electrons is called the electron gun. As a thin beam of electrons shoots through the length of the tube, it is controlled by two sets of parallel deflecting plates. The plates of each set are oppositely charged. Depending on the kind and amount of charge on the plates, the electron beam is pushed up and down or from side to side. Electromagnets can be used instead of plates to control the electron beam.

After passing between the deflecting plates, the cathode ray hits the end of the tube and causes a special mineral-coated screen to glow, or fluoresce. The horizontal deflecting plates control the side-to-side motion of the beam across the face of the screen. The vertical deflecting plates control the up-and-down motion of the beam.

The television tube is a form of cathode-ray tube. In order to understand the connection between television and cathode-ray emission, you must first understand how color television is transmitted and received. Light entering the television camera from the studio scene is separated into the three primary colors of light: red, green, and blue. Three separate signals, one for each color, are coded. These visual signals are then combined in the transmitter with the audio signal from the microphone. The combined signal is broadcast from the antenna on the station's assigned carrier wave.

The receiving antenna picks up TV signals from the broadcast antenna. The viewer uses the TV's tuner to select the signals that correspond to the station the viewer has chosen. The combined audio and visual signal is separated, and the visual signal is then changed into the three signals that represent the three primary colors. The color picture tube has three electron guns, producing one electron beam for each of the primary colors. The strength of each of the electron beams determines the brightness of the light it produces.

Each electron beam in the picture tube sweeps back and forth in a scanning motion across a screen on the inner face of the picture tube. In 1 sec, an electron beam makes approximately 16,000 strokes across the screen. This screen is coated with three different minerals, arranged in a series of tiny dots called phosphor dots. Each electron gun scans only one color of dots. Because the dots are so small, they blend together to reproduce the same color that was picked up by the camera in the studio. This process is illustrated in Figure 28-9. If you look closely at a TV screen with a magnifying glass, you can see the dots.

Color TV programs may be received by black and white sets. However, black and white sets have no colored phosphor dots on the screen and only one electron gun. Therefore, the picture appears as gradations of light rather than different colors.

Television signals can travel no farther than 100 km to 150 km from the sending station. The high-frequency waves of TV do not reflect from the ionosphere. Therefore, for long-distance transmission, the program is either carried by cables, or is sent by a series of relay towers or by earth satellite stations.

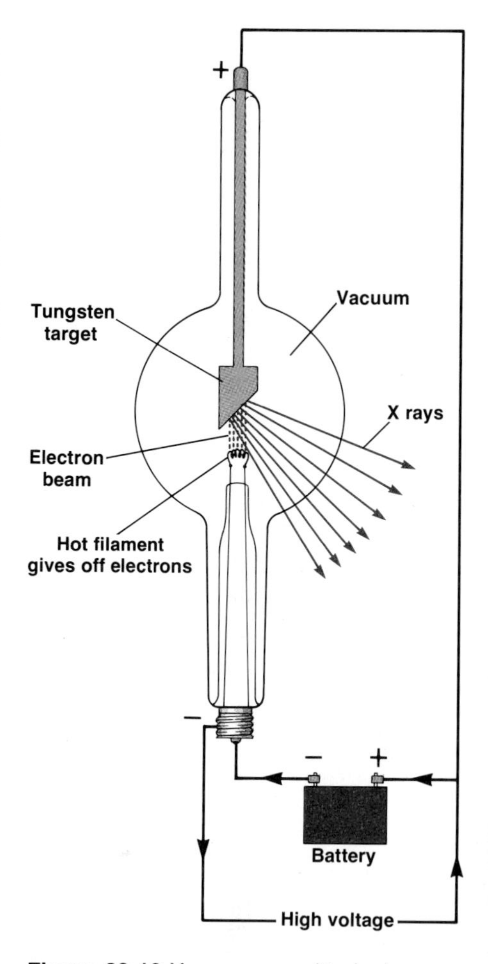

Figure 28-10 X-rays are emitted when an electron beam strikes the metal target.

Work with cathode rays led to the discovery of X-rays. While working with cathode rays in 1895, Wilhelm Roentgen (RENT-*gen*) noticed that a nearby fluorescent mineral was glowing from the effects of his cathode-ray tubes. He found that this ray went right through many kinds of matter. Not knowing what it was, Roentgen called it simply X-ray. Scientists now know that X-rays are electromagnetic radiation similar to light, but with a much shorter wavelength (see also Chapter 23).

X-rays are formed when an electron beam strikes a metal target, as shown in Figure 28-10. If a higher voltage is used, X-rays of higher frequency are produced. More than a million volts will produce X-rays that can penetrate several centimeters of steel. Such radiation can be used to check metal castings and welds for hidden flaws.

X-rays are widely used in diagnostic medicine to provide information about the body. For example, they are used to detect broken bones, cavities in teeth, tumors, and diseases of the lungs.

Intense doses of X-rays can destroy body tissue. Many early workers suffered serious burns and even death before the danger of exposure to X-rays was recognized. Studies have indicated that, in addition to causing burns and killing body cells, exposure to X-rays may produce unwanted mutations. Mutations are changes in the factors that control heredity. Extreme care should be taken to protect the body from an overdose of X-rays.

As soon as scientists found that X-rays could destroy body cells, they put this knowledge to use. They found that certain types of cancer could be destroyed or controlled by the proper use of X-rays. Today, X-rays are commonly used to treat cancer. This treatment is usually called radiation therapy.

28.4 DEVICES USING PHOTOTUBES

The photoelectric effect is used in solar batteries. The **phototube**, or photoelectric cell, sometimes called an electric eye, is an application of the electron device in which electron flow is initiated by the energy of light. The phototube is based on the photoelectric action of light discussed in Chapter 22.

Recall that when metals, such as cesium, are struck by light, some electrons are released, leaving the metal with a positive charge. A single photon with the right amount of energy can detach an electron from its atom. This reaction is called the photoelectric effect.

The photoelectric effect is used in sound films and TV cameras. The photoelectric effect is used to produce sound in movies. If you examine a section of sound movie film, you will notice an uneven dark streak along one edge of the film. This dark streak is the record of the sound, called the sound track.

A tiny beam of light shining through this sound track strikes a phototube. The amount of light that passes through the film to the phototube varies with the changing density of the sound track. This changing light causes the phototube to produce a changing current. This current is amplified and then changed to sound in loudspeakers located near the movie screen.

The television camera also depends upon the photoelectric effect. In the TV camera (Figure 28-11), light forms an

Figure 28-11 What causes the flow of electrons in a television camera?

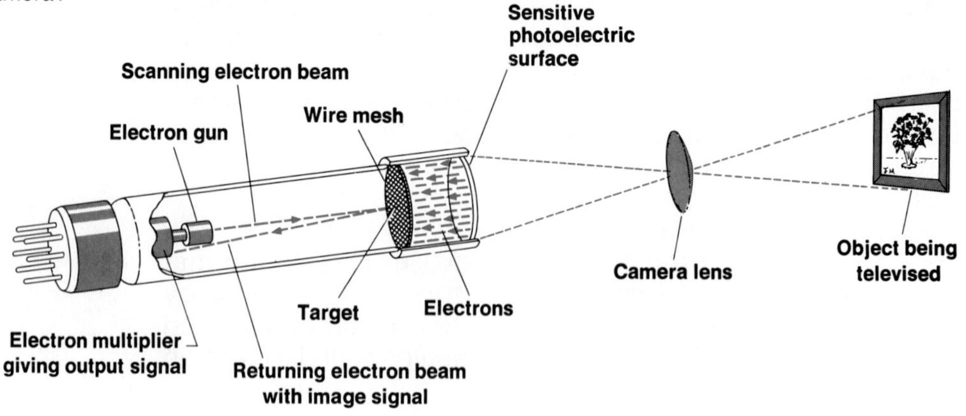

image on a surface that is coated with thousands of particles of a photoelectric substance. Each particle of this substance acts like a tiny photoelectric cell. The substance emits a varying charge, depending upon the brightness of the light that strikes it. An electron beam within the tube scans this photoelectric surface at great speed. The beam produces an electric current that varies in the same way as the pattern of light from the scene being televised. This varying current is then amplified and broadcast.

There are many other uses of the photoelectric effect. Recall from Chapter 22 that a phototube circuit can be arranged in such a way that a counting device operates each time a beam of light is broken. These interruptions might be caused by traffic on the highway or packages on a moving belt. Also, doors can be opened and closed by phototubes when someone walks through a light beam. Phototubes can be used to turn on street and building lights as darkness approaches and to control the foul lights on bowling alleys. Even burglar alarms can be operated by phototubes. When a burglar breaks an unseen beam of infrared light, the alarm is set off.

28.5 COMPUTERS

There are two types of computers—analog and digital. Computers in use today can be divided into two basic groups: analog and digital. Computers in each group can solve a wide range of complex problems in math and logic. Some modern computers operate mechanically and others are powered electrically, although all of the earliest computers, such as the abacus (Figure 28-12), were mechanical.

An analog computer solves problems by measuring things. It may measure the expansion or contraction of a solid, liquid, or gas; the increase or decrease of voltage or current in an electric circuit; or a change in the loudness of a sound.

A mercury thermometer is a simple analog computer. As it gets warmer, the mercury column rises and as it gets colder, the mercury drops. A slide rule is another example of an analog computer. You can multiply or divide with a slide rule by making it longer or shorter.

Figure 28-12 The abacus was the first computer. It was originally used in China as a counting device approximately 3,000 years ago. Addition and subtraction are performed on an abacus by simply moving beads with certain values toward or away from the center board.

TABLE 28-1

Decimal Digit	Binary Code
0	0
1	1
2	10
3	11
4	100
5	101
6	110
7	111
8	1000
9	1001

A digital computer provides solutions to problems by counting things. These machines process data and do other math-related tasks. Examples of digital computers are adding machines, cash registers, and desk calculators. The following list indicates some of the functions of modern electronic computers:

1. Digital computers are amazingly accurate and fast. In microseconds, a computer can accurately divide 29.2473051 by 8.90677162, whereas a human would probably round off to 29.2 divided by 8.9 and take several seconds to one minute to complete the task.
2. Electronic computers are automatic. Once the human operator provides the machine with a set of step-by-step instructions, or a program, and supplies the needed data, the machine operates automatically.
3. Unlike the human memory, the electronic computer never forgets. It can recall facts and instructions almost instantly.

Electronic computers use the binary, or base-two, number system. The operation of a digital electronic computer depends on the thousands of electronic circuits it contains. The circuits themselves operate like ordinary electrical switches that can be switched on or switched off. To understand a computer's operation, think of it as a box of electrical switches that are constantly switching on and off. The computer's information is stored in this pattern of on-and-off activity. By assigning a number to each switch position, namely 0 for off and 1 for on, information can be coded, or represented numerically. Most digital electronic computers are based on a **binary**, or base-two, code or number system. This system uses only two numerals, 1 and 0, to carry all data and do all arithmetic processes.

The binary system is quite different from our common decimal, or base-ten, system. The decimal system has ten numerals as digits: 0, 1, 2, 3, 4, 5, 6, 7, 8, 9. The base is the total number of digits used in the numbering system. The highest number that can be represented by a single digit is one less than the base. In the decimal, or base-ten, system the highest numeral is 9; in the binary, or base-two, system the highest numeral is 1.

The decimal and binary systems are compared in Table 28-1. The English alphabet can also be expressed in a binary code as indicated in Table 28-2.

Computers have three basic functions. All digital computers have three common functions: input, processing, and output. How these functions are related is shown in Figure 28-13.

TABLE 28-2

Alphabetic Characters and Their Binary Codes		
A 01000001	J 01001010	S 01010011
B 01000010	K 01001011	T 01010100
C 01000011	L 01001100	U 01010101
D 01000100	M 01001101	V 01010110
E 01000101	N 01001110	W 01010111
F 01000110	O 01001111	X 01011000
G 01000111	P 01010000	Y 01011001
H 01001000	Q 01010001	Z 01011010
I 01001001	R 01010010	

The input section receives data to be processed by the computer. Coded data are fed into this section in a variety of ways, including the use of keyboards, punched cards, magnetic tape, disks, touch screens, joy sticks, and scanners. The data to be processed include human-produced coded data and step-by-step instructions about what to do with these data.

The function of processing includes control, memory, and arithmetic. Control makes sure that the data and instructions are sent to the right places. Once in the computer, the data and instructions are sent to the memory unit to await recall by the control unit whenever needed. The arithmetic section does the needed calculations as directed by the control function. This section is capable of working problems in logic.

The computer processes data in the form of electric signals only a computer can use. Output devices then translate the electric signals into records that people can read. Data or solutions to a problem may appear as a binary code on magnetic tape, decoded into numerals and letters of the alphabet, printed into words, depicted as images, or in some other form.

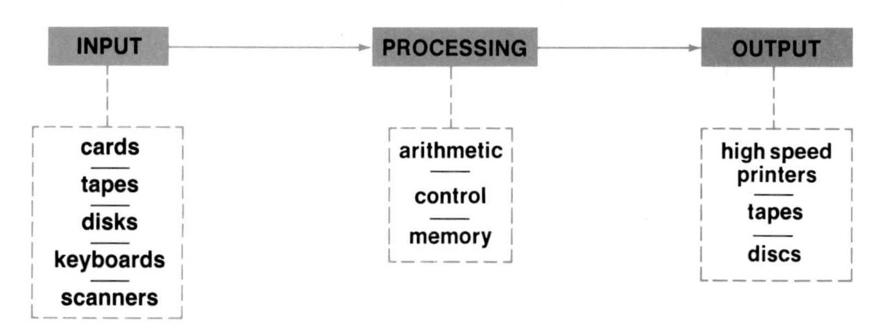

Figure 28-13 This flow chart shows the basic functions of a digital computer and how these functions are related.

BIOGRAPHY

GEORGE R. CARRUTHERS

astrophysicist

George R. Carruthers, a senior astrophysicist with the Naval Research Laboratory in Washington D.C., has spent his professional life looking outward—beyond the earth. Dr. Carruthers designs and builds electronic instruments that detect ultraviolet radiation from sources in both nearby and outer space. These instruments have been used in Apollo, Skylab, and Space Shuttle missions.

One of Dr. Carruthers' many accomplishments was the first detection of interstellar molecular hydrogen (H_2). He did this by noticing that ultraviolet radiation from stars was missing certain frequencies. These frequencies were exactly those that hydrogen molecules would absorb. He reasoned that the ultraviolet light he was observing must pass through clouds of molecular hydrogen on its journey through space. By identifying and studying substances in outer space, scientists can understand how stars, like our sun, and accompanying planets are formed from interstellar hydrogen and dust.

Before he entered high school, Carruthers became interested in outer space from reading science fiction comic books and textbooks on astronomy. While in high school he participated in science fairs and clubs concerned with astronomy, and space flight. After receiving his Ph.D. in 1964, Carruthers joined the Naval Research Laboratory where he has worked since.

The instruments that Dr. Carruthers has designed make use of many different techniques to detect ultraviolet radiation. His latest method is proposed for use in satellite and deep space missions. It uses a special photocathode that emits electrons when exposed to ultraviolet radiation, and what is known as a *charge coupled device*. Charge coupled devices are specially prepared semiconductors which can store and move electrons on their surface. They are used in most television cameras on the market today.

When operating, the instrument is aimed at a particular region of space. Radiation from that region hits the photocathode. The photocathode then emits electrons depending on how much ultraviolet radiation it absorbs. The charge coupled device collects these electrons, which create a measurable voltage on the device. Voltage measurements can then be used to infer the amount of ultraviolet radiation coming from a region of space in a given time.

As you have learned, electronics is used to build things such as computers and televisions. Thanks to scientists like Dr. Carruthers, however, electronics can also be used to help us understand the universe we live in.

CHAPTER REVIEW

SUMMARY

1. Modulation is the process of tailoring radio waves so that they carry a specific pattern of audio waves.
2. Radio stations are assigned certain radio frequencies for carrier waves so that the stations do not interfere with each other.
3. Radio receivers can be tuned to one station by a variable capacitor.
4. Radar sends out high-frequency waves and detects the signals reflected from objects in order to determine the location of the objects.
5. Electron devices make use of the Edison effect, in which electrons are emitted from a cathode and attracted to a positive plate.
6. The original diode and triode are electron devices that were made using vacuum bulbs. Modern diodes and triodes, more commonly called transistors, are made of semiconductors.
7. The cathode-ray tube produces a stream of electrons (a cathode ray) in low-pressure gases.
8. Cathode-ray tubes are used in TV sets and X-ray machines.
9. Phototubes, or photoelectric cells, use light energy to generate a flow of electrons.
10. Analog computers solve problems by measuring.
11. Digital computers solve problems by counting.
12. Electronic computers use the binary (or base-two) number system.
13. Electronic computers perform the functions of input, processing, and output.

VOCABULARY

Match the item in the left column with the best answer in the right column. Do not write in this book.

1. binary
2. cathode rays
3. diode
4. modulation
5. phototube
6. radar
7. transistor
8. triode

a. applying audio-wave patterns to carrier waves
b. base-two number system
c. device that uses light energy to generate electron flow
d. electron device containing three electrodes
e. electron device containing two electrodes
f. streams of high-speed electrons
g. device used to detect moving or fixed objects
h. semiconductor device that functions like a vacuum tube triode

REVIEW QUESTIONS

1. What is meant by tuning in a radio station?
2. What is meant by modulation?
3. Compare the two methods of modulating radio broadcasts.
4. What effect does the ionosphere have on AM radio waves?
5. How does radar work?
6. What is the Edison effect? How was it discovered?
7. How does a diode convert AC to DC?
8. Name the electrodes in a vacuum tube triode. What are their functions?
9. Explain how the grid is used to help amplify the current in a triode.
10. In what ways are transistors better than vacuum tubes?
11. To which electrode of a transistor is the antenna current fed? Explain.
12. What is an integrated circuit?
13. What parts of a transistor act like the cathode, plate, and grid of a vacuum tube?
14. Discuss the operation of a cathode-ray tube.
15. How is the electron beam in a TV set controlled?
16. Describe how color is formed on the screen of a color TV set.
17. Can a black and white TV set receive color TV signals? What does the viewer see on the screen?
18. Why are relay stations necessary for long-distance TV transmissions?
19. How are X-rays produced?
20. In what ways is exposure to X-rays dangerous?
21. How is the phototube used in a motion-picture projector?
22. Explain how a TV camera works.
23. What are some uses of the photoelectric effect?
24. Into what basic types can computers be classified? Give two examples of each type.
25. In what ways do the two basic types of computers solve problems?
26. What is the binary system and how is it used by electronic computers?
27. What three basic functions are common to digital computers?

CRITICAL THINKING

28. Radar waves travel at the speed of light (about 300,000 km/sec). Explain how you would determine the distance to an object if it takes 2 sec for a radar wave to travel from the transmitter to the receiver.
29. Study the binary number system shown in Table 28-1 and write the binary numbers for the base-ten numbers 10, 11, 12, 13, 14, and 15.
30. Using the binary code for the alphabet in Table 28-2, decode the following word(s):

    01000011 01001111 01001101 01010000 01010101
    01010100 01000101 01010010 01010011.

FURTHER READING

Bender, Alfred. *Science Projects with Electrons and Computers*. New York: Arco Publishing, Inc. 1978.

Corn, Joseph J. *Imagining Tomorrow: History, Technology, and the American Future*. Cambridge, Mass.: MIT Press, 1988.

Englebardt, S. L. *Miracle Chip: The Microelectronic Revolution*. New York: Lothrop, Lee, and Shepard Books, 1979.

Heller, R. S., and C. D. Martin. *Bits 'n Bytes about Computing: A Computer Literacy Primer*. Rockville, Md.: Computer Science, 1982.

Leblanc, Wayne J., and Alden R. Carter. *Modern Electronics*. New York: Franklin Watts, Inc., 1986.

Renmore, C. D. *Silicon Chips*. New York: Beaufort, 1980.

How the Sciences Work Together

The Process in Physical Science: Making a Computer Chip.

A good computer will take a second to do calculations an average person would spend weeks doing. Computerized controls on an airplane measure speed and position in heavy turbulence and make hundreds of adjustments per second to maintain a smooth ride. The brains behind these and other devices such as calculators and video games are integrated circuits—or computer chips.

Video games have become more sophisticated with improved computer chips.

Computer chips help maintain a smooth ride in today's airplanes.

Computer chips, often no larger than a few millimeters on a side, can contain thousands of transistors. Each transistor can process information by switching the flow of electrons on and off.

Most chips are made from silicon, which is a semiconductor. To make a chip, high purity silicon is mixed with impurities that either donate or accept electrons within the silicon crystal. These impurities are called "dopants." Silicon treated this way can be made to act like a conductor whose resistance depends on the voltage applied. This property can be used to amplify electrical signals or switch them on and off. Switching signals on and off is a key function of computer circuits.

Since their development in the early 1960's, chip circuitry has gone through a rapid miniaturization. This has reduced costs and increased performance. A tiny electronic circuit is cheaper to mass produce than a larger one that does the same things. Miniature circuits provide more computing power on a given surface area. Also, information is processed more quickly in smaller circuits since electrons have shorter distances to travel.

Businesses rely on information processed by computer chips.

The Connection to Physics

Transistors were developed by physicists looking for ways to amplify (strengthen) electrical signals. They found that a small input current applied to two layers of doped silicon would create an output current through the adjoining layers. The output current was about equal to the input current but it flowed through a greater resistance. In effect, resistance was transferred from one circuit to another—thus the name "transistor" for transferred resistance. Since power is proportional to the resistance through which a current flows, the power in the output circuit was greater than in the input circuit. The energy for this increased power was obtained from voltage applied across certain layers of the transistor.

An engineer inspects a large scale circuit drawing used to design a computer chip.

The Connection to Chemistry

Several chemical processes are involved in making today's computer chips. First, extremely pure crystalline silicon is obtained by melting silicon dioxide (sand). Rods of pure silicon are then formed and sliced into wafers. Thousands of chips can then be manufactured on each wafer.

Next, dopants are combined within the surface layer of pure silicon by heating the wafer in the presence of a gaseous dopant or bombarding the wafer with ion beams of the dopant.

Precise patterns and layers of differently doped silicon are then formed by applying a film of light sensitive material onto the wafer surface. An image of the microscopic circuit is then projected onto the wafer. The light sensitive film hardens where light strikes it. Lines of the soft film and underlying layers of doped silicon are removed by washing the wafer in chemical baths.

A pencil eraser is large compared to this computer chip.

Once this is done, another chemical bath washes away the hardened portion of the film, exposing the rest of the chip surface. The chip is now ready for the process to be repeated until the required pattern of doped silicon layers is laid down.

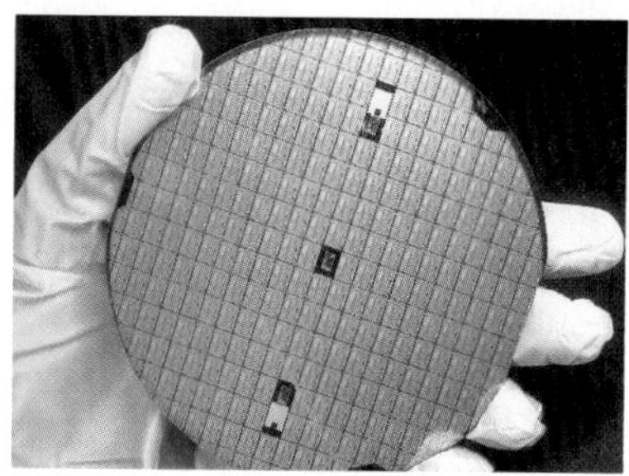

Each square on this silicon wafer contains a computer chip.

Continuing innovation in design and production has recently enabled chip manufacturers to put one million transistors on a 1.27 cm^2 chip. In the early 1960's, only four or five transistors could have been squeezed onto an area that small.

APPENDIX CONTENTS

KEY FORMULAS AND EQUATIONS

Section	Subject
1.2	Volume volume = length $\times$ width $\times$ height
2.1	Density $\text{density} = \frac{\text{mass}}{\text{volume}}$
6.3	Einstein's Mass-Energy Relationship $E = mc^2$ E is energy; m is mass; c is speed of light
15.1	Acceleration $a = \frac{v_f - v_i}{t}$ a is average acceleration; v_f is final velocity; v_1 is initial velocity; t is elapsed time
15.2	Displacement of Free-falling Object $s = \frac{at^2}{2}$ s is displacement; a is uniform acceleration; t is elapsed time
15.2	Torque torque = force $\times$ distance from fulcrum
15.3	Newton's Second Law of Motion $F = ma$ F is force; m is mass; a is acceleration
15.3	Momentum momentum = mass $\times$ velocity
15.3	Newton's Third Law of Motion (Action-Reaction) $m_A v_A = m_B v_B$ m_A is mass of object A; v_A is velocity of object A; m_B is mass of object B; v_B is velocity of object B
16.1	Efficiency $\text{Efficiency} = \frac{\text{output work}}{\text{input work}} = \frac{AMA}{IMA}$ AMA is actual mechanical advantage; IMA is ideal mechanical advantage
17.2	Boyle's Law $P_1 V_1 = P_2 V_2$ P_1 is original pressure; V_1 is original volume; P_2 is new pressure; V_2 is new volume
17.2	Fluid Pressure $\text{pressure} = \text{depth} \times \frac{\text{weight}}{\text{volume}}$
17.2	Pressure $\text{pressure} = \frac{\text{force}}{\text{area}}$

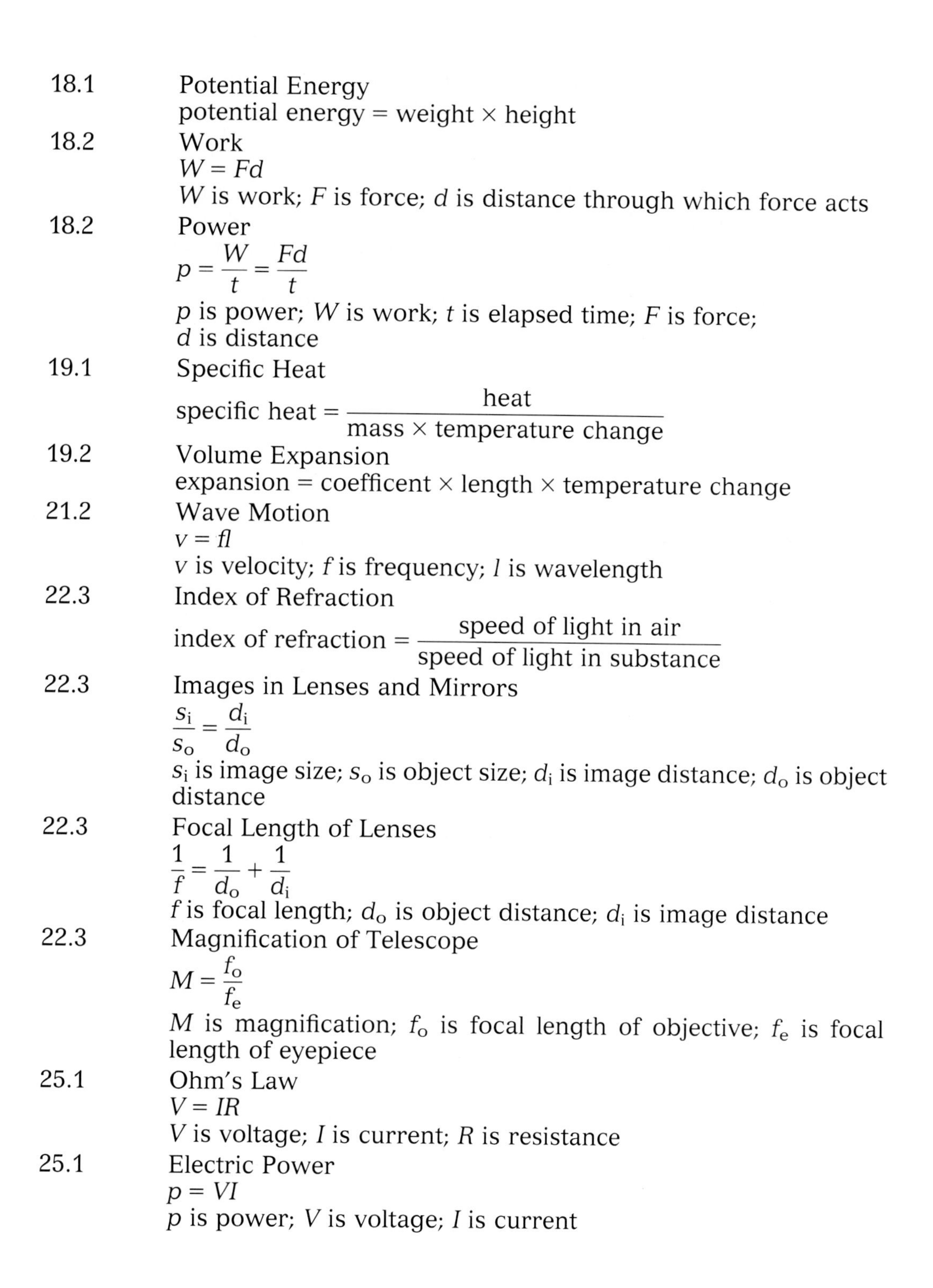

18.1 Potential Energy

potential energy = weight × height

18.2 Work

$W = Fd$

W is work; F is force; d is distance through which force acts

18.2 Power

$$p = \frac{W}{t} = \frac{Fd}{t}$$

p is power; W is work; t is elapsed time; F is force; d is distance

19.1 Specific Heat

$$\text{specific heat} = \frac{\text{heat}}{\text{mass} \times \text{temperature change}}$$

19.2 Volume Expansion

expansion = coefficent × length × temperature change

21.2 Wave Motion

$v = fl$

v is velocity; f is frequency; l is wavelength

22.3 Index of Refraction

$$\text{index of refraction} = \frac{\text{speed of light in air}}{\text{speed of light in substance}}$$

22.3 Images in Lenses and Mirrors

$$\frac{s_i}{s_o} = \frac{d_i}{d_o}$$

s_i is image size; s_o is object size; d_i is image distance; d_o is object distance

22.3 Focal Length of Lenses

$$\frac{1}{f} = \frac{1}{d_o} + \frac{1}{d_i}$$

f is focal length; d_o is object distance; d_i is image distance

22.3 Magnification of Telescope

$$M = \frac{f_o}{f_e}$$

M is magnification; f_o is focal length of objective; f_e is focal length of eyepiece

25.1 Ohm's Law

$V = IR$

V is voltage; I is current; R is resistance

25.1 Electric Power

$p = VI$

p is power; V is voltage; I is current

METRIC MEASUREMENTS AND CONVERSIONS

The metric system is a logical system of measurement, suited to the needs of scientists. Metric measurements use the decimal system, in which each unit is 10 times larger than the next smaller unit. Although many people in the United States still use British measurements (inches, feet, miles, and so on), metric measurements are easy to use and universally accepted in scientific communities.

Measurement	Units (abbreviations)	Equivalents/ Conversions
Mass/weight	milligram (mg)	1 mg = 0.001 g
	gram (g)	1 g = 0.001 kg
	kilogram (kg)	1 kg = 1,000 g
Length/ distance	millimeter (mm)	1 mm = 0.001 m
	centimeter (cm)	1 cm = 10 mm
	meter (m)	1 m = 100 cm
	kilometer (km)	1 km = 1,000 m
Area	square millimeter (mm^2)	
	square centimeter (cm^2)	1 cm^2 = 100 mm^2
	square meter (m^2)	1 m^2 = 10,000 cm^2
	square kilometer (km^2)	1 km^2 = 1,000,000 m^2
Volume	milliliter (mL)	1 mL = 1 cm^3
	liter (L)	1 L = 1,000 mL
	cubic centimeter (cm^3)	1,000 cm^3 = 1 L
	cubic meter (m^3)	1 m^3 = 1,000,000 cm^3
Time	second (sec)	
	minute (min)	1 min = 60 sec
	hour (hr)	1 hr = 60 min
Temperature	degree Celsius (°C)	
	Kelvin (K)	

Metric Unit Prefix	Symbol	Multiplication Factor
mega-	M	1,000,000
kilo-	k	1,000
hecta-	h	100
deka-	da	10
deci-	d	0.1
centi-	c	0.01
milli-	m	0.001
micro-	μ	0.0000001

TABLE OF IMPORTANT ELEMENTS

Name	Symbol	Atomic Number	Atomic Mass*
Aluminum	Al	13	26.9815
Barium	Ba	56	137.34
Bromine	Br	35	79.909
Calcium	Ca	20	40.08
Carbon	C	6	12.01115
Chlorine	Cl	17	35.453
Copper	Cu	29	63.54
Fluorine	F	9	18.9984
Gold	Au	79	196.967
Helium	He	2	4.0026
Hydrogen	H	1	1.00797
Iodine	I	53	126.9044
Iron	Fe	26	55.847
Lead	Pb	82	207.19
Lithium	Li	3	6.939
Magnesium	Mg	12	24.312
Nickel	Ni	28	58.71
Nitrogen	N	7	14.0067
Oxygen	O	8	15.9994
Phosphorus	P	15	30.9738
Silicon	Si	14	28.086
Silver	Ag	47	107.870
Sodium	Na	11	22.9898
Sulfur	S	16	32.064
Uranium	U	92	238.03
Zinc	Zn	30	65.37

*Atomic masses shown represent the average mass of an element's isotopes.

LABORATORY TECHNIQUES AND SAFETY

Proper precautions are an important part of any science program, in schools as well as in research institutions. They are intended to protect both people and property.

Most chapters in this textbook include activities that relate to the scientific concepts you are studying. The activities may involve using chemicals or equipment that can be hazardous or dangerous if used improperly. Some of the hazards you may encounter in the laboratory include caustic or corrosive substances, electrical shock, toxic gases, hot surfaces, and sharp edges. Your teacher will review and discuss all aspects of laboratory safety. To prevent injury whenever you are performing these activities, you must follow instructions exactly as given by your teacher.

When you are working in the science classroom or laboratory, you must assume the responsibility for your own safety. The following guidelines set forth what is expected of you as you perform the activities. If you do not understand any of these points, ask your teacher for an explanation.

1. Perform only those activities assigned or authorized by your teacher. Carefully follow all printed and verbal instructions. Be alert to potential problems introduced by your teacher and particularly to precautions presented in the activities. Carefully read and follow all statements preceded by the word CAUTION in the activities.

2. Remember that the laboratory is a place to work and learn; it is not a place to play.

3. As directed by your teacher and the activity procedure, wear approved safety goggles for eye protection, gloves to protect against skin irritants, and laboratory aprons for clothing protection. Contact lenses should not be worn while in the science laboratory.

4. Do not wear loose, bulky clothing or long, heavy jewelry in the science laboratory. Tie back long hair.

5. Become familiar with the health and safety hazards of all items of equipment, supplies, and chemicals.

6. Tell the teacher immediately if there are spills, breakages, or injuries. Be prepared to file a detailed report on what happened and how the incident could have been avoided.

7. Learn the location and operation of first aid kits, safety showers, eyewash stations, fire extinguishers, fire blankets, and other emergency facilities. Know the school's evacuation procedures in case of fire or other disaster.

8. Never taste anything in the laboratory. Never remove glassware from the laboratory or use this glassware for eating and drinking. Never store lunch in the science classroom or laboratory.

9. Never mix, touch, heat, or inhale chemicals unless you are told to do so by your teacher.

10. Do not continue an activity if the teacher must leave the laboratory.

11. Never leave a lighted gas burner or hot objects unattended. Do not point a lighted burner or the open end of a container being heated at anyone.

12. Never use, or return to storage, glassware that is broken, chipped, or cracked. Use only Pyrex glassware for heating procedures and use a wire gauze between the glass and the gas burner. Use a dust pan and broom to clean up broken glass. Dispose of broken or damaged glassware in the container designated for broken glass.

13. Read the labels on all bottles before using the chemicals in the bottles. Take only the amount needed for the activity and place the chemical in a labeled container. Unused chemicals must not be returned to their original containers.

14. Handle dry ice, radioactive materials, and all hot objects with tweezers, tongs, clamps, or gloves.

15. Follow all electrical safety guidelines as provided by your teacher. Do not turn on the power until the teacher has checked and approved the circuits you have connected.

16. Do not handle electrical equipment with wet hands or when standing in wet or damp areas.

17. Always observe proper disposal techniques for all waste chemicals and equipment.

18. If the flame of a gas burner goes out, immediately turn off gas burners by turning off the gas outlet valves, not by turning off the gas adjustment valve of the gas burner.

19. Keep work areas clean, dry, and free of clutter.

20. After you have completed the activity, clean the glassware and equipment and return them to their proper storage areas.

21. Wash your hands thoroughly with soap and water at the end of every activity period.

These safety guidelines for working in a school laboratory with science materials may be supplemented with more information from your teacher and other sources. Once you understand your responsibilities, you are ready to learn by performing the activities.

SOLVING MATHEMATICAL PROBLEMS

Mathematical problems are like all problems in science: they are best solved by a logical, well-thought-out approach. In *Modern Physical Science,* all Sample Problems use a three-step approach. Using this approach should help you to arrive at, and understand, solutions to the problems posed in *Modern Physical Science.* The following Sample Problem explains this three-step approach to problem solving.

SAMPLE PROBLEM

What is the index of refraction of glass through which light travels at 200,000 km/sec?

SOLUTION

Step 1: Analyze

First, "Analyze" the problem by listing all the given and known information that may help you solve the problem. Also, be sure to write down what you are asked to find—the "unknown."

You are given that the speed of light in glass is 200,000 km/sec. The index of refraction for the glass is the unknown.

Step 2: Plan

In the second step, "Plan" your solution out by organizing the given and known information so that it leads to the value of the unknown. This usually involves writing one or more equations that translate the "unknown" into the "known."

You can derive the following equation from the definition of the index of refraction:

$$\text{index of refraction} = \frac{\text{speed of light in vacuum or air}}{\text{speed of light in substance}}$$

Since you know the speed of light in air and in the substance (glass), you can use the equation to calculate the index of refraction.

Step 3: Compute and Check

In the third step, "Compute and Check," do the necessary arithmetic and find an answer. Be sure to use all units of measure in your calculations. Check your answer. Does it have the right units? Is the size of the number what you might expect from the given information? If your answer to these questions is yes, then you have completed your solution.

Substitute all values in the equation.

$$\text{index of refraction} = \frac{300{,}000 \text{ km/sec}}{200{,}000 \text{ km/sec}}$$
$$= 1.5$$

Since your answer is approximately equal to the index of refraction given for crown glass in Table 22-2, your solution is complete.

If you find out later from your teacher that your answer is incorrect, be sure to determine where you made your mistake. Did you make an incorrect assumption in any of the steps? Were the equations you used the correct ones? Each time you go through these steps, you will become a better problem solver. Note the application of the same three steps in the following sample problem.

SAMPLE PROBLEM

A piece of brass has a mass of 32.0 g and a volume of 3.8 cm^3. Find the density of brass.

SOLUTION

Step 1: Analyze

You are given that the piece of brass has a mass of 32.0 g and a volume of 3.8 cm^3. Its density is the unknown.

Step 2: Plan

Since you know the mass and volume, you can use the following equation for density:

$$\text{density} = \frac{\text{mass}}{\text{volume}}$$

Step 3: Compute and Check

Substitute all values in the equation.

$$\text{density} = \frac{32.0 \text{ g}}{3.8 \text{ cm}^3}$$
$$= 8.4 \text{ g/cm}^3$$

Since density in solids is most often expressed in grams per cubic centimeter, your solution is complete.

PROCESS SKILLS

You may have noticed the phrase *process skills* used in the activities followed by a sentence that tells you to *observe* or *classify* for instance. What are process skills and why are they important in science?

To answer this question think about times when you have been confronted with a problem. For instance, how do you get a wide chair through a narrow doorway? Or how do you turn a recipe for six people into one for eight? The solution may come in a moment of unexplained inspiration. Sometimes the solution does not come at all. Usually, however, the best approach to finding a solution is to have a logical, well-thought-out plan. Scientists solve problems for a living. They use a logical, well-thought-out approach to solving problems that is often referred to as a *scientific method.*

When applying a scientific method scientists do not use the same approach to every problem. They do, however, draw repeatedly from a number of different process skills. The process skills used in *Modern Physical Science* are listed and explained in the following paragraphs. You probably use many of these processes in your everyday life. In studying physical science you will use some of them as you carry out the activities, read the book, work the problems, or answer the questions.

The following 13 process skills are referred to in *Modern Physical Science* activities: *observing, collecting data, measuring, hypothesizing, experimenting, organizing data, classifying, analyzing data, drawing conclusions, inferring, predicting, modeling,* and *communicating.*

To help understand these process skills and how they are applied in a scientific method we can use a hypothetical investigation into the buoyancy of objects. The question will be, "why do some objects float in water whereas other objects sink?"

1. The skill of *observing* involves using your senses of sight, touch, taste, hearing, and smell, all of which enable you to gather information. Scientists make observations with the help of such devices as satellites, microscopes, telescopes, and microphones.

In our buoyancy investigation it would be helpful to observe the buoyancy of as many different objects as possible. We would observe that many objects made of wood and plastic, for instance, float in water whereas objects made of solid rock or metal sink.

2. *Collecting data* involves gathering and recording information based on observations. Devices that can be used to collect data are pen and paper, cameras, tape recorders, computers, and so on.

Pen and paper would probably be the best method to collect data about the buoyancy of various objects.

3. In *measuring* you determine the size of an object, the number of objects in a group, the duration of an event, the mass of an object, or other characteristics in precise units.

Measuring the mass and volume of objects is helpful in studying their buoyancy. Making these measurements would require a scale, and, for solid objects with simple shapes, some type of ruler or tape measure. The ratio of an object's mass to its volume is called its *density*. For comparison, the density of a given sample of water could be measured and found to be 1 g/mL or 1 g/cm^3.

4. *Hypothesizing* involves making testable statements about available data and information. Hypotheses provide the basis for experimenting.

A possible hypothesis about the buoyancy of objects is, "objects the density of which are less than that of water (1 g/cm^3) will float whereas objects the density of which are greater than 1 g/cm^3 will sink."

5. *Experimenting* involves designing and carrying out procedures that test hypotheses.

To test the hypothesis in part 4 we might determine the buoyancy of six objects of the following densities: 0.25 g/cm^3, 0.50 g/cm^3, 0.75 g/cm^3, 1.5 g/cm^3, 2.0 g/cm^3, and 2.5 g/cm^3.

6. *Organizing data* is arranging data in an orderly way so that they are easy to understand and interpret. This may involve creating tables, lists, graphs, maps, or illustrations.

The data from the buoyancy experiment could be arranged in a table using the headings "floats" and "does not float."

7. *Classifying* is the process of grouping objects or phenomena into categories.

Classification is usually used to categorize vast numbers of objects and phenomena. We can categorize any object as either one that floats in water or one that does not float in water.

8. *Analyzing data* is the process that determines whether data collected are reliable and whether or not they support or refute the hypothesis.

In the buoyancy experiment we would analyze the data to see whether or not the hypothesis stated in part 4 has been supported or refuted. We would also check to see that the data make sense, have been recorded properly and are therefore considered reliable.

9. *Drawing conclusions* involves making judgments based solely on the data at hand.

We will very likely conclude, from the data in the buoyancy experiment, that the hypothesis in part 4 is supported.

10. *Inferring* is the process of drawing conclusions based on experience but without direct observation or measurement.

From the results of our buoyancy experiment, we could infer that *any* object with a density less than water will float whereas any object with a density greater than water will not float. This statement is an inference since we have not measured the density of *all* possible objects and observed whether or not they will float.

11. *Predicting* involves stating in advance the result of a measurement or observation.

Given an object with a density less than water, we could predict that it would float if placed in water.

12. *Modeling* involves creating a physical, mathematical, or verbal representation of an object. Models can show relationships among data.

One model that represents an object that would be tested in part 5 is the mathematical equation: density = mass/volume.

13. By *communicating* the results of experiments in books, journals, and at scientific meetings, other scientists can learn of their colleagues' work and build on it by forming new lines of investigation and hypothesis.

If you wrote a report on the results of the buoyancy experiment and handed it out to your classmates, some of them might be puzzled. They might recall that many objects, boats for instance, float but are often made of material, such as steel, that is denser than water. How can boats made of steel float? Further investigation and experimenting would answer this question.

Obviously shipbuilders, and a lot of other people, already know why boats made of steel float. However, the application of process skills described in this hypothetical investigation represents the way scientists handle questions that have not yet been answered.

GLOSSARY

(Note: Glossary terms in **boldface** appear as boldface vocabulary terms on the page indicated.)

absolute zero Temperature at which the motion of molecules is at a minimum; established at − 273° C. 374

acceleration Measure of any change in speed or direction. 287

acid Substance that ionizes in water and gives hydrogen ions (protons) to water molecules to form hydronium ions. 78

acid dye Dye made with acids and used for wool and silk. 253

acid rain Precipitation that is acidic due to reactions between water and oxides of sulfur and nitrogen occurring as air pollutants. 142

acoustics Qualities of a building or place that provide clarity of sound. 425

actual mechanical advantage Amount a machine multiplies the effort force if there is friction. 315

addition reactions Chemical reactions that add hydrogen to unsaturated hydrocarbons. 189

adhesion Force of attraction between molecules of two or more different substances. 337

aeration Process of purifying water by spraying it into the air, such that oxygen dissolves in the water from the mechanical action of spraying. 147

ailerons Used to turn an airplane by imparting a rolling motion. 325

air Mixture of gases that make up the earth's atmosphere. 121

alcohol Organic compound in which one or more hydrogen atoms are replaced by hydroxyl (OH) groups. 191

alkali Strong soluble base. 85

alkane series Simplest and most abundant hydrocarbons having straight or branched chains and single bonds between the carbon atoms; characterized by the general formula C_nH_{2n+2}. 182

alkene series Hydrocarbons having straight or branched chains and double bonds between the carbon atoms; characterized by the general formula C_nH_{2n}. 186

alkyne series Hydrocarbons having straight or branched chains and triple bonds between the carbon atoms; characterized by the general formula C_nH_{2n-2}. 187

alloy Substance consisting of two or more metals melted together. 174

alpha particles Helium nuclei that can be emitted during radioactive decay. 99

alternating current (AC) Flow of electrons that changes direction in a circuit. 507

ampere Unit of electric current defined as one coulomb per second. 506

amplify To strengthen electric currents. 566

amplitude Maximum distance that a particle of wave medium moves up or down from its rest position to the top of a crest or to the bottom of a trough. 412

amplitude modulation (AM) Mode of broadcasting in which the amplitude of the carrier wave is changed to match the pattern of the sound wave. 562

analgesic Drug used to relieve pain. 215

analog computer Computer that solves problems by measuring things. 571

angle of incidence Angle at which light strikes an object, measured between the light beam and the perpendicular to the object's surface. 450

angle of reflection Angle at which light is reflected from an object, measured between the reflected light beam and the perpendicular to the surface of the object. 450

annealing Process in which heated steel is cooled slowly, first in a furnace and then in air, in order to soften the steel for shaping. 164

anode Positive (+) terminal of an electrochemical cell; also the positive electrode or plate of an electron device. 507

anthracite Bituminous coal that has been subjected to increased pressure and has lost most of its moisture and volatile matter; known as hard coal. 224

antibiotic Chemical originally produced by living organisms that destroys or inhibits the growth of infection-producing organisms. 216

antiparticles Mirror images or opposite counterparts of ordinary atomic particles. 111

antiseptic Chemical safe to use on body tissues that retards or stops the growth of bacteria. 213

asbestos Mineral fiber that occurs in some types of rock; characterized by a resistance to heat and chemicals, and a potential to cause health problems when inhaled. 250

atmosphere Envelope of air surrounding the earth. 121

atom Smallest unit of an element that can combine chemically with other elements; made up of protons and, except for hydrogen, neutrons, in a nucleus surrounded by electrons at various energy levels. 25

atomic mass Average mass of all isotopes of an element reflecting the proportions in which those isotopes occur in nature. 43

atomic mass unit Value that compares an element's mass to a standard, namely the mass of carbon-12 (12 atomic mass units). 43

atomic number Number of protons contained in the nucleus of one atom of an element. 41

atomic theory Theory proposed by John Dalton to explain the nature and properties of matter and that matter consists of atoms. 37–38

background radiation Naturally occurring radiation from the earth and space. 274

barometer Device that measures atmospheric air pressure. 349

base Substance that ionizes in water and accepts hydrogen ions (protons) from water to form hydroxide ions. 84

battery Combination of two or more electrochemical cells connected together in series or in parallel. 526

beats Variations in loudness caused by interference of sound waves. 427

benzene series Hydrocarbons that contain rings similar to the benzene molecule; characterized by the general formula C_nH_{2n-6}. 188

Bernoulli principle When the speed of a fluid (liquid or gas) is increased, its internal pressure is decreased. 322

beta particles High-speed electrons that can be given off during radioactive decay. 99

binary (number system) Base-two number system used by most digital computers to carry and process all data; uses only the two numbers 0 and 1. 572

biodegradable Substances that can be broken down into simpler compounds by bacteria, sunlight, or other natural processes. 272

bituminous coal Lignite that has been subjected to heat and pressure; third stage in the formation of coal; known as soft coal. 224

bleaching Procedure by which color is wholly or partially removed from a colored material. 251

block and tackle System of pulleys used to raise a heavy object with a small amount of effort. 319

boiling point Temperature at which the surface molecules of a liquid have sufficient energy to overcome the attractive forces of other liquid molecules and escape as a gas. 64

Boyle's law States that the pressure of a dry gas varies inversely with its volume if temperature remains constant. 350

branched-chain hydrocarbons Hydrocarbon compounds in which the carbon atoms are linked together with branches attached to the main string of carbon atoms. 184

breeder reactor Device that produces new fissionable material at a rate greater than that needed to use up fuel. 108

buoyancy Upward force exerted by fluids on objects immersed in them. 345

bypass jet (fanjet) Jet engine characterized by a stream of cool air that bypasses the combustion chamber and goes directly into the exhaust gases. 402

calorie Unit of heat (work); amount of heat required to raise the temperature of one gram of water one Celsius degree. 374

candela Standard light unit of brightness. 444

capacitor Device that stores static charges; consists of metal plates separated by an insulator. 499

capillary action Rise or fall of liquids in small-diameter tubes due to adhesion. 339

catalyst Substance that changes the rate of a chemical reaction but is not itself changed. 125

catalytic converter Device in modern motor vehicles that changes hydrocarbons and carbon monoxide in exhaust to water vapor and carbon dioxide, and the harmful oxides of nitrogen to nitrogen and oxygen gases. 266

cathode Negative (−) terminal of an electrochemical cell; also the electrode that emits electrons in an electron device. 507; 565

cathode rays Streams of high-speed electrons emitted by a negative electrode. 567

Celsius scale Scale used to measure temperature that establishes the freezing point of water at 0 degrees and the boiling point of water at 100 degrees. 372

center of gravity Point where all the weight of an object may be considered to be concentrated. 298

chain reaction Any self-sustaining nuclear reaction. 107

chemical bonds Forces that hold two or more atoms together in a compound or molecule. 49

chemical change Change in which a new substance with different properties is produced. 21

chemical equation Expression that, using formulas and symbols, tells the kind and number of each reactant and product in a chemical reaction. 27

chemical formula Designation used to identify a compound by showing the kind and the number of atoms present. 26

chemical properties Properties that describe the way a substance can be changed into a different substance or the behavior of a substance in reactions with other substances. 21

chemical symbol One- or two-letter abbreviation used to represent an element. 26

chlorofluorocarbons Class of synthetic organic compounds containing chlorine and fluorine that are believed to be harmful to the earth's protective ozone layer. 268

chord Musical tones sounded together in harmony. 430

coefficient of linear expansion Increase per unit length of a solid per degree rise in temperature. 379

coefficient of volume expansion Increase in volume of a substance per degree rise in temperature. 378

coherent waves Light waves in which the crests and troughs are all in step; property of laser light. 447

cohesion Tendency for the same kind of particles of matter to stick together. 337

coke Hard, porous, solid fuel produced by the destructive distillation of bituminous coal. 227

color Property of visible light that is determined by its wavelength. 472

combustion See rapid oxidation.

complementary colors Two colored lights that produce white light when mixed. 477

complete combustion Occurs when a fuel joins with enough oxygen to complete the burning process, yielding the products carbon dioxide, water, and energy. 233

complex machine Device made up of two or more simple machines. 314

compound Substance composed of two or more different elements that are chemically combined (in fixed, whole-number ratios). 25

compression Part of a wave in which the distance between vibrating particles in the medium is less than the distance between particles when the medium is at rest. 414

compression ratio Ratio of the volume within the cylinder of an engine when the piston is at the bottom of its stroke to the volume within the cylinder when the piston is at the top of its stroke. 399

concave Describes any object, particularly a mirror or lens, the surface of which curves in or recesses. 450

concentrated (solution) Solution with a relatively large amount of solute in the solvent. 68

condensation Process of changing from a gas to a liquid. 64

conduction Transfer of heat through a substance from molecule to molecule as they collide. 384

conductor Substance that permits heat or electricity to move through it readily. 46

cones Three types of cells in the retina that register color, each one of which is sensitive to one of the primary colors. 479

conservation of energy, law of States that energy cannot be made or destroyed, but can only be changed from one form to another. 362

conservation of matter, law of States that matter cannot be made or destroyed by ordinary chemical means. 29

constant proportions, law of States that every compound always contains the same proportion by mass of the elements of which it is formed. 29

contact charging Process by which objects become equally charged after touching each other. 495

control rod Rod of neutron-absorbing material used to regulate the reaction in a nuclear reactor. 108

convection Transfer of heat by the movement of liquids and gases. 384

convex Describes any object, particularly a mirror or lens, the surface of which curves out. 450

cornea Tough, transparent layer over the eye through which light enters and is refracted. 462

cosmic rays High energy radiation or particles from outer space. 100

cotton Fiber of nearly pure cellulose produced by the cotton plant; characterized by its moisture-absorbing qualities. 249

Cottrell device Industrial filter that uses charged plates to remove particulate pollutants from the air. 271

coulomb Unit equal to the quantity of electrical charge found on 6.25 billion billion electrons. 506

covalent bond Chemical bond formed when atoms share one or more pairs of electrons. 51

crest Highest point of a wave. 412

critical mass Smallest mass of material needed for a chain reaction of nuclear fission to occur. 107

crystal Solid containing atoms that are arranged in a regular pattern. 65

daughter product Product of radioactive decay. 100

decibel Unit of sound intensity or loudness; measure of sound-wave amplitude. 268

decomposition reaction Reaction in which a complex substance is broken down into two or more simpler substances. 87

density Mass per unit volume of a substance; measure of how close particles of matter are packed together. 20

destructive distillation Process of breaking down material, such as coal, in the absence of air. 227

detergent Substance that helps remove dirt from various materials by dissolving the oily covering of dirt particles. 209

diesel engine Internal combustion engine in which the fuel is ignited in the cylinder by the heat of compression of air. 398

differential Part of a motor vehicle containing gears that permits the rear wheels to turn at different speeds when the vehicle is making a turn. 400

diffraction Spread of waves beyond the edge of a barrier. 423

diffusion Movement of molecules from an area of higher concentration to an area of lower concentration; also scattering of light rays. 63; 450

digital computer Computer that solves problems by counting things. 572

dilute (solution) Solution with a relatively small amount of solute in the solvent. 68

diode Electron device that contains two electrodes. 565

direct current (DC) Flow of electrons in one direction in a circuit. 507

direct dye Dye that the fabric takes and holds easily; used for all natural fibers. 253

disinfectant Chemical that retards or stops the growth of bacteria on non-living surfaces. 213

distillation Process of evaporating a liquid and then condensing the vapors in a separate container. 144

domain Microscopic region composed of atoms whose magnetic fields are aligned in the same direction. 546

Doppler effect Rise or fall of pitch that is due to relative motion between the observer and the source of sound; shift in color towards the blue or red end of the spectrum due to relative movement. 423

double replacement reaction Chemical reaction in which substances change places. 83

drag Force caused by the friction of the air that is pushed aside by a moving airplane. 322

drug Chemical that alters normal life processes. 215

dyes Chemicals used to color fabrics and other substances. 252

Edison effect Flow of current from a hot cathode to a positively charged plate. 565

efficiency Ratio of work that goes into a machine (input work) to work that comes out (output work); also a comparison of actual mechanical advantage with ideal mechanical advantage. 315

elastic limit Point of stretching beyond which matter will not return to its former shape. 336

elasticity Property that allows matter to return to its original shape after a distorting force is removed. 336

electric circuit Complete conducting path along which an electric current travels. 505

electric current Movement of electrons within a conductor due to a potential difference in the circuit. 505

electric field Region in which an electric force acts on a charge placed in the region. 492

electric generator Device that changes mechanical energy to electrical energy. 499

electric motor Device that changes electrical energy into mechanical energy. 550

electrochemical cell Device that changes chemical energy into electrical energy; consists of two electrodes and an electrolyte. 525

electrochemical series List of elements that rates their relative activity compared to hydrogen, which has a value of zero. 527

electrodes Two poles or different metals in an electrochemical cell. 525

electrolysis Chemical change produced when an electric current passes through a solution that contains ions. 22

electrolytes Substances that conduct an electric current when dissolved or melted. 71

electromagnet Temporary magnet produced when a direct electric current flows through a coil of wire that surrounds a soft iron core. 547

electromagnetic induction Electric current produced by relative movement between a magnet and a conducting circuit. 536

electromagnetic spectrum Representation of the range of wavelengths of electromagnetic emissions in a logarithmic scale. 472

electromagnetic waves Transverse waves that have both electric and magnetic components. 443

electron Negatively charged subatomic particle. 38

electronics Study of the behavior and effects of electrons and of devices that control the flow of their energy. 560

electroscope Device used to detect an electric charge or indicate how much charge is present. 494

electrostatics Study of charges at rest. 490

element Substance that cannot be broken down into simpler substances by chemical means. 23

elevators Control up and down movements of an airplane's nose. 324

emulsion Temporary mixture of immiscible liquids. 70

energy Ability to do work; that which produces change in matter. 14

energy level One of up to seven regions around a nucleus within which electrons are found. 39

engine Any machine that uses heat energy produced by burning a fuel to do work. 393

eutrophication Process by which the dissolved oxygen of lakes and streams is depleted due to the addition of nutrients from untreated sewage in runoff. 153

external combustion engine Engine such as a steam engine or steam turbine where the fuel burns outside the engine itself. 396

farsightedness Condition resulting from eyeballs that are too short or lenses that are too flat, producing a fuzzy image of nearby objects. 463

fast dye Dye that clings to the fabric, does not lose its color, and does not wash away. 253

fermentation Chemical reaction in the absence of oxygen that produces carbon dioxide gas. 212

fiber optics Field of science based on the transmission of information in the form of light through thin fibers of pure glass or plastic. 449

field Region around a charged object within which a force can be detected. 492

first-class lever Class of lever in which the fulcrum is located between the resistance and the effort. 317

fission, nuclear Splitting of a heavy nucleus into two or more lighter nuclei. 106

flaps Surfaces on an airplane that help control lift, speed, and turns, and are used as air brakes during landing. 325

fluorescent Describes substances that produce light when exposed to radiation. 443

focal length Distance from the center of a lens to the principal focus. 458

force Any push or pull on an object; that which can change the state of rest or of motion of a body. 12

forced vibrations Transfer of wave energy by contact between objects. 431

fossil fuel Coal and other fuels that formed from organic matter. 224

four-stroke engine Most common gasoline engine in which the piston makes four movements during one full cycle. 396

fractional distillation Process of separating a mixture of liquids having different boiling points; used for the initial refining of petroleum. 230

freeze separation Process of extracting pure water from salt water by freezing. 149

frequency Number of waves that pass a given point in one second. 412

frequency modulation (FM) Mode of broadcasting in which the frequency of the carrier wave is changed to indicate the pattern of the sound wave. 562

friction Resistance to motion caused by surfaces in contact with each other. 289

fuel Substance that can be burned to generate heat at a reasonable cost. 223

fuel cell Very efficient type of electromagnetic cell in which chemical energy is changed to electrical energy by a continuous supply of fuel (generally hydrogen). 532

fulcrum Point of support on which a lever pivots. 296

fundamental Lowest frequency that can be made by a musical sound source; the tone produced when a string vibrates as a whole (instead of in segments). 429

fungicide Chemical compound that destroys fungi. 212

fuse Safety device placed in an electric circuit that melts and breaks the circuit when the current increases beyond a safe load. 516

fusion, nuclear Merging of two or more lighter nuclei to yield a heavier nucleus and great amounts of energy. 110

gamma rays Very high frequency radiation that can be emitted during radioactive decay. 473

gas Matter that does not have definite shape or volume. 63

germicide Chemical too strong to use on body tissues that retards or stops the growth of bacteria. 213

gram Unit of mass in the metric system. 13

gravity Force of attraction of the earth for a given object, or of one mass for another. 12

greenhouse effect Trapping of the sun's energy by the earth's atmosphere caused by increased amounts of carbon dioxide in the air. 130

grid Element of a vacuum tube that controls the flow of electrons moving from the cathode to the plate. 565

grounding Absorption of an object's induced charge by contact with a much larger object. 495

half-life Time needed for half of the atoms of a given radioactive sample to break down. 99

hard water Water that contains dissolved minerals and causes soap to precipitate. 150

harmony Effect of two or more tones sounded together to produce a pleasing combination. 430

heat Total amount of energy that can be transferred from one object to another because of a difference in their temperatures; produced by the motion of molecules of a substance. 372

heat of fusion Heat that is absorbed in the melting process. 382

heat of vaporization Heat needed to change a liquid into a gas. 382

heating value Amount of heat produced by a given amount of fuel when it is burned; usually expressed in units of calories per gram. 226

holograms Lifelike, three-dimensional images produced by laser light. 448

homogenization Process by which particles in an emulsion are broken up to the extent that they do not separate. 70

horsepower Unit of power (work done per unit time) used to rate the strength of engines. 364

hydraulic (machine) Device that works by putting pressure on a liquid. 344

hydrocarbons Compounds that are composed only of hydrogen and carbon. 182

hydroelectric energy Energy obtained from falling water. 361

hydrogen bond Weak bond formed by the attraction between the positive hydrogen end of one molecule and the negative end of another molecule. 141

hydrogenation Process by which hydrogen combines chemically with other substances. 134

hydrometer Device that measures the specific gravity of liquids. 348

ideal mechanical advantage Amount a machine can multiply effort force if there is no friction. 315

image Reproduction of an object formed with mirrors or lenses. 449

immiscible Not soluble in each other. 70

incandescent Describes any object heated until it emits light. 442

incomplete combustion Occurs when there is insufficient oxygen for the burning of a fuel, yielding the products carbon monoxide, water, and energy. 234

inclined plane Ramp or slanting surface used as a simple machine. 320

indicator Substance that changes color in the presence of a certain ion; used to indicate the pH value of a solution. 78

induction Process by which a charged object produces the opposite charge on a nearby object. 495

induction coil Device that uses an iron core and a current in a primary coil to induce a current in a secondary coil. 552

inertia Resistance to any change of motion or change in direction. 13

infrared rays Electromagnetic emissions with frequencies below those of red light, referred to as radiant heat. 473

input work Product of the effort force and the distance through which it travels; work that goes into a machine. 315

insecticide Chemical compound that kills insects. 211

insulator (electrical) Material that does not readily carry an electric current. 507

insulator (heat) Substance that is a poor conductor of heat. 384

interference Effect caused by two or more waves superposed. 426

internal combustion engine Engine in which the fuel burns inside the cylinders. 396

inverse-square rule Mathematical relationship wherein if one component changes by some amount, the other component changes by the reciprocal of the square of that amount. 292

ion Atom (or group of atoms) that has lost or gained electrons such that it becomes an electrically charged particle. 47

ion exchange Process of mixing two ionic compounds to yield new compounds, one of which may precipitate and be separated from solution. 149

ionic bond Chemical bond formed by the attraction between ions of opposite charge and resulting from the transfer of electrons from one of the bonding atoms to another atom. 49

ionized Said of substances that lose or gain electrons such that they become ions; said of substances that react with water to form ions. 71–72

isomers Compounds whose molecules have the same number and kind of atoms but with a different arrangement and different properties. 185

isotopes Atoms that contain the same number of protons but a different number of neutrons. 42

jet engine Aircraft engine that works on the basis of Newton's action-reaction principle, such that the thrust that drives the craft forward comes from the reaction force directed back on the engine by escaping gas. 401

joule Unit of work done when one newton of force acts through a distance of one meter. 357

Kelvin scale Scale used to measure temperature that establishes absolute zero at 0 degrees, the freezing point of water at 273 degrees, and the boiling point of water at 373 degrees. 374

kilowatt Unit of power in the metric system, equal to 1,000 watts. 364

kilowatt–hour Unit of electrical energy equivalent to using 1,000 watts of power for one hour. 510

kindling temperature Lowest temperature at which a material begins to burn. 129

kinetic energy Energy of motion. 61

kinetic theory Used to explain how matter behaves and states that matter is made up of particles such as atoms, ions, and molecules that are in constant motion. 62

kite effect Way in which moving air striking the underside of an airplane causes the airplane to be pushed upward. 322

laser Device that amplifies light as it passes through energized material and produces monochromatic, coherent light that travels in a plane wave front; acronym for light amplification by stimulated emission of radiation. 447

latex Milky fluid, or sap, from the rubber tree that consists of 30 percent to 35 percent rubber. 240

leavening agent Substance that causes dough to rise. 212

left-hand rule Guide for identifying a magnet's north pole, which states that the outstretched thumb of the left hand indicates the north pole of the coil if the fingers are wrapped around the electromagnet pointing in the direction of the current flow. 548

lens Transparent substance, having at least one curved surface, that is used to refract light. 457

lever Simple machine that is a rigid bar free to turn about its point of support, the fulcrum. 296

lift Upward force on an airplane caused as air pressure builds up under the wing until it lifts the airplane off the runway. 322

light Form of energy produced when electrons fall from one energy level to a lower level; the visible portion of the electromagnetic spectrum. 440

lightning Light caused when air is heated by discharge of static electricity between clouds or between a cloud and the earth. 497

lignite Decomposed peat that has lost most of its fiber; second stage in the formation of coal, known as brown coal. 224

linen Fiber of nearly pure cellulose produced from the flax plant. 249

lines of force Lines representing the strength and direction of a magnetic, electric, or gravitational field. 543

liquid Matter that takes up a definite amount of space but has no definite shape. 63

lodestone Naturally occurring magnetic mineral. 542

longitudinal wave Wave in which the particles of the medium move back and forth parallel to the direction of the motion of the wave. 413–414

loudness Effect of the intensity of sound energy on the ears. 424

lumen Measure of the rate at which light energy is given off by a source. 445

machine Device used to change the size, direction, or speed of application of a force. 313

magnet Piece of iron, nickel, cobalt, or alloy that attracts other pieces of like materials. 542

magnetic field Space within which magnetic force is felt. 543

magnetic pole Part of a magnet where the magnetic force is strongest. 542

magnification Apparent enlargement of an object by an optical instrument. 459

mass Measure of the amount of matter in an object; measure of an object's inertia. 13

mass number Sum of the number of protons and neutrons in the nucleus of an atom. 42

matter Anything that occupies space and has weight. 13

mechanical advantage Ratio that tells how a machine multiplies effort force. 315

medium Material that carries a wave or transfers energy. 412

melting Process of changing a substance from a solid to a liquid. 64

metal fatigue Gradual weakening of a metal caused by continued stress on that metal. 177

metallurgy Science of taking useful metals from their ores, refining the metals, and preparing them for use. 160

metals Elements grouped at the left side of the periodic table that have luster, are usually solid at room temperature, have relatively high densities, are malleable, are good conductors of heat and electricity, and tend to lose electrons in chemical reactions. 46

meter Unit of length in the metric system. 9

mirage Optical illusion caused by bending of light rays as they pass through air layers of different densities. 454

miscible Soluble in each other. 70

mixture Material containing different substances that have not been chemically combined. 24

moderator Material used to slow the movement of neutrons in a fission reaction. 108

modulation Process of shaping or coding a radio station's carrier waves with the pattern of audio waves; accomplished by changing the amplitude or the frequency of the carrier wave. 562

molecule Smallest unit of a substance that has all the properties of that substance. 25

momentum Product of the mass and velocity of a body. 305

monomer Single unit of a whole molecule that, when linked together with other units, forms a polymer. 241

mordant Chemical that combines with dye molecules and fixes them within cotton fibers, thus enabling dyes to adhere. 253

motion Movement of an object from one place to another. 285

narcotic Drug that relieves pain but may become habit-forming. 215

natural frequency One fundamental frequency at which an object vibrates when disturbed. 431

natural gas Fuel made up almost entirely of methane, but may also include propane and butane. 232

natural rubber Elastic, waterproof, non-conducting substance extracted from the sap of the rubber tree. 239–240

nearsightedness Condition resulting from eyeballs that are too long or lenses that are too thick, producing a fuzzy image of distant objects. 463

negative charge Electric charge resulting from a gain of electrons. 38

neutralization reaction Reaction between the hydronium ions of an acid and the hydroxide ions of a base to form water. 88

neutron Neutral subatomic particle found in the nucleus of an atom. 38

newton Unit of force in the metric system. 12

Newton's first law of motion States that a body at rest or in uniform motion will remain at rest or in the same motion unless acted upon by some external force. 301

Newton's second law of motion States that the acceleration of a body is directly proportional to the net force acting on the body and inversely proportional to the mass of the body. 303

Newton's third law of motion States that the force exerted by one body on a second body is equal in magnitude and opposite in direction to the force exerted by the second body on the first; action causes reaction. 304

nitrogen fixation Process whereby bacteria produce nitrogen compounds from nitrogen in the air. 130

noble gases Relatively inactive elements, sometimes called inert or rare gases. 130

noise Sounds characterized by irregular vibrations. 427

nonelectrolytes Substances that do not conduct an electric current when dissolved. 72

nonmetals Elements grouped at the right side of the periodic table that lack luster, are poor conductors of heat and electricity, vary in color, are not malleable, have low densities, and tend to gain electrons in chemical reactions. 46

nonpolar (molecule) Molecule with a symmetrical distribution of electrical charge. 140

nuclear reactor Device in which controlled fission of radioactive material produces new radioactive substances and energy; also referred to as an atomic pile. 107

nucleus Central core of an atom containing protons and neutrons, and having a positive charge equal to the number of protons. 38

objective Large convex lens with a long focal length placed at the end of a telescope tube; small convex lens with a short focal length placed at the lower end of a microscope tube. 465

octane number Rating that expresses the ability of a gasoline to resist knocking and reflects a gasoline's relative proportion of branched-chain isomers. 185

ohm Unit that measures electrical resistance. 508

opaque Describes a substance that does not transmit light waves. 444

ore Rock or mineral from which a metal can be obtained profitably. 160

organic compounds Compounds that contain carbon. 181

osmosis Diffusion of water across a semipermeable membrane from an area of greater concentration of water to an area of lesser concentration of water. 149

output work Product of the resistance force and the distance through which it acts; work that comes out of a machine. 315

overtones High-pitched tones produced by a musical sound source as it vibrates in segments; the frequencies are whole-number multiples of the fundamental frequency. 429

oxidation Chemical reaction in which an element loses one or more electrons. 127

oxidation number Charge an atom has in a particular molecule or ion when it loses, acquires, or shares electrons during a chemical reaction. 47

oxidizer Substance that combines with rocket fuel to supply energy; permits the rocket engine to operate outside the earth's atmosphere. 402

parallel circuit Type of circuit that contains two or more branches through which a current can flow. 513

Pascal's law States that pressure applied to a confined liquid acts equally in all directions. 344

peat Soft, brown, spongy material composed of plant matter that has been changed through the action of heat and pressure; first stage in the formation of coal. 224

period Horizontal row of elements in the periodic table. 46

periodic table Chart that groups elements according to their increasing atomic numbers and their properties. 46

petroleum Solid, liquid, or gaseous fuel found beneath the earth's surface composed mostly of carbon and hydrogen. 229

ph Value that indicates how acidic or basic a solution is. 87

phases (or states) of matter Physical property of matter, which can be either solid, liquid, gas, or plasma. 63

photochemical smog Formed when unburned hydrocarbons react with oxides of nitrogen in the presence of ozone and sunlight. 266

photoelectric effect Emission of electrons from a metal surface that is hit by a beam of light. 440

photon Particle of light energy. 440

photosynthesis Food-making process in green plants; using sunlight, water from soil, and CO_2 from air to make sugars and starches. 130

phototube Photoelectric cell that uses light energy to produce an electric current. 570

physical change Change that does not produce a new substance. 21

physical property Property of a substance that can be easily observed or measured without altering the identity of that substance. 20

piezoelectric effect Electric current produced when certain types of crystals are subjected to mechanical stress; change of mechanical energy to electrical energy. 535

pigment Matter that has color because it reflects light of only certain wavelengths. 479

piston Device inside the cylinder of an engine that is moved back and forth by the pressure of a fluid that transmits motion. 395

pitch Distance on a screw from one thread to the next; also the effect of the frequency of sound waves on the ears. 321; 419

plasma State of matter similar to a gas that exists only at very high temperatures and consists of charged particles. 63

plastics Synthetic products made from coal, oil, or related raw materials that have been shaped into final forms by heat and pressure. 244–245

pneumatic (machine) Device that is operated by compressed gas. 344

polar (molecule) Molecule with an uneven distribution of electric charge. 140

polarization Screening process that permits the passage of light waves vibrating only in one plane. 442

pollutant Any substance added to a natural system in larger amounts than can be disposed of by that system such that it becomes harmful to living organisms. 259

polymer Large molecule formed from the linking of many monomers, or small, single units. 241

porous Said of solids that have very small holes, or openings, through which gases or liquids can pass. 63

positive charge Electric charge resulting from a loss of electrons. 38

positron Electron with a positive charge; the antiparticle of an electron. 110

potable water Water that is fit to drink. 142

potential difference Difference in electrical potential energy between two points in an electric circuit or electric field. 507

potential energy Stored energy that is due to position or condition; equal to the work a body can do or to the work done on a body to lift it. 358

power Rate at which work is done; work done per unit of time. 364

precious metals Metals that are useful but scarce and costly. 171

precipitate Insoluble solid substance that separates from an aqueous solution due to a chemical reaction. 29

pressure Force acting on a unit area of surface. 340

primary colors of light Three colors of light (red, green, and blue) that produce white light when mixed; by changing their relative amounts, they can be mixed to produce all colors of the spectrum. 477

primary pigments Three pigment colors (red, blue, and yellow) that produce black when mixed; by changing their relative amounts, they can be mixed to produce all colors of the spectrum. 480

principal focus (or focal point) Point at which the rays parallel to the principal axis converge after passing through a convex lens. 458

prism Three-sided piece of glass that separates light into its colors. 457

product Substance that results from a chemical reaction. 28

proton Positively charged subatomic particle found in the nucleus of an atom. 38

quenching Process by which steel is heated and then cooled suddenly in order to make it strong, hard, and brittle. 164

radar Electronic instrument used to locate moving or fixed objects by sending out pulses of high-frequency waves and picking up wave reflections from those objects. 563

radiation Transfer of energy in waves through space. 385

radioactivity Process by which energy is given off in the form of alpha particles, beta particles, or gamma radiation as a result of disintegration of the nucleus. 98

radioisotope Unstable isotope of a radioactive element. 98

radon Naturally occurring radioactive gas considered to be a pollutant when found in buildings at high levels of concentration. 274

ramjet Simplest of jet engines that is used after the aircraft has been boosted to a high speed. 401

rapid oxidation Oxidation that occurs quickly and gives off heat and light. 127

rarefaction Part of a wave in which the distance between vibrating particles in the wave medium is greater than the distance between particles when the medium is at rest. 414

reactant Substance that enters into a chemical reaction. 28

real image Inverted image formed by light rays that converge at one location. 452

reciprocating engine Engine that converts heat energy to mechanical energy by the back-and-forth motion of a piston in a cylinder. 395

rectifier Device that changes alternating current to direct current. 552

recycling Process of reusing wastes to make new products. 273

red shift Doppler effect that indicates a light source is moving away from the viewer. 482

reducing agent Substance that removes oxygen from a compound. 161

reduction Chemical reaction in which an element gains one or more electrons. 168

refining Process of separating a metal from the other materials in its ore. 161

reflecting telescope Telescope that uses concave mirrors to focus the light and magnify the object. 453

reflection Return movement of a wave from the surface of a different medium. 422

reflection, law of States that the angle of reflection equals the angle of incidence. 450

refracting telescope Telescope that uses a lens or combination of lenses to focus light. 465

refraction Bending of waves as they pass at an angle from one medium to another. 423

relay Electromagnetic switch. 548

REM Unit that measures the effect of radiation on the human body. 274

resistance Opposition of a conductor to the flow of an electric current; also, any weight to be moved. 508; 315

resonance Induced vibration at an object's own natural frequency by similar vibration of another object. 431

respiration Use of oxygen to provide energy for living things. 126

resultant Combined effect of two or more forces acting together. 299

retina Light-sensitive layer in back of the eyeball that transmits images to the brain via the optic nerve. 462–463

reverse osmosis Process by which potable water is made from sea water, where sea water is forced across a semipermeable membrane from an area of lesser concentration of water (salt water) to an area of greater concentration of water (fresh water). 149

ring compound Compound in which the ends of a chain of carbon atoms are linked together, forming a ring. 187

roasting Refining process in which an ore is heated in oxygen-enriched air to change it into an oxide. 168

rocket engine Device propelled by exhaust gases that can operate beyond the atmosphere because it does not require oxygen. 402

rolling friction Force that is less than or equal to sliding friction, produced when an object is rolled rather than slid. 289

rudder Moves nose of an airplane to the left or right. 324

salt Compound formed when the positive ions of a base and the negative ions of an acid react. 88

sanitary landfill Landsite for the regulated disposal of wastes. 273

saturated (hydrocarbon) Molecule that contains only single bonds between carbon atoms. 186

saturated (solution) Solution that contains the maximum amount of solute that can be dissolved under the existing conditions. 68

science Orderly search for answers to questions about nature. 5

scientific law General statement that describes the orderly behavior of nature. 29

scientific method Logical and systematic approach to collecting information and applying knowledge. 6

screw Inclined plane wrapped in a spiral around a cylinder; type of simple machine. 314

second-class lever Class of lever in which the resistance is located between the fulcrum and the effort. 317

semiconductor Material having an ability to carry a current that is between that of a conductor and an insulator. 508

series circuit Type of circuit in which current has only one path to follow and along which current is the same everywhere. 512

short circuit Sudden and large drop in resistance that increases the current and overloads a circuit with dangerous results. 514

SI (International System of Units) Set of measurement standards called the metric system that is based on decimals. 9

silk Fiber that is produced by the silkworm caterpillar when it spins its cocoon. 248

simple machine Class of devices used to change the force that is used to do work; characterized by the lever, pulley, inclined plane, wedge, screw, and wheel and axle. 314

single replacement reaction Reaction in which one element replaces another in a compound. 79

slag Waste product from the smelting of ores. 161

sliding friction Force that is less than or equal to starting friction and is produced when contacting surfaces slide past one another. 289

slow oxidation Oxidation process such as decay or rusting that produces some heat but no light. 128

soft water Water that is relatively free of dissolved minerals and easily makes suds with soaps. 150

solid Matter that takes up a definite amount of space and has a definite shape. 63

solute Component of a solution that is dissolved in the solvent. 67

solution Uniform mixture of a solute and a solvent. 67

solvent Component of a solution in which the solute is dissolved. 67

sonic boom Loud noise caused by a jump in air pressure that results when planes fly faster than the speed of sound. 325

sound Form of energy produced by the vibration of matter. 415

specific gravity Value that compares the density of a substance with the density of water; found by dividing the weight of an object by the weight of an equal volume of water. 347

specific heat Number of calories needed to warm up 1 g of any substance by 1 C° (Celsius degree). 375

spectroscope Device used to analyze light from objects, particularly stars and other distant objects; includes a prism and system of lenses. 480

speed Measure of the distance an object moves in a certain amount of time. 286

spinneret Small metal plate with tiny holes used for forming strands of synthetic fibers. 250

spontaneous combustion Combustion resulting from the accumulation of heat from slow oxidation. 129

stabilizers Equipment on a plane that keeps the plane steady in flight. 323

static electricity Accumulation of positive or negative charges on an object. 492

static friction Force that prevents motion until the contacting surfaces begin to slide. 289

steam turbine Engine containing a set of rotating blades that produce a steady, constant motion. 395

storage battery Series of electrochemical cells that can be repeatedly recharged. 529

straight-chain hydrocarbons Hydrocarbon compounds in which the carbon atoms are linked together in long, straight chains. 184

structural formula Formula that shows the arrangement of atoms in a molecule. 183–184

sublimate, or sublime To change phase directly from a solid to a gas. 65

substitution reaction Chemical reaction that replaces one atom or group of atoms with another atom or group of atoms. 189

superconductor Substance with little or no resistance to electric current. 176

supermagnet Very powerful electromagnet made from a superconductor. 548

superposition, principle of States that when the displacements produced by two sound waves traveling through the same medium are combined, the resultant wave is their algebraic sum. 426

supersonic Faster than the speed of sound. 325

surface tension Cohesion between molecules on the surface of a liquid. 338

suspension Mixture consisting of a liquid and large solid particles. 142

synthesis reaction Reaction in which one or more substances combine to form a more complex substance. 80

synthetic fibers Class of fibers made from cellulose or petrochemicals, of which rayon, acetate, nylon, acrylic, and polyester are examples. 250

synthetic (substance) Substance that does not occur naturally but is produced artificially by assembling chemically simpler substances. 244

temperature Measure of the average kinetic energy (motion) of the molecules of a substance. 372

temperature inversion Atmospheric condition characterized by a layer of cool air close to the earth topped by a layer of warm air; creates a stagnant air mass that traps pollutants and creates a health hazard. 270

tempering Process of reheating and recooling quenched steel to produce the degree of hardness desired. 164

tesla Unit that measures the strength of a magnetic field. 548

thermal pollution Increase in heat in lakes and rivers caused by discharge of hot water from power plants or factories; drives oxygen from the water, altering the ecology of the water environment. 262

thermocouple Device that measures temperature difference by use of the thermoelectric effect. 534

thermoelectric effect Production of electric current in a circuit composed of two different metals when the two junctions of the metals are maintained at different temperatures. 534

thermoplastic Substance that results when chains of molecules are formed by head-to-tail addition of monomer units; melts easily and can be reshaped. 245

thermosetting plastic Substance that results when bonds are formed between chains of molecules and monomer units; takes a permanent shape when heat and pressure are applied during the forming process. 245

thermostat Device containing a bimetal bar that utilizes differences in expansion during temperature changes to turn furnaces on and off. 380

third-class lever Class of lever in which the effort is located between the resistance and the fulcrum, and the mechanical advantage is less than one. 317

thrust Force that pushes an airplane forward. 321

torque Twisting movement caused by one or more forces. 296

transformer Device that changes the voltage of an alternating current. 554

transistor Solid-state semiconductor device that functions like vacuum tubes. 566

translucent Describes a substance that scatters the light waves it transmits. 444

transmission Part of a motor vehicle that is made up of gears that change its speed and direction; allows the driver to shift gears to get the best mechanical advantage. 400

transparent Describes a substance that transmits light waves. 444

transverse wave Wave in which the particles of the medium move at right angles to the direction of the motion of the wave. 413

triode Electron device that consists of three electrodes. 565

trough Lowest point of a wave. 412

turbine Rotary engine that makes electricity from the force of a gas or liquid acting against its vanes or blades. 361

turbojet Jet engine consisting of turbine blades in the rear and air-compressor blades in front; escaping gases give the engine its forward thrust. 401

ultrasonic Sound with a frequency above 20,000 vps. 421

ultraviolet light Electromagnetic emissions with frequencies greater than those of violet light. 473

unsaturated (hydrocarbon) Molecule that contains double or triple bonds between carbon atoms. 186

vacuum Space that contains no gas, liquid, or solid. 349

vector Any quantity, such as a force, that has both size and direction. 299

velocity Measure of the speed and direction of an object. 286

virtual images Images formed by plane mirrors that are upright and the same size as the object, but do not exist where they appear to be, behind the mirror. 450

visible light Part of the electromagnetic spectrum that can be seen by the human eye. 472

visible spectrum Band of colors, arranged in order of increasing wavelength from red to violet, produced when white light travels through a diffracting medium such as a prism. 474

volt Unit that measures potential electrical difference (electrical pressure). 507

volume Measure of the amount of space occupied by an object. 10

vulcanization Process by which sulfur joins with rubber polymer chains to produce rubber products with desirable properties. 241

water gas Fuel composed of hydrogen and carbon monoxide that is produced when steam passes through burning coke. 233

watt Unit of power (work done per unit of time) in the metric system. 364

wave Disturbance that propagates through a medium or space. 411

wavelength Distance measured from the crest of one wave to the crest of the next wave, or from trough to trough. 412

wedge Double-inclined plane used to split or separate objects. 314

weight Measure of the pull (force of gravitational attraction) between two objects. 12

wool Fiber that comes from the fleece of sheep, goat, llama, alpaca, or camel. 248

work Results from a force acting through a distance such that there is motion or displacement. 357

xerography Process of copying printed material using static charges and dry powdered inks. 500

X-rays High-energy electromagnetic radiation produced when an electron beam strikes a metal target. 473

INDEX

(Note: Page numbers in **boldface** type include illustrations.)

PHOTO CREDITS:

Table of Contents: Page iii, N.A.S.A.; iv, Ralph P. Earlandson/Hillstrom Stock Photos; v, Courtesy of The Silver Dome, Pontiac, Michigan; vii, M.P.L. Fogden/Bruce Coleman, Inc.; ix, Dr. E. R. Degginger/Bruce Coleman, Inc.; x, Karl Sims, MIT Media Laboratory/Dan McCoy/Rainbow; xii, Courtesy of The Silver Dome, Pontiac, Michigan; xiii, John Shaw/Tom Stack & Associates; xv, Courtesy Department of Chemistry, University of California at Berkeley; xvii, Bob Peterson/FPG International; Dr. Harold Rose/Science Photo Library/Photo Researchers, Inc.; Robert Kristofik/The Image Bank; Michael Philip Manheim/The Stock Market; Comstock, Inc.; and Adam Hart-Davis/Science Photo Library/Photo Researchers, Inc.

Unit 1: Pages 2–3, N.A.S.A.

Chapter 1: Pages 4–5, Steve Krasemann/DRK Photo; 7(t), Courtesy of Ford Motor Company; 7(b), Nancy Wolff/Imagery; 8, Pam Hasegawa/Taurus Photos; 15, Courtesy of AIP Niels Bohr Library, Hebrew University of Jerusalem.

Chapter 2: Pages 18–19, Jeff Foott/Tom Stack & Associates; 21(t), HRW Photo by Ken Karp; 21(b), Gainesville Sun Photo by Tom Kennedy/Black Star; 25(t,b), Chem Study Film: Molecular Motion; 32, Will McIntyre/Photo Researchers, Inc.

Chapter 3: Pages 36–37, Jane Burton/Bruce Coleman, Inc.; 46(t–b), HRW Photo by Dr. E.R. Degginger, HRW Photo by John King, HRW Photo by John King, HRW Photo by Dr. E.R. Degginger; 53, W. F. Meggers Collection/AIP Neils Bohr Library.

Intra-Science One: Page 56(l), Shostal Associates/Superstock International; 56(r), Bob & Ira Spring; 57(l), Dallas and John Heaton/The Stock Shop; 57(r), Ray Nelson/Phototake.

Unit 2: Pages 58–59, Ralph P. Earlandson/Hillstrom Stock Photos.

Chapter 4: Pages 60–61, Robert Iscar/Photo Researchers, Inc.; 65, Runk/Schoenberger/Grant Heilman; 66(t), L. V. Bergman & Associates; 66(b), Breck Kent/Earth Scenes; 70(t), HRW Photo by Russell Dian; 70(bl,br), St. Regis, C.P. Division; 73, Imagery.

Chapter 5: Pages 76–77, Thomas Styczynski/Tony Stone Worldwide/Click/Chicago Ltd.; 80, Courtesy of Monsanto Enviro-Chem; 81, Courtesy of Robert L. Pundy/U.S. Steel Corporation; 87, Courtesy of Micro Essential Laboratory; 92, Courtesy of AIP Niels Bohr Library.

Chapter 6: Pages 96–97, Courtesy of Los Alamos National Laboratory; 98(t), Matt Grimaldi and Mark Shupack; 98(b), Hank Morgan/Rainbow; 104(c), Roger Tully/Medichrome/The Stock Shop; 104(b), John Marmaras/Woodfin Camp & Associates; 105, Dan McCoy/Rainbow; 107,109, Courtesy of U.S. Department of Energy; 111,112, Dan McCoy/Rainbow; 113, The Bettmann Archive.

Intra-Science Two: Page 116(l), Larry Mulvehill/Science Source/Photo Researchers; 116(r), Alexander Tsiaras/Science Source/Photo Researchers; 117(bl), T. Qing/FPG International; 117(tl), Petit Format/Guigoz/Steiner/Science Source/Photo Researchers; 117(tc), Howard Sochurek/Medichrome/The Stock Shop; 117(tr), David C. London/Tom Stack & Associates; 117(br), Gould Electronics/Peter Arnold, Inc.

Unit 3: Pages 118–119, Courtesy of The Silver Dome, Pontiac, Michigan.

Chapter 7: Pages 120–121, Phil Degginger/Dr. E.R. Degginger; 123, The Bettmann Archive; 126, Spencer Swanger/Tom Stack & Associates; 127, Dick Durrance II/Woodfin Camp & Associates; 128, Courtesy of EIMCO, Fairmont, West Virginia; 130, Breck P. Kent; 132, Tim Eagan/Woodfin Camp & Associates; 133, John Colwell/Grant Heilman; 135, Courtesy of Department of Chemistry, University of California at Berkeley.

Chapter 8: Pages 138–139, Dr. E.R. Degginger; 140, HRW Photo by Russell Dian; 146, Courtesy of Elizabeth Town Water Co., Westfield, New Jersey; 147, Runk/Schoenberger/Grant Heilman; 149, John Zoiner; 150(l,r), HRW Photos by Ken Karp; 152, Grant Heilman; 153, Ted Spiegel/Black Star; 154, Tom Tracy/Black Star; 155, Gary Milburn/Tom Stack & Associates.

Chapter 9: Pages 158–159, Lee Boltin Picture Library; 160, 168(t), Courtesy of Kennecott Copper Co.; 163, 168(b), Courtesy of Charles Rotkin, P.F.I.; 171, Courtesy of American Smelting & Refinery Co.; 175, Courtesy of Clevite Corp.; 177, Hank Morgan/Rainbow.

Chapter 10: Pages 180–181, John Shaw/Tom Stack & Associates; 186, Dr. E.R. Degginger; 187, Harvey Lloyd/Peter Arnold; 192, Dr. E.R. Degginger; 195, Ralph Crane/Globe Photos.

Chapter 11: Pages 198–199, Marie Ueda/Tony Stone Worldwide/Click/Chicago Ltd.; 200(t), Courtesy of Atlantic Cement Co.; 200(b), Adam Woolfitt/Woodfin Camp & Associates; 201, Dr. E.R. Degginger; 202, Imagery; 203, 205, Courtesy of PPG Industries, Inc.; 204, Foto Messerschmidt/The Stock Shop; 208, Dion Ogust/The Image Works; 210, Courtesy of Pfizer, Inc.; 211, Fred Ward/Black Star; 213, HRW Photo. 216, Visuals Unlimited; 217, George Holtin/Photo Researchers, Inc.; 218, Culver Pictures.

Chapter 12: Pages 222–223, Photo Researchers, Inc.; 224, Breck P. Kent; 227, Courtesy of Bethlehem Steel Corp. 229(t), 232, Courtesy of Exxon; 229(b), Drake Well Museum; 230, Courtesy of Champlen Petroleum Co.; 233(t), Thomas Hovland/Grant Heilman; 233(b), Courtesy of Cities Service Co.; 235, Spencer Swanger/Tom Stack & Associates.

Chapter 13: Pages 238–239, Stuart L. Craig, Jr./Bruce Coleman, Inc.; 240(t), 242(t,b), 244, 246, Courtesy of Goodyear Tire & Rubber Co.; 240(b), Shostal Associates/Superstock International; 249, Courtesy of U.S.D.A.; 250, Courtesy of Monsanto; 251, Courtesy of DuPont; 252, HRW Photo by Russell Dian; 254, Martin Rogers/Tony Stone Worldwide/Click/Chicago Ltd.

Chapter 14: Pages 258–259, Dr. E.R. Degginger; 260, EPA/Documerica; 261, Gerhard Gscheidle/The Image Bank; 263, Runk/Schoenberger/Grant Heilman; 266, Tom Stack/Tom Stack & Associates; 270, Grant Heilman; 271(t,b), Courtesy of Joy Manufacturing Co.; 273, Wil Blanche/Design Photographers International; 277, Dan McCoy/Rainbow.

Intra-Science Three: Page 280(l), David R. Frazier Photolibrary; 280(r), Tom Tracy/The Stock Shop; 281, Courtesy of Denver Board of Water Commissioners.

Unit 4: Pages 282–283, M.P.L. Fogden/Bruce Coleman, Inc.

Chapter 15: Pages 284–285, T.J. Florian/Rainbow; 286, Dr. E.R. Degginger; 289, Focus on Sports; 296, Dr. E.R. Degginger; 303, Focus on Sports; 308, Courtesy of Ford Motor Company.

Chapter 16: Pages 312–313, Nancy Simmerman/Bruce Coleman, Inc.; 314, Dr. E.R. Degginger; 321, Imagery; 329, N.A.S.A.; 330, The Bettmann Archive.

Chapter 17: Pages 334–335, Tony Stone Worldwide/Click/Chicago Ltd.; 344, Brian Parker/Tom Stack & Associates; 347, Imagery; 352, The Granger Collection.

Chapter 18: Pages 356–357, Bob Peterson/FPG International; 361, Courtesy of the Tennessee Valley Authority; 366, Harald Sund.

Chapter 19: Pages 370–371, Dr. R. Clark and M. Goff/Photo Researchers, Inc.; 379, Jon Riley/The Stock Shop; 386, N.A.S.A.; 388, Culver Pictures.

Chapter 20: Pages 392–393, N.A.S.A.; 394, Grant Heilman; 400, Woodfin Camp & Associates; 402, N.A.S.A.; 403, Don & Pat Valenti/DRK Photos.

Intra-Science Four: Page 406, Eric Carle/Shostal Associates/Superstock International; 407(l), Ken Lax/The Stock Shop; 407(r), S.I.U./Photo Researchers, Inc.

Unit 5: Pages 408–409, Dr. E. R. Degginger/Bruce Coleman, Inc.

Chapter 21: Pages 410–411, Mark Newman/Tom Stack & Associates; 412, Fred Anderson/Photo Researchers; 426, John Gair/Tom Stack & Associates; 428, Imagery; 429, Tana Hoban/Design Photographers International; 434, Alan F. Blumenthal.

Chapter 22: Pages 438–439, Stuart L. Craig, Jr./Bruce Coleman, Inc.; 442(t,c), Martin Adler Levic/Black Star; 442(b), Linda Lindrath; 448, Courtesy of John Goeller, New York Eye and Ear Infirmary; 466, The Granger Collection.

Chapter 23: Pages 470–471, John Cancalosi/Tom Stack & Associates; 483, Jeff March/Tom Stack & Associates.

Intra-Science Five: Page 486, N.A.S.A.; 487(t), Earth Satellite Corp./Science Photo Library/Photo Researchers, Inc.; 487(bl), Courtesy of Ocean Sciences Division/NOAA; 487(br), N.A.S.A./The Stock Shop; 487(bc), Carl Purcell/Photo Researchers, Inc.

Unit 6: Pages 488–489, Karl Sims, MIT Media Laboratory/Dan McCoy/Rainbow.

Chapter 24: Pages 490–491, Spencer Swanger/Tom Stack & Associates; 501, The Bettmann Archives.

Chapter 25: Pages 504–505, Jean-Marie Truchet/Tony Stone Worldwide/Click/Chicago Ltd.; 511, Courtesy of Con Edison; 520, Courtesy of the U.S. Department of the Interior, National Park Service, Edison National Historic Site.

Chapter 26: Pages 524–525, Greg L. Ryan/Sally A. Beyer; 533(t), Jeffrey E. Blackman/The Stock Shop; 533(b), Courtesy of Medtronic, Inc.; 535, Equipment provided by Wabash Instrument Corp./Photo by Carolina Biological Supply Co; 537, Culver Pictures.

Chapter 27: Pages 540–541, Courtesy of Argonne National Laboratory; 542,543, Courtesy of International Minerals & Chemicals; 554, Harald Sund/The Image Bank; 556, Culver Pictures.

Chapter 28: Pages 560–561, Alfred Pasieka/Bruce Coleman, Inc.; 571, HRW Photo by Ken Karp; 574, Courtesy of Naval Research Laboratory.

Intra-Science Six: Page 578(c), Bob Peterson/FPG International; 578(t), Dr. Harold Rose/Science Photo Library/Photo Researchers, Inc.; 578(b), Robert Kristofik/The Image Bank; 579(bl), Michael Philip Manheim/The Stock Market; 579(t), Comstock, Inc.; 579(br), Adam Hart-Davis/Science Photo Library/Photo Researchers, Inc.